Since $D \gg d$ w use the st energy equation between sur funnel to exit from pipe (pressure $= P_a$, ... $= 0$)

$$g z_0 = \tfrac{1}{2} v^2 + g z_2 + g h_e \quad (\alpha = 1)$$

$$g(z_0 - z_2) = \tfrac{1}{2} v^2 + \tfrac{1}{2} v^2 \left(K_1 + K_2 + f \tfrac{L}{d}\right)$$

$$v^2 = \frac{2g(z_0 - z_2)}{1 + K_1 + K_2 + f \tfrac{L}{d}}$$

$$V = \sqrt{80 g d}$$

Diagram labels: D, P_a, O, $40d$, K_1 — 1, $L = 1000d$, d, $-K_2$, 2, $V \epsilon P_a$, $\alpha = 1$

Diagram labels: ②, $D = 1.25$, $L = 300$, ①, $\dot{q} = .1$

$$\frac{P_1}{\rho} + \frac{\alpha v^2}{2} + g z_1 = \frac{P_2}{\rho} + \frac{\alpha v^2}{2} + g z_2 + g h_e$$

$$\frac{P_1 - P_2}{\rho} = g(z_2 - z_1) + g h_e$$

$$= g(z_2 - z_1) + f \frac{L}{D} \frac{v^2}{2}$$

(if all pressure losses due to friction)

$$P = \text{Power required} = \Delta P \times \dot{q} = \left(g(z_2 - z_1) + f \frac{L}{2D} \left(\frac{\dot{q}^2}{(\tfrac{\pi}{4} D^2)^2}\right) \right) \times \rho \dot{q}$$

$$P = \rho \dot{q} \left(g(z_2 - z_1) + \frac{8 \dot{q}^2 L}{\pi^2 D^5} f \right)$$

to find f, we need $Re = \frac{VD}{\nu} = \frac{4 \dot{q}}{\pi D \nu}$

internal flow— Consider flow through circular pipe of constant X-Section assume fully developed — apply control vol analysis

laminar flow

$r = R$

$r = 0$

$$T_w = \left| \mu \frac{du}{dr} \right|$$

$$u = u_{max} \left(1 - \left(\frac{r^2}{R} \right) \right)$$

shear stress

viscosity

Wall Shear Stress? $\quad T_w = \left| u \frac{du}{dr} \right|$

$$P_1 A_1 - P_2 A_3 = A_3 T_w$$

$$(P_1 - P_2) A = (2 \pi R L) T_w$$

$$A = \pi R^2$$

$$P_1 - P_2 \pi R^2 = 2 \pi R L T_w$$

$$\frac{P_1 - P_2}{L} = \frac{2}{R} T_w$$

$$F_v = (2 \pi R L) T_w$$

Area

$$\rightarrow \dot{q} = .06$$

Control volume

$\eta = 87\%$ $D = .2$ $90°$ Bend

$$\frac{dE}{dt} + m\left(\frac{v^2}{2} + gz + \frac{P}{\rho}\right)_1 - m\left(\frac{v^2}{2} + gz + \frac{P}{\rho}\right)_2 + W_{in} = losses$$

The losses include those due to viscous dissapation, (Major)
geo of piping system (Minor) and those in the pump

Assumptions: Steady flow of water at $20°C$. The pump recieves & delivers
water at atmospric pressure.

By neglecting changes in kinetic & potential energy
the energy equation is given:

$$m\left(\frac{P_1 - P_2}{\rho}\right) + W_a = losses$$

$W_a =$ actual Power

mass flow rate is $m = \rho Q$

the required power is higher than what is
required under ideal conditions, when there
are no losses.

efficiency of pump $\eta = \frac{W_t}{W_a} < 1$

energy losses at the pump are equal to $(W_a - W_t)$
or losses $= W_a - \eta W_a = (1-\eta) W_a$

$$W_a = \frac{\rho Q}{\eta}\left(\frac{P_2 - P_1}{\rho}\right)$$

$$\left(\frac{v^2}{2} + gz + \frac{P}{\rho}\right) - \left(\frac{v^2}{2} + gz + \frac{P}{\rho}\right) = losses_{2 \rightarrow 3}$$

at Inlet   at exit
$z_2 = 0$ $z_3 = 50m$ $V_2 = V_3$ (mass Conservation)

$$\left(\frac{P_2 - P_3}{\rho}\right) = h_{major} + h_{minor} + gz_3$$

$$h_{major} = f\frac{L}{D}\frac{v^2}{2}$$

$$Re = \frac{VD}{\nu} = \frac{4\dot{q}}{\pi D \nu} \Rightarrow 380,450$$

Commercial Steel pipe $\frac{e}{D} = .0002$ $f = .0158$

$$h_{major} = f\frac{L}{D}\frac{v^2}{2} = 43.22$$

—minor losses are those due to the geometry
in this case minor losses occur at a $90°$ &'s
for each elbow $\frac{L_e}{D} = 30$

$$h_{minor} = 2f\frac{L_e}{D}\frac{v^2}{2} \Rightarrow 1.729$$

Pressure Drop

$$\left(\frac{P_2 - P_3}{\rho}\right) = (43.22 + 1.729 + 9.81)(50)$$

Power Input to Pump

$$W_a = \frac{\rho \dot{q}}{\eta}\left(\frac{P_2 - P_1}{\rho}\right) \quad \frac{998.2}{.87}(.06)(5354)$$

$$= 36.86$$

A PHYSICAL INTRODUCTION
TO FLUID MECHANICS

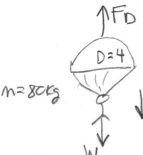

By having man & parachute as control volume,
when control volume drops with constant terminal
velocity, the acceleration vanishes $(m\frac{dU}{dt}=c)$

linear momentum: $m\frac{dU}{dt} = \Sigma F = 0$

assume density variations are negligble

forces acting on CV; $W = -mg\,\hat{\jmath}$

$$F_D = F_0\,\hat{\jmath}$$

$$F_D = C_D \frac{1}{2}\rho U^2 A$$

A = frontal Area

$A = \frac{\pi D^2}{4}$

$C_D = 1.4$

Substituting we get

$$C_D \frac{1}{2}\rho U^2 A = mg$$

$$U = \sqrt{2\frac{mg}{C_D \rho A}} = \boxed{8.901}$$

A Boundary layer is a region close to a solid surface where the
flow over the surface has strong velocity gradients, from zero at
the wall to the free stream value U_e a short distance away
(both velocities are measured relative to the surface). Within the
Boundary layer, viscous stresses are important. Outside the
boundary layer, they are not,

a) Displacement thickness $\delta^* = \int_0^\infty \left(1-\frac{u}{U_e}\right)dy$

The mass flux passing through a distance δ^* in the absense of a boundary layer equals the deficit in Mass flux due to the presense of a boundary layer. From the point of view of the free stream flow, the presence of the boundary layer appears to displace the streamlines by a distance equal to δ^*

b) Momentum thickness $\theta = \int_0^\infty \frac{u}{U_e}\left(1-\frac{u}{U_e}\right)dy$

the momentum flux passing through a distance θ in the absense of a boundary layer is the same as the deficit in momentum flux due to the presense of the boundary layer

c) Shape factor $H = \frac{\delta^*}{\theta}$! smaller values of H indicate a fuller velocity profile

---

68) for boundary layer velocity profile $\frac{u}{U_e} = \frac{y}{\delta}$ $\left(\frac{y}{\delta} \le 1\right)$

find wall shear stress
Skin friction coefficient
displacement thickness
momentum thickness

a) $T_w = \mu \frac{du}{dy}\Big|_{y=0} = \mu \frac{U_e}{\delta}$

b) $C_f = \frac{T_w}{\frac{1}{2}\rho U_e^2} = \frac{2\mu U_e}{\delta} \frac{1}{\rho U_e^2} = \frac{2\mu}{\rho U_e \delta}$

c) $\delta^* = \int_0^\infty \left(1-\frac{u}{U_e}\right)dy$

$\approx \int_0^\delta \left(1-\frac{u}{U_e}\right)dy = \int_0^\delta \left(1-\frac{y}{\delta}\right)dy = \left[y - \frac{y^2}{2\delta}\right]_0^\delta = \frac{\delta}{2}$

D) $\theta = \int_0^\infty \frac{u}{U_e}\left(1-\frac{u}{U_e}\right)dy \approx \int_0^\delta \frac{y}{\delta}\left(1-\frac{y}{\delta}\right)dy = \left[\frac{y^2}{2\delta} - \frac{y^3}{3\delta^2}\right]_0^\delta = \frac{\delta}{6}$

A PHYSICAL INTRODUCTION TO FLUID MECHANICS

ALEXANDER J. SMITS
Department of Mechanical Engineering
Princeton University

JOHN WILEY & SONS, INC.
New York / Chichester / Weinheim / Brisbane / Singapore / Toronto

ACQUISITIONS EDITOR	Bill Zobrist
MARKETING MANAGER	Katherine Hepburn
SENIOR PRODUCTION EDITOR	Robin Factor
ILLUSTRATION COORDINATOR	Gene Aiello
SENIOR DESIGNER	Harold Nolan
COVER DESIGNER	Suzanne Noli

Photo Credits: Figure 1.1: Austin Post/U.S. Department of the Interior, U.S. Geological Survey, David A. Johnston Cascades Volcano Observatory, Vancouver, WA. Figure 1.8: Corbis/Ralph A. Clevenger. Figure 1.10: Courtesy United Airlines. Figure 11.9: From Siemens, with permission. Figure 13.27: Courtesy U.S. Department of Energy.

This book was set in New Times Roman by Bi-Comp, Inc. and printed and bound by Hamilton Printing. The cover was printed by Phoenix Color.

This book is printed on acid-free paper. ∞

Library of Congress Cataloging in Publication Data
Smits, Alexander J.
    A physical introduction to fluid mechanics / Alexander J. Smits.
        p.   cm.
    ISBN 0-471-25349-9 (alk. paper)
    1. Fluid mechanics.   I. Title.
    TA357.S517   1999
    620.1'06—DC21   99-16027
                                                CIP

Printed in the United States of America

10 9 8 7 6 5 4 3 2

CONTENTS

CHAPTER 13   *TURBOMACHINES*   401

CHAPTER 14   *ENVIRONMENTAL FLUID MECHANICS*   433

PREFACE

The purpose of this book is to summarize and illustrate basic concepts in the study of fluid mechanics. Although fluid mechanics is a challenging and complex field of study, it is based on a small number of principles, which in themselves are relatively straightforward. The challenge taken up here is to show how these principles can be used to arrive at satisfactory engineering answers to practical problems. The study of fluid mechanics is undoubtedly difficult, but it can also become a profound and satisfying pursuit for anyone with a technical inclination, and I hope the book conveys that message clearly.

The scope of this introductory material is rather broad, and many new ideas are introduced. It will require a reasonable mathematical background, and those students who are taking a differential equations course concurrently sometimes find the early going a little challenging. The underlying physical concepts are highlighted at every opportunity to try to illuminate the mathematics. For example, the equations of fluid motion are introduced through a reasonably complete treatment of one-dimensional, steady flows, including Bernoulli's equation, and then developed through progressively more complex examples. This approach gives the students a set of tools that can be used to solve a wide variety of problems, as early as possible in the course. In turn, by learning to solve problems, students can gain a physical understanding of the basic concepts before moving on to examine more complex flows. Dimensional reasoning is emphasized, as well as the interpretation of results (especially through limiting arguments). Throughout the text, worked examples are given to demonstrate problem-solving techniques. They are grouped at the end of major sections to avoid interrupting the text as much as possible. Historical references are given throughout, and some brief biographical sketches are collected near the end of the text. I hope they add to the fabric of the book, and that they will stimulate further reading in the history of fluid mechanics.

The book is intended to provide students with a broad introduction to the mechanics of fluids. The material is sufficient for two quarters of instruction. For a one-semester course only a selection of material should be used. A typical one-semester course, might consist of the material in Chapters 1 to 10, not including Chapter 7. If time permits, one of Chapters 11 to 14 may be included. For a course lasting two quarters, it is possible to cover Chapters 1 to 6, and 8 to 10, and select three or four of the other chapters, depending on the interests of the class. The sections marked with asterisks may be omitted without loss of continuity. Although some familiarity with thermodynamic concepts is assumed, it is not a strong prerequisite. Omitting the sections marked by a single asterisk, and the whole of Chapter 12, will leave a curriculum that does not require a prior background in thermodynamics.

A limited number of Web sites are suggested to help enrich the written material. In particular, a number of Java-based programs are available on the Web to solve specific fluid mechanics problems. They are especially useful in areas where traditional methods limit the number of cases that can be explored. For example, the programs designed to solve potential flow problems by superposition and the programs that handle compressible flow problems, greatly expand the scope of the examples that can be solved in a limited amount of time, while at the same time dramatically reducing the effort involved. A listing of current links to sites of interest to students and researchers in fluid dynamics may be found at *http://www. princeton.edu/~gasdyn/fluids.html.* In an effort to keep the text as current as possible, additional problems, illustrations and Web resources, as well as a Corrigendum and Errata may be found at *http://www.princeton.edu/~asmits/fluidmechanics.html.*

In preparing this book, I have had the benefit of a great deal of advice from my colleagues. One persistent influence that I am very glad to acknowledge is that of Professor Sau-Hai Lam of Princeton University. His influence on the contents and tone of the writing is profound. Also, my enthusiasm for fluid mechanics was fostered as a student by Professor Tony Perry of the University of Melbourne, and I hope this book will pass on some of my fascination with the subject.

Many other people have helped to shape the final product. Professor David Wood of Newcastle University in Australia provided the first impetus to start this project. Professor George Handelman of Rensselaer Polytechnic Institute, Professor Peter Bradshaw of Stanford University, and Professor Robert Moser of the University of Illinois Urbana-Champaign were very helpful in their careful reading of the manuscript and through the many suggestions they made for improvement. Professor Victor Yakhot of Boston University test-drove an early version of the book, and provided a great deal of feedback, especially for the chapter on dimensional analysis. My wife, Louise Handelman, gave me wonderfully generous support and encouragement, as well as advice on improving the quality and clarity of the writing. I would like to dedicate this work to the memory of my brother, Robert Smits (1946–1988), and to my children, Peter and James, who represent the future.

Alexander J. Smits
Princeton, New Jersey, USA

A PHYSICAL INTRODUCTION TO FLUID MECHANICS

CHAPTER 1

INTRODUCTION

Fluid mechanics is the study of the behavior of fluids under the action of applied forces. Typically, we are interested in finding the force required to move a solid body through a fluid, or the power necessary to move a fluid through a system. The speed of the resulting motion, and the pressure, density, and temperature variations in the fluid, are also of great interest. To find these quantities, we apply the principles of dynamics and thermodynamics to the motion of fluids, and develop equations to describe the conservation of mass, momentum, and energy.

As we look around, we can see that fluid flow is a pervasive influence on all parts of our daily life. To the ancient Greeks, the four fundamental elements were Earth, Air, Fire, and Water; and three of them, Air, Fire, and Water, involve fluids. The air around us, the wind that blows, the water we drink, the rivers that flow, and the oceans that surround us, affect us daily in the most basic sense. In engineering applications understanding fluid flow is necessary for the design of aircraft, ships, cars, propulsion devices, pipe lines, air conditioning systems, heat exchangers, clean rooms, pumps, artificial hearts and valves, spillways, dams, and irrigation systems. It is essential to the prediction of weather, ocean currents, pollution levels, and greenhouse effects. Not least, all life-sustaining bodily functions involve fluid flow since the transport of oxygen and nutrients throughout the body is governed by the flow of air and blood. Fluid flow is, therefore, crucially important in shaping the world around us, and its full understanding remains one of the great challenges in physics and engineering.

What makes fluid mechanics challenging is that it is often very difficult to predict the motion of fluids. In fact, even to observe fluid motion can be difficult. Most fluids are highly transparent, like air and water, or they are of a uniform color, like oil, and their motion only becomes visible when they contain some type of particle. Snowflakes swirling in the wind, dust kicked up by a car along a dirt road, smoke from a fire, or clouds scudding in a stiff breeze, help to mark the underlying fluid motion (Figure 1.1). It is clear that this motion can be very complicated. By following a single snowflake in a snowstorm, for example, we see that it traces out a complex path, and that each flake follows a different path. Eventually, all the flakes end up on the ground, but it is difficult to predict where and when a particular snowflake ends up. The fluid that carries the snowflake on its path experiences similar contortions, and generally the velocity and acceleration of a particular mass of fluid vary with time and location. This is true for all fluids in motion: the position, velocity, and acceleration of a fluid is, in general, a function of time and space.

To describe the dynamics of fluid motion, we need to relate the fluid acceleration to the resultant force acting on it. For a rigid body in motion, such as a satellite in orbit, we can follow a fixed mass, and only one equation (Newton's second law of motion, $F = ma$) is required. Fluids can also move in rigid-body motion, but more commonly one part of the fluid is moving with respect to another part (there

FIGURE 1.1   The eruption of Mt. St. Helens, May 18, 1980. Austin Post/U.S. Department of the Interior, U.S. Geological Survey, David A. Johnston, Cascades Volcano Observatory, Vancouver, WA.

is *relative* motion), and then the fluid behaves more like a huge collection of particles. Each snowflake, for example, marks one small mass of fluid (a fluid *particle*) and to describe the dynamics of the entire flow requires a separate equation for each fluid particle. The solution of any one equation will depend on every other equation because the motion of one fluid particle depends on its neighbors, and solving this set of simultaneous equations is obviously a daunting task. It is such a difficult task, in fact, that for almost all practical problems the exact solution cannot be found, even with the aid of the most advanced computers. It seems likely that this situation will continue for many years to come, despite the projected developments in computer hardware and software capabilities.

To make any progress in the understanding of fluid mechanics and the solution of engineering problems, we usually need to make approximations and use simplified flow models. But how do we make these approximations? Physical insight is often necessary. We must determine the crucial factors that govern a given flow, and to identify the factors that can safely be neglected. This is what sometimes makes fluid mechanics difficult to learn and understand: physical insight takes time and familiarity to develop, and the reasons for adopting certain assumptions or approximations are not always immediately obvious.

To help develop this kind of intuition, this book starts with the simplest types of problems and progressively introduces higher levels of complexity, while at the same time stressing the underlying principles. We begin by considering fluids that are in rigid-body motion, then fluids where relative motions exist under the action of simple forces, and finally more complex flows where viscosity and compressibility are important. At each stage, the simplifying assumptions will be discussed, although the full justification is sometimes postponed until after the later material is understood. By the end of the book, the reader should be able to solve basic problems in fluid mechanics, while understanding the limitations of the tools used in their solution.

Before starting along that path, we need to consider some fundamental aspects of fluids and fluid flow. In this chapter, we discuss the differences between solids and fluids, and introduce some of the distinctive properties of fluids such as density, viscosity, and surface tension. We will also consider the type of forces that can act on a fluid, and its deformation by stretching, shearing, and rotation. We begin by describing how fluids differ from solids.

1.1 THE NATURE OF FLUIDS

Almost all the materials we see around us can be described as solids, liquids, or gases. Many substances, depending on the pressure and temperature, can exist in all three states. For example, H_2O can exist as ice, water, or vapor. Liquids and gases are both called fluid states, or simply fluids.

Fluids behave differently from solids in two respects. The most obvious property of fluids that is not shared by solids is the ability of fluids to flow and change shape; fluids do not hold their shape independent of their surroundings, and they will flow spontaneously within their containers. In this respect, liquids and gases respond somewhat differently in that gases fill a container fully, whereas liquids occupy a definite volume. When a gas and a liquid are both present, an interface forms between the liquid and the gas called a *free surface* (Figure 1.2). At a free surface, surface tension may be important, and waves can form. Gases can also be dissolved in the liquid, and when the pressure changes bubbles can form, as when a soda bottle is suddenly opened.

The most distinctive property of fluids, however, is its response to an applied force or an applied stress (stress is force per unit area). For example, when a shear stress is applied to a fluid, it experiences a continuing and permanent distortion. Drag your hand through a basin of water and you will see the distortion of the fluid (that is, the flow that occurs in response to the applied force) by the swirls and eddies that are formed in the free surface. This distortion is permanent in that the fluid does not return to its original state after your hand is removed from the

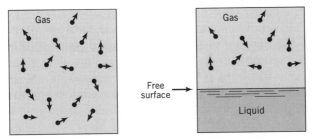

FIGURE 1.2 Gases fill a container fully (left), whereas liquids occupy a definite volume, and a free surface can form (right).

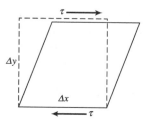

FIGURE 1.3 When a shear stress τ is applied to a fluid element, the element distorts. It will continue to distort as long as the stress acts.

fluid. Also, when a fluid is squeezed in one direction (that is, a normal stress is applied), it will flow in the other two directions. Squeeze a hose in the middle and water will issue from its ends. If such stresses persist, the fluid continues to flow. Fluids cannot offer permanent resistance to these kinds of loads. This is not true for a solid; when a force is applied to a solid it will deform only as much as it takes to accommodate the load, and then the deformation stops.

> *Thus, a fluid can be defined unambiguously as a material that deforms continuously and permanently under the application of a shearing stress, no matter how small. This definition does not address the issue of how fast the deformation occurs and as we shall see later this rate is dependent on many factors including the properties of the fluid itself. The inability of fluids to resist shearing stress gives them their characteristic ability to change shape or to flow; their inability to support tension stress is an engineering assumption, but it is a well-justified assumption because such stresses, which depend on intermolecular cohesion, are usually extremely small. . . .*
>
> *Because fluids cannot "support" shearing stresses, it does not follow that such stresses are nonexistent in fluids. During the flow of real fluids, the shearing stresses assume an important role, and their prediction is a vital part of engineering work. Without flow, however, shearing stresses cannot exist, and compression stress or pressure is the only stress to be considered."*[1]

So we see that the most obvious property of fluids, their ability to flow and change their shape, is precisely a result of their inability to support shearing stresses (Figure 1.3). Flow is possible without a shear stress, since differences in pressure will cause a fluid to experience a resultant force and an acceleration, but when the shape of the fluid is changing, shearing stresses must be present.

With this definition of a fluid, we can recognize that certain materials that look like solids are actually fluids. Tar, for example, is sold in barrel-sized chunks, which appear at first sight to be the solid phase of the liquid that forms when the tar is heated. However, cold tar is also a fluid. If a brick is placed on top of an open barrel of tar, we will see it very slowly settle into the tar. It will continue to settle as time goes by—the tar continues to deform under the applied load—and eventually the brick will be completely engulfed. Even then it will continue to move downwards until it reaches the bottom of the barrel. Glass is another substance that appears to be solid, but is actually a fluid. Glass flows under the action of its own weight. If you measure the thickness of a very old glass pane you would find

[1] *Elementary Fluid Mechanics,* 7th edition, by R.L. Street, G.Z. Watters, and J.K. Vennard, John Wiley & Sons, 1996.

it to be larger at the bottom of the pane than at the top. This deformation happens very slowly because the glass has a very high viscosity, which means it does not flow very freely, and the results can take centuries to become obvious. However, when glass experiences a large stress over a short time, it behaves like a solid and it can crack. Silly putty is another example of a material that behaves like an elastic body when subject to rapid stress (it bounces like a ball) but it has fluid behavior under a slowly acting stress (it flows under its own weight).

1.2 STRESSES IN FLUIDS

In this section, we consider the stress distributions that occur within the fluid. To do so, it is useful to think of a fluid particle, which is a small volume of fluid of fixed mass.

   The stresses that act on a fluid particle can be split into normal stresses (stresses that give rise to forces acting normal to the surface of the fluid particle) and tangential, or shearing, stresses (stresses that produce forces acting tangential to its surface). Normal stresses tend to compress or expand the fluid particle without changing its shape. For example, a rectangular particle will remain rectangular, although its dimensions may change. Tangential stresses shear the particle and deform its shape: a particle with an initially rectangular cross-section will become lozenge-shaped.

   What role do the properties of the fluid play in determining the level of stress required to obtain a given deformation? In solids, we know that the level of stress required to compress a rod depends on the Young's modulus of the material, and that the level of tangential stress required to shear a block of material depends on its shear modulus. Young's modulus and the shear modulus are properties of solids, and fluids have analogous properties called the bulk modulus and the viscosity. The bulk modulus of a fluid relates the normal stress on a fluid particle to its change of volume. Liquids have much larger values for the bulk modulus than gases since gases are much more easily compressed (see Section 1.3.3). The viscosity of a fluid measures its ability to resist a shear stress. Liquids typically have larger viscosities than gases since gases flow more easily (see Section 1.4). Viscosity, as well as other properties of fluids such as density and surface tension, are discussed in more detail later in this chapter. We start by considering the nature of pressure and its effects.

1.3 PRESSURE

When a gas is held in a container, the molecules of the gas move around and bounce off its walls. When a molecule hits the wall, it experiences an elastic impact, which means that its momentum magnitude and energy are conserved. However, its direction of motion changes, so that the wall must have exerted a force on the gas molecule. Therefore, an equal and opposite force is exerted by the gas molecule on the wall during impact. If the piston in Figure 1.4 was not constrained in any way, the continual impact of the gas molecules on the piston surface would tend to move the piston out of the container. To hold the piston in place, a force must be applied to it, and it is this force (per unit area) that we call the gas pressure.

   If we consider a very small area of the surface of the piston, so that over a short time interval Δt very few molecules hit this area, the force exerted by the

FIGURE 1.4 The piston is supported by the pressure of the gas inside the cylinder.

molecules will vary sharply with time as each individual collision is recorded. When the area is large, so that the number of collisions on the surface during the interval Δt is also large, the force on the piston due to the bombardment by the molecules becomes effectively constant. In practice, the area need only be larger than about $10 \times \ell_m^2$, where the mean free path ℓ_m is the average distance traveled by a molecule before colliding with another molecule. Pressure is therefore a *continuum* property, by which we mean that for areas of engineering interest, which are almost always much larger than areas measured in terms of the mean free path, the pressure does not have any measurable statistical fluctuations due to molecular motions.[2]

We make a distinction between the *microscopic* and *macroscopic* properties of a fluid, where the microscopic properties relate to the behavior on a molecular scale (scales comparable to the mean free path), and the macroscopic properties relate to the behavior on an engineering scale (scales much larger than the mean free path). In fluid mechanics, we are concerned only with the continuum or macroscopic properties of a fluid, although we will occasionally refer to the underlying molecular processes when it seems likely to lead to a better understanding.

1.3.1 Pressure: Direction of Action

Consider the direction of the force acting on a flat solid surface due to the pressure exerted by a gas at rest. On a molecular scale, of course, a flat surface is never really flat. On average, however, for each molecule that rebounds with some amount of momentum in the direction along the surface, another rebounds with the same amount of momentum in the opposite direction, no matter what kind of surface roughness is present (Figure 1.5). The average force exerted by the molecules on the solid in the direction along its surface will be zero. We expect, therefore, that the force due to pressure acts in a direction which is purely normal to the surface.

Furthermore, the momentum of the molecules is randomly directed, and the magnitude of the force due to pressure should be independent of the direction of the surface on which it acts. For instance, a thin flat plate in air will experience no resultant force due to air pressure since the forces due to pressure on its two sides have the same magnitude and they point in opposite directions. This result is independent of the orientation of the plate. We say that pressure is *isotropic* (based on Greek words, meaning "equal in all directions," or more precisely, "independent of direction"). We will prove that this is so using an argument based on macroscopic or continuum concepts in Section 1.3.2.

---

[2] The mean free path of air molecules in the atmosphere at sea level is about $10^{-7}\,m$, which is about 1000 times smaller than the thickness of a human hair.

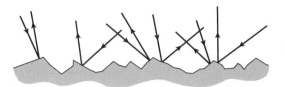

FIGURE 1.5 Molecules rebounding off a macroscopically rough surface.

In summary, pressure is a stress, and it is a normal stress since it produces a force that acts in a direction normal to the surface on which it acts. That is, the direction of the force is given by the orientation of the surface, as indicated by a unit normal vector **n** (Figure 1.6). The force has a magnitude equal to the average pressure times the area of contact. By convention, a force acting to compress the volume is positive, but for a closed surface the vector **n** always points outward (by definition). So

> The force due to a pressure p acting on one side of a small element of surface dA defined by a unit normal vector **n** is given by $-p\mathbf{n}dA$.

In some textbooks, the surface element is described by a vector **dA**, which has a magnitude dA and a direction defined by **n**, so that $\mathbf{dA} = \mathbf{n}dA$. We will not adopt that convention, and the magnitude and direction of a surface element will always be indicated separately.

For a fluid at rest, the pressure is the normal component of the force per unit area. What happens when the fluid is moving? The answer to this question is somewhat complicated.[3] For the flows considered in this text, this description of pressure is a very good approximation, even for fluids moving at very high speeds. Furthermore, this definition is consistent with the concept of pressure as used in thermodynamics.

1.3.2 Forces Due to Pressure

Pressure is given by the normal force per unit area, so that even if the force itself is moderate, the pressure can become very large if the area is small enough. This effect makes skating possible: the thin blade of the skate combined with the weight of the skater produces intense pressures on the ice, melting it and producing a thin film of water that acts as a lubricant and reduces the friction to very low values.

It is also true that very large forces can be developed by small fluid pressure differences acting over large areas. Rapid changes in air pressure, such as those produced by violent storms, can result in small pressure differences between the inside and the outside of a house. Since most houses are reasonably airtight to save air conditioning and heating costs, pressure differences can be maintained for some time. Even small pressure differences can produce very large forces when they act over the large interior surfaces of a house. When the outside air pressure is lower than that inside the house, as is usually the case when the wind blows, the forces produced by these small pressure differences can be large enough to cause the house to explode. Example 1.2 illustrates this phenomenon.

[3] See, for example, I.G. Currie, "Fundamental Fluid Mechanics," McGraw-Hill, 1974.

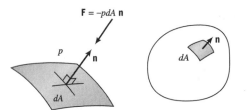

FIGURE 1.6   The vector force **F** due to pressure p acting on an element of surface area dA.

This effect can be demonstrated with a simple experiment. Take an empty metal container and add a small amount of water in the bottom. Heat the water so that it boils. The water vapor that forms displaces some of the air out of the container. If the container is then sealed, and allowed to cool, the water vapor inside the container condenses back to liquid, and now the mass of air in the container is less than at the start of the experiment. The pressure inside the container is therefore less than atmospheric (since fewer molecules of air hit the walls of the container). As a result, strong crushing forces develop which can cause the container to collapse, providing a dramatic illustration of the large forces produced by small differential pressures. More common examples include the slamming of a door in a draft and the force produced by pressure differences on a wing to lift an airplane off the ground.

Similarly, to drink from a straw requires creating a pressure in the mouth that is below atmospheric, and a suction cup relies on air pressure to make it stick. In one type of suction cup, a flexible membrane forms the inside of the cup. To make it stick, the cup is pressed against a smooth surface, and an external lever is used to pull the center of the membrane away from the surface, leaving the rim in place as a seal. This action reduces the pressure in the cavity to a value below atmospheric, and the external air pressure produces a resultant force that holds the cup onto the surface.

When the walls of the container are curved, pressure differences will also produce stresses within the walls. In Example 1.3, we calculate the stresses produced in a pipe wall by a uniform internal pressure. The force due to pressure acts radially outward on the pipe wall, and this force must be balanced by a circumferential force acting within the pipe wall material, so that the fluid pressure acting normal to the surface produces a tensile stress in the solid.

1.3.3 Pressure Is Isotropic

In Section 1.3.1, we gave an argument based on molecular dynamics to show that pressure is isotropic, and that it produces a force that has a direction normal to the surface on which it acts. Here we take a macroscopic approach to show the same result, somewhat more rigorously.

Consider the small wedge-shaped fluid element shown in Figure 1.7, with a volume $\frac{1}{2}dxdydz$. The fluid element is in equilibrium under the action of the forces due to pressure and its own weight. Let p_1, p_2, and p_3 be the average values of the pressure on the three surfaces. The element is not accelerating, so that the resultant force acting on the element must be zero. That is, $\Sigma\,\mathbf{F} = 0$. Resolving forces in the

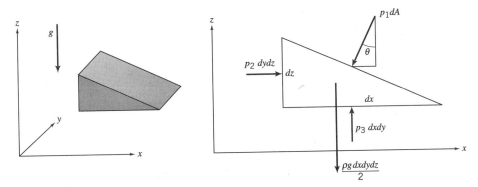

FIGURE 1.7 Pressures acting on a wedge-shaped fluid element.

x-direction gives

$$F_x = p_2\, dydz - p_1\, dA \sin\theta = 0$$

That is

$$p_2\, dydz = p_1\, dy\, \frac{dz}{\sin\theta}\sin\theta$$

and therefore

$$p_2 = p_1 \tag{1.1}$$

In the z-direction we have

$$F_z = p_3\, dydx - \tfrac{1}{2}\rho g\, dxdydz - p_1\, dA \cos\theta = 0$$

That is,

$$p_3\, dydx = \tfrac{1}{2}\rho g\, dxdydz + p_1\, dy\, \frac{dx}{\cos\theta}\cos\theta$$

and

$$p_3 = p_1 + \tfrac{1}{2}\rho g\, dz$$

As the volume decreases in size, the contribution from the weight of the fluid decreases rapidly as $dz \to 0$, and it becomes negligible when the volume becomes infinitesimally small. Hence,

$$p_3 = p_1$$

Since we showed that $p_2 = p_1$ (Equation 1.1),

$$p_3 = p_2 = p_1 \tag{1.2}$$

Therefore the pressure at a point is independent of the orientation of the surface passing through the point. In other words, the pressure is isotropic.

The pressure at a point in a fluid is independent of the orientation of the surface passing through the point. Pressure is a scalar, and it always acts at right angles to a given surface.

This result was obtained because the body force due to weight becomes negligible compared to the forces due to pressure as the fluid element becomes very small. Therefore, it will hold regardless of what body forces exist and no matter what the acceleration of the fluid is, even if the fluid were moving with an acceleration many times that of gravity.

1.3.4 Bulk Stress and Fluid Pressure

In the interior of a fluid, away from the walls of the container, each fluid particle feels the pressure due to its contact with the surrounding fluid. This is very similar to having a solid body such as a cube suspended and completely immersed in a fluid. In that case, the body experiences a *bulk* strain and a *bulk* stress since the fluid exerts a pressure on all surfaces of the body. Similarly, a fluid particle experiences a bulk compression due to the pressure exerted by the surrounding fluid on its surface.

When the pressure is uniform throughout the fluid, the forces due to pressure acting over each surface of a fluid particle all have the same magnitude. The force acting on any one face of the particle acts normal to that face with a magnitude equal to the pressure times the area. The force acting on the top face of a rectangular fluid particle, for example, is cancelled by an opposite but equal force acting on its bottom face. This will be true for all pairs of opposing faces. Therefore, the resultant force acting on that particle is zero. This result will also hold for a spherical fluid particle (an element of surface area on one side will always find a matching element on the opposite side), and, in fact, it will hold for a body of any arbitrary shape. Therefore there is no resultant force due to pressure acting on a body if the pressure is uniform in space, regardless of the shape of the body. Resultant forces due to pressure will appear only if there is a pressure variation within the fluid, that is, when pressure gradients exist.

The force due to pressure acts to compress the fluid particle. This type of strain is called a bulk strain, and it is measured by the fractional change in volume, $d\forall/\forall$, where $\forall$ is the volume of the fluid particle. The change pressure dp required to produce this change in volume is linearly related to the bulk strain by the bulk modulus, K. That is,

$$dp = -K\frac{d\forall}{\forall} \tag{1.3}$$

The minus sign indicates that an increase in pressure causes a decrease in volume (a compressive pressure is taken to be positive). We can write this in terms of the fractional change in density, where the density of the fluid ρ is given by its mass divided by its volume (see Section 1.3.5). Since the mass m of the particle is fixed,

$$\rho = \frac{m}{\forall}$$

so that

$$\forall = \frac{m}{\rho}$$

and

$$\frac{d\forall}{\forall} = \frac{d(m/\rho)}{(m/\rho)} = \rho d\left(\frac{1}{\rho}\right) = -\frac{d\rho}{\rho}$$

Equation 1.3 becomes

$$dp = K\frac{d\rho}{\rho} \tag{1.4}$$

This compressive effect is illustrated in Example 1.5(*a*). Note that the value of the bulk modulus depends on how the compression is achieved; the bulk modulus for isothermal compression (where the temperature is held constant) is different from its adiabatic value (where there is no heat transfer allowed) or its isentropic value (where there is no heat transfer and no friction).

1.3.5 Density and Specific Gravity

Density is defined as the mass per unit volume, and it is measured in kg/m^3, or *slugs*. The usual symbol is ρ. At 4°C, water has a density of 1000 kg/m^3, so that at this temperature one cubic meter contains 1000 kg of water. At 20°C, water has a density of 998.2 kg/m^3. In contrast, air has a density of 1.204 kg/m^3 at atmospheric pressure and 20°C, so that its density is about 830 times smaller than water (see Tables 1.1 and A–C.7).

It is common practice to express the density of other liquids relative to that of water. This is called the *specific gravity*. Formally, the specific gravity (SG) of a material is the ratio of its density to that of water, that is,

$$SG = \frac{\text{density of substance}}{\text{density of water}}$$

Since the density is mass per unit volume, an equivalent definition is

$$SG = \frac{\text{mass of a given volume of substance}}{\text{mass of an equal volume of water}}$$

as long as the temperatures of the substance and the water are equal. Therefore air has a specific gravity of $1.204/998.2 = 0.001206$ at 20°C. One type of alcohol widely used in manometers (and alcoholic beverages) is ethanol, which has a density of 789 kg/m^3 at 20°C, so that its specific gravity is $789/998.2 = 0.790$. Ethanol, therefore, floats on water. Steel, on the other hand, has a density of 7850 kg/m^3, so that its specific gravity is 7.86, and it will not float on water, except possibly through the action of surface tension (see Section 1.8).

TABLE 1.1 Density and Viscosity of Some Common Fluids (at 20°C and 1 *atm*)

	$\rho\ (kg/m^3)$	$\mu\ (N \cdot s/m^2)$
Air	1.204	18.2×10^{-6}
Water	998.2	1.002×10^{-3}
Sea water	1025	1.07×10^{-3}
SAE 30 motor oil	917	0.290
Honey	≈ 1430	≈ 1.4
Mercury	13,550	1.56×10^{-3}

FIGURE 1.8   Iceberg, above and below the sea surface. Corbis/Ralph A. Clevenger.

Ice will float on water since it has a specific gravity of 0.917 (at $20°C$). For an iceberg floating in sea water (which has a specific gravity of 1.025), this means that only about 10% of its bulk will be visible above the sea surface (Figure 1.8). This result is demonstrated in Example 2.9.

1.3.6   Ideal Gas Law

Take another look at the piston and cylinder example shown in Figure 1.4. If we double the number of molecules in the cylinder, the density of the gas will double. If the extra molecules have the same speed (that is, the same temperature) as the others, the number of collisions will double, to a very good approximation. Since the pressure depends on the number of collisions, we expect the pressure to double also, so that at a constant temperature the pressure is proportional to the density.

On the other hand, if we increase the temperature without changing the density, so that the speed of the molecules increases, the impact of the molecules on the piston and walls of the cylinder will increase. The pressure therefore increases with temperature, and by observation we know that the pressure is very closely proportional to the absolute temperature.

These two observations are probably familiar from basic physics, and they are summarized in the ideal gas law, which states that

$$p = \rho RT \tag{1.5}$$

where R is the gas constant. Gas constants for a number of different gases are given in Table A–C.8. For air, $R = 287.03 \ m^2/s^2K = 1716.4 \ ft^2/s^2R$.

Equation 1.5 is an example of an equation of state, in that it relates several thermodynamic properties such as pressure, temperature, and density. Most gases obey equation 1.5 to a good approximation, except under conditions of extreme pressure or temperature where more complicated relationships must be used.

1.3.7 Compressibility in Fluids

All fluids are compressible. However under some range of conditions, it is often possible to make the approximation that a fluid is incompressible. This is particularly true for liquids. Water, for example, only changes its volume very slightly under extreme pressure [see, for instance, Example 1.5(*a*)]. Other liquids behave similarly, and under commonly encountered conditions of pressure and temperature, we generally assume that liquids are incompressible.

Gases are much more compressible. The compressibility of air, for example, is part of our common experience. By blocking off a bicycle pump and pushing down on the handle, we can easily decrease the volume of the air by 50% (Figure 1.9), so that its density increases by a factor of two (the mass of air is constant). If we assume that the temperature remains constant (somehow), we know from the ideal gas law (equation 1.5) that the pressure must also increase by a factor of two. If the initial air was at atmospheric pressure, the pressure will rise by one atmosphere (14.696 *psi*, or $1.01325 \times 10^5 \ Pa$). This experiment suggests a bulk modulus for air

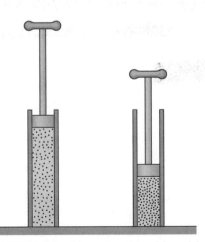

FIGURE 1.9 Air compressed in a bicycle pump.

of about $2 \times 10^5 Pa$, which is close to the exact value. Also, if we assume a pump diameter of 1.25 $in.$, then a force of $[14.7\ psi \times \pi(1.25\ in.)^2/4]\ lb_f = 18.1\ lb_f$ will be required. This is not a large force, so that doubling the pressure of ambient air is easily demonstrated.

Even though gases are much more compressible than liquids (by perhaps a factor of 10^4), small pressure differences cause only small changes in gas density. For example, a 1% change in pressure at constant temperature will change the density by 1%. In the atmosphere, a 1% change in pressure corresponds to a change in altitude of about 85 meters, so that for changes in height of the order of tall buildings we can usually assume air has a constant pressure and density [see Example 1.5(b)].

Velocity changes will also affect the fluid pressure and density. When a fluid accelerates from velocity V_1 to velocity V_2 at a constant height, the change in pressure Δp that occurs is given by

$$\Delta p = -\tfrac{1}{2}\rho(V_2^2 - V_1^2) \tag{1.6}$$

as long as its total energy is conserved (see Section 4.2). The pressure decreases as the velocity increases, and vice versa. The quantity $\tfrac{1}{2}\rho V^2$ turns up in many flow problems, and it is called the *dynamic pressure* (see Section 4.3 for further details). It represents the change in pressure due to a change in fluid velocity.

As long as the velocity variations are small, the pressure variations are small, and the fluid can be assumed to have constant density. A common yardstick is to compare the flow velocity V to the speed of sound a. This ratio is called the Mach number, M, so that

$$M = \frac{V}{a} \tag{1.7}$$

The Mach number is a nondimensional parameter since it is defined as the ratio of two velocities. That is, it is just a number, independent of the system of units used to measure V and a (see Section 1.9 and Chapter 8). It is named after Ernst Mach, who was an early pioneer in studies of sound and compressibility (see Section 15.15).

When the Mach number is less than about 0.3, the flow is usually assumed to be incompressible. To see why this is so, consider air held at 20°C as it changes its speed from zero to 230 mph (114 m/s). The speed of sound in an ideal gas is given by

$$a = \sqrt{\gamma R T} \tag{1.8}$$

where T is the absolute temperature, R is the gas constant, and γ is the ratio of specific heats ($\gamma = 1.4$ for air). At 20°C, the speed of sound in air is 343 m/s = 1126 ft/s = 768 mph. Therefore, at this temperature, 230 mph corresponds to a Mach number of 0.3. At sea level, according to equation 1.6, the pressure will decrease by about 7800 Pa at the same time, which is less than 8% of the ambient pressure. If the air temperature is kept constant, the density will decrease by the same amount. We see that relatively high speeds are required for the density to change significantly. However, when the Mach number approaches one, compressibility effects become very important. Passenger transports, such as the Boeing 747 shown in Figure 1.10, travel at a Mach number of about 0.8, and the compressibility of air is a very important factor affecting its aerodynamic design.

FIGURE 1.10 A Boeing 747 cruising at 35,000 *ft* at a Mach number of about 0.82. Courtesy of United Airlines.

1.3.8 Pressure: Transmission Through a Fluid

An important property of pressure is that it is transmitted through the fluid. For example, when an inflated bicycle tube is squeezed at one point, the pressure will increase at every other point in the tube. Measurements show that the increase is (almost) the same at every point and equal to the applied pressure; if an extra pressure of 5 *psi* were suddenly applied at the tube valve, the pressure would increase at every point in the tube by almost exactly this amount. Small differences will occur due to the weight of the air inside the tube (see Chapter 2), but in this particular case the contribution is very small. This property of transmitting pressure undiminished is a well established experimental fact, and it is a property possessed by all fluids.

However, the transmission does not occur instantaneously. It depends on the speed of sound in the medium and the shape of the container. The speed of sound is important because it measures the rate at which pressure disturbances propagate (sound is just a small pressure disturbance traveling through a medium). The shape of the container is important because pressure waves refract and reflect off the walls, and this process increases the distance and time the pressure waves need to travel. The phenomenon should be familiar to anyone who has experienced the imperfect acoustics of a poorly designed concert hall.

1.3.9 Hydraulic Presses and Hoists

A hydraulic press uses the transmissibility of fluid pressure to produce large forces. A simple press consists of two connected cylinders of significantly different size, each fitted with a piston and filled with either oil or water (see Figure 1.11). The

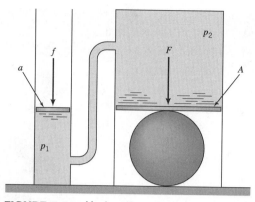

FIGURE 1.11 Hydraulic press.

pressures produced by hydraulic devices are typically hundreds or thousands of *psi*, so that hydrostatic forces due to height differences are usually neglected. If there is an unobstructed passage connecting the two cylinders, $p_1 \approx p_2$. Since pressure = (magnitude of force)/area,

$$\frac{F}{A} = \frac{f}{a}, \quad \text{so that} \quad F = \frac{A}{a} f$$

The pressure transmission amplifies the applied force; a hydraulic press is simply a hydraulic lever.

A hydraulic hoist is basically a hydraulic press that is turned around. In a typical garage hoist, compressed air is used (instead of an actuating piston) to force oil through the connecting pipe into the cylinder under a large piston, which supports the car. A lock-valve is usually placed in the connecting pipe, and when the hoist is at the right height the valve is closed, holding the pressure under the cylinder and maintaining the hoist at a constant height.

A similar application of the transmissibility of pressure is used in hydraulic brake systems. Here, the force is applied by a foot pedal, increasing the pressure in a "master" cylinder, which in turn transmits the pressure to each brake or "slave" cylinder (Figure 1.12). A brake cylinder has two opposing pistons, so that when the pressure inside the cylinder rises the two pistons move in opposite directions. In a drum brake system, each brake shoe is pivoted at one end, and attached to one of the pistons of the brake cylinder at the other end. As the piston moves out, it forces the brake shoe into contact with the brake drum. Similarly, in a disk brake

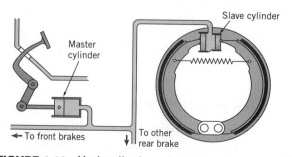

FIGURE 1.12 Hydraulic drum brake system.

system, there are two brake pads, one on each side of the disk, and the brake cylinder pushes the two brake pads into contact with the disk.

EXAMPLE 1.1 *Finding the Pressure in a Fluid*

Consider the piston and cylinder illustrated in Figure 1.4. If the piston had a mass of 1 *kg*, and an area $A = 0.01 \ m^2$, what is the pressure p of the gas in the cylinder? Atmospheric pressure p_a acts on the outside of the container.

Solution: The piston is not moving so that it is in equilibrium under the force due to its own weight, the force due to the gas pressure inside of the piston (acting up), and the force due to air pressure on the outside of the piston (acting down). The weight of the piston = mass × gravitational acceleration. The force due to pressure = pressure × area of piston. Therefore

$$pA - p_a A = 1 \ kg \times 9.8 \ m^2/s = 9.8 \ N$$

That is,

$$p - p_a = \frac{9.8}{0.01} \frac{N}{m^2} = 980 \ Pa$$

where $Pa = pascal = N/m^2$.

What is this excess pressure in *psi* (pounds per square inch)? A standard atmosphere has a pressure of 14.7 $lb_f/in.^2$, or 101,325 *Pa* (see Table 2.1). Therefore, 980 *Pa* is equal to 14.7 × 980/101,325 *psi* = 0.142 *psi*. ∎

EXAMPLE 1.2 *Force Due to Pressure*

In the center of a hurricane, the pressure can be very low. Find the force acting on the wall of a house, measuring 10 *ft* by 20 *ft*, when the pressure inside the house is 30.0 *in.* of mercury, and the pressure outside is 26.3 *in.* of mercury. Express the answer in lb_f and *N*.

Solution: A mercury barometer measures the local atmospheric pressure. A standard atmosphere has a pressure of 14.7 $lb_f/in.^2$, or 101,325 *Pa* or N/m^2 (see Table 2.1). When a mercury barometer measures a standard atmosphere, it shows a reading of 760 *mm* or 29.92 *in.* To find the resultant force, we need to find the difference in pressure on the two sides of the wall and multiply it by the area of the wall, A. That is,

$$\text{Force on wall, acting out} = (p_{inside} - p_{outside})A$$

$$= \left(\frac{30.0 - 26.3}{29.92} \times 14.7\right) \frac{lb_f}{in.^2} \times 10 \ ft \times 20 \ ft \times 12 \frac{in.}{ft} \times 12 \frac{in.}{ft}$$

$$= 52,354 \ lb_f$$

$$= 52,354 \times 4.448 \frac{N}{lb_f}$$

$$= 232,871 \ N$$

We see that the force acting on the wall is very large, and without adequate strengthening, the wall can explode outward. ∎

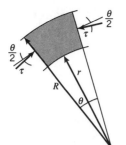

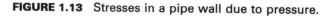

FIGURE 1.13 Stresses in a pipe wall due to pressure.

EXAMPLE 1.3 *Stresses in a Pipe Wall*

Consider a section through a pipe of outside radius R and inside radius r, as shown in Figure 1.13. If the yield stress of the material is τ_y, what is the maximum gauge pressure p_{max} that can be contained in the pipe?

Solution: For a uniform pressure inside the pipe, the force dF acting radially outward on a segment of length dz with included angle θ is

$$dF = \text{pressure} \times \text{area}$$
$$= pr\theta \, dz$$

Since the pipe is in static equilibrium (it is not tending to move), this force is exactly counterbalanced by the forces set up within the pipe material. If the stress is assumed to be uniform across the pipe wall thickness, then

$$dF = \text{wall stress} \times \text{area}$$
$$= \tau(R - r) \sin \tfrac{1}{2}\theta \, dz$$

Therefore,

$$pr\theta \, dz = (R - r)\sin \tfrac{1}{2}\theta \, dz$$

For small angles, $\sin \tfrac{1}{2}\theta \approx \tfrac{1}{2}\theta$, so that

$$p = \frac{(R - r)}{2r} \tau$$

and

$$p_{max} = \frac{t}{2r} \tau_y$$

where t is the wall thickness. In practice, the maximum allowable stress is considerably lower because of prescribed factors of safety, allowances for the way the pipe was manufactured and heat treated, possible corrosion effects, and the addition of fittings and joints, which all tend to weaken the pipe. ∎

EXAMPLE 1.4 *Density and Specific Gravity*

(*a*) Find the density of a rectangular block with dimensions 300 *mm* × 100 *mm* × 25 *mm*, of mass 10 *kg*.

TABLE 1.2 Units and Dimensions

Dimension		Units	
		SI	BG
Mass	M	kilogram	slug
Length	L	meter	foot
Time	T	second	second
Velocity	LT^{-1}	m/s	ft/s
Acceleration	LT^{-2}	m/s²	ft/s²
Velocity gradient (strain rate)	T^{-1}	s^{-1}	s^{-1}
Density	ML^{-3}	kg/m³	slug/ft³
Force	MLT^{-2}	newton	lb_f
Energy	ML^2T^{-2}	joule	$ft \cdot lb_f$
Power	ML^2T^{-3}	watt	$ft \cdot lb_f/s$
Stress	$ML^{-1}T^{-2}$	pascal (N/m²)	psi
Viscosity	$ML^{-1}T^{-1}$	Pa·s (N·s/m²)	slug/ft·s (= $lb_f \cdot s/ft^2$)
Kinematic viscosity	L^2T^{-1}	m²/s	ft²/s

(*b*) Find the density of a rectangular block with dimensions 12 *in.* × 4 *in.* × 1 *in.*, of mass 20 *lb$_m$*.

Solution: For part (*a*)

$$\rho = \frac{\text{mass}}{\text{volume}} = \frac{10}{300 \times 100 \times 25 \times 10^{-9}} \, kg/m^3 = 13{,}333 \, kg/m^3$$

For part (*b*), the unit *lb$_m$* is not part of the engineering system of units (see Table 1.2), so we first convert it to *slugs,* where

$$\text{mass in } slugs = \frac{\text{mass in } lb_m}{32.1739}$$

Therefore

$$\rho = \frac{20 \times 12 \times 12 \times 12}{32.2 \times 12 \times 4 \times 1} \, slug/ft^3 = 22.36 \, slug/ft^3$$

From Table A–C.7, we see that this material has a density somewhere between lead and gold. Its specific gravity is equal to its density divided by the density of water at 20°C. For part (*a*), therefore, the specific gravity = 13,333/998.2 = 13.36. ∎

EXAMPLE 1.5 *Bulk Modulus and Compressibility*

(*a*) Calculate the fractional change in volume of a fixed mass of seawater as it moves isentropically from the surface of the ocean to a depth of 5000 *ft.*
(*b*) Calculate the fractional change in density of a fixed mass of air as it moves isothermally from the bottom to the top of the Empire State building (a height of 350 *m,* equivalent to a change in pressure of about 4100 *Pa*).

Solution: For part (*a*), from equation 1.3,

$$dp = -K\frac{d\forall}{\forall}$$

so that the fractional change in volume is given by

$$\frac{d\forall}{\forall} = -\frac{dp}{K}$$

For seawater, the isentropic bulk modulus $K_v = 2.34 \times 10^9 \, N/m^2$ (see Table A–C.9). The change in pressure due to the change in depth may be found as follows. One standard atmosphere is equal to $101,325 \, N/m^2$, but it can also be expressed in terms of an equivalent height of water, equal to $33.90 \, ft$ (see Table 2.1). So, if the fluid moves isentropically,

$$\frac{d\forall}{\forall} = -\frac{dp}{K} = \frac{\frac{5000}{33.90} \times 101,325 \, N/m^2}{2.34 \times 10^9 \, N/m^2} = 0.0064 = 0.64\%$$

We see that seawater is highly incompressible.

For part (*b*), we use the ideal gas law (equation 1.5)

$$p = \rho RT$$

with $R = 287.03 \, m^2/s^2K$ for air. When the temperature is constant, we obtain

$$dp = RTd\rho$$

so that

$$\frac{dp}{p} = \frac{d\rho}{\rho}$$

Therefore

$$\frac{d\rho}{\rho} = \frac{4100}{101,325} = 0.0405 = 4.05\%$$

We see that air is much more compressible than seawater. ∎

EXAMPLE 1.6   *Dynamic Pressure and Mach Number*

(*a*) Calculate the change in pressure of water as it increases its speed from 0 to 30 *mph* at a constant height.

(*b*) What air speed V corresponds to $M = 0.6$ when the air temperature is 270 K?

Solution: For part (*a*), using equation 1.6, the change in pressure due to a change in speed is given by:

$$\Delta p = -\tfrac{1}{2}\rho(V_2^2 - V_1^2)$$

In this case,

$$\Delta p = -\tfrac{1}{2}\rho V_2^2$$

where 1000 kg/m^3 and $V_2 = 30 \, mph = (30/2.28) \, m/s$ (Appendix B). Hence,

$$\Delta p = -\tfrac{1}{2} \times 1000 \times \left(\frac{30}{2.28}\right)^2 Pa = 86,565 \, Pa$$

That is, the change in pressure is a little less than one atmosphere.

For part (b), from equation 1.8

$$a = \sqrt{\gamma R T}$$

For air, $R = 287.03 \ m^2/s^2K$, and $\gamma = 1.4$. Since $M = V/a$,

$$V = Ma = 0.6 \times \sqrt{1.4 \times 287.03 \times 270} \ m/s = 197.6 \ m/s \qquad \blacksquare$$

1.4 VISCOUS STRESSES

As indicated earlier, when there is no flow the stress distribution is completely described by its pressure distribution, and the bulk modulus relates the pressure to the fractional change in volume (the compression strain). When there is flow, however, shearing stresses become important, and additional normal stresses also come into play. The magnitude of these stresses depends on the fluid viscosity. Viscosity is a property of fluids, and it is related to the ability of a fluid to flow freely. Intuitively, we know that the viscosity of motor oil is higher than that of water, and the viscosity of water is higher than that of air (see Section 1.4.4 for more details). To be more precise about the nature of viscosity, we need to consider how it gives rise to viscous stresses.

1.4.1 Viscous Shear Stresses

When a shear stress is applied to a solid, it deforms by an amount that can be measured by an angle called the shear angle $\Delta\gamma$ (Figure 1.14). We can also apply a shear stress to a fluid particle by confining the fluid between two parallel plates, and moving one plate with respect to the other (see Fig. 1.20). We find that the shear angle in the fluid will grow indefinitely if the shear stress is maintained. The shear stress τ is not related to the magnitude of the shear angle, as in solids, but to the rate at which the angle is changing. For many fluids, the relationship is linear, so that

$$\tau \propto \frac{d\gamma}{dt}$$

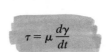

That is

$$\tau = \mu \frac{d\gamma}{dt}$$

where the coefficient of proportionality μ is called the *dynamic viscosity* of the fluid, or simply the fluid *viscosity*.

Imagine an initially rectangular fluid particle of height Δy, where the tangential force is applied to the top face, and its base is fixed (Figure 1.15). In time Δt, the

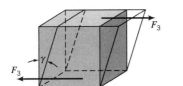

FIGURE 1.14 Solid under shear.

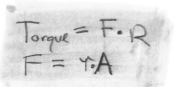

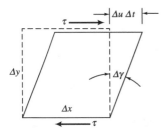

FIGURE 1.15   A fluid in shear.

top face of the particle moves a distance $\Delta u \Delta t$ relative to the bottom face, where Δu is the velocity of the top face relative to the bottom face. Since

$$\Delta \gamma \Delta y \approx \Delta u \Delta t, \quad \text{we have} \quad \frac{\Delta \gamma}{\Delta t} \approx \frac{\Delta u}{\Delta y}$$

and in the limit of a very small cube we have

$$\frac{d\gamma}{dt} = \frac{du}{dy}$$

so that

$$\tau_{yx} = \mu \frac{du}{dy} \tag{1.9}$$

The subscript yx indicates that the shear stress in the x-direction is associated with a strain rate in the y-direction. Note that du/dy has dimensions of a strain rate (time^{-1}).

Fluids which obey a linear relationship between stress and strain rate such as that given in equation 1.9 are called *Newtonian* fluids. Most common fluids are Newtonian, including air and water over very wide ranges of pressures and temperatures. The density and viscosity of some common fluids are shown in Table 1.1. A more complete listing appears in Appendix A.

Not all fluids obey a Newtonian stress–strain relationship. An enormous variety of non-Newtonian or *visco-elastic* fluids exists that obey more complicated relationships between stress and strain rate. The relationships can be nonlinear, similar to the plastic deformation of solids, or demonstrate history effects, where the stress history needs to be known before the deformation can be predicted. Such fluids are commonly encountered in the plastics and chemical industries.

1.4.2   Energy and Work Considerations

For the velocity profile shown in Figure 1.16, the local shear stress at any distance from the surface is given by equation 1.9. To overcome this viscous stress, work must be done by the fluid. If no further energy were supplied to the fluid, all motion would eventually cease because of the action of viscous stresses. For example, after we have finished stirring our coffee we see that all fluid motion begins to slow down and finally come to a halt. Viscous stresses *dissipate* the energy associated with the fluid motion. In fact, we often say that viscosity gives rise to a kind of friction within the fluid.

Viscosity also causes a drag force on a solid surface in contact with the fluid. In particular, the viscous stress at the wall, $\tau_w = (du/dy)_w$, transmits the fluid drag to the surface of the body, as illustrated in Figure 1.17.

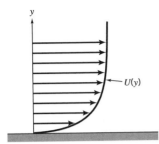

FIGURE 1.16 Velocity profile in the region near a solid surface.

Example 1.8 shows how the drag force on a solid surface can be found from the velocity profile by evaluating the velocity gradient at the wall and integrating the local surface stress over the area of the body. If the stress is constant over the area, the viscous force F_v is simply given by the shear stress at the wall times the area over which it acts. If the body moves at a constant velocity U_b, it must do work to maintain its speed. If it moves a distance Δx in a time Δt, it does work equal to $F_v\Delta x$, and it expends power equal to $F_v\Delta x/\Delta t$, that is, F_vU_b.

1.4.3 Viscous Normal Stresses

Viscosity is also important when normal stress differences occur. Consider a length of taffy. Taffy behaves a little like a fluid in that it will continue to stretch under a constant load, and it has very little elasticity, so that it does not spring back when the load is removed. Imagine that we pull lengthwise on the taffy (call it the x-direction). Its length will increase in the direction of the applied load, while its cross-sectional area will decrease. Normal stress differences are present. For example, as the taffy stretches, the point in the center will remain in its original location, but all other points move outward at speeds that increase with the distance from the center point. If the velocity at any point is u, we see that u varies with x, and that a velocity gradient or strain rate du/dx exists. The resistance to the strain rate depends on the properties of the taffy.

A fluid behaves somewhat similarly to taffy, in that fluids offer a viscous resistance to stretching or compression. The magnitude of the stress depends on a fluid property called the *extensional* viscosity. For Newtonian fluids, the viscosity is isotropic, so that the shear viscosity and the extensional viscosity are the same. Then, for a fluid in simple extension or compression, the normal stress is given by

$$\tau_{xx} = \mu\frac{du}{dx} \tag{1.10}$$

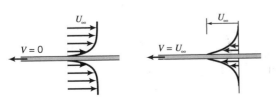

FIGURE 1.17 A long flat plate moving at constant speed in a viscous fluid. On the left is shown the velocity distributions as they appear to a stationary observer, and on the right they are shown as they appear to an observer moving with the plate.

The subscript *xx* indicates that the stress in the *x*-direction is associated with a strain rate in the *x*-direction.

1.4.4 Viscosity

We have seen that when fluids are in relative motion, shear stresses develop which depend on the viscosity of the fluid. Viscosity is measured in units of $Pa \cdot s$, $kg/(m \cdot s)$, $lb_m/(ft \cdot s)$, $N \cdot s/m^2$, or *Poise* (a unit named after the French scientist Poiseuille—see Section 15.11).

Since a viscous stress is developed (that is, a viscous force per unit area), we know from Newton's second law that the fluid must experience a rate of change of momentum. Sometimes we say that momentum "diffuses" through the fluid by the action of viscosity. To understand this statement, we need to examine the basic molecular processes that give rise to the viscosity of fluids. That is, we will take a microscopic point of view.

A flowing gas has two characteristic velocities: the average molecular speed $\bar{v}$, and the speed at which the fluid mass moves from one place to another, called the *bulk velocity, V*. For a gas, $\bar{v}$ is equal to the speed of sound. Consider a flow where the fluid is maintained at a constant temperature, so that $\bar{v}$ is the same everywhere, but where V varies with distance as in Figure 1.18. As molecules move from locations with a low bulk velocity to locations where the bulk velocity is larger (from B to A in Figure 1.18), the molecules will interact and exchange momentum with their faster neighbors. The net result is to reduce the local average bulk velocity. At the same time, molecules from regions of higher velocity will migrate to regions of lower velocity, interact with the surrounding molecules, and increase the local average velocity (from A to B in Figure 1.18).

We see that the exchange of momentum on a microscopic level tends to smooth out, or diffuse, the velocity differences in a fluid. On a macroscopic scale, we see a change in momentum in the bulk fluid and infer the action of a stress—we call it the viscous shear stress, with a diffusion coefficient called the viscosity.

Since the interactions occur within a distance comparable to the mean free path, the viscosity must depend on the average molecular speed $\bar{v}$, the number density ρ and the mean free path ℓ_m. Viscosity is therefore a property of the fluid, and for a gas it can be estimated from molecular gas dynamics. Typically, the dynamic viscosity is very small. Air at 20°C, for example, has a viscosity of $\mu = 18.2 \times 10^{-6} N \cdot s/m^2$ (see Table A–C.1). Nevertheless, the stress is given by the product of the viscosity and the velocity gradient, and viscous stresses can become very important when the magnitude of the velocity gradient is large, even when the viscosity itself is very small.

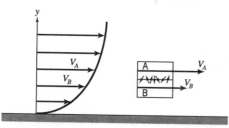

FIGURE 1.18 Momentum exchange by molecular mixing.

The molecular interpretation of viscosity also helps us to know what to expect when the temperature of the gas increases. Since the number of collisions will increase, enhancing the momentum exchange among molecules, the viscosity should increase. Figure A–C.1 confirms this expectation.

The opposite behavior is found for liquids, where the viscosity decreases with temperature. This is because liquids have a much higher density than gases, and intermolecular forces are more important. As the temperature increases, the relative importance of these bonds decreases, and therefore the molecules are more free to move. As a consequence, the viscosity of liquids decreases as their temperature increases (see Figure A–C.1).

Finally, we note that it is sometimes convenient to use a parameter called the *kinematic viscosity, ν,* defined as the dynamic viscosity divided by the density,

$$\nu = \frac{\mu}{\rho}$$

The dimensions of kinematic viscosity are length^2/time, and common units are m^2/s or ft^2/s.

1.5 MEASURES OF VISCOSITY

We have seen that viscosity is associated with the ability of a fluid to flow freely. For example, honey has about 1000 times the viscosity of water, and it is obvious that it takes a much greater force to stir honey than water (at the same rate). The fact that viscosity and flow rate are directly connected is sometimes used to measure the fluid viscosity. Motor oils, for example, are rated by the Society of Automobile Engineers in terms of the time it takes for 60 mL of the oil to flow through a calibrated hole in the bottom of a cup under the action of gravity. An oil rated as SAE 30 will take about 60 seconds, and a SAE 120 oil will take 240 seconds, and so forth. The instrument is called a Saybolt viscometer (Figure 1.19).

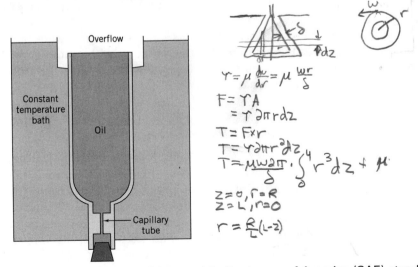

FIGURE 1.19 Society of Automobile Engineers of America (SAE) standard viscosity test. With permission, "Fluid Mechanics," Streeter & Wylie, 8th ed., published by McGraw-Hill, 1985.

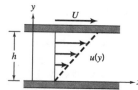

FIGURE 1.20   Linear Couette flow.

Alternatively, it is possible to use the relationship given in equation 1.9 to measure the viscosity. If we consider two plates, separated by a small gap h, and fill the gap with a fluid of viscosity μ, the force required to move one plate with respect to the other is a measure of the fluid viscosity. If the top plate moves with a velocity U relative to the bottom plate, and if the gap is small enough, the velocity profile becomes linear, as shown in Figure 1.20. This flow is called linear Couette flow. By measuring τ, the force per unit area required to move the top plate at a constant speed, the viscosity can be found as shown in Example 1.8.

It is also possible to use a circular geometry, where the fluid fills the annular gap between two concentric cylinders, and the two cylinders move at different speeds. This flow is called circular Couette flow. Another common "viscometer" uses a cone rotating in a cup (see Figure 8.7).

EXAMPLE 1.7   *Dynamic and Kinematic Viscosity*

What are the dynamic and kinematic viscosities of air at $20°C$? $60°C$? What are the values for water at these temperatures?

Solution: From Table A–C.1, we find that for air at $20°C$, $\mu = 18.2 \times 10^{-6}$ *N s/m²* and $\nu = 15.1 \times 10^{-6}$ *m²/s*. At $60°C$, $\mu = 19.7 \times 10^{-6}$ *N s/m²* and $\nu = 18.6 \times 10^{-6}$ *m²/s*.

From Table A–C.3, we find that for water at $20°C$, $\mu = 1.002 \times 10^{-3}$ *N s/m²* and $\nu = 1.004 \times 10^{-6}$ *m²/s*. At $60°C$, $\mu = 0.467 \times 10^{-3}$ *N s/m²* and $\nu = 0.475 \times 10^{-6}$ *m²/s*.

Note that: (a) the viscosity of air increases with temperature, while the viscosity of water decreases with temperature; (b) the variation with temperature is more severe for water than for air; and (c) the dynamic viscosity of air is much smaller than that of water, but its kinematic viscosity is much larger.   ∎

EXAMPLE 1.8   *Forces Due to Viscous Stresses*

Consider linear Couette flow, as shown in Figure 1.20. Note that if there is no force acting on the top plate, it will slow down and eventually stop because of the viscous stress exerted by the fluid on the plate. That is, if the plate is to keep moving at a constant speed, the viscous force set up by the shearing of the fluid must be exactly balanced by the applied force. Find the force per unit area τ to keep the top plate moving at a constant speed in terms of the fluid viscosity μ, the velocity of the moving plate U, and the gap distance h, and the power required.

Solution: The stress τ_w required to keep the plate in constant motion is equal to the viscous stress exerted by the fluid on the top plate. For a Newtonian fluid, we have

$$\tau_w = \mu \left.\frac{du}{dy}\right|_{wall}$$

so that

$$\tau_w = \mu \frac{U}{h}$$

We see that as the velocity and the viscosity increase, the stress increases, as expected, and as the gap size increases, the stress decreases, again as expected. Also,

$$\mu = \frac{h\tau_w}{U}$$

We can use this result to determine the viscosity of a fluid by measuring τ_w at a known speed and gap width.

The power required to drive the top plate is given by the applied force times the plate velocity. The force is equal to the stress τ_w times the area of the plate A, so that

$$\text{power} = \tau_w A U = \mu A \frac{U^2}{h}$$

which shows that the power is proportional to the velocity squared.   ■

1.6 BOUNDARY LAYERS

Viscous effects are particularly important near solid surfaces, where the strong interaction of the molecules of the fluid with the molecules of the solid reduces the relative velocity between the fluid and the solid to zero. For a stationary surface, therefore, the fluid velocity in the region near the wall must reduce to zero (Figure 1.21).

We see this effect in nature when a dust cloud driven by the wind moves along the ground. Not all the dust particles are moving at the same speed; close to the ground they move more slowly than further away. If we were to look in the region very close to the ground, we would see that the dust particles there are almost stationary, no matter how strong the wind. Right at the ground, the dust particles do not move at all, indicating that the air has zero velocity at this point. We call this phenomenon the *no-slip condition*, in that there is no relative motion allowed between the air and the ground at their point of contact. It follows that the flow velocity varies with distance from the wall; from zero at the wall to its full value some distance away. In most cases, these regions of significant velocity gradient are thin (compared to a typical body dimension), and they are called *boundary*

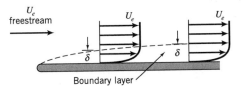

FIGURE 1.21 Growth of a boundary layer along a stationary flat plate.

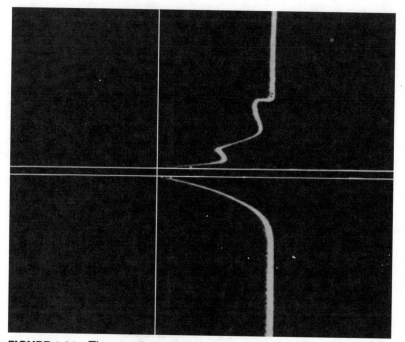

FIGURE 1.22 The no-slip condition in water flow past a thin plate. Flow is from left to right. The upper flow is turbulent, and the lower flow is laminar. With permission, *Illustrated Experiments in Fluid Mechanics,* (*The NCMF Book of Film Notes,* National Committee for Fluid Mechanics Films, Education Development Center, Inc., © 1972).

layers. Within the boundary layer, strong velocity gradients can occur, and therefore viscous stresses can become important (as indicated by equation 1.9). Generally, the thickness of the boundary layer will grow with distance along the surface as the molecular mixing diffuses the momentum differences between adjacent layers of fluid.

The no-slip condition is illustrated in Figure 1.22. Here, water is flowing above and below a thin flat plate. The flow is made visible by forming a line of hydrogen bubbles in the water (the technique is described in Section 3.3). The line was originally straight, but as the flow sweeps the bubbles downstream (from left to right in Figure 1.22), the line changes its shape since bubbles in regions of faster flow will travel further in a given time than bubbles in regions of slower flow. The hydrogen bubbles therefore make the velocity distribution visible. Bubbles near the surface of the plate will move slowest of all, and *at* the surface they are stationary with respect to the surface because of the no-slip condition. The upper and lower flows in Figure 1.22 are different (the upper one is turbulent, and the lower one is laminar; see Section 1.7), but the no-slip condition applies to both.

EXAMPLE 1.9 *Viscous Stress in a Boundary Layer*

A particular laminar boundary layer velocity profile is given by

$$\frac{u}{U_e} = 2\left(\frac{y}{\delta}\right) - \left(\frac{y}{\delta}\right)^2$$

for $y \leq \delta$ so that at $y = \delta$, $u = U_e$. Find the shear stress τ as a function of the distance from the wall, y.

Solution: We have

$$\tau = \mu \frac{du}{dy}$$

so that

$$\tau = \mu \frac{d}{dy}\left[U_e \left(2\left(\frac{y}{\delta}\right) - \left(\frac{y}{\delta}\right)^2 \right)\right]$$

$$= \mu U_e \left[\frac{2}{\delta} - \frac{2y}{\delta^2} \right] = \frac{2\mu U_e}{\delta}\left(1 - \frac{y}{\delta}\right)$$

At the wall, where $y = 0$, $\tau = \tau_w$, and so

$$\tau_w = \frac{2\mu U_e}{\delta}$$

∎

1.7 LAMINAR AND TURBULENT FLOW

We have described boundary layer flow as the flow in a region close to a solid surface where viscous stresses are important. When the layers of fluid inside the boundary layer slide over each other in a very disciplined way, the flow is called *laminar* (the lower flow in Figure 1.22). Whenever the size of the object is small, or the speed of the flow is low, or the viscosity of the fluid is large, we observe laminar flow. However, when the body is large, or it is moving at a high speed, or the viscosity of the fluid is small, the entire nature of the flow changes. Instead of smooth, well-ordered, laminar flow, irregular eddying motions appear, signaling the presence of *turbulent* flow (the upper flow in Figure 1.22).

Turbulent flows are all around us. We see it in the swirling of snow in the wind, the sudden and violent motion of an aircraft encountering "turbulence" in the atmosphere, the mixing of cream in coffee, and the irregular appearance of water issuing from a fully opened faucet. Turbulent boundary layers are seen whenever we observe the dust kicked up by the wind, or look alongside the hull of a ship moving in smooth water, where swirls and eddies are often seen in a thin boundary layer close to the hull. Inside the eddies and between them, fluid layers are in relative motion, and viscous stresses are causing energy dissipation. Because of the high degree of activity associated with the eddies and their fluctuating velocities, the viscous energy dissipation inside a turbulent flow can be very much greater than in a laminar flow.

We suggested that laminar flow is the state of fluid flow found at low velocities, for bodies of small scale, for fluids with a high kinematic viscosity. In other words, it is the flow state found at *low Reynolds number*, which is a nondimensional ratio defined by

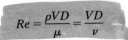

$$Re = \frac{\rho V D}{\mu} = \frac{VD}{\nu}$$

where V is the velocity and D is a characteristic dimension (the body length, the tube diameter, etc.). Turbulent flow is found when the velocity is high, on large bodies, for fluids with a low kinematic viscosity. That is, turbulent flow is the state of fluid flow found at *high* Reynolds number. Since the losses in turbulent flow are much greater than in laminar flow, the distinction between these two flow states is of great practical importance.

We can give a physical interpretation to the Reynolds number by noting that we can write it as

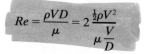

$$Re = \frac{\rho V D}{\mu} = 2 \frac{\frac{1}{2}\rho V^2}{\mu \frac{V}{D}}$$

so that it can be thought of as (twice) the ratio of the dynamic pressure to a typical viscous stress. From Section 1.3.6, we also note that the dynamic pressure measures the rate of change in momentum produced by a pressure difference, so that another interpretation for the Reynolds number is to think of it as the ratio of a typical inertia force (mass times acceleration) to a typical viscous force.

EXAMPLE 1.10   *Reynolds Number*

(*a*) Find the Reynolds number of a water flow at an average speed of 20 *ft/s* at 60°*F* in a 6 *in.* diameter pipe.
(*b*) If the flow is laminar at Reynolds numbers below 2300, what is the maximum speed at which we can expect to see laminar flow?

Solution: For part (*a*), we have the Reynolds number for pipe flow

$$Re = \frac{\overline{V}D}{\nu}$$

where D is the pipe diameter and $\overline{V}$ is the average flow velocity. Using Table A–C.4, we find that for water at 60°*F*, $\nu = 1.21 \times 10^{-5}$ *ft²/s*, so that

$$\frac{\overline{V}D}{\nu} = \frac{20\,ft/s \times 0.5\,ft}{1.21 \times 10^{-5}\,ft^2/s} = 8.26 \times 10^5$$

For part (*b*), we have

$$V_{max} = \frac{Re_{max} \times \nu}{D} = \frac{2300 \times 1.21 \times 10^{-5}\,ft^2/s}{0.5\,ft} = 0.056\,ft/s$$

so that the flow is laminar for velocities less than 0.056 *ft/s* or 0.67 *in./s*. This is a very slow flow. In most practical applications, such as in domestic water supply systems, we would therefore expect to see turbulent flow.   ∎

1.8   **SURFACE TENSION

At the free surface that forms between a gas and a liquid, a fluid property called *surface tension* becomes important.

By observation, we know that the surface of a liquid tends to contract to the smallest possible area, behaving as though its surface were a stretched elastic membrane. For example, small drops of liquid in a spray become spherical, since a sphere has the smallest surface area for a given volume. Spherical lead shot (as found in shotgun pellets) used to be made by dropping molten lead from a high tower. The drops of liquid lead would take a spherical form because of surface

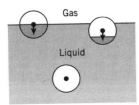

FIGURE 1.23   Surface tension and "spheres of molecular attraction."

tension, and maintain this form as they cooled and solidified during their fall. Also, when a brush is wet, the hairs cling together, since the films between them tend to contract. Some insects can walk on water, and a steel needle placed carefully on a water surface will float.

These surface tension phenomena are due to the attractive forces that exist between molecules. The forces fall off quickly with distance, and they are appreciable only over a very short distance (of the order of 5×10^{-6} m, that is, 5 μm). This distance forms the radius of a sphere around a given molecule, and only molecules contained in this sphere will attract the one at the center (Figure 1.23). For a molecule well inside the body of the liquid, its "sphere of molecular attraction" lies completely in the liquid, and the molecule is attracted equally in all directions by the surrounding molecules, so that the resultant force acting on it is zero. For a molecule near the surface, where its sphere of attraction lies partially outside the liquid, the resultant force is no longer zero; the surrounding molecules of the liquid tend to pull the center molecule into the liquid, and this force is not balanced by the attractive force exerted by the surrounding gas molecules because they are fewer in number (the gas has a much lower density than the liquid). The resultant force on molecules near the surface is inward, tending to make the surface area as small as possible.

The coefficient of surface tension σ of a liquid is the tensile force per unit length of a line on the surface. Common units are lb_f/ft or N/m. Some typical values at $20°C$ are

air-water: $\sigma = 0.0050 \ lb_f/ft = 0.073 \ N/m$

air-mercury: $\sigma = 0.033 \ lb_f/ft \ = 0.48 \ N/m$

Generally, dissolving an organic substance in water, such as grease or soap, will lower the surface tension, whereas inorganic substances raise the surface tension of water slightly. The surface tension of most liquids decreases with temperature, and this effect is especially noticeable for water.

We will now describe some particular phenomena due to surface tension, including the excess pressure in a drop or bubble, the formation of a meniscus on a liquid in a small-diameter tube, and *capillarity*.

1.8.1   Drops and Bubbles

The surfaces of drops or bubbles tend to contract due to surface tension, which increases their internal pressure. When the drop or bubble stops growing, it is in equilibrium under the action of the forces due to surface tension and the excess pressure Δp (the difference between the internal and external pressures).

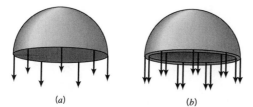

FIGURE 1.24 Equilibrium of (a) drop and (b) bubble, where the excess pressure is balanced by surface tension.

Figure 1.24(a) shows one half of a spherical drop of radius r. The resultant upward force due to the excess pressure is $\pi r^2 \Delta p$. Because the drop is assumed to be static equilibrium (that is, it is not accelerating, and it is not growing), this force must be balanced by the surface tension force $2\pi r\sigma$, acting around the edge of the hemisphere (we neglect its weight). Therefore, for a drop

$$\pi r^2 \Delta p = 2\pi r\sigma$$

That is,

$$(a) \quad \Delta p = \frac{2\sigma}{r}$$

In the case of a bubble [Figure 1.24(b)] there are two surfaces to be considered, inside and out, so that for a bubble

$$(b) \quad \Delta p = \frac{4\sigma}{r}$$

1.8.2 Forming a Meniscus

The free surface of a liquid will form a curved surface when it comes in contact with a solid. Figure 1.25(a) shows two glass tubes, one containing mercury and the other water. The free surfaces are curved, convex for mercury, and concave for water. The angle between the solid surface AB and the tangent BC to the liquid surface at the point of contact [Figure 1.25(b)] is called the *angle of contact, θ*. For liquids which *wet* the surface (for example, water on glass), the angle is less than 90°, and for liquids which do not wet the surface (such as mercury on glass), the angle is greater than 90°. For pure water on clean glass, the angle of contact is

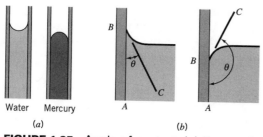

Water   Mercury

(a)

(b)

FIGURE 1.25 Angle of contact. (a) Free surface shape of water and mercury in glass tubes. (b) A wetting, and a non-wetting liquid.

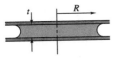

FIGURE 1.26 A drop of liquid squeezed between two glass plates.

approximately zero, and for various metal surfaces it lies between 3° and 11°. For mercury on glass, its value is 130° to 145°.

Water "wets" glass because the attractive forces between the water molecules and the glass molecules exceed the forces between water molecules, and the reverse holds true for mercury. The fact that the contact angle depends on the nature of the surface is clearly illustrated by the behavior of water droplets. On a clean glass plate, $\theta \approx 0$, and a drop of water will spread out to wet the surface. However, on a freshly waxed surface such as a car hood, drops will "bead up," showing that $\theta > 0°$.

Consider also a drop of water pressed between two plates a small distance t apart (Figure 1.26). The radius of the circular spot made by the drop is R. The pressure inside the drop is less than the surrounding atmosphere by an amount depending on the tension in the free surface. To pull the plates apart, a force F is required. The force due to surface tension is given by $4\pi R\sigma \cos \theta$, which is balanced by the reduced pressure acting on the circumferential area $2\pi Rt$ (only the radial component needs to be taken into account). That is,

$$4\pi R\sigma \cos \theta = 2\pi Rt\, \Delta p$$

In turn, the force required to pull the plates apart is given by the reduced pressure times the area of the spot, which is approximately equal to πR^2. Hence

$$F = \pi R^2 \Delta p = \frac{2\sigma \pi R^2 \cos \theta}{t} \tag{1.11}$$

This force can be quite large if the film thickness t is small. For example, when a pot sits on a wet kitchen bench, the gap between the pot and the bench fills with water, and the pot can be remarkably difficult to pull off the bench. This force is due to surface tension.

1.8.3 Capillarity

Another phenomenon due to surface tension is capillarity. When a clean glass tube of radius r is inserted into a dish of water, the water will rise inside the tube a distance h above the surface (Figure 1.27). This happens because the attraction between glass and water molecules is greater than that between water molecules themselves, producing an upward force. The liquid rises until the weight of

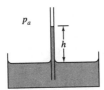

FIGURE 1.27 Water in a glass tube: a demonstration of capillarity.

the liquid column balances the upward force due to surface tension. If θ is the angle of contact, the upward component of surface tension is $\sigma \cos \theta$ [see Figure 1.25 (b)], and it produces an upward force of $2\pi r\sigma \cos \theta$ around the inside perimeter of the tube. If we neglect the contribution of the curved surface to the height of the column, then the weight of the column of liquid is equal to its volume $(\pi r^2 h)$ times the fluid density ρ times the acceleration due to gravity g, that is, $\rho g \pi r^2 h$. Hence,

$$\rho g \pi r^2 h = 2\pi r\sigma \cos \theta$$

and

$$h = \frac{2\sigma \cos \theta}{\rho g r}$$

The capillary rise h is therefore inversely proportional to the tube radius. For water on clean glass, $\theta \approx 0$, and

$$h = \frac{2\sigma}{\rho g r}$$

For mercury in a glass tube, $\theta > 90°$ and h is negative, so that there is a capillary depression.

1.9 UNITS AND DIMENSIONS

As a final topic in this chapter, we will say a few words about units and dimensions. Whenever we solve a problem in engineering or physics, it is extremely important to pay strict attention to the units used in expressing the forces, accelerations, material properties, and so on. The two systems of units used in this book are the SI system (Système Internationale) and the BG (British Gravitational) system. To avoid errors, it is absolutely essential to correctly convert from one system of units to another, and to maintain strict consistency within a given system of units. There are no easy solutions to these difficulties, but by using the SI system whenever possible, many unnecessary mistakes can often be avoided. A list of commonly used conversion factors is given in Appendix B.

It is especially important to make the correct distinction between mass and force. In the SI system, mass is measured in kilograms, and force is measured in newtons. A mass m in kilograms has a weight in newtons equal to mg, where g is the acceleration due to gravity $(= 9.8 \ m/s^2)$. There is no such quantity as "kilogram-force," although it is sometimes (incorrectly) used. What is meant by kilogram-force is the force required to move one kilogram mass with an acceleration of 9.8 m/s^2, and it is equal to 9.8 N.

In the BG system, mass is measured in slugs, and force is measured in pound-force (lb_f). A mass m in slugs has a weight in lb_f equal to mg, where g is the acceleration due to gravity $(= 32.1739 \ ft/s^2)$. When the quantity "pound-mass" (lb_m) is used, it should always be converted to slugs first by dividing lb_m by the factor 32.1739. The force required to move 1 lb_m with an acceleration of 1 ft/s^2 is

$$1 \ lb_m \ ft/s^2 = \frac{1}{32.2} \ slug \ ft/s^2 = \frac{1}{32.2} \ lb_f$$

Remember that $1 \ lb_f = 1 \ slug \ ft/s^2$.

It is also necessary to make a distinction between units and dimensions. The *units* we use depend on whatever system we have chosen, and they include quantities like feet, seconds, newtons, and pascals. In contrast, a *dimension* is a more abstract notion, and it is the term used to describe concepts such as mass, length, and time. For example, an object has a quality of "length" independent of the system of units we choose to use. Similarly, "mass" and "time" are concepts that have a meaning independent of any system of units. All physically meaningful quantities, such as acceleration, force, stress, and so forth, share this quality.

Interestingly, we can describe the dimensions of any quantity in terms of a very small set of what are called *fundamental* dimensions. For example, acceleration has the dimensions of length/(time)^2 (in shorthand, LT^{-2}), force has the dimensions of mass times acceleration (MLT^{-2}), density has the dimensions of mass per unit volume (ML^{-3}), and stress has the dimensions of force/area ($ML^{-1}T^{-2}$) (see Table 1.2).

A number of quantities are inherently nondimensional, such as the numbers of counting. Also, ratios of two quantities with the same dimension are dimensionless. For example, the bulk strain $d\forall/\forall$ is the ratio of two quantities with the dimension of volume, and it is nondimensional. The shear strain γ is another example. It is measured in terms of an angle, and angles are usually measured in radians. Since a radian is the ratio of an arc-length to a radius, that is, the ratio of two lengths, it is nondimensional. Specific gravity and Mach number are also the ratio of two quantities with the same dimensions (density and velocity, respectively), and therefore they are also nondimensional. Finally, we have introduced the Reynolds number, which is a combination of quantities such that the combination is nondimensional. Nondimensional quantities are independent of the system of units as long as the units are consistent, that is, if the same system of units is used throughout. They are widely used in fluid mechanics, as we shall see.

EXAMPLE 1.11 *Units and Converting Between Units*

Consider a rectangular block with dimensions 300 *mm* × 100 *mm* × 25 *mm*, of mass 10 *kg*, resting on a surface (Figure 1.28). The pressure acting over the area of contact can be found as follows.

Solution: When in position (*a*), the force exerted on the table is equal to the weight of the block (= mass × gravitational acceleration = 98 *N*), but the average pressure over the surface in contact with the table is

$$\frac{98}{100 \times 25 \times 10^{-6}} \, N/m^2 = 39,200 \, Pa$$

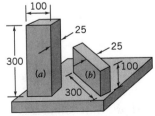

FIGURE 1.28 Pressure exerted by a weight resting on a surface. All dimensions in millimeters.

where $Pa = pascal = N/m^2$. Since pressure is a stress, it has dimensions of force per unit area.

In position (b), the force exerted on the table is still equal to 98 N, but the average pressure over the surface in contact with the table is reduced to $98/(300 \times 25 \times 10^{-6})$ N/m^2, that is, 13,067 Pa.

We can repeat this example using engineering units. Let's take a rectangular block, made of a different material with dimensions 12 $in. \times 4\ in. \times 1\ in.$, of mass 20 lb_m (this is similar to the case shown in Figure 1.28). Find the pressure acting over the area in the BG system.

Solution: The unit lb_m is not part of the engineering system of units (see Table 1.2), so we first convert it to *slugs,* where

$$\text{mass in } slugs = \frac{\text{mass in } lb_m}{32.1739}$$

So 20 $lb_m = 0.622$ *slug.*

When in position (a), the force exerted on the table is equal to the weight of the block (= mass $\times$ gravitational acceleration = 20 lb_f), but the average pressure over the surface in contact with the table is $20/(4 \times 1)$ $lb_f/in.^2$, that is, 5 *psi.* In position (b), the force exerted on the table is still equal to 20 lb_f, but now the average pressure over the surface in contact with the table is $20/(12 \times 1)$ $lb_f/in.^2$, that is, 1.67 *psi.* ∎

PROBLEMS

1.1   A body requires a force of 400 N to accelerate at a rate of 1.0 m/s^2. Find the mass of the body in

(a) kilograms

(b) slugs

(c) lb_m

1.2   What volume of fresh water at 20°C will have the same weight as a cubic foot of lead? Oak? Seawater? Air at atmospheric pressure and 100°C? Helium at atmospheric pressure and 20°C?

1.3   What is the weight of one cubic foot of gold? Pine? Water at 40°F? Air at atmospheric pressure and 60°F? Hydrogen at 7 atmospheres of pressure and 20°C?

1.4   What is the weight of 5000 *liters* of hydrogen gas at 20°C and 100 *atm* on earth? On the moon?

1.5   What is the specific gravity of gold? Aluminum? Seawater? Air at atmospheric pressure and −20°C? Argon at atmospheric pressure and 20°C?

1.6   When the temperature changes from 20 to 30°C, at atmospheric pressure

(a) by what percentage does the viscosity μ of air change?

(b) of water?

(c) by what percentage does the kinematic viscosity v of air change?

(d) of water?

1.7   A scuba diving tank initially holds 0.25 ft^3 of air at 3000 *psi.* If the diver uses the air at a rate of 0.05 kg/min, approximately how long will the tank last? Assume that the temperature of the gas remains at 20°C.

1.8 A hollow cylinder of 30 *cm* diameter and wall thickness 20 *mm* has a pressure of 100 *atm* acting on the inside and 1 *atm* acting on the outside. Find the stress in the cylinder material.

1.9 What is the change in volume of 1 kilogram of fresh water kept at 5°C as it moves from a depth of 1 *m* to a depth of 100 *m*. Take the isothermal bulk modulus of water to be 2×10^4 *atm*.

1.10 A square submarine hatch 60 *cm* by 60 *cm* has 3.2 *atm* pressure acting on the outside and 2.6 *atm* pressure acting on the inside. Find the resultant force acting on the hatch.

1.11 A Boeing 747 airplane has a total wing area of about 500 m^2. If its weight in level cruise is 500,000 *lb_f*, find the average pressure difference between the top and bottom surfaces of the wing, in *psi* and in *Pa*.

1.12 Find the force acting on the wall of a house in the eye of a hurricane, when the pressure inside the house is 1,000 *mbar,* and the pressure outside is 910 *mbar*. The wall measures 3.5 *m* by 8 *m*. Express the answer in N and *lb_f*.

1.13 The pressure at any point in the atmosphere is equal to the total weight of the air above that point, per unit area. Given that the atmospheric pressure at sea level is about $10^5 Pa$, and the radius of the earth is 6370 *km*, estimate the mass of all the air contained in the atmosphere.

1.14 Estimate the change in pressure that occurs when still air at 100°*F* and atmospheric pressure is accelerated to a speed of 100 *mph* at constant temperature. Neglect compressibility effects.

1.15 For the previous problem, estimate the error in the calculation of the pressure change due to compressibility. Assume an isothermal acceleration.

1.16 Compute the Mach number of a flow of air at a temperature of −50°*C* moving at a speed of 500 *mph*.

1.17 Compute the Mach number of a bullet fired into still air at a temperature of 60°*F* at a speed of 1500 *ft/s*.

1.18 A boogy board of area 1 m^2 slides at a speed of 3 *m/s* over the beach supported on a film of water 3 *mm* thick. If the velocity distribution in the film is linear, and the temperature of the water is 30°*C*, find the force applied to the board.

1.19 The viscosity of a Newtonian liquid can be estimated using the instrument shown in Figure P1.19. The outer cylinder rotates at an angular speed ω. If the torque (dimensions of force times length) required to keep the inner cylinder stationary is *T*, and the velocity distribution in the gap is linear, find an expression for the viscosity μ in terms of *T*, ω, *H*, δ, and *R*.

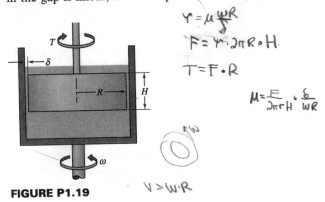

$$\tau = \mu \frac{\omega R}{\delta}$$

$$F = \tau \cdot 2\pi R \cdot H$$

$$T = F \cdot R$$

$$\mu = \frac{E}{2\pi r H} \cdot \frac{\delta}{\omega R}$$

$$V = \omega \cdot R$$

FIGURE P1.19

1.20 A Taylor-Couette apparatus consists of an inner cylinder and an outer cylinder that can rotate independently. The gap between the cylinders is filled with a Newtonian fluid. For a particular experiment at 20°*C*, we find that when the inner cylinder is held stationary and a force of 0.985 N is applied to the outer cylinder, the outer cylinder rotates at a speed of 1 Hz. Find the viscosity of the fluid, assuming that the velocity profile in the gap is linear. The inner and outer cylinders have radii of 200 mm and 150 mm, respectively, and the cylinders are 300 mm long. Can you suggest which fluid is being tested?

1.21 A 10 cm cube of mass 2 kg lubricated with SAE 10 oil at 20°C (viscosity of 0.104 $N \cdot s/m^2$) slides down a 10° inclined plane at a constant velocity. Estimate the speed of the body if the oil film has a thickness of 1 mm, and the velocity distribution is linear.

1.22 When water at 40°F flows through a channel of height h, width W, and length L at low Reynolds number, the flow is laminar and the velocity distribution is parabolic, as shown in Figure P1.22. When h = 2 $in.$, W = 30 $in.$, L = 10 ft, and the maximum velocity is 1 ft/s:

(a) Calculate the Reynolds number based on channel height and maximum velocity.

(b) Find the viscous stress at the wall τ_w, and the total viscous force acting on the channel, assuming τ_w is constant.

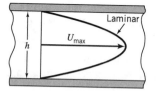

FIGURE P1.22

1.23 Given that the mean free path of a given gas is inversely proportional to its density, and that the mean free path of air at sea level is 0.089 μm, find the mean free path at 30 km altitude, and at 50 km altitude. Estimate the altitude where the mean free path is 5 mm. Can we apply the continuum approximation under these conditions?

1.24 Estimate the Reynolds number for the following:

(a) a professionally pitched baseball

(b) a shark swimming at full speed

(c) a passenger jet liner at cruise speed

(d) the wind flow around the Empire State building

(e) methane (CH_4) flowing in a 2 m diameter pipeline at 40 kg/s;

(f) a mosquito flying in air.

Make reasonable estimates of characteristic speeds and lengths, and state them.

1.25 Find the characteristic Reynolds number of a submarine of length 120 m moving at 10 m/s in water. If a 12:1 model is to be tested at the same Reynolds number, describe the test conditions if you were to use

(a) water

(b) air at standard temperatures and pressure

(c) air at 200 atmospheres and room temperature (assume that the viscosity is not a function of pressure)

1.26 The largest artery in the body is the aorta. If the maximum diameter of the aorta is 2 cm, and the maximum average velocity of the blood is 20 cm/s, determine if the flow in the aorta is laminar or turbulent (assume that blood has the same density as water but three times its viscosity—blood, after all, is thicker than water).

1.27 Find the height to which water at 300°K will rise due to capillary action in a glass tube 3 mm in diameter.

1.28 A liquid at $10°C$ rises to a height of 20 *mm* in a 0.4 *mm* glass tube. The angle of contact is 45°. Determine the surface tension of the liquid if its density is 1200 kg/m^3.

1.29 What is the pressure inside a droplet of water 0.1 *mm* diameter if the ambient pressure is atmospheric?

1.30 A beer bubble has an effective surface tension coefficient of 0.073 *N/m*. What is the overpressure inside the bubble if it has a diameter of 1.0 *mm*?

1.31 What is the maximum diameter of a thin glass ring for it to continue to float on a water surface at 20°C? The mass of the ring is 5 g.

Prob 1.3

$r\Omega_2 = V_2$

$\delta = \Delta y = r_2 - r_1$

$F = (2\pi r_2 L)\Upsilon$

Area $\nearrow$

$\Upsilon = \dfrac{F}{2\pi r_2 L}$

$\Upsilon = \mu_1 \dfrac{du}{dy} = \mu \dfrac{(V_2 - 0)}{\delta}$

$\mu = \dfrac{\Upsilon \delta}{V_2} = \boxed{\dfrac{F}{2\pi r_2 L} \cdot \dfrac{(r_2 - r_1)}{r_2 \Omega_2}}$

Prob 1.21

$F = W \sin\theta$

$W = mg$

$\Upsilon_w = \mu \dfrac{du}{dy} = \mu \dfrac{V}{\delta}$

$\Upsilon_w = \dfrac{Fx}{A} = \dfrac{mg \sin\theta}{10 cm^2}$

$V = \dfrac{\Upsilon_w \delta}{\mu}$

$\Upsilon = \mu \dfrac{V_\theta}{\delta} = \mu \dfrac{r\Omega}{.5-x}$ one side

$= \mu \dfrac{r\Omega}{.5+x}$ otherside

$V_\theta = r\Omega$

$dA = r\, dr\, d\theta$

$A = \int_0^{2\pi} \int_\delta^R r\, dr\, d\theta$

$= 2\pi \cdot \dfrac{R^2}{2} = \pi R^2$

$F = \Upsilon\, dA$

$dM = r\, dF = \Upsilon r\, dA$

$T_R = \int_0^{2\pi} \int_0^R \Upsilon r\, dA$

$= \int\int \left(\dfrac{\mu r\Omega}{.5-x}\right)(r)(r\, dr\, d\theta)$

$T_R = \left(\dfrac{\mu\Omega}{.5-x}\right) 2\pi \left(\dfrac{R^4}{4}\right)$

$T_0 = \left(\dfrac{\mu\Omega}{.5+x}\right) 2\pi \left(\dfrac{R^4}{4}\right)$

$T_+ = \left(\dfrac{\pi}{2}\mu\Omega R^4\right)\left(\dfrac{1}{.5-x} + \dfrac{1}{.5+x}\right)$

CHAPTER *2*

FLUID STATICS

In this chapter, we consider fluids in static equilibrium. To be in static equilibrium, the fluid must be at rest, or it must be moving in such a way that there is no relative motion between adjacent fluid particles. There can be no velocity gradients, and consequently, there will be no viscous stresses. A fluid in static equilibrium, therefore, is acted on only by forces due to pressure and its own weight, and perhaps by additional body forces due to externally imposed accelerations.

The simplest case occurs when the fluid is at rest. For example, if a liquid is poured into a bucket and left to stand until all relative motions have died out, the fluid is then in static equilibrium. At this point, there is no resultant force acting on the fluid.

It is also possible to have a moving fluid in static equilibrium, as long as no part of the fluid is moving with respect to any other part. This is called *rigid-body* motion. When the fluid and its container are moving at constant speed, for instance, it reaches an equilibrium where the forces due to pressure and its own weight are balanced. However, when this system is accelerating, the inertia force needs to be taken into account, as we shall see in Section 2.11.

2.1 THE HYDROSTATIC EQUATION

We begin by considering a fluid at rest. We choose a small fluid element, that is, a small fixed volume of fluid located at some arbitrary point in the liquid with dimensions of δx, δy, and δz in the x-, y-, and z-directions, respectively. The z-axis points in the direction opposite to the gravitational vector (see Figure 2.1) so that the positive direction is vertically up.

The only forces acting on the fluid element are those due to gravity and pressure differences. Since there is no resultant acceleration of the fluid element, these forces must balance. The force due to gravity acts only in the vertical direction, and we see immediately that the pressure cannot vary in the horizontal plane. In the x-direction, for example, the force due to pressure acting on the left face of the element (*abef*) must cancel the force due to pressure acting on the right face of the element (*cdgh*), since there is no other force acting in the horizontal plane. The pressures on these two faces must be equal, and so the pressure cannot vary in the x-direction. Similarly, it cannot vary in the y-direction.

For the vertical direction, we use a Taylor-series expansion to express the pressure on the top and bottom faces of the element in terms of pressure at the center of the element and its derivatives at that point (see Section A–A.8.4). That is,

$$p_{top} = p_0 + \frac{\delta z}{2} \frac{dp}{dz}\bigg|_0 + \frac{1}{2!} \left(\frac{\delta z}{2}\right)^2 \frac{d^2p}{dz^2}\bigg|_0 - \cdots \qquad (2.1)$$

(the positive sign on the first derivative reflects the fact that when we move from the center of the cube to the top face, we move in the positive z-direction). Similarly,

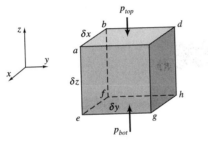

FIGURE 2.1 Static equilibrium of a small element of fluid under the action of gravity and fluid pressure.

on the bottom face of the cube

$$p_{bot} = p_0 - \frac{\delta z}{2} \frac{dp}{dz}\Big|_0 + \frac{1}{2!} \left(\frac{\delta z}{2}\right)^2 \frac{d^2p}{dz^2}\Big|_0 + \cdots \tag{2.2}$$

As the volume becomes infinitesimally small, the resultant force due to pressure is given by

$$F_{press} = (p_{bot} - p_{top})\, dx\, dy = \frac{dp}{dz}\Big|_0 dx\, dy\, dz$$

which acts in the positive z-direction (opposite to the direction of $\mathbf{g}$). The second-order terms cancel exactly, and the third- and higher-order terms become negligible as the volume of the element becomes very small.

To find the weight of the element, we will assume that the density (like the pressure) is only a function of z, since we expect a direct relationship between pressure and density. Hence,

$$F_{weight} = \int_{top}^{bot} \rho(z)g\, dx\, dy\, dz = g\, dx\, dy \int_{top}^{bot} \rho(z)\, dz$$

$$= g\, dx\, dy\, \tfrac{1}{2}\left(\rho_{top} + \rho_{bot}\right) dz$$

The densities at the top and bottom of the cube are given by equations similar to equations 2.1 and 2.2, respectively, with the pressure p replaced by the density ρ. The weight of the fluid element is then given by

$$F_{weight} = (\rho_0 + \cdots)\, g\, dx\, dy\, dz$$

For static equilibrium, $F_{weight} + F_{press} = 0$, and so we obtain the *hydrostatic equation* describing the variation of pressure in a fluid under the action of gravity

$$\boxed{\frac{dp}{dz} = -\rho g,} \tag{2.3}$$

Remember that z points in the direction opposite to that of $\mathbf{g}$ (vertically up). The subscript identifying the center of the element has been dropped because the final result applies to any point in the fluid.

For a different coordinate system, where z points in the same direction as the gravitational vector,

$$\boxed{\frac{dp}{dz} = \rho g} \tag{2.4}$$

We can also use vector notation to write the hydrostatic equation in a form that is independent of the coordinate system.

$$\nabla p = \rho \mathbf{g},$$

(2.5)

where the symbol ∇ is the gradient operator (see Section A–A.4).

2.2 ABSOLUTE AND GAUGE PRESSURE

Consider a vessel filled to a depth h with a liquid of constant density ρ. The top of the container is open to the atmosphere, so that the pressure at the surface of the liquid is equal to atmospheric pressure, p_a (Figure 2.2). The z-direction is positive going up, so we use equation 2.3. This equation is a first-order, linear, ordinary differential equation, so we need only one boundary condition to solve it. We have that at $z = h$, $p = p_a$, so that

$$p - p_a = -\rho g(z - h) = \rho g(h - z)$$

(2.6)

which shows that the pressure increases with depth (given by $d = h - z$), with a constant slope equal to ρg. That is,

> For a constant density fluid, the pressure increases linearly with depth.

This is the single most important observation regarding the behavior of constant density fluids in static equilibrium. In solving problems in fluid statics, this is the first result you should consider using.

It is always useful to check relationships by examining their behavior "in the limit." When using the hydrostatic equation, for example, it is important to check that the depth is expressed correctly since it is easy to get the sign wrong. Remember that the pressure should have its maximum value at the maximum depth. We see that equation 2.6 has the right behavior: when $z = 0$, the depth is h and the pressure takes its maximum value.

We can derive equation 2.6 directly by considering a column of fluid extending from the free surface to a depth $d\ (= h - z)$(see Figure 2.2). If the cross-sectional area of the column is δA, the weight of the fluid column is $\rho g d \delta A$. If p_a is the atmospheric pressure at the free surface, and p is the pressure at depth d, the

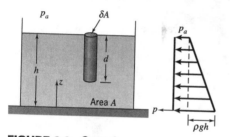

FIGURE 2.2 Container open to atmosphere. The graph on the right is a plot of the pressure as a function of depth.

$$P_g \approx P - P_a$$

upward thrust on the column is given by $p\delta A - p_a\delta A$. Since the column is in static equilibrium,

$$p\delta A - p_a\delta A = \rho g d\delta A$$

or

$$p - p_a = \rho g(h - z) = \rho g d$$

Here p is the *absolute* pressure, since it is measured relative to an absolute vacuum. Absolute pressure is the pressure that appears in the ideal gas law (equation 1.5). When the pressure is measured relative to atmospheric pressure, such as $p - p_a$, it is called *gauge* pressure. From equation 2.7, we see that the gauge pressure at depth d is given by the weight (per unit area) of a column of fluid extending from the surface to a depth d.

Gauge pressure is often useful when pressures above (or below) atmospheric are of interest. Most pressure instruments measure gauge pressures. A tire gauge, for instance, measures the pressure in a tire over and above the local atmospheric pressure. A vacuum gauge, in contrast, will measure the pressure below atmospheric (in common engineering usage a *vacuum* is any pressure lower than the ambient atmospheric pressure).

Air pressure can often be assumed to be constant. Consider a simple container like that shown in Figure 2.2. The pressure difference measured from the top to the bottom of the liquid is given by $\rho g h$. Outside the container, atmospheric pressure acts, and the air pressure changes by $\rho_a g h$ over the same distance h, where ρ_a is the density of air. The pressure change for air relative to water over the same distance is therefore given by the ratio

$$\frac{\rho_a g h}{\rho g h} = \frac{\rho_a}{\rho}$$

If the liquid was water, this ratio is equal to about $\frac{1.2}{1000} = 0.0012$, and for alcohol it would be about $\frac{1.2}{800} = 0.0015$. So the relative change of pressure in air compared to a liquid over the same depth is always very small, simply because the density of a gas is always much smaller than the density of a liquid. In many problems involving the pressure change due to a change in depth of a liquid, therefore, the change in air pressure over the same distance can be neglected and the air pressure can be assumed to be constant.

When this is the case, the air pressure can sometimes be neglected altogether, and the problem can be solved using gauge pressure. We will see some additional examples later, but for now we consider the resultant force exerted on the bottom of the container shown in Figure 2.2. We will assume that the air pressure can be taken to be constant everywhere.

First, we use absolute pressures. The absolute pressure acting on the bottom of the container from the outside, where air pressure acts, is simply p_a, and the absolute pressure exerted on the bottom of the container from the inside by the liquid is p_b, where

$$p_b = p_a + \rho g h$$

If the area of the bottom is A, then the resultant force F acting on the bottom surface is

$$F = (p_a + \rho gh)A - p_a A = \rho ghA$$

Second, we use gauge pressures. The gauge pressure at the bottom of the container is $p_b - p_a = \rho gh$, and so

$$F = \rho ghA$$

as before. In this case, the air simply adds a constant pressure to the pressure developed hydrostatically in the liquid. Since the same pressure acts on the bottom of the container from the outside in the opposite direction, the forces due to air pressure cancel out and they do not contribute to the resultant force.

2.3 APPLICATIONS OF THE HYDROSTATIC EQUATION

2.3.1 Pressure Variation With Height and Depth

Equation 2.6, obtained for constant density fluids, will apply in a broad sense to variable density fluids as well. For example, we can think of atmospheric pressure as being equal to the weight (per unit area) of all the air above our point of measurement. As we increase our height above sea level, this weight decreases, and therefore pressure decreases with altitude. However, the pressure will not be given by a simple relationship such as $\rho g \times depth$ because the density varies with altitude. We need to go back to equation 2.3 or 2.4 and find the solution from first principles.

We choose to use equation 2.3 where the z-direction is positive going up, so that

$$\frac{dp}{dz} = -\rho g$$

The variation of temperature, density, and, pressure with altitude is usually described in terms of a Standard Atmosphere (see Tables A–C.5 and A–C.6, and Fig. 14.1). For the first 10,000 m, the temperature variation is almost linear, and it is closely approximated by

$$T = T_0 + m(z - z_0).$$

From Table A–C.5 we see that, with $T_0 = 288.16°K$ ($15.0°C$) and $z_0 = 0$, $m = -0.0065°K/m$. From the hydrostatic equation and the ideal gas law (equation 1.5),

$$\frac{dp}{dz} = -\frac{p}{RT}g$$

That is,

$$\frac{dp}{p} = -g\frac{dz}{RT}$$

$$= -\frac{g}{R}\left(\frac{dz}{T_0 + m(z - z_0)}\right)$$

$$= -\frac{g}{mR}\left(\frac{m\,dz}{T_0 + mz}\right)$$

By integration

$$\ln\left(\frac{p}{p_0}\right) = -\frac{g}{mR}\ln\left(\frac{T_0 + mz}{T_0}\right)$$

That is, the pressure varies with altitude according to

$$\frac{p}{p_0} = \left(\frac{T_0 + mz}{T_0}\right)^{-g/mR}$$

What about the variation of pressure with depth in the ocean? We saw earlier that water is slightly compressible, and that this leads to a small increase in density with increasing depth. The density varies almost linearly with pressure, and the coefficient is given by the bulk modulus K. With z measured from the surface downward, and substituting for dp using equation 1.4,

$$\frac{dp}{dz} = \frac{K}{\rho}\frac{d\rho}{dz} = \rho g$$

That is,

$$\frac{d\rho}{dz} = \frac{g}{K}\rho^2$$

so that

$$\frac{d\rho}{\rho^2} = \frac{g}{K}dz$$

If we assume that K remains constant,

$$\rho = \frac{\rho_0}{1 - \frac{\rho_0 g z}{K}}$$

where ρ_0 is the density at the point where $z = 0$. Seawater has a bulk modulus of $K = 2.34 \times 10^9\,Pa$ at $20°C$ (see Table A–C.9). If we neglect the variation of K with temperature, then at a depth of $1000\,m$ the density of seawater is different from its value at the surface by about 0.5%.

2.3.2 Manometers

Manometers are used to find the pressure difference between two points. A simple manometer can be made using a U-shaped tube (Figure 2.3) filled to some depth with a liquid. Water is easy, but many manometers use alcohol to prevent the growth of algae or bacteria. Alcohol can also be easily dyed to make it clearly visible. Tubes are used to connect the pressure ports to the legs of the manometer.

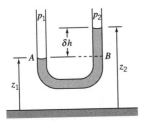

FIGURE 2.3   A simple U-tube manometer.

The pressures may come from the upper and lower part of a wing, for example. In that case, since the wing supports the weight of the aircraft, we expect the pressure at some point on the bottom of the airfoil, p_1, to be greater than the pressure at a point on the top, p_2. Using the hydrostatic equation we obtain

$$p_1 - p_2 = \rho_m g(z_2 - z_1)$$

that is,

$$p_1 - p_2 = \rho_m g \Delta h$$

where ρ_m is the density of the manometer fluid. Therefore, we can find the pressure difference $p_1 - p_2$ by measuring the deflection of the manometer fluid, Δh.

In deriving this result, we assumed that the pressure at points A and B are equal. This assumption is perfectly correct, since the pressure in a fluid under the action of gravity alone does not vary in the horizontal plane. In other words, any horizontal line is an *isobar*, which is a line connecting points of equal pressure. This is a general principle, as long as the path from A to B can be drawn within the same liquid. Points A and B are then called *simply connected*. If, for example, there was a slug of different fluid (an air pocket, perhaps) along the path between A and B, they are no longer simply connected, and the pressures at A and B would be different. For the manometer filled with three different fluids shown in Figure 2.4, $p_1 = p_6 = p_a$ and $p_3 = p_4$, but $p_2 \neq p_5$. A manometer problem similar to this one is worked out in detail in Example 2.1.

When a pressure is transmitted to a manometer through a tube, it is important to wait long enough for the pressure to be felt fully by the manometer, and this will take longer for smaller diameter tubes. Again, this occurs because pressure disturbances travel slowly along tubes of small diameter since they reflect and refract off the walls of the tube. If we wait long enough, however, the full pressure will always be felt at the manometer, because of the transmissibility of pressure in a fluid (see Section 1.3.7).

2.3.3 Barometers

Barometers are devices that measure the atmospheric pressure relative to a complete vacuum. In a complete vacuum, the absolute pressure is zero, so that barometers measure the absolute atmospheric pressure.

A mercury barometer consists of a vertical tube that contains mercury. One end is in contact with a vacuum where the absolute pressure is zero, and the other end is exposed to atmospheric pressure. A simple way to make a mercury barometer is to take a glass tube about 1 m long, close it at one end and fill it with mercury. Invert the tube and put the open end below the surface of a pool of mercury in a

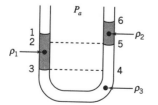

FIGURE 2.4 A U-tube manometer with multiple manometer fluids, open to atmospheric pressure.

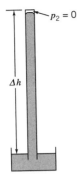

FIGURE 2.5 Mercury barometer as devised by Torricelli.

bowl. This will create a near vacuum in a small space at the closed end of the tube. The level of the mercury in the tube will settle at a height of about 760 *mm* or 30 *in.* above the level of the mercury in the bowl (see Figure 2.5). This is exactly what the Italian scientist Evangalista Torricelli did in 1643. He explained this observation by saying that the atmosphere must be exerting a pressure on the free mercury surface and that this pressure must equal to that exerted by the mercury column. For the barometer shown in Figure 2.5, with the very good approximation that $p_2 = 0$,

$$p_a = \rho g \Delta h = \text{absolute pressure}$$

One of the units of pressure, the *torr,* is named in honor of Torricelli for his observations on the nature of pressure, and his contributions to the development of barometers to measure atmospheric pressure (see Section 15.3).

When the barometer fluid is mercury, $\Delta h \approx 30$ *in.*, and when it is water, $\Delta h \approx 34$ *ft.* It is possible, therefore, to express atmospheric pressure in terms of an equivalent height of a column of fluid. This is common practice for all kinds of pressures, so we can talk about a pump developing a *head* of 60 feet of water, which is just a way of saying that it develops a pressure equal to that found at the bottom of 60 feet of water, that is, a pressure of about 1.8 atmospheres, or 26 *psi.* More exact equivalences are given in Table 2.1.

The type of mercury barometer most widely used in fluid mechanics laboratories is called a Fortin barometer (see Figure 2.6). The scale is fixed and the level of the mercury in the bowl is adjustable. Before making the reading, the free

**TABLE 2.1   Common
Equivalent Expressions for
Atmospheric Pressure**

1 atmosphere =	101,325 N/m^2 (= Pa)
=	1.01325 *bar*
=	14.70 *psi* (= lb_f/in^2)
=	2,116 *psf* (= lb_f/ft^2)
=	29.92 *in. Hg*
=	760.0 *mm Hg*
=	760.0 *torr*
=	10.33 *m* H_2O
=	33.90 *ft* H_2O

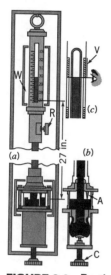

FIGURE 2.6   Fortin barometer. From Martin and Connor, *Basic Physics,* 8th ed., published by Whitcombe & Tombs Pty. Ltd., Melbourne, Australia, 1962, with permission.

mercury surface is brought to the scale zero by turning the knob *C* until the free surface just touches the tip of the pointer *A*. The vernier scale *V* is then moved up and down by means of a rack and pinion *R*, until the top of the mercury column and the lower front and back edges are seen in the same straight line [Figure 2.6(*c*)]. A white surface *W* behind the tube helps to make this adjustment. The reading needs to be adjusted for ambient temperature since the length of the mercury column and the length of the scale both increase with temperature. Also, moist air is less dense than dry air at the same temperature and pressure, since the molecular weight of water is less than that of air (18 compared to 28.96), and so the barometer reading decreases as the moisture of the atmosphere increases.

Another type of barometer is the aneroid barometer, which consists of an evacuated metal box (Figure 2.7). The air pressure acting on the outside of the box causes the box to deform, and this deformation is amplified by a series of levers and springs to move a pointer over a graduated scale.

The height of the mercury column, or the reading of the aneroid barometer,

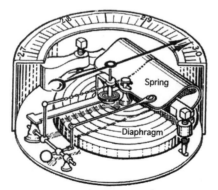

FIGURE 2.7   Aneroid barometer. From Martin and Connor, *Basic Physics,* 8th ed., published by Whitcombe & Tombs Pty. Ltd., Melbourne, Australia, 1962, with permission.

will vary with altitude, since the atmospheric pressure decreases with height. If it is known how the pressure varies with altitude, this property can be used to find the height above sea level. A modified aneroid barometer is commonly used as an airplane altimeter.

EXAMPLE 2.1 *Manometers*

Consider the manometer shown in Figure 2.4. Let z_1 be the height of point 1 above a horizontal reference level, z_2 be the height of point 2, and so on. If $\rho_3/\rho_1 = 2$, and $z_6 = 12$ *in.*, $z_1 = 10$ *in.*, $z_2 = 8$ *in.*, and $z_3 = 6$ *in.*, find the ratio ρ_2/ρ_3.

Solution: We know that $p_1 = p_6 = p_a$ and $p_3 = p_4$. Therefore, if we equate pressures at height z_3, we get

$$\rho_1 g(z_1 - z_3) = \rho_3 g(z_6 - z_2) + \rho_2 g(z_2 - z_3)$$

That is,

$$\rho_2 g(z_2 - z_3) = 2\rho_1 g(z_6 - z_2) - \rho_1 g(z_1 - z_3)$$
$$\frac{\rho_2}{\rho_1} = \frac{2(z_6 - z_2) - (z_1 - z_3)}{(z_2 - z_3)} = 2$$

and $\rho_2 = \rho_3$ so that liquids 2 and 3 have the same density. ∎

2.4 VERTICAL WALLS OF CONSTANT WIDTH

So far, we have used the hydrostatic equation to find the variation with pressure due to height. We will now consider the forces exerted by pressure differences on the walls of a container.

In Chapter 1 we showed that pressure is an isotropic stress. That is, at a given point in the fluid, the pressure that acts on a vertical surface passing through the point has the same value as it has on a horizontal surface passing through the same point. In a container filled with water, the hydrostatic pressure acting on the bottom of the vessel is equal to the hydrostatic pressure acting on the side wall at the same depth. The pressure on the side wall decreases with decreasing depth according to the hydrostatic equation, and so the side wall feels a linearly varying pressure distribution.

Consider the vertical wall of a reservoir, with water of depth h on one side (see Figure 2.8). A plot of the pressure as a function of depth is shown on the left.

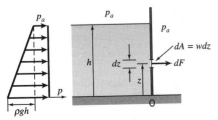

FIGURE 2.8 Vertical wall with water pressure acting on one side, atmospheric pressure on the other. The graph on the left is a plot of the pressure as a function of depth.

At the free surface, the pressure is atmospheric and below the surface, the pressure increases linearly with depth to a maximum value of $p_a + \rho g h$. What is the resultant force due to the water pressure acting over the area of the reservoir wall, and where does it act? That is, if we were to replace the distributed force due to pressure by a single resultant force, what is its magnitude and direction, and where must it be placed? Air pressure is assumed to be acting everywhere, on the free surface as well as on the outside of the wall. The wall is of constant width, w (into the page), and the coordinate z is chosen to be positive in the direction of increasing altitude (that is, it is in the direction opposite to the gravitational force). The force acting on a local element of area, dA, depends only on the depth, that is, the distance below the free surface, given here by $(h - z)$. The total force due to pressure depends on the area over which the pressure acts. In the case shown here, w is constant (that is, independent of z), so that we can use $dA = w\,dz$. In other words, the elemental area we consider is a strip of width w and height dz.

2.4.1 Solution Using Absolute Pressures

First, we use absolute pressures. From the hydrostatic equation (equation 2.3), the pressure at any depth is given by

$$p - p_a = \rho g(h - z)$$

(remember, z is positive going up). On the side of the wall where the water acts, the pressure p acts on element $dA = w\,dz$ to produce a force dF_x that acts to the right. So

$$dF_x = (\text{pressure}) \times (\text{area}) = p\,dA$$
$$= pw\,dz$$
$$= [p_a + \rho g(h - z)]w\,dz$$

On the other side of the wall, air pressure acts, and we will assume that this pressure is constant (remember that the density of water is about 800 times the density of air, so that the relative variations of water and air pressure over the same distance is different by a factor of about 800). The resultant force, dF, acting on the area $w\,dz$ is therefore given by

$$dF = dF_x - p_a w\,dz$$

and it acts to the right. That is,

$$dF = [p_a + \rho g(h - z)]w\,dz - p_a w\,dz = \rho g(h - z)w\,dz \tag{2.8}$$

Integrating equation 2.8 from $z = 0$ to $z = h$ gives the resultant force

$$F = \int_0^h \rho g(h - z)w\,dz$$

so that

$$F = \tfrac{1}{2}\rho g w h^2$$

2.4.2 Solution Using Gauge Pressures

If instead we use gauge pressure (p_g), we have

$$p_g = \rho g(h - z)$$

and

$$dF = p_g \, dA = \rho g(h - z)w \, dz \tag{2.9}$$

just as before, but much more simply. The force F is formed by integration, in the same way.

2.4.3 Moment Balance

Where should the resultant force F be placed so that it exerts the same moment as the distributed force due to pressure? For this we need to take moments. A convenient place to take moments is the point where $z = 0$. Any other point will do, but this one happens to make the problem simpler. The moment of the resultant force dF acting on the element $w \, dz$ about the axis passing through the point $z = 0$ is called dM_0, and it is given by

$$dM_0 = z \times dF$$

where z is the *moment arm* of the force dF_x about the axis through $z = 0$, and it is measured *perpendicular* to the line of action of the force. This moment is clockwise about the origin. Substituting for dF from equations 2.8 or 2.9, we obtain

$$dM_0 = z \times dF = \rho g(h - z)wz \, dz$$

Integrating from $z = 0$ to $z = h$ gives the resultant moment

$$M_0 = \tfrac{1}{6}\rho gwh^3$$

By definition, the resultant force times its moment arm must be equal to the total moment exerted by the pressure acting on the wall, so that

$$M_0 = \bar{z} \times F$$

where $\bar{z}$ is the moment arm for the resultant force, as measured from the axis through $z = 0$. Finally,

$$\bar{z} = \tfrac{1}{3}h$$

To summarize:

> When atmospheric pressure acts everywhere, the resultant force exerted by a fluid of depth h acting on a vertical wall of constant width is given by
>
> $$F = \tfrac{1}{2}\rho gwh^2 \tag{2.10}$$
>
> and it acts a distance $\tfrac{1}{3}h$ from the bottom of the reservoir.

In finding the resultant force and moment in statics problems, there are many similarities with methods that are usually taught in mechanics of solids. The problem just considered is very similar to finding the resultant force acting on a cantilever beam under a distributed load, where the load per unit area of beam is p, varying from its minimum value of zero at $z = h$ to its maximum value of ρgh at $z = 0$ (see Figure 2.9). The resultant force is given by the integral of the distributed force over the length of the beam. That is, it is given by the area of the triangle describing the load distribution. The point at which the force acts will be at the centroid of the triangle, which is of course one-third the distance from the point where the load is a maximum. In fluid statics, the point through which the resultant force acts is called the *center of pressure* (see also Section 2.8).

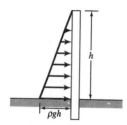

FIGURE 2.9 Cantilever under the action of a distributed load similar to that produced by hydrostatic pressure.

2.4.4 Gauge Pressure or Absolute Pressure?

For the problem we just solved, air pressure acted on top of the water surface so that it added a constant pressure p_a to the pressure at any point inside the water. Since we assumed that the atmospheric pressure was constant everywhere outside the reservoir, it also acts with the same value of p_a at any position on the outside of the wall. Since air pressure acts on the inside and the outside of the wall with the same value, it cancels out. The only pressure that contributes to the resultant force is the "excess" pressure, that is, the gauge pressure. It is always a good idea to look at a problem before starting its solution to see if the ambient pressure by itself is going to lead to a resultant force. If not, the solution can be made considerably simpler.

To give an example where this is not possible, consider the closed gas container shown in Figure 2.10, which is similar to the kind of propane container used in a gas grill. Here, the container happens to have a square cross-section with dimensions $w \times w$, and a height H. The liquid propane fills the container to a depth h. The rest of the container is filled with propane vapor at a pressure p_v, equal to its vapor pressure at the ambient temperature. Since the density of the gaseous phase of propane is much less than the density of the liquid phase (this is true for any fluid), the vapor pressure can be taken to be constant throughout the container. Similarly, air pressure acts with a constant value everywhere outside the gas container. What is the resultant force due to all the fluid pressures on a side wall of the container and where does it act?

We can think of the total pressure acting on the inside of the container as the sum of two parts: a constant part equal to p_v, and a varying part p_l due to the hydrostatic pressure exerted by the liquid. This allows us to split up the problem into two parts also, starting with the pressures that are constant; on the inside we

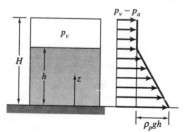

FIGURE 2.10 Closed propane container where the gaseous phase exerts a pressure on the liquid phase.

have p_v and on the outside p_a. The resultant force due to these pressures is simply $(p_v - p_a)wH$, acting to the right. Within the liquid, the pressure p_l varies according to its depth $p_l = \rho_p g(h - z)$ (where ρ_p is the density of the liquid propane). From our earlier work, we know this gives rise to a resultant force $\frac{1}{2}\rho_p gwh^2$, acting to the right. So the total resultant force F acting to the right due to all the fluid pressures is

$$F = (p_v - p_a)wH + \tfrac{1}{2}\rho_p gwh^2$$

To find out where this resultant force acts, we can again split up the problem into two parts. The resultant force due to the constant pressures p_v and p_a will act at $z = \frac{H}{2}$, for the same reason the resultant force on a beam due to a constant distributed load acts at the center of the beam. We found earlier that the resultant force due to the liquid acts at a distance $\frac{1}{3}h$ from the bottom. When we take moments about the axis through the point $z = 0$, we have

$$F \times \bar{z} = (p_v - p_a)wH \times \frac{H}{2} + \frac{1}{2}\rho_p gwh^2 \times \frac{h}{3}$$

which can then be solved for $\bar{z}$.

EXAMPLE 2.2 Forces and Moments on Vertical Walls

A cubic tank of dimension h contains water of density ρ. It has a pipe open to atmosphere located in its top surface. This pipe contains water to a height L, as shown in Figure 2.11. A square vent of size D in the top surface is just held closed by a lid of mass M.

(a) Find the height L in terms of M, D, and ρ.
(b) Find the force F due to the water acting on a vertical face of the tank in terms of ρ, g, h, and L.
(c) Find where this force acts in terms of h and L.

Solution: For part (a), the weight of the lid Mg is just balanced by the water pressure acting on the area of the lid. Hence,

$$\rho g L D^2 = Mg$$

and so

$$L = \frac{M}{\rho D^2}$$

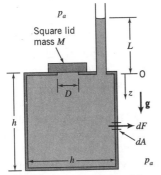

FIGURE 2.11 Tank with stand-pipe.

For part (b), we identify an element of area dA on a vertical wall of the tank. Atmospheric pressure acts everywhere, so dF, the force due to water pressure acting on dA is given by

$$dF = p_g \, dA = \rho g(depth) \, dA = \rho g(L + z) \, dA$$

Note that the depth is measured from the free surface, not the top of the tank. Since the wall is of constant width $(= h)$, $dA = h \, dz$, and

$$dF = \rho g(L + z)h \, dz$$

To find F we integrate

$$F = \int_0^h \rho g(L + z)h \, dz$$

$$= \rho g h \left[Lz + \frac{z^2}{2} \right]_0^h$$

so that

$$F = \rho g h^2 \left(L + \frac{h}{2} \right)$$

For part (c), we take moments about the top of the tank side wall (where $z = 0$). If dM is the moment due to dF,

$$dM = z \, dF = \rho g(L + z)zh \, dA$$

Integrating, we obtain

$$M = F \times \bar{z} = \int_0^h \rho g(L + z)zh \, dz$$

$$= \rho g h \left[L\frac{z^2}{2} + \frac{z^3}{3} \right]_0^h$$

$$= \rho g h^2 \left(L\frac{h}{2} + \frac{h^2}{3} \right)$$

Therefore, the resultant force acts at a distance $\bar{z}$ below the top of the tank, where

$$\bar{z} = \frac{M}{F} = \frac{L\frac{h}{2} + \frac{h^2}{3}}{L + \frac{h}{2}} = \frac{L + \frac{2h}{3}}{1 + \frac{2L}{h}}$$

∎

EXAMPLE 2.3 *Equilibrium of a Hinged Gate*

Sometimes we have a problem where the equilibrium of a solid body depends on the sum of the moments. For instance, if a hinged gate is placed in a vertical wall, the water pressure will try to open the gate unless a moment of sufficient strength is exerted on the gate to keep it shut. Consider a simple case where the whole wall serves as a gate. The gate is vertical and rectangular in shape, of width w and height h [Figure 2.12(a)]. The top of the gate is level with the surface of the water, where it is supported by a frictionless hinge H. Halfway down the gate, an arm sticks out horizontally and a weight W hangs at a distance a from the gate. The bottom of the gate rests against a stop. Atmospheric pressure acts everywhere. Neglect the weight of the gate and the arm. What is the minimum value of W necessary to keep the gate shut?

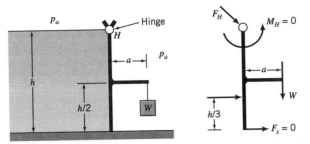

FIGURE 2.12 Left: simple gate, hinged at the level of the free surface. Right: free body diagram of gate.

Solution: Consider the free body diagram for the gate [Figure 2.12(b)]. This is a diagram of the gate showing the forces and moments acting on it when it is at the point of opening, so that the reaction exerted on the gate by the stop at the foot of the gate is zero. The force F exerted by the water pressure on the gate acts from the left in the horizontal direction, and it tends to rotate the gate in a counterclockwise direction. We know from our previous work that it has a magnitude $F = \frac{1}{2}\rho g w h^2$ and that it acts at a distance $\frac{2}{3}h$ from the top of the gate. The weight W acts at a distance a horizontally out from the gate, and it tends to rotate the gate in a clockwise direction. There is also a force exerted by the hinge on the gate, but no moment since the hinge is frictionless (a hinge without friction cannot exert a moment). Since the gate is not moving, it is in static equilibrium under the action of these forces and moments.

How do we find the critical value of W, where the gate is just on the point of opening? We know that $\Sigma\mathbf{F} = 0$ and $\Sigma\mathbf{M} = 0$. If we use $\Sigma\mathbf{F} = 0$, we see from the free body diagram that there is a force exerted at the hinge F_H that needs to be found separately before the force balance can be solved for W. We may be able to find F_H using the moment equation $\Sigma\mathbf{M} = 0$, but instead we can use the moment equation to find W directly. If we choose the moment axis to coincide with the top of the gate, then the hinge force F_H exerts no moment about this axis and we need not consider it any further. We simply balance the moments about the hinge exerted by the weight W and the force due to water pressure.

$$W \times a - \frac{1}{2}\rho g w h^2 \times \frac{2h}{3} = 0$$

and so we find

$$W = \frac{\rho g w h^3}{3a} \qquad \blacksquare$$

EXAMPLE 2.4 *Another Hinged Gate*

In Figure 2.13, a gate is shown, hinged at the top using a frictionless hinge H, located at the same level as the water surface. There is a rectangular overhang in the wall that sticks out a distance a horizontally from the gate. There is a stop at the bottom of the wall to resist the force of the water pressure on the inside of the gate, and to prevent it opening in a counterclockwise direction. However, as a is increased, there will come a point where the weight of the water in the overhang

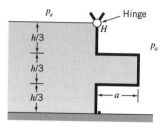

FIGURE 2.13 Gate with overhang, hinged at the level of the free surface.

will be large enough to cause the gate to move away from the stop with a clockwise rotation. What is this critical value of a?

Solution: If we were to draw the free body diagram of this gate, we would see that it is best to take moments about the hinge line since the unknown force exerted by the hinge on the gate has no moment about this axis.

We could solve the problem by considering each vertical part of the wall separately, and find the moments about the hinge exerted by the forces acting on each surface (these moments are all counterclockwise), and then find the moment exerted by the forces acting on the horizontal parts of the overhang (these moments are clockwise). For equilibrium, the sum of the moments must be zero, and so we could find a.

However, there is a simpler way. The total horizontal force is equal to the sum of the forces acting on all the vertical parts of the gate, and therefore it is equal to the force acting on a vertical wall of the same height, given by $\frac{1}{2}\rho g w h^2$. It acts at a distance $\frac{2}{3}h$ from the top of the gate, and so its moment can be found directly.

As for the overhang, there are two approaches. First, we work in terms of the pressures acting on the top and bottom surfaces of the overhang. We can find the pressure acting on the bottom surface, multiply by the area of the bottom surface to find the force (since the bottom surface is at a constant depth, the pressure is constant over the area), and then multiply by its moment arm to find its (clockwise) moment. The moment arm is equal to $\frac{1}{2}a$, since the loading on the bottom surface is uniformly distributed. If we say that clockwise moments are positive, then, for the bottom surface of the overhang,

$$\text{pressure on bottom surface} = \tfrac{2}{3}\rho g h$$
$$\text{force on bottom surface} = \tfrac{2}{3}\rho g h w a$$
$$\text{moment due to force on bottom surface} = \tfrac{2}{3}\rho g h w a \times \tfrac{1}{2}a$$

Similarly, for the top surface of the overhang, we find for the clockwise moment.

$$\text{pressure on top surface} = \tfrac{1}{3}\rho g h$$
$$\text{force on top surface} = \tfrac{1}{3}\rho g h w a$$
$$\text{moment due to force on top surface} = -\tfrac{1}{3}\rho g h w a \times \tfrac{1}{2}a$$

Therefore the total clockwise moment exerted by the overhang is $\frac{1}{3}\rho g h w a \times \frac{1}{2}a$.

Second, we note that the moment produced by the overhang is due to the weight of water contained in it, and this weight is the volume multiplied by the

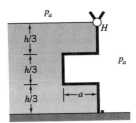

FIGURE 2.14 Gate with cut-out.

density, that is, $\frac{1}{2}\rho ghwa$. The moment arm of this weight is located at the centroid of the volume, at a point $\frac{1}{2}a$ out from the hinge, so that the total clockwise moment exerted by the overhang is $\frac{1}{3}\rho ghwa \times \frac{1}{2}a$, as before.

The sum of all the moments is given by the clockwise moment due to the weight of water contained by the overhang, plus the counterclockwise moment due to the water acting on the vertical portions of the gate. That is, a can be found from:

$$\tfrac{1}{3}\rho ghwa \times \tfrac{1}{2}a - \tfrac{1}{2}\rho gwh^2 \times \tfrac{2}{3}h = 0$$

That is

$$a = \sqrt{2h}$$ ∎

EXAMPLE 2.5 *A Final Hinged Gate*

What happens if the overhang in Example 2.4 was negative, as shown in Figure 2.14? That is, if instead of an overhang there was a cut-out of the same dimensions?

Solution: The moment due to the water acting on the vertical parts of the gate is the same as in Example 2.4: about the hinge, it is counterclockwise, of magnitude $\frac{1}{2}\rho gwh^2 \times \frac{2}{3}h$. For the cut-out, the moment due to pressure acting on the bottom surface is $\frac{2}{3}\rho ghwa \times \frac{1}{2}a$ in the clockwise direction, and for the top surface is $\frac{1}{3}\rho ghwa \times \frac{1}{2}a$ in the counterclockwise direction. The resultant moment produced by the horizontal surfaces of the cut-out is therefore $\frac{1}{3}\rho ghwa \times \frac{1}{2}a$ in the clockwise direction, exactly the same as that found in Example 2.4. So the moment equilibrium of the gate shown in Figure 2.14 is the same as for the gate shown in Figure 2.13, and the value of a is also the same. ∎

2.5 SLOPING WALLS OF CONSTANT WIDTH

What happens when the wall is sloping? That is, consider a straight wall of constant width leaning over at an angle of θ to the horizontal (Figure 2.15). The pressure acts in a direction normal to the surface, so that the resultant force due to water pressure will now have a horizontal and a vertical component (F_x and F_z, respectively). What is the magnitude of the resultant force F, and where does it act? Atmospheric pressure acts everywhere.

We will solve this problem in a number of different ways. We begin by finding

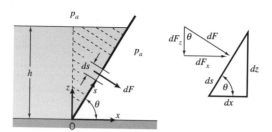

FIGURE 2.15   Pressure acting on a sloping wall of constant width.

F_x and F_z separately, and obtain the resultant force using $F = \sqrt{F_x^2 + F_z^2}$. We will then find F directly.

2.5.1 Horizontal Force

Since atmospheric pressure acts everywhere, we can use gauge pressure, p_g. Then, at any depth below the surface,

$$p_g = \rho g(h - z)$$

This pressure acts normal to the wall, so that the force dF on the element dA is given by

$$dF = p_g \, dA$$

Here, $dA = w \, ds$, where s is the coordinate along the direction of the wall, measured from the origin, where $s = 0$ and $z = 0$. From the geometry, we have

$$dz = ds \sin \theta$$

and

$$dz = ds \cos \theta$$

The horizontal component of the force acting on dA is given by

$$
\begin{aligned}
dF_x &= p_g \, dA \sin \theta \\
&= p_g w \, ds \sin \theta \\
&= p_g w \, \frac{dz}{\sin \theta} \sin \theta
\end{aligned}
$$

That is,

$$dF_x = p_g w \, dz = \rho g(h - z)w \, dz$$

This is the same result found for a vertical wall (equation 2.8), and we find that the horizontal component of the force acting on a straight wall of constant width is independent of its slope, so that $F_x = \frac{1}{2}\rho g w h^2$, as before.

> The horizontal component of the force due to hydrostatic pressure on an inclined wall is equal to that on a vertical wall at the same depth, as long as the wall has a constant width.

2.5.2 Vertical Force

Consider now the vertical component of the force acting on dA. This is given by

$$dF_z = p_g \, dA \cos \theta$$
$$= p_g w \, ds \cos \theta$$
$$= p_g w \, dx$$

The equation to the line describing the shape of the wall is

$$z = x \tan \theta$$

so that $dz = dx \tan\theta$ (θ is a constant), and

$$F_z = \int_0^{h/\tan \theta} \rho g (h - z) w \, dx$$

The integration is with respect to x, so that the limits have to be x-limits corresponding to the lowest point on the wall (where $x = 0$) and the top of the water level (where $z = h$ and $x = h/\tan \theta$). In addition, z appears in the integrand, and needs to be expressed in terms of x. Hence

$$F_z = \int_0^{h/\tan \theta} \rho g (h - x \tan \theta) w \, dx = \frac{\rho g w h^2}{2 \tan \theta}$$

which acts downwards if θ is positive.

If instead we consider the weight of the fluid "supported" by the wall, (the hatched area in Figure 2.15) we get

$$\text{weight of fluid} = \frac{1}{2} (\rho g w h) \times \frac{h}{\tan \theta} = F_z$$

as we might expect.

2.5.3 Resultant Force

The resultant force can now be found.

$$F = \sqrt{F_z^2 + F_x^2}$$

and therefore

$$F = \frac{\rho g w h^2}{2 \sin \theta} \tag{2.11}$$

We can arrive at the same result by considering the force dF directly, rather than splitting it up into horizontal and vertical components. Here we use the s-coordinate system rather than the $[x, y]$ system. Then,

$$dF = p_g \, dA$$

That is,

$$dF = \rho g (h - z) w \, ds$$
$$= \rho g (h - z) w \, \frac{dz}{\sin \theta} \tag{2.12}$$

So

$$F = \int_0^h - \rho g(z - h)w \, \frac{dz}{\sin \theta}$$

and therefore

$$F = \frac{\rho g w h^2}{2 \sin \theta}$$

as before. Choosing a particular coordinate system can often make the problem simpler.

2.5.4   Moment Balance

To find where this force acts we need to take moments. In order to take moments, we must specify the axis about which the moment is taken, and determine the lever arm or moment arm, which is the minimum distance between the moment axis and the line of action of the force. Let M_0 be the moment about the axis through the origin due to the force exerted by the water pressure, so that

$$M_0 = \bar{s} \times F$$

where $\bar{s}$ is the moment arm of F about the origin, and where a clockwise moment is taken to be positive. We can find M_0 by considering the moment dM_0 due to dF about the origin. That is,

$$dM_0 = s \times dF = \rho g(h - z)sw \, ds$$

where we have used the result for dF from equation 2.12. Integration gives

$$M_0 = \int \rho g(h - z)sw \, ds$$

With $z = s \sin \theta$,

$$\bar{s} = \frac{M_0}{F} = \frac{1}{F} \int_0^{h/\sin \theta} \rho g w (h - s \sin \theta)s \, ds$$

$$= \frac{2 \sin \theta}{\rho g w h^2} \times \rho g w \left[\frac{s^2 h}{2} - \frac{s^3 \sin \theta}{3} \right]_0^{h/\sin \theta}$$

Hence

$$\bar{s} = \frac{h}{3 \sin \theta}$$

We can check the solution by examining its behavior "in the limit." For example, if $\theta \to \pi/2$, the wall becomes vertical and we find $\bar{s} = \frac{h}{3}$, as we should. In the other limit, where $\theta \to 0$, the wall becomes horizontal, and the result is not so useful.

We can also obtain $\bar{s}$ by evaluating the moment arms for the two components of the resultant force, since $\bar{s} = \sqrt{\bar{x}^2 + \bar{z}^2}$, where $\bar{x}$ is the moment arm of F_z, and $\bar{z}$ is the moment arm of F_x, both taken about the origin. That is,

$$\bar{z} = \frac{1}{F_x} \int z \, dF_x$$

$$= \frac{1}{F_x} \int_0^h \rho g w (h - z)z \, dz = \frac{2}{3h^2} \left[\frac{hz^2}{2} - z^3 \right]_0^h$$

Hence

$$\bar{z} = \frac{h}{3}$$

as expected. Also

$$\bar{x} = \frac{1}{F_z} \int x \, dF_z$$

$$= \frac{1}{F_z} \int_0^{h/\tan\theta} \rho g w (h - x) x \, dx$$

$$= \frac{2 \tan\theta}{3h^2} \left[\frac{lx^2}{2} - x^3 \tan\theta \right]_0^{h/\tan\theta}$$

Hence

$$\bar{x} = \frac{h}{3 \tan\theta}$$

which is also the location of the centroid of the fluid supported by the wall (we will show later that this is a general result).

Finally,

$$\bar{s} = \sqrt{\bar{x}^2 + \bar{z}^2} = \frac{h}{3} \sqrt{1 + \frac{1}{\tan^2\theta}}$$

and, as before,

$$\bar{s} = \frac{h}{3 \sin\theta}$$

EXAMPLE 2.6   *Weight and Forces Due to Pressure*

Consider the two containers shown in Figure 2.16. They have a width w, and they are filled with water to the same height, h. Assume the weight of each container is negligible. From the hydrostatic equation, we know that the pressure at depth h will be the same for both vessels. Therefore, if the bottom area A is the same, the force exerted on the bottom of the container will be equal, that is, $F_1 = F_2$, in spite of the obvious difference in the total weight of liquid contained. Is there a paradox?

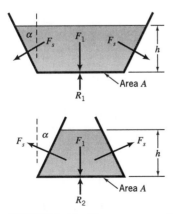

FIGURE 2.16 Pressure exerted on the bottom of different shaped containers.

Solution: We must be careful to consider all the forces, including the forces acting on the side walls, the bottom of the container, and the reaction from the surface it is resting on. Note that the forces on the side walls (F_s) act at right angles to the wall, and that the horizontal components of the side-wall forces cancel out since they act in opposing directions.

For the container shown in the upper part of Figure 2.16, the vertical components of F_s act downward, and for static equilibrium

$$R_1 = 2F_s \sin \alpha + \rho g h A$$

The force F_s is given by equation 2.11, where $\alpha = \frac{\pi}{2} - \theta$. Hence,

$$F_s = \frac{\rho g w h^2}{2 \cos \alpha}$$

and

$$F_s \sin \alpha = \frac{\rho g w h^2}{2} \frac{\sin \alpha}{\cos \alpha} = \tfrac{1}{2} \rho g w h^2 \tan \alpha$$

That is,

$$R_1 = \rho g w h^2 \tan \alpha + \rho g h A$$

which is exactly equal to the weight of the water in the container, as it should be.

For the container shown in the lower part of Figure 2.16, the force components in the vertical direction act upward, and for static equilibrium,

$$R_2 = -2F_s \sin \alpha + \rho g h A$$

and we can show that R_2 is equal to the weight of the water in this particular container. So when the walls slope inward, the force on the base due to the liquid is greater than the weight of the liquid contained because of the forces exerted by the side walls on the liquid. In this case, the sum of the vertical components of these forces, plus the weight of the fluid, equals the force exerted on the base.

In 1646, the French scientist Blaise Pascal gave an interesting illustration of this principle (see Section 15.4). He placed a long vertical pipe in the top of a barrel filled with water and found that by pouring water into the pipe he could burst the barrel even though the weight of water added in the pipe was only a small fraction of the force required to break the barrel (Figure 2.17). ∎

FIGURE 2.17   Pascal's rain barrel. From Martin and Connor, *Basic Physics*, 8th ed., published by Whitcombe & Tombs Pty. Ltd., Melbourne, Australia, 1962, with permission.

2.6 HYDROSTATIC FORCES ON CURVED SURFACES

So far we have considered only plane surfaces. In many cases, the surfaces are curved. For example when building a dam, the bottom of the dam needs to be stronger than the top because the pressure increases with depth. To achieve this with minimum material, the thickness of the dam must increase with depth, so that the strength of the dam matches the increasing force and moments due to pressure.

To see how the hydrostatic forces acting on a curved surface can be found, consider the parabolic wall shown in Figure 2.18. Atmospheric pressure acts everywhere. We will find the resultant hydrostatic force acting on the wall, and its point of action, from first principles.

2.6.1 Resultant Force

We begin by defining a coordinate system so that the shape of the wall can be expressed as simply as possible. For the system shown, the equation describing the shape of the wall is $z = \frac{h}{a^2} x^2$. The gauge pressure at height z is p_g, so that

$$p_g = \rho g(h - z)$$

Therefore

$$dF = \rho g(h - z)\, dA$$

where dF is the net force due to the water pressure acting on the element dA, and $dA = w\, ds$, where the coordinate s is measured along the surface of the wall, and w is the width of the wall. For curved walls it is probably simpler to find the horizontal and vertical components of the resultant force separately, starting with the horizontal component F_x. Now,

$$dF_x = dF \sin \theta$$

so

$$dF_x = \rho g(h - z)w\, ds \sin \theta$$

Since

$$ds \sin \theta = dz$$

we have

$$dF_x = \rho g(h - z)w\, dz$$

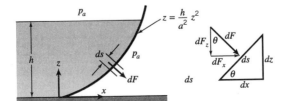

FIGURE 2.18 Pressure acting on a parabolic wall of constant width.

By integration

$$F_x = \rho g \int_0^h (h - z)w\, dz = \rho g w \left[hz - \frac{z^2}{2} \right]_0^h$$

so that

$$F_x = \tfrac{1}{2}\rho g w h^2$$

This is the same result found for the horizontal force acting on a sloping wall (see Section 2.5). It is, in fact, a general result: the horizontal force acting on a wall of constant width is independent of the slope of the wall.

Now find the vertical component F_z.

$$dF_z = dF \cos \theta$$

So

$$dF_z = \rho g (h - z)w\, ds \cos \theta$$

Since

$$ds \cos \theta = dx$$

we have

$$dF_z = \rho g (h - z)w\, dx$$

By integration

$$F_z = \rho g \int_{-a}^0 \left(h - \frac{h}{a^2}x^2 \right) w\, dx = \rho g h w \left[x - \frac{x^3}{3a^2} \right]_{-a}^0$$

so that

$$F_z = \tfrac{2}{3}\rho g w a h$$

We can also show that this vertical force is equal to the weight of fluid supported by the wall.

The total force may be found by vector addition.

$$F = \sqrt{F_x^2 + F_z^2} = \rho g h w \sqrt{\frac{h^2}{4} + \frac{4a^2h^2}{9}}$$

Therefore

$$F = \tfrac{1}{2}\rho g w h^2 \sqrt{\left(1 + \frac{16a^2}{9h^2} \right)}$$

We could also try to find this resultant force directly by using

$$dF = \rho g (h - z)w\, ds$$

and

$$F = \int \rho g (h - z)w \frac{dz}{\sin \theta}$$

For a curved wall, θ is not a constant. In fact, $\tan\theta$ is the slope of the wall at any point, where

$$\tan \theta = \frac{dz}{dx} = \frac{2h}{a^2}x$$

We need to substitute for $\sin \theta$ in the expression for dF. With some trigonometry, we obtain

$$\sin \theta = \sqrt{\frac{\tan^2 \theta}{1 + \tan^2 \theta}} = \sqrt{\frac{4h^2 x^2}{a^4 + 4h^2 x^2}}$$

Clearly, this integration is not straightforward, and so we will not proceed. If it is possible to get the resultant force directly, it can save some effort, but in most curved-wall cases, it is easier to deal with the two components of the resultant force separately.

2.6.2 Line of Action

To find where the resultant force acts, we take each component in turn. First, we take moments about the y-axis, which is the line passing through the origin of the coordinate system pointing into the page. The moment arm for F_x is $\bar{z}$, so that

$$F_x \times \bar{z} = \rho g \int_0^h z(h - z)\, dz$$

and we can show that

$$\bar{z} = \tfrac{1}{3}h$$

Again, we could have anticipated this result from our earlier analysis; the horizontal component of the resultant force on a plane wall of constant width acts at a point one-third from the bottom of the wall, regardless of the inclination of the wall.

Next, we take moments about the y-axis. The moment arm for F_z is $\bar{x}$, so that

$$F_z \times \bar{x} = \rho g \int_a^h (h - z)x\, dx = \rho g \int_a^h \left(h - \frac{x^2}{a^2} \right) x\, dx$$

and we find that

$$\bar{x} = \tfrac{3}{8}a$$

This is also the position of the centroid of the volume of water supported by the parabolic wall, which may not be obvious since this is probably not a commonly known fact. We obtained a similar conclusion for a plane, inclined wall. In fact, this is a general result: the vertical component of the resultant force on any surface acts at the centroid of the volume of fluid it supports (usually called the *displaced volume*), regardless of the shape of the wall. This observation will be considered further in Section 2.9.

2.7 TWO-DIMENSIONAL SURFACES

The next level of complexity in hydrostatic problems is the case where the surface is flat, but it has a width that varies. This is called a two-dimensional surface. The force on the surface depends on its depth and its width. An example using a triangular plate is given in Figure 2.19 where a side view and a plan view are shown. The plate is symmetrical, with a height ℓ and a base width $2a$. Its apex is located a depth $L \sin \theta$ below the surface of the water. The surface of the water, and the outer face of the plate, are open to the atmosphere, so that we can use gauge pressure. What is the resultant force due to the water pressure, and where does it act?

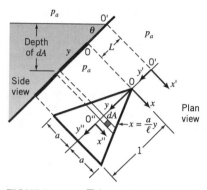

FIGURE 2.19 Triangular plate, with water pressure on one side, and air pressure on the other. Side and plan views.

We begin by defining a coordinate system. A good choice is to put the origin O at the apex, so that the y-coordinate lies along the centerline of the plate, pointing toward the base, and the x-coordinate points out of the page in the side view shown near the top of Figure 2.19. Next, we mark an element of area dA on the surface of the plate, where $dA = dx\,dy$. The depth of the element dA is $(y + L)\sin\theta$, and therefore the gauge pressure acting on dA is $p_g = \rho g(y + L)\sin\theta$, and the force acting on dA is

$$dF = p_g\,dA$$
$$= \rho g(y + L)\sin\theta\,dA$$
$$= \rho g(y + L)\sin\theta\,dx\,dy$$

so that

$$F = \iint \rho g(y + L)\sin\theta\,dx\,dy$$

This is a double integral, with one integration in x and one in y. We can choose which one to do first, and since the pressure depends only on depth, that is, it depends only on y and not on x, it is probably better to perform the x integration first. We can write this as

$$F = \int \rho g(y + L)\sin\theta\left(\int dx\right)dy$$

The x-integration is done by holding y constant, so that the limits of integration must go from one side of the plate to the other side, at a constant y. The x-value on the positive edge of the plate is ay/ℓ, and on the negative edge it is $-ay/\ell$. The x-integration expands the original square element dA into a strip of width dy and length $2ay/\ell$. The subsequent y-integration expands this strip into the whole area of the plate, so that the limits on y are 0 and ℓ. That is,

$$F = \int_0^\ell \rho g(y + L)\sin\theta\left(\int_{-\frac{ay}{\ell}}^{\frac{ay}{\ell}} dx\right)dy$$

$$= \int_0^\ell \rho g(y + L)\sin\theta\left(\frac{2ay}{\ell}\right)dy$$

Completing the y-integration gives

$$F = \rho g a \ell \left(L + \frac{2\ell}{3} \right) \sin \theta \qquad (2.13)$$

To find the point of action of F, we need to take moments. The area is symmetrical and therefore the point of action lies somewhere along the y-axis. To find exactly where, we will take moments about the x-axis.

The moment arm of the force dF about the x-axis is given by y, so that dM, the moment of dF about the x-axis, is given by

$$dM = y \times dF = \rho g \sin \theta \, y(y + L) \, dA$$

The total moment M can be found by integration. M is also given by the resultant force F multiplied by its moment arm $\bar{y}$ about the x-axis, so that

$$M = F \times \bar{y} = \rho g \sin \theta \int y(y + L) \, dA$$

Hence,

$$\bar{y} = \frac{1}{a\ell(L + \frac{2\ell}{3})} \int_0^l y(y + L) \left(\int_{-\frac{ay}{\ell}}^{\frac{ay}{\ell}} dx \right) dy$$

and

$$\bar{y} = \frac{2\ell}{3} \left(\frac{L + \frac{3\ell}{4}}{L + \frac{2\ell}{3}} \right). \qquad (2.14)$$

To demonstrate the consequences of choosing a different moment axis, we choose the moment axis through the point O', where O' is located at the point where the y-axis intersects the water surface (see Figure 2.19). We still measure y and $\bar{y}$ from the point O. Then

$$dM = (y + L) \times dF = \rho g \sin \theta (y + L)^2 \, dA$$

$$F \times (\bar{y} + L) = \rho g \sin \theta \int (y + L)^2 \, dA$$

$$\bar{y} + L = \frac{1}{a\ell(L + \frac{2\ell}{3})} \int_0^l (y + L)^2 \left(\frac{2ay}{\ell} \right) dy$$

$$= \frac{2}{\ell^2(L + \frac{2\ell}{3})} \int_0^l (y^3 + 2y^2 L + yL^2) \, dy$$

$$= \frac{2}{\ell^2(L + \frac{2\ell}{3})} \left[\frac{y^4}{4} + \frac{2y^3 L}{3} + \frac{y^2 L^2}{2} \right]_0^l$$

$$= \frac{2}{(L + \frac{2\ell}{3})} \left(\frac{\ell^2}{4} + \frac{2\ell L}{3} + \frac{L^2}{2} \right)$$

$$= \frac{1}{6(L + \frac{2\ell}{3})} (3\ell^2 + 8\ell L + 6L^2)$$

Hence,

$$\bar{y} = \frac{1}{6(L + \frac{2\ell}{3})} (3\ell^2 + 8\ell L + 6L^2 - 6L^2 - 4\ell L)$$

and

$$\bar{y} = \frac{2\ell}{3} \left(\frac{L + \frac{3\ell}{4}}{L + \frac{2\ell}{3}} \right) \tag{2.15}$$

We obtain the same answer as before (see equation 2.14), but the calculations were more complicated. It is always a good idea to think a little before choosing the coordinate system for a particular problem. The "right" choice will help to reduce the overall algebraic complexity and the likelihood of making a mistake.

To summarize:

(A) To find the resultant force.

Step 1: Choose a coordinate system. It is best to choose a system which makes the task of expressing the shape of the surface as straightforward as possible. Clearly show the origin and direction of your coordinate system.

Step 2: Choose an element of area dA on the surface.

Step 3: Find the depth of dA, that is, its distance below the surface, measured vertically down.

Step 4: Determine if it is possible to use gauge pressure instead of absolute pressure. The gauge pressure acting on dA is p_g, where $p_g = p - p_a = \rho g \times depth$. If gauge pressure can be used, the force acting on dA is $dF = p_g \times dA$.

Step 5: Integrate to find F. For a double integral, do the integral at a constant depth first. The shape of the surface sets the limits of integration.

(B) To find the point of action, take moments.

Step 1: Look for symmetry, since this will always lead to simplifications. For instance, in the example given above, F will act on y-axis so that $\bar{x} = 0$.

Step 2: Choose the axis about which to take moments (the x-axis in the example given above). Then $dM = y \times dF$, where y is the moment arm of dF about the x-axis, and $F \times \bar{y} = \int dM = \int y \times dF$. When gauge pressure can be used, $F \times \bar{y} = \int yp_g \, dA$.

Step 3: Integrate to find M. For a double integral, do the integral at a constant depth first. The shape of the surface sets the limits of integration.

EXAMPLE 2.7 *Choosing Moment Axes*

Consider a rectangular tank filled with water, with a triangular gate located in its side-wall (Figure 2.20). The top edge of the gate is level with the surface of the water. Atmospheric pressure acts everywhere outside the tank. The gate is held on by three bolts. Find the force in each bolt.

Solution: Since the gate is in static equilibrium, the sum of the forces must be zero. That is, in the horizontal direction,

$$F_1 + F_2 + F_3 - F = 0$$

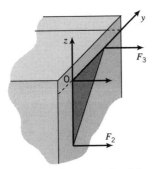

FIGURE 2.20 Wall with triangular gate.

where F is the force exerted by the water pressure on the area of the gate. It is clear that we need additional information to solve for F_1, F_2, and F_3. This information will come from the moment equation; since the gate is in static equilibrium, the sum of the moments must also be zero. This is true for any axis we choose, but some axes are better than others. For example, if we choose the z-axis, F_1 and F_2 need not be considered since they have no moment about the z-axis (their moment arm about the z-axis is zero). Therefore the moment exerted by F_3 about the z-axis must balance the moment exerted by F about the z-axis, and F_3 can be found directly. Similarly, F_2 can be found by taking moments about the y-axis, and so, together with $\Sigma \mathbf{F} = 0$, we have three equations for three unknowns. ■

EXAMPLE 2.8 *Complex Two-Dimensional Surfaces*

What is the force on the gate shown in Figure 2.21 due to the hydrostatic pressure?

Solution: The shape of the gate is rather complex, and it is best to treat it in two parts, where the forces acting on the left hand side and the right hand side of the gate are found separately, and the resultant force is found by simple addition. We will not give the full solution here, only an outline of how the problem may be solved. Here is the basic result for the left hand side.

$$F_1 = \int p \, dy \, dz = \int_{-a}^{+a} p(z) \left(\int_{-\sqrt{a^2-z^2}}^{0} dy \right) dz$$

Because p is a function of z (only) it is best to do the integration with respect to y first. The integration can be completed using a table of standard integrals.

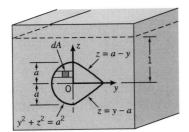

FIGURE 2.21 Gate with circular section on one side, and triangular section on the other.

To find the force acting on the right hand side, we need to subdivide the area further: the top half (A_{2t}) is described by the equation $z = a - y$, that is, $y = a - z$, and the bottom half (A_{2b}) is described by the equation $z = y - a$, that is, $y = a + z$. For the top half

$$F_{2t} = \int p\, dA_{2t} = \int_0^{+a} p(z) \left(\int_0^{a-z} dy \right) dz$$

and for the bottom half

$$F_{2b} = \int p\, dA_{2b} = \int_{-a}^{0} p(z) \left(\int_0^{a+z} dy \right) dz$$

∎

2.8 **CENTERS OF PRESSURE, MOMENTS OF AREA

There are other ways to solve these types of problems. One way that is popular in many textbooks is to note that hydrostatics has a lot in common with the mechanics of solids. In particular, for the problem with the triangular plate (Figure 2.19), we know that the centroid of the triangle is located at a depth $(L + \frac{2\ell}{3}) \sin \theta$ below the water surface. The gauge pressure at this point is $\rho g(L + \frac{2\ell}{3}) \sin \theta$, so that from equation 2.13 we can see that the resultant force on the triangular plate is equal to the pressure at the centroid times the area of the plate $a\ell$. This is a general result for flat surfaces, so that

> The resultant force acting on a flat plate is given by the pressure at the centroid, p_c, times the area of the plate. That is,
>
> $$F = p_c A. \tag{2.16}$$

This observation is only useful if we know where the centroid is located.

Another interesting result can be obtained by considering the sum of the moments. If we choose a new coordinate system $[x', y']$ located at the point O' (see Figure 2.19), and we measure the moment arm $\bar{y}'$ from the x'-axis, then, taking moments about the x'-axis

$$dM = y \times dF = y'(\rho g \sin \theta\, y')\, dA$$

That is

$$F \times \bar{y}' = \rho g \sin \theta \int y'^2\, dA$$
$$= \rho g \sin \theta I_{x'} dA$$

where $I_{x'} = \int y'^2\, dA$. This integral can be recognized as the second moment of the area with respect to the x'-axis.

We can continue by adding another coordinate system $[x'',y'']$ with its origin located at the centroid O'', located a distance $[x_c, y_c]$ from O'. Hence, $y' = y'' + y_c$, and

$$I_{x'} = \int y'^2\, dA = \int (y''^2 + 2y_c y'' + y_c^2)\, dA$$
$$= \int y''^2\, dA + \int 2y_c y''\, dA + \int y_c^2\, dA$$

The integral $\int 2y_c y'' \, dA = 2y_c \int y'' \, dA$, and $\int y'' \, dA$ is the first moment of area about an axis passing through the centroid, which is identically zero by the definition of a centroid. So

$$I_{x'} = \int y''^2 \, dA + \int y_c^2 \, dA = \int y''^2 \, dA + y_c^2 \int dA$$

or:

$$I_{x'} = I_{xc} + Ay_c^2$$

This is the *theorem of parallel axes,* where $I_{x'}$ is the second moment of area with respect to an arbitrary x'-axis, I_{xc} is the second moment of area with respect to the x''-axis passing through the centroid, and y_c is the distance between the x'-axis and the x''-axis. Finally,

$$F \times \bar{y}' = \rho g \sin \theta I_{x'}$$

so that

$$\bar{y}' = \frac{\rho g \sin \theta I_{x'}}{F} = \frac{\rho g \sin \theta I_{x'}}{p_c A} = \frac{\rho g \sin \theta I_{x'}}{\rho g y_c \sin \theta A} = \frac{I_{x'}}{y_c A}$$

> Therefore, if the position of the centroid is known, and the moment of area about the centroid is known, the line of action can be found using
>
> $$\bar{y}' = \frac{I_{xc}}{y_c A} + y_c \tag{2.17}$$

This result also shows that the point of action of the resultant force is always lower than the position of the centroid. To illustrate this point more physically, consider a simple rectangular plate oriented vertically. The pressure acting over the area increases with depth so that it has a higher value at greater depth, and therefore the resultant force will always act below the middle of the plate, that is, it acts below the centroid of the area.

Methods based on centroids and moments of inertia are one way to solve problems in hydrostatics. They require you to know the position of the centroid and the moment of inertia for the shape in question, and although many common shapes are tabulated in handbooks, irregular shapes still need to be evaluated from first principles. In addition, unless great care is taken, shifting axes to find moments of area about axes other than the standard ones is a process open to many errors. Finally, these approaches tend to obscure the basic physics underlying the problem. For all these reasons, these short cut methods are not recommended. Instead, the fundamental methods outlined in Section 2.7 should be used, in a careful step-by-step approach.

2.9 ARCHIMEDES' PRINCIPLE

We saw in Sections 2.5 and 2.6 that the vertical force acting on a sloping or curved wall was equal to the weight of the fluid supported by the wall, and that it acted through the centroid. More can be done with this result.

Consider a V-shaped "boat" floating in water, where the weight of the boat is W (Figure 2.22). The lowest point of the boat is a distance h below the surface of the water. The boat is in equilibrium, so that the resultant force acting on it is

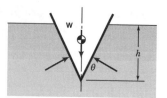

FIGURE 2.22 A V-shaped boat floating in water.

zero (the resultant moment must be zero also). The horizontal component of the force due to water pressure acting on the left half of the hull acts to the right, and on the right half it acts to the left. Since they are of the same magnitude and they act at the same depth, they cancel. As for the vertical forces, we know that the weight of the boat is balanced by the sum of the two vertical forces due to water pressure acting on the two halves of the hull. The vertical components both act up, and since they are of the same magnitude, they add. Using the result found in Section 2.5,

$$W = 2 \times \frac{\rho g w h^2}{2 \tan \theta}$$

(In the limit, we see that when $\theta \to \pi/2$, the weight that can be supported by the water pressure goes to zero, which is the correct limiting behavior.)

Note that the volume of the displaced fluid is $wh^2/\tan \theta$, and its weight W is given by $\rho g w h^2/\tan \theta$. In other words, the weight of the boat is balanced by a *buoyancy* force equal to the weight of water displaced by the boat. Since there is no resultant moment (all moments cancel), the resultant force must act through the centroid of the displaced fluid, in line with the weight force. These observations can be written in a general way, known as Archimedes principle.

> The buoyancy force on a solid is equal to the weight of liquid it has displaced, and it acts through the centroid of the displaced volume.

The centroid of the displaced fluid is also known as the *center of buoyancy*.

For a different example involving buoyancy forces, consider a steel barge floating in a small pond. The barge contains a number of steel bars, and therefore displaces a certain amount of water (the weight of the fluid displaced must equal the weight of the barge plus the weight of the steel bars). The maximum depth of the water is h_1 (Figure 2.23). An accident upsets the barge so that all the steel bars fall in the water, and since they have a greater density than water, they sink to the bottom. The maximum depth of the water is now h_2. How does h_1 compare to h_2? There are three possible answers: $h_1 > h_2$, $h_1 = h_2$, or $h_1 < h_2$. Which one is right?

Initially, the total weight of water displaced equals the weight of the steel barge plus the weight of the steel bars. Since the density of steel is greater than

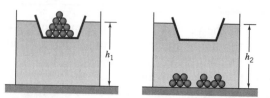

FIGURE 2.23 Floating barge, containing steel bars (left), and empty (right).

that of water (steel has a density of 7850 kg/m^3, compared to water, which has a density of 998 kg/m^3, so that the specific gravity of steel is 7.86), the volume of water displaced is much greater than the volume of steel. After the barge is upset and the steel bars have sunk to the bottom, the water displaced by the barge is such that the weight of the displaced water equals the weight of the barge. Water is also displaced by the steel bars, but since they are not floating they only displace a volume of water equal to their own volume. This volume is much less than the volume displaced by the steel bars when they float, so the correct answer is $h_1 > h_2$.

EXAMPLE 2.9   *The Tip of an Iceberg*

Consider an iceberg floating in seawater. Find the fraction of the volume of the iceberg that shows above the sea surface.

Solution: If the volume of the iceberg is $\forall$, and the fraction showing above the surface is $\Delta \forall$, the buoyancy force acting up on the iceberg is given by $\rho_{sw} g$ $(\forall - \Delta \forall)$, where ρ_{sw} is the density of sea water. For static equilibrium, this must equal the weight of the iceberg, which is given by $\rho_{ice} g \forall$, where ρ_{ice} is the density of ice. That is,

$$\rho_{sw} g(\forall - \Delta \forall) = \rho_{ice} g \forall$$

so that

$$\frac{\Delta \forall}{\forall} = 1 - \frac{\rho_{ice}}{\rho_{sw}}$$

Ice has a density of 920 kg/m^3, and seawater has a density of 1025 kg/m^3 (Table A–C.7). Hence

$$\frac{\Delta \forall}{\forall} = 0.102$$

so that only about 10% of the bulk of an iceberg is visible above the surface (see Figure 1.8). ∎

2.10 **STABILITY OF FLOATING BODIES

Floating bodies are in equilibrium under body forces (the weight of the vessel) and buoyancy forces (the weight of the displaced fluid). The lines of action of these forces determines the stability of the body (Figure 2.24). The body force W acts through the the center of gravity of the body, CG, and the buoyancy force F_B acts

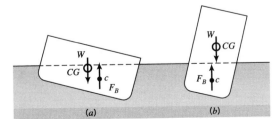

FIGURE 2.24 Stability of floating bodies. (*a*) Stable. (*b*) Unstable.

through the centroid of the displaced fluid volume, that is, the center of buoyancy c. The body is *neutrally stable* if the lines of action are co-linear, and *stable* if they produce a moment which tends to right the vessel [Figure 2.24(a)]. It is *unstable* if the moment tends to capsize the vessel [Figure 2.24(b)].

2.11   **FLUIDS IN RIGID-BODY MOTION

We noted earlier that a fluid in motion can also be in static equilibrium, as long as all parts of the fluid are moving together as a rigid body. In *rigid-body motion*, there can be no relative motion within the liquid; no part of the fluid can be moving with respect to any other part. When the fluid and its container are moving at constant speed, such a system is in simple translation and there are no additional forces acting. However, when this system is accelerating, the inertia force needs to be taken into account and a new analysis is required.

2.11.1   Vertical Acceleration

Consider a container that is moving in the vertical direction with an acceleration a_z. The fluid is in rigid body motion. Newton's second law states that for an accelerating fixed mass of fluid:

{mass of fluid × acceleration} = {forces due to pressure differences} + {weight of fluid}

For a small element of fluid of volume $\delta x \delta y \delta z$, where the positive z-direction is vertically down as in Figure 2.25,

$$(\rho \delta x \delta y \delta z)a_z = \left(p - \frac{dp}{dz}\frac{\delta z}{2}\right)\delta x \delta y - \left(p + \frac{dp}{dz}\frac{\delta z}{2}\right)\delta x \delta y + (\rho \delta x \delta y \delta z)g$$

where we have used a Taylor series expansion (Section A–A.8.4) to express the pressure on the top and bottom faces in terms of the pressure at the center of the element. Therefore,

$$\rho a_z = -\frac{dp}{dz} + \rho g$$

That is,

$$\frac{dp}{dz} = \rho(g - a_z) \tag{2.18}$$

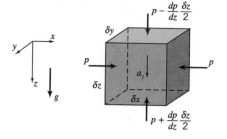

FIGURE 2.25   Static equilibrium of a fluid element under the action of gravity and a constant vertical acceleration.

What happens when the container is falling freely under gravity? In this case, $a_z = g$ since z is positive going down, and the hydrostatic pressure variations in the container go to zero. This is true for all fluids in free fall. For example, a jet of water issuing from a tank will be in free fall under gravity, and there will be no hydrostatic pressure variations within the jet. When in space, where there is zero gravity because the spacecraft is essentially in free fall, the pressure in any fluid will be constant, and the only thing keeping a fluid together in an open container is surface tension.

In contrast, when the container is accelerating upwards, the pressure variations in the fluid increase above their static values. Astronauts accelerating into space at high "g-levels" need to lie at right angles to the acceleration vector so that the pressure differences across their bodies are minimized, and their hearts can cope with the extra load imposed by the increased pressure differences.

2.11.2 Vertical and Horizontal Accelerations

Consider a container that has an acceleration in the horizontal and vertical directions, a_x and a_z, respectively (Figure 2.26). The equation of motion in the z-direction (equation 2.19) becomes

$$\frac{\partial p}{\partial z} = \rho(g - a_z) \tag{2.19}$$

Partial derivatives are needed because the pressure is now a function of x and z. The equation of motion in the x-direction gives

$$(\rho\delta x\delta y\delta z)a_x = \left(p - \frac{dp}{dx}\frac{\delta x}{2}\right)\delta y\delta z - \left(p + \frac{dp}{dx}\frac{\delta x}{2}\right)\delta y\delta z$$

Therefore,

$$\frac{\partial p}{\partial x} = -\rho a_x \tag{2.20}$$

Equations 2.19 and 2.20 are a pair of first-order partial differential equations, which can be solved as follows. Integrating equation 2.19 with respect to z we obtain

$$p = \rho(g - a_z)z + f(x) + C_1 \tag{2.21}$$

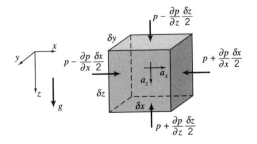

FIGURE 2.26 Static equilibrium of a fluid element under the action of gravity and constant vertical and horizontal accelerations.

where $f(x)$ is an unknown function of x (only), and C_1 is a constant of integration. Integrating equation 2.20 with respect to x we obtain

$$p = -\rho a_x x + g(z) + C_2 \qquad (2.22)$$

where $g(z)$ is an unknown function of z (only), and C_2 is another constant of integration. From equations 2.21 and 2.22 we see that

$$p = \rho(g - a_z)z - \rho a_x x + C \qquad (2.23)$$

(When solving simultaneous partial differential equations, it is always recommended to check that the original equations, in this case equations 2.19 and 2.20, are recovered by differentiating the solution.)

For a surface of constant pressure (an *isobaric* surface), the left hand side of equation 2.23 is constant. The slope of an isobaric surface can be found by differentiating this equation while holding p constant. That is, the slope of an isobaric surface is given by

$$\frac{dz}{dx} = \frac{a_x}{g - a_z} \qquad (2.24)$$

At the free surface, the pressure is constant and equal to the atmospheric pressure, so that equation 2.24 gives the slope of the free surface. In fact, since a_x and a_z are constant throughout the fluid, all isobaric surfaces have the same slope so that they are all parallel to the free surface.

2.11.3 Rigid-Body Rotation

A similar analysis will apply to a volume of fluid that is rotating at a constant angular speed, as shown in Figure 2.27. In that case, an element of fluid experiences a radial acceleration V^2/r, where V is the circumferential component of velocity, and r is the distance from the axis of rotation. There is also an acceleration in the vertical direction due to gravity, so that the pressure is a function of r and z. If the rate of rotation is ω *rad/s*, the equation of motion in the r-direction becomes

$$\frac{\partial p}{\partial r} = \rho \frac{V^2}{r} = \rho r \omega^2 \qquad (2.25)$$

For the z-direction (where z is positive down), we get the usual hydrostatic variation

$$\frac{\partial p}{\partial z} = \rho g \qquad (2.26)$$

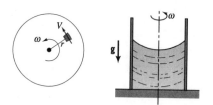

FIGURE 2.27 Fluid in rigid-body motion under rotation.

Integrating equation 2.25 with respect to r we obtain

$$p = \tfrac{1}{2}\rho r^2 \omega^2 + f'(z) + C_3 \tag{2.27}$$

and integrating equation 2.26 with respect to z we obtain

$$p = \rho g z + g'(x) + C_4 \tag{2.28}$$

From equations 2.27 and 2.28

$$p = \rho g z + \tfrac{1}{2}\rho r^2 \omega^2 + C' \tag{2.29}$$

To find C', we need to know the pressure at a given point. At the free surface, the pressure is equal to atmospheric pressure, so that $p = p_a$ at $r = z = 0$, and

$$z = \frac{p - p_a}{\rho g} - \frac{r^2 \omega^2}{2g} \tag{2.30}$$

On the axis, where $r = 0$, the radial acceleration is zero, and the free surface is horizontal. Further from the axis of rotation, the surface is inclined to the horizontal, and we see that the free surface actually forms a parabola of revolution, as do all other isobaric surfaces (Figure 2.27). Question: For a cylindrical container, where is the maximum pressure found?

EXAMPLE 2.10   *Rigid-Body Motion*

For the case shown in Figure 2.28, find the horizontal acceleration that would make the water spill out of the container.

Solution: From equation 2.24, we know that

$$\frac{dz}{dx} = \frac{a_x}{g - a_z} = \frac{a_x}{g}$$

since $a_z = 0$. The fluid will spill out of the container when

$$\frac{dz}{dx} = \frac{\frac{h}{3}}{\frac{3h}{2}} = \frac{2}{9}$$

which requires a horizontal acceleration

$$a_x = \tfrac{2}{9}g$$

$\blacksquare$

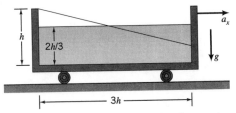

FIGURE 2.28 Fluid in rigid-body motion under the action of gravity and a constant horizontal acceleration.

PROBLEMS

2.1   Express the following pressures in *psi:*
(a) $2.5 \times 10^5 \, Pa$
(b) 4.3 *bar*
(c) 31 *in.* Hg
(d) 20 *ft* H_2O

2.2   Express the following pressures in *Pa:*
(a) 3 *psia*
(b) 4.3 *bar*
(c) 31 *in.* Hg
(d) 8 *m* H_2O

2.3   Express the following absolute pressures as gauge pressures in SI and BG units:
(a) 3 *psia*
(b) $2.5 \times 10^5 \, Pa$
(c) 31 *in.* Hg
(d) 4.3 *bar*
(e) 20 *ft* H_2O

2.4   In a hydraulic press, a force of 200 *N* is exerted on the small piston (area 10 cm^2). Determine the force exerted by the large piston (area 100 cm^2), if the two pistons were at the same height.

2.5   For the hydraulic press described in the previous problem, what is the the force produced by the large piston if it were located 2 *m* above the small piston? The density of the hydraulic fluid is 920 kg/m^3.

2.6   A device consisting of a circular pipe attached to a rectangular tank is filled with water as shown in Figure P2.6. Neglecting the weight of the tank and the pipe, determine the total force on the bottom of the tank. Compare the total weight of the water with this result and explain the difference.

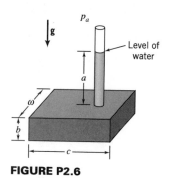

FIGURE P2.6

2.7   The gauge pressure at the liquid surface in the closed tank shown in Figure P2.7 is 4.0 *psi.* Find *h* if the liquid in the tank is
(a) water
(b) kerosene
(c) mercury

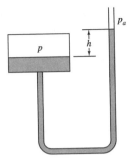

FIGURE P2.7

2.8 Find the maximum possible diameter of the circular hole so that the tank shown in Figure P2.8 remains closed. The lid has a mass of 50 *kg*.

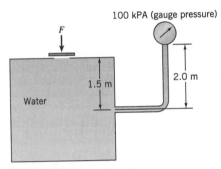

FIGURE P2.8

2.9 A hollow cylinder of diameter 1 *m* has a closed bottom, which is pushed into a swimming pool to a depth of 3 *m*. The cylinder is open to atmosphere at the top, and the swimming pool is at sea level.

(a) Find the force acting on the bottom of the cylinder when the pool contains fresh water, and the air pressure is taken to be constant everywhere.

(b) How does the answer to part **(a)** change when the pool contains sea water?

(c) How does the force found in part **(a)** change when the variation in air pressure inside the cylinder is taken into account?

(d) Do the answers to parts **(a)**, **(b)**, and **(c)** change if the swimming pool is moved to the top of a 5000 *m* mountain?

2.10 For the manometer shown in Figure P2.10, both legs are open to the atmosphere. It is filled with liquids A and B as indicated. Find the ratio of the liquid densities.

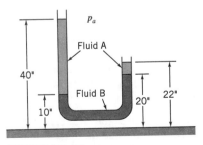

FIGURE P2.10

2.11   Find the pressure at an elevation of 3000 *m* if the temperature of the atmosphere decreases at a rate of 0.006° *K* per *m*. The ground-level temperature is 15° *C*, and the barometer reading is 29.8 *in*. Hg. (The gas constant for air is = 287.03 $m^2 s^2 K$.)

2.12   At a particular point in the Pacific Ocean, the density of sea water increases with depth according to $\rho = \rho_0 + mz^2$, where ρ_0 is the density at the surface, *z* is the depth below the surface, and *m* is a constant. Develop an algebraic equation for the pressure as a function of depth.

2.13   If the specific gravity of concrete is 2.4, find the vertical reactions R_1 and R_2 per unit width of the concrete dam shown in Figure P2.13.

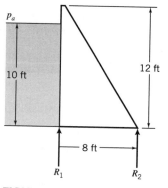

FIGURE P2.13

2.14   The gate shown in Figure P2.14 has a width *w* and a height *H* and it is pivoted on a frictionless hinge at a point *z** below the surface of the water. The top of the gate is level with the surface of the water. The water is of density ρ, and outside the tank the pressure is uniform everywhere and equal to the atmospheric pressure.

(a) Find the magnitude of the resultant force *F* on the gate, in terms of ρ, *g*, *w*, and *H*.

(b) Find the location of the pivot line *z** below the top of the gate, so that there is no resultant moment about the hinge tending to open the gate.

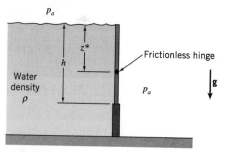

FIGURE P2.14

2.15   A gate 3 *ft* square in a vertical dam is exposed to air at atmosphere pressure on one side and water on the other. The resultant force acts 2 *in*. below the center of the gate. How far is the top of the gate below the water surface?

2.16   A gate of width *w* stands vertically in a tank, as shown in Figure P2.16, and it is connected to the bottom of the tank by a frictionless hinge. On one side the tank is filled to a depth h_1 by a fluid density ρ_1; on the other side it is filled to a depth h_2 by a fluid density ρ_2. Find h_2/h_1 in terms of ρ_2/ρ_1 if the gate is in static equilibrium.

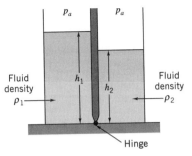

FIGURE P2.16

2.17 The tank shown in Figure P2.17 has a gate which pivots on a vertical, frictionless hinge. Find the ratio of the depths of water h_1/h_2 in terms of the densities ρ_1 and ρ_2 when the gate is in static equilibrium.

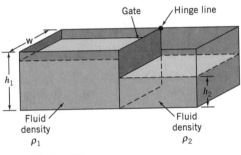

FIGURE P2.17

2.18 A rectangular gate is located in a dam wall as shown in Figure P2.18. The reservoir is filled with a heavy fluid of density ρ_2 to a height equal to the height of the gate, and topped with a lighter fluid of density ρ_1.

(a) Write down the variation of pressure for $z \leq D$, and for $D \leq z \leq D + L$.

(b) Find the resultant force acting on the gate due to the presence of the two fluids.

(c) Find the point where this resultant force acts.

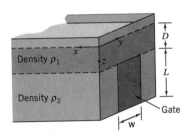

FIGURE P2.18

2.19 A primitive safety valve for a pressure vessel containing liquids is shown in Figure P2.19. The liquid is water, of density ρ, and it has a constant depth H. The gauge pressure exerted on the surface of the water is p_w, and the pressure outside the vessel is atmospheric. The gate is rectangular, of height B and width W, and there is a spring at the hinge which exerts a constant clockwise moment M_h. Find p_w for which the gate is just on the point of opening.

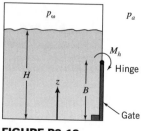

FIGURE P2.19

2.20   A square gate of dimension b separates two fluids of density ρ_1 and ρ_2, as shown in Figure P2.20. The gate is mounted on a frictionless hinge. As the depth of the fluid on the right increases, the gate will open. Find the ratio ρ_2/ρ_1 for which the gate is just about to open in terms of H_1, H_2, and b.

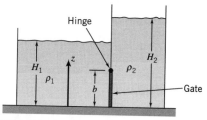

FIGURE P2.20

2.21   The symmetric trough shown in Figure P2.21 is used to hold water. Along the length of the trough steel wires are attached to support the sides at distances w apart. Find the magnitude of the resultant force exerted by the water on each side, and the magnitude of the tension in the steel wire, neglecting the weight of the trough.

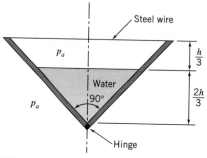

FIGURE P2.21

2.22   A rigid uniform thin gate of weight Mg and constant width b is pivoted on a frictionless hinge as shown in Figure P2.22. The gate makes an angle θ with the ground. The water depth on the left hand side of the gate is H and remains constant. On the right hand side

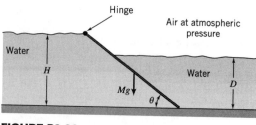

FIGURE P2.22

of the gate the level of water is slowly decreased until the gate is just about to open. Find the depth D at which this occurs.

2.23 The rectangular gate shown in Figure P2.23 (of width w and length L) is made of a homogeneous material and it has a mass M. The gate is hinged without friction at point B. Determine the mass required to hold the gate shut when the water depth at the point B is H.

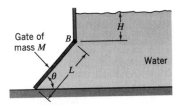

FIGURE P2.23

2.24 A gate of constant width w is hinged at a frictionless hinge located at point O and rests on the bottom of the dam at the point A, as shown in Figure P2.24. Find the magnitude and direction of the force exerted at the point A due to the water pressure acting on the gate.

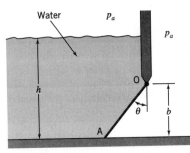

FIGURE P2.24

2.25 Figure P2.25 shows a very delicate balancing act. On one side of a weightless wedge there is fluid of density ρ_1, and on the other side there is fluid of density ρ_2. If the wedge is just balanced, find the ratio ρ_2/ρ_1 in terms of H_1, H_2, and θ.

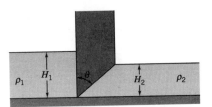

FIGURE P2.25

2.26 A rectangular window of width w is set into the sloping wall of a swimming pool, as shown in Figure P2.26. Find the point of action of the resultant force acting on the window.

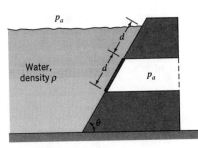

FIGURE P2.26

2.27 A triangular trough contains two fluids, with densities ρ_1 and ρ_2, and depths h_1 and h_2, respectively, as shown in Figure P2.27. Find the resultant hydrodynamic force acting on the divider, and where it acts.

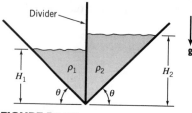

FIGURE P2.27

2.28 A rigid, weightless, two-dimensional gate of width w separates two liquids of density ρ_1 and ρ_2, respectively, as shown in Figure P2.28. The gate pivots on a frictionless hinge and it is in static equilibrium. Find the ratio ρ_2/ρ_1 when $h = b$.

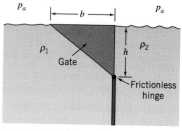

FIGURE P2.28

2.29 A rectangular gate 1 m by 2 m, is located in a wall inclined at 45°, as shown in Figure P2.29. It is held shut by a force F. If the gate has a mass of 100 kg, find F.

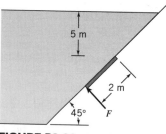

FIGURE P2.29

2.30 A beach ball of weight Mg and diameter D is thrown into a swimming pool. If the ball just floats, what is the diameter of the ball?

2.31 Determine what fraction of the volume of an ice cube is visible above: (a) the surface of a glass of fresh water, and (b) the surface of a glass of ethanol. Do these answers change if we were on the surface of the moon?

2.32 A 1 m^3 of aluminum, of specific gravity 2.7, is tied to a piece of cork, of specific gravity 0.24, as shown in Figure P2.32. What volume of cork is required to keep the aluminum block from sinking in water if both masses are completely submerged?

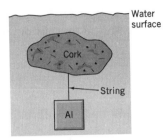

FIGURE P2.32

2.33 A concrete block rests on the bottom of a lake. The block is a cube, 1 *ft* on a side. Calculate the force required to hold the block at a fixed depth. The density of concrete is 2400 kg/m^3.

2.34 A two-dimensional symmetrical prism floats in water as shown in Figure P2.34. The base is parallel to the surface. If the specific gravity of the prism material is 0.25, find the ratio a/d.

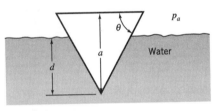

FIGURE P2.34

2.35 A rectangular barge of length L floats in water (density ρ_w) and when it is empty, it is immersed to a depth D, as shown in Figure P2.35. Oil of density ρ_o is slowly poured into the barge until it is about to sink. Find a relationship for the depth of oil at this point in terms of H, D, ρ_w, and ρ_o.

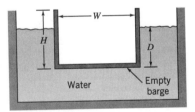

FIGURE P2.35

2.36 For the example shown in Figure 2.23, what will happen if the steel cylinders are replaced by logs of wood?

2.37 A cork float with dimensions $A \times A \times A$ and specific gravity 0.24 is thrown into a swimming pool with a water surface area $2A \times 2A$, and an initial depth of $2A$. Derive an expression for the pressure at the bottom of the pool.

2.38 A circular cylinder of length 3 *ft* and diameter 6 *in.* floats vertically in water so that only 6 *in.* of its length protrudes above the water level. If it was turned to float horizontally, how far would its longitudinal axis be below the water surface?

2.39 A rigid, helium-filled balloon has a total mass M and volume $\forall$, and it is floating in static equilibrium at a given altitude in the atmosphere. Using Archimedes' principle, describe what happens when ballast of mass m is dropped overboard. How does this answer change if the balloon is no longer rigid but it is allowed to stretch?

2.40 The circular cylinder shown in Figure P2.40 has a specific gravity of 0.9.
(a) If the system is in static equilibrium, find the specific gravity of the unknown liquid.
(b) Do you think the system is stable?

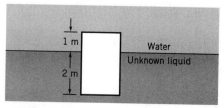

FIGURE P2.40

2.41 A rectangular body of specific gravity 0.8 has length and width a, and height b, as shown in Figure P2.41, where $a > b$. Find the position of the body relative to the interface between the fluids.

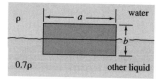

FIGURE P2.41

2.42 A float and lever system is used to open a drain valve, as shown in Figure P2.42. The float has a volume $\forall$ and a density ρ_f. The density of the water is ρ_w. Find the maximum force available to open the drain valve, given that the hinge is frictionless. Hint: first consider the forces acting on the float, then consider the free-body diagram of the gate.

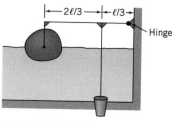

FIGURE P2.42

2.43 A rectangular cork float is attached rigidly to a vertical gate as shown in Figure P2.43. The gate can swing about a frictionless hinge. Find an expression for the depth of water D where the gate will just open, in terms of a, b, L, and h. The cork float has the same width w as the gate, and the specific gravity of cork is 0.24.

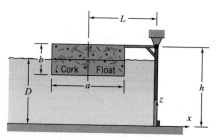

FIGURE P2.43

2.44 The two-dimensional gate shown in Figure P2.44 is arranged so that it is on the point of opening when the water level reaches a depth H. If the gate is made of a uniform material that has a weight per unit area of mg, find an expression for H.

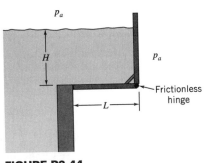

p_a

H

p_a

Frictionless hinge

L

FIGURE P2.44

2.45 A rectangular gate of width w and height h is placed in the vertical side wall of a tank containing water. The top of the gate is located at the surface of the water, and a rectangular container of width w and breadth b is attached to the gate, as shown in Figure P2.45. Find d, the depth of the water required to be put into the container so that the gate is just about to open, in terms of h and b. The top and sides of the tank and container are open to the atmosphere. Neglect the weight of the container.

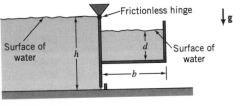

Frictionless hinge

$\downarrow g$

Surface of water

h

d

Surface of water

b

FIGURE P2.45

2.46 A rigid gate of width w is hinged without friction at a point H above the water surface, as shown in Figure P2.46. Find the ratio b/H at which the gate is about to open. Neglect the weight of the gate.

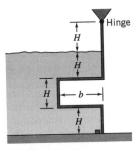

Hinge

H

H

H

b

H

FIGURE P2.46

2.47 If the weightless gate shown in Figure P2.47 is just on the point of opening, find an expression for B in terms of h.

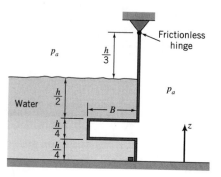

FIGURE P2.47

2.48 A certain volume of water is contained in the square vessel shown in Figure P2.48. Where the sealing edges of the inclined plate come into contact with the vessel walls, the reactive force normal and parallel to the wall is zero.

(a) Find the magnitude and direction of the single force F required to hold the plate in position. The weight of the plate may be neglected.

(b) Where does this force F act?

(c) If the inclined plate is replaced by a horizontal one, find the relative position of the new plate if the force used to maintain its position has the same magnitude as before. The volume of fluid remains the same as before.

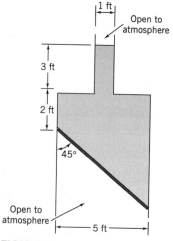

FIGURE P2.48

2.49 Figure P2.49 shows a vessel of constant width w that contains water of density ρ. Air is introduced above the surface of the water at pressure p_1. There is a rectangular blow-off

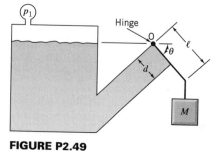

FIGURE P2.49

safety lid of width w and length ℓ hinged without friction at point O which is located at the same height as the water surface. At what pressure will the lid open? Express the answer in terms of w, d, ℓ, Mg, θ, and ρ.

2.50 A vessel of constant width w is filled with water, and the surface is open to atmospheric pressure. On one side, a relief valve is located, as shown in Figure P2.50. The relief valve has the same width as the vessel, and it has a length D. It is hinged without friction at point A. At point B, a mass M is connected to the gate to keep it shut. Find M in terms of H, D, θ, and ρ, where ρ is the density of water.

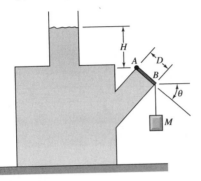

FIGURE P2.50

2.51 The rectangular submarine escape hatch shown in Figure P2.51 (of width w and length L) will open when the constant pressure inside the chamber, p_i, exceeds a critical value p_{ic}. The hatch has negligible mass and it is hinged without friction at point A, which is a depth D below the surface. Find p_{ic} in terms of w, L, D, ρ, g, and θ.

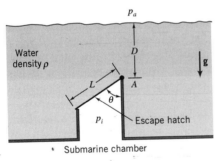

FIGURE P2.51

2.52 The gate AB shown in Figure P2.52 is rectangular with a length L and a width w into the page. The gate has a frictionless hinge at point A and it is held against a stop at point B by a weight Mg. Neglecting the weight of the gate, derive an expression for the water height h at which the gate will start to move away from the stop.

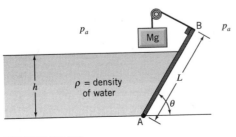

FIGURE P2.52

2.53 The gate shown in Figure P2.53 has a constant width w into the page. The gate has a frictionless hinge at point A and it is held against a stop at point B. Neglecting the weight of the gate, find the magnitude of the resultant force exerted on the gate by the water, and the dimension b for which there is no force on the gate at point B.

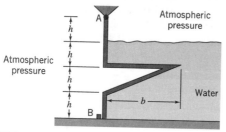

FIGURE P2.53

2.54 A tank of water of density ρ has a symmetric triangular gate of height H and maximum width $2a$, as shown in Figure P2.54. Calculate the force F exerted by the water on the gate, and where it acts, in terms of ρ, g, h, H, and a. The air pressure outside the tank is uniform everywhere.

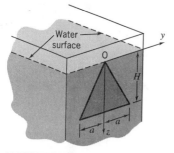

FIGURE P2.54

2.55 A rigid, weightless, two-dimensional gate of width w separates two liquids of density ρ_1 and ρ_2 respectively. The gate has a parabolic face, as shown in Figure P2.55, and it is in static equilibrium. Find the ratio ρ_2/ρ_1 when $h = \ell$.

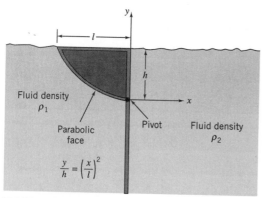

FIGURE P2.55

2.56 A rigid, submerged, triangular gate is hinged as shown in Figure P2.56.
(a) Find the magnitude and direction of the total force exerted on the gate by the water.
(b) A weight Mg on a rigid lever arm of length L is meant to keep this gate shut. What is the depth of water D when this gate is just about to open? You can neglect the weight of the gate itself, and the weight of the lever arm.

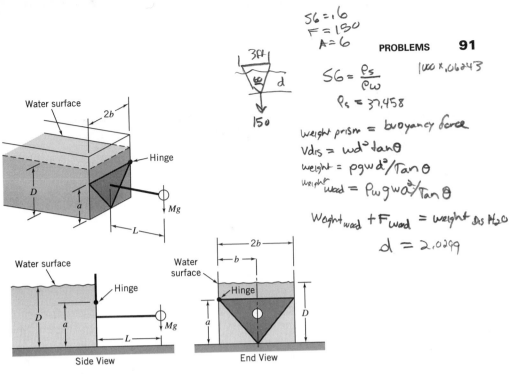

Handwritten notes:
$S6 = 1.6$
$F = 130$
$A = 6$

$S6 = \dfrac{\rho_s}{\rho_w}$ $\qquad 100 \times .06243$

$\rho_s = 37.458$

weight prism = buoyancy force
$V_{dis} = wd^2 \tan\theta$
weight $= \rho g w d^2 / \tan\theta$
$weight_{wad} = \rho_w g w a^2 / \tan\theta$

$Weight_{wad} + F_{wad} = weight \; dis \; H_2O$

$d = 2.0299$

FIGURE P2.56

2.57 A circular gate of radius R is mounted halfway up a vertical dam face as shown in Figure P2.57. The dam is filled to a depth h with water, and the gate pivots without friction about a horizontal diameter.

(a) Determine the magnitude of the force due to the water pressure acting on the gate.

(b) Determine the magnitude and sign of the force F required to prevent the gate from opening.

FIGURE P2.57

2.58 An elevator accelerates vertically down with an acceleration of $\frac{1}{4}g$. What is the weight of a 120 lb_f person, as measured during the acceleration?

2.59 A rocket accelerating vertically up with an acceleration of a carries fuel with a density ρ in tanks of height H. The top of the tank is vented to the atmosphere. What is the pressure at the bottom of the tank?

2.60 A car accelerates at a constant rate from 0 *mph* to 60 *mph* in 10 s. A U-tube manometer with vertical legs 2 *ft* apart is partly filled with water and used as an accelerometer.

(a) What is the difference in height of the water level in the two legs?

(b) Starting from rest, how fast would the car be going at the end of 10 s if the difference in level were 1.0 *in.* larger?

2.61 The cart shown in Figure 2.28 is now moving upward on a 5° incline at a constant acceleration a_i. Find the value of a_i that would make the water spill out of the container.

2.62 A 10 *cm* diameter cylinder initially contains 10 *cm* of water. It is then spun about its axis at an angular speed of ω. Find the value of ω where the bottom of the container just becomes exposed.

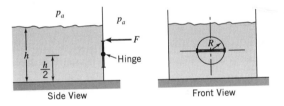

INTRODUCTION TO FLUID MOTION I

3.1 INTRODUCTION

The principal aim of the next two chapters is to introduce the basic principles governing fluid motion: mass conservation, the momentum, and the energy equation. These principles will be illustrated by examples of one-dimensional, steady flows. In this chapter, we consider the mass conservation and momentum equations. Before doing so, we introduce some tools for describing fluid motion, including the concepts of pathlines, streamlines, fluid particles, fluid elements, and control volumes.

3.2 FLUID PARTICLES AND CONTROL VOLUMES

In Chapter 2 we considered fluids in rigid-body motion. But a fluid rarely moves as a rigid body, and generally there exist relative motions among different parts of the fluid. How are we to describe the displacement, velocity, and acceleration of these parts? We can either use a *fluid particle* approach, or a *control volume* approach. In the fluid particle approach, we identify and follow small, fixed masses of fluid as illustrated in Figure 3.1(*a*). This is also called a *Lagrangian* approach. In the control volume approach, we draw an imaginary box around a fluid system. This is also called an *Eulerian* approach. The box can be large or small, and it can be stationary or in motion. Generally, there will be flow in and out through the surface of the control volume, and the fluid inside the control volume is changing all the time [Figure 3.1(*b*)].

3.2.1 Lagrangian System

In the Lagrangian system we use *fluid particles,* which are small fluid elements of fixed mass. They are called particles in analogy with the dynamics of solid bodies. We follow an individual fluid particle as it moves through the flow, and the particle is identified by its position at some instant and the time elapsed since that instant. If we have a velocity described by

$$\mathbf{V} = u\mathbf{i} + v\mathbf{j} + w\mathbf{k}$$

where $u\mathbf{i} + v\mathbf{j} + w\mathbf{k}$ are the unit vectors in a Cartesian $[x, y, z]$ coordinate system, then in Lagrangian terms the velocity of a fluid particle that was located at point $[x_0, y_0, z_0]$ at time $t = t_0$, is given by $\mathbf{V} = \partial(\mathbf{x} - \mathbf{x_0})/\partial t$, and its acceleration is given by $\partial \mathbf{V}/\partial t$. This is the system used in the dynamics of rigid bodies because the particles tend to be few in number and easily identified. To describe a fluid flow where there is relative motion, however, we need to follow many particles, and to

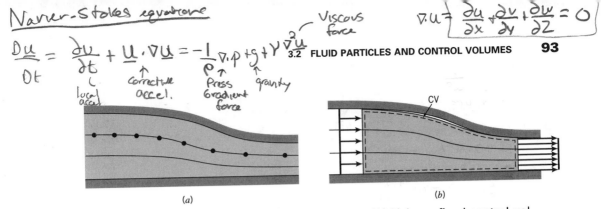

(handwritten annotations above figure:)

Navier-Stokes equations

$\nabla \cdot \underline{u} = \dfrac{\partial u}{\partial x} + \dfrac{\partial v}{\partial y} + \dfrac{\partial w}{\partial z} = 0$

$\dfrac{D\underline{u}}{Dt} = \underbrace{\dfrac{\partial \underline{u}}{\partial t}}_{\substack{\text{local} \\ \text{accel}}} + \underbrace{\underline{u} \cdot \nabla \underline{u}}_{\substack{\text{convective} \\ \text{accel.}}} = -\dfrac{1}{\rho} \underbrace{\nabla \cdot p}_{\substack{\text{Press} \\ \text{gradient} \\ \text{force}}} + \underbrace{g}_{\text{gravity}} + \gamma \underbrace{\nabla^2 \underline{u}}_{\substack{\text{Viscous} \\ \text{force}}}$

FIGURE 3.1 (a) Following a fluid particle in time. (b) Using a fixed control volume (CV).

resolve the smallest details of the flow we may need to follow a huge number of particles. The motion of each particle is described by a separate ordinary differential equation (*ODE*), such as Newton's second law, and each equation is coupled to all the others (that is, the solution of one equation depends on the solution of all others). The solution of these coupled *ODE*s is usually very difficult to find so that the Lagrangian approach is not widely used in fluid mechanics, although it can be very useful for some particular kinds of problems.

3.2.2 Eulerian System

In the Eulerian system we try to find a description which gives the details of the entire flow field at any position and time. Instead of describing the fluid motion in terms of the movement of individual particles, we look for a "field" description. In other words, for the particle that happens to be at the position [x, y, z] at a time t, we search for a description that gives its velocity, acceleration, momentum, and energy at any other position and time. For example, if we were given a Cartesian velocity field described by $\mathbf{V} = 2x^2\mathbf{i} - 3t\mathbf{j} + 4xy\mathbf{k}$, we would know its velocity at every point in the flow, at any time.

At first sight, this approach appears to be very straightforward. However, we are no longer explicitly following fluid particles of fixed mass; at a given point in the flow, new particles are arriving all the time. This makes it difficult to apply Newton's second law since it applies only to particles of fixed mass. We therefore need a relationship that gives the acceleration of a fluid particle in terms of the Eulerian system, and this proves to be somewhat complicated, as we shall see.

3.2.3 Fluid Elements

In deriving the equations of motion, we often choose fixed, infinitesimally small control volumes, similar to the volume element that was used to demonstrate that pressure is an isotropic stress (see Figure 1.7). Such "fluid elements" are control volumes that are small enough so that the variation in all fluid properties over its volume, such as the variation in density and temperature, as well as velocity and stress levels, is approximately linear.[1] Fluid elements are different from fluid particles.

> A fluid element has a fixed volume and it is fixed in space, whereas a fluid particle has a fixed mass and moves with the flow.

---

[1] That is, we need to take only the first two terms in a Taylor series expansion.

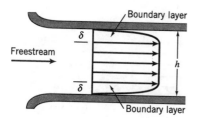

FIGURE 3.2   Flow in a channel, showing boundary layers near the walls.

There are upper and lower limits on the size of the element. The characteristic dimension of the element, ℓ (its height, width, or length), should always be large compared to the mean free path ℓ_m so that the continuum approximation holds ($\ell \gg \ell_m$). At the other end of the scale, ℓ should be small enough so that the changes in fluid velocity or fluid pressure across the element can be expressed linearly. In other words, the fluid element needs to be small compared to the characteristic scale of the flow field. For example, if we have a duct of height h, then we should certainly make sure that $\ell \ll h$. In addition, near solid walls, thin boundary layers can form. To be able to say something about the flow inside the boundary layer, we need $\ell \ll \delta$, where δ is a measure of the boundary layer thickness (see Figures 3.2 and 3.3).

3.2.4 Large Control Volumes

Instead of using small fluid elements as control volumes, we can choose a control volume that is comparable in size to the size of the system [for instance, the whole of the duct, as in Figure 3.1(b)]. Whatever their size, control volumes allow us to think in terms of overall balances of mass, momentum, and energy. For example, mass must be conserved in any fluid flow. That is, we need to account for all the mass of fluid entering and leaving the control volume, as well as the rate of change of the amount of mass contained inside the volume. A piping system, for instance, will have a number of places where fluid enters, and a number of other places where fluid exits. If the amount of mass entering over a given time exceeds the amount of mass leaving during the same time, we know that mass is accumulating somewhere inside the system.

We can think in terms of a large box, which encloses the entire fluid flow system; this box is an example of a large control volume. The mass, momentum, or energy budget can then be applied in terms of the entering and exiting flow, as well as the rate of change of the mass, momentum, or energy contained in the

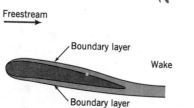

FIGURE 3.3   Flow over an airfoil, showing boundary layers near the surface, and the formation of a wake.

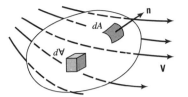

FIGURE 3.4 Fixed control volume for derivation of the integral form of the momentum equation.

volume. The amorphous shape shown in Figure 3.4 illustrates a general control volume, with a surface area element dA, and a volume element $d\forall$.

In this book, we will generally use fixed control volumes, large and small, although it is sometimes useful to follow a fluid particle or use a moving control volume. Broadly speaking, the control volume approach is one of the most valuable tools for the study of fluid flow. When we use a large control volume to consider the momentum balance, it can help us to find the forces that accelerate the fluid, and the forces that act on a solid body, such as an airplane wing. The control volume method can give many insights into fluid flow problems, without solving for all the detailed aspects of the fluid behavior inside the box.

On the other hand, it is sometimes essential to know the details in the box. For example, the aerodynamic performance of an airplane wing depends critically on its shape, and using a large control volume that encloses the entire wing cannot give any guidance as to its correct shape. To design the wing shape, we need to use sufficiently small control volumes or fluid elements, since this is the only way to obtain detailed information on the pressure distribution and velocity field generated by the wing. Remember: all physical laws are originally stated in terms of a *system*. A system is a collection of matter you have picked. For example, the law of conservation of mass says the mass of a system does not change. Newton's second law says that the force acting on a system equals the time rate of change of momentum of the system. In fluid mechanics, we normally choose an arbitrary mass of fluid as our system.

Now, a control volume is a fixed volume in space you have selected to be your control volume. To apply the physical laws to a control volume, we need to consider the system that occupies the control volume at the moment we do our analysis. Since the system properties may be changing in time, and the system itself is generally moving, we need to relate the system properties to the control volume properties. For infinitesimally small control volumes (fluid elements), this relationship is called the *total derivative* (see Section 6.1). For large control volumes, it is called the *Reynolds transport theorem* (see Section 5.4).

3.2.5 Steady and Unsteady Flow

Another very important concept is the idea of a *steady* flow. If we have a large, fixed control volume, it is possible that the inflow and outflow conditions do not change with time. If the fluid properties inside the control volume are also independent of time, we say that the flow is steady. However, if we follow an individual fluid particle, we might see that its velocity changes as it moves through the system, and therefore from its point of view, the flow is unsteady. Whether the flow is steady or unsteady often depends on how we choose to look at the system. In the case of the large control volume, we might see a variation of velocity, momentum, energy, and so on, in space, but no variation in time, whereas when we move with a fluid particle we would see only a variation in time.

It is sometimes possible to change an unsteady flow into a steady flow by changing the point of view of the observer. For instance, if you stand by the side of the road as a car is approaching, you will feel no air flow at first, then a sudden rush as the car goes by, and then, later, nothing again. Standing by the side of the road, you experience an unsteady flow. However, if you were travelling with the car, and the car was moving at a constant speed, the flow relative to your new vantage point would not vary with time: it is steady. Steady flows are much more easily analyzed than unsteady flows, and it is always useful if a coordinate transformation can be found whereby an unsteady flow becomes steady.

3.3   STREAMLINES AND STREAMTUBES

Although we have examined how fluid flows may be analyzed, we have not considered how fluid motion can be described, or how it can be visualized. The most common methods identify the streamline pattern, the particle paths, or the streaklines.

3.3.1   Streamlines

Velocity is a vector, and therefore it has a magnitude and a direction. A *streamline* is defined as a line that is everywhere tangential to the instantaneous velocity direction. To visualize a streamline in a flow, we could imagine the motion of a small marked particle of fluid. For example, we could mark a drop of water with fluorescent dye and illuminate it using a laser so that it fluoresces. If we took a short exposure photograph as the drop moves according to the local velocity field (where the exposure needs to be short compared to the time it takes for the velocity to change appreciably), we would see a short streak, with a length $V \Delta t$, and with a direction tangential to the instantaneous velocity direction. If we mark many drops of water in this way, the streamlines in the flow would become visible.

Alternatively, we can find the shape of the streamlines if we know the velocity field. From the definition of a streamline, it follows that for a flow in the $[x, y]$-plane, the slope of a streamline, dy/dx, must be equal to v/u, where u and v are the instantaneous velocity components in the x- and y-directions. For a three-dimensional flow,

$$\frac{dy}{dx} = \frac{v}{u}, \quad \frac{dy}{dx} = \frac{v}{w}, \quad \frac{dx}{dz} = \frac{u}{w}$$

The procedure is illustrated in Examples 3.1 and 3.2.

Since the velocity at any point in the flow has a single value, streamlines cannot cross (the flow cannot go in more than one direction at the same time).[2]

---

[2] *Singular points* can occur in the flow at points where the velocity magnitude is zero. Their nature depends strongly on the velocity of the observer. At a critical point, the velocity direction is indeterminate, and streamlines can meet. One example of a singular point is a stagnation point—see Figure 3.8 and Section 4.3.

FIGURE 3.5 Surface tufts for surface-flow visualization. From *Race Car Aerodynamics,* J. Katz, Robert Bentley Publishers, 1995. Originally from MIRA. With permission.

Surface streamlines are the streamlines followed by the flow very close to a solid surface. For example, small, flexible tufts of yarn are sometimes attached to the body, and because of the drag on the tufts by the flow, the tufts will align themselves in the local flow direction (see, for example, Figure 3.5).

3.3.2 Pathlines

There are other ways to make the flow visible. For example, we can trace out the path followed by our fluorescent drop using a long-exposure photograph. This line is called a *pathline,* and it is similar to what you see when you take a long-exposure photograph of car lights on a freeway at night. It is possible for pathlines to cross, as you can imagine from the freeway analogy; as a car changes lanes, the pathline traced out by its lights might cross another pathline traced out by another vehicle at an earlier time.

A pathline is the line traced out by a fluid particle as it moves through the flowfield. Since we are following fluid particles along a pathline, it is a Lagrangian concept. For a pathline in a two-dimensional flow, we use the fact that $u = dx/dt$, and $v = dy/dt$. If we know how u and v depend on time, and a sufficient number of boundary conditions are given, we can integrate u and v with respect to time to find particle path coordinates x and y as a function of time. By eliminating time from the expressions for x and y, the particle path in $[x, y]$-space can be found (see Examples 3.1 and 3.2).

3.3.3 Streaklines

Another way to visualize flow patterns is by *streaklines*. A streakline is the line traced out by the particles that pass through a particular point. For instance, if we

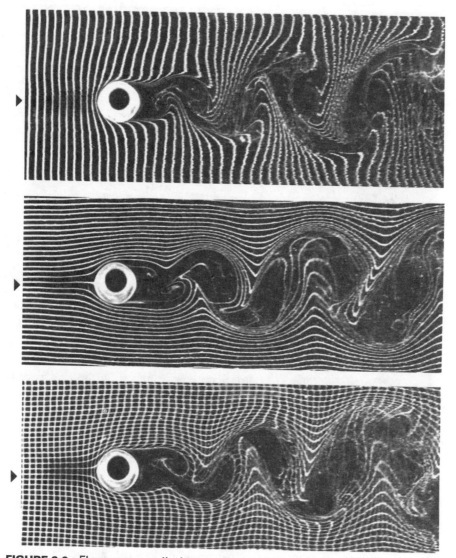

FIGURE 3.6   Flow over a cylinder at a Reynolds number of 170 made visible using streaklines and timelines generated by the hydrogen bubble technique. From *Visualized Flow,* Japan Society of Mechanical Engineers, Pergamon Press, 1988.

emit dye continuously from a fixed point, the dye makes up a streakline as it moves downstream. To use the freeway analogy again, a streakline is the line made up of the lights on all the vehicles that passed through the same toll booth. If they all follow the same path (a steady flow), a single line results, but if succeeding cars follow different paths (unsteady flow), it is possible for lines to cross over each other as well as themselves. As an example, Figure 3.6(*b*) shows how streaklines can be used to visualize the flow over a circular cylinder.

Surface streaklines are the streaklines followed by the flow very close to a solid surface. For example, raindrops falling on a windscreen will tend to trace out a path representing a surface streakline.

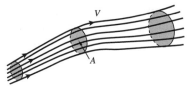

FIGURE 3.7 Streamlines forming a streamtube.

In unsteady flow, streamlines, pathlines, and streaklines are all different, but it may be shown that:

> In steady flow, streamlines, pathlines, and streaklines are identical.

3.3.4 Streamtubes

Imagine a set of streamlines starting at points on a closed loop (Figure 3.7). These streamlines form a tube that is impermeable since the walls of the tube are made up of streamlines, and there can be no flow normal to a streamline (by definition). This tube is called a *streamtube*. In Section 3.5, we will see that for a steady, one-dimensional flow, the mass-flow rate ρAV is constant along a streamtube. If the density is constant, and the area decreases, the velocity increases. The cross-sectional area of the streamtube therefore gives information on the local velocity.

3.3.5 Timelines

In Figure 1.22 boundary layer flows were made visible by forming a line of hydrogen bubbles in the water. To form the bubbles, a thin wire is mounted across the flow and connected to the cathode of a voltage source. An anode is placed in the water at some other location. By passing a current between the anode and cathode, hydrogen is generated by electrolysis at the wire in the form of small bubbles. If a short pulse of current is used, a line of bubbles is formed at the wire that is swept downstream by the flow. If the hydrogen bubbles are small enough, the buoyancy force on them can be neglected (it is proportional to the volume of the bubbles), and the bubbles follow the flow. The line formed by the bubbles is an example of a *timeline*. A timeline is a line made up of fluid particles that were all marked at the same time. Examples of timelines, separately, and in combination with streaklines, are shown in (Figure 3.6).

EXAMPLE 3.1 *Streamlines and Pathlines in Steady Flow*

Consider the flow field given by

$$\mathbf{V} = x\mathbf{i} - y\mathbf{j}$$

This flow is steady (independent of time) and two-dimensional (it depends on two space coordinates, x and y).
(*a*) Describe the velocity field.
(*b*) Find the shape of the streamlines.
(*c*) Find the shape of the pathlines.

Solution: For part (a), the vector velocity makes an angle θ with the x-axis, where

$$\tan\theta = \frac{v}{u} = -\frac{y}{x}$$

Along the x-axis, $y = 0$, and $\theta = 0°$ (or $180°$), and **V** has the same sign and magnitude as x, so that the velocity points directly along the x-axis and increases in magnitude with distance from the origin. Along the y-axis, $x = 0$, and $\theta = 90°$ (or $270°$). Here, **V** has the opposite sign but the same magnitude as y, so that the velocity points along the negative y-axis and increases in magnitude with distance from the origin. At $y = x$, $\theta = -45°$, and so the entire velocity field can be built up by successively finding the velocity direction and magnitude at all points in the flow. The flowfield represents a stagnation point flow, as illustrated in Figure 3.8.

For part (b), the shape of the streamlines is given by the solution of

$$\frac{dy}{dx} = \frac{v}{u} = -\frac{y}{x}$$

The variables can be separated and integrated to give

$$\int \frac{dy}{y} = -\int \frac{dx}{x}$$

or

$$\ln y = -\ln x + \text{constant}$$

This can be written as

$$xy = C$$

where C is a constant. That is, the streamlines are hyperbolae in the x-y plane.

For part (c), we can find the pathlines by using

$$u = \frac{dx}{dt} = x, \quad \text{and} \quad v = \frac{dy}{dt} = -y$$

That is,

$$\frac{dx}{x} = -\frac{dy}{y}$$

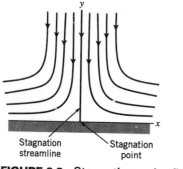

Stagnation
streamline

Stagnation
point

FIGURE 3.8 Stagnation point flow.

Multiplying through by xy gives

$$y \, dx + x \, dy = d(xy) = 0$$

The equation to the pathline is then

$$xy = C$$

the same result obtained for the equation to the streamline. This is expected, since streamlines and pathlines coincide in steady flow. ∎

EXAMPLE 3.2 *Streamlines and Pathlines in Unsteady Flow*

Consider the unsteady flow field given by

$$\mathbf{V} = x\mathbf{i} + yt\mathbf{j}$$

This flow is unsteady (it depends on time) and two-dimensional (it depends on two space coordinates, x and y). Find the shape of
(a) The streamline and;
(b) The pathline passing through the point [1, 1] at time $t = 0$.

Solution: For part (a), the shape of the streamlines is given by the solution of

$$\frac{dy}{dx} = \frac{v}{u} = \frac{yt}{x}$$

The variables can be separated and integrated to give

$$\ln y = t \ln x + \text{constant}$$

or

$$y = C_1 x^t$$

For the streamline passing through the point [1, 1] at $t = 0$, $C_1 = 1$, so for this particular streamline,

$$y = x^t \tag{3.1}$$

For part (b), to find the pathlines (the location of particles as a function of time), we use the fact that $u = dx/dt$, and $v = dy/dt$. For this problem,

$$\frac{dx}{dt} = x,$$

and

$$\frac{dy}{dt} = yt$$

Integrating,

$$x = C_2 e^t$$

and

$$y = C_3 e^{t^2/2}$$

For the particle located at the point [1, 1] at $t = 0$, $C_2 = C_3 = 1$, so for this particular pathline,

$$x = e^t$$

and

$$y = e^{t^2/2}$$

Eliminating time gives

$$2 \ln y = (\ln x)^2 \tag{3.2}$$

Equations 3.1 and 3.2 are different because in an unsteady flow, streamlines and pathlines do not coincide.   ∎

3.4   DIMENSIONALITY OF A FLOW FIELD

The dimensionality of a flow field is governed by the number of space dimensions needed to define the flow field completely. For example, the flow parameters describing a one-dimensional flow can only vary in one direction. The direction might coincide with a coordinate axis such as x, or it may be directed along a streamline, as in Figure 3.9(a). A two-dimensional flow varies along the flow direction as well as across it [Figure 3.9(b)], and a three-dimensional flow varies in all three space directions.

The flows considered in the rest of this chapter are all one-dimensional; the fluid velocity, pressure, and density may change as the fluid travels from the inlet to the outlet of a control volume, but they are constant over the inlet and outlet areas. Two- and three-dimensional flows will be analyzed in Chapter 5.

Perfectly one-dimensional flows are not found in nature. For example, if the duct or streamtube diverges so that its cross-sectional area increases with distance, the flow will also diverge [Figure 3.9(b)]. The flow must develop a cross-stream component of the velocity that varies across the duct. In addition, frictional stresses will cause the fluid near the wall to slow down, and the no-slip condition guarantees that the flow velocity at the wall must be zero (see Section 1.6 and Example 3.5). Nevertheless, the concept of a one-dimensional flow, or a *quasi-one-dimensional* flow, is still a very useful approximation in many cases.

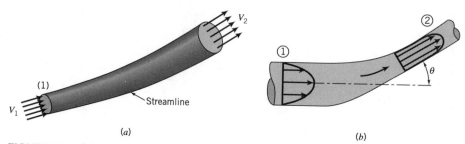

FIGURE 3.9   (a) One-dimensional flow. (b) Two-dimensional flow.

3.5 MASS CONSERVATION

We are now ready to consider the first basic principle of fluid motion: mass conservation, which requires that when a fluid is in motion, it moves in such a way that mass is conserved. To see how mass conservation places restrictions on the velocity field, consider the steady flow of fluid through a duct or a streamtube. We will use a large control volume enclosing the duct (CV in Figure 3.10). The inflow and outflow are one-dimensional, so that the velocity V and density ρ distributions are constant over the inlet and outlet areas.

Mass conservation in a steady flow requires that whatever mass flows into the control volume must leave the control volume at the same rate. If the mass inflow was greater than the outflow, for example, mass would accumulate inside the control volume and the flow could not be steady. Over a short time interval Δt, therefore,

$$\text{volume flow in over } A_1 = A_1 V_1 \Delta t$$
$$\text{volume flow out over } A_2 = A_2 V_2 \Delta t$$

That is

$$\text{mass in over } A_2 = \rho_1 A_1 V_1 \Delta t$$
$$\text{mass out over } A_2 = \rho_2 A_2 V_2 \Delta t$$

So

$$\rho_1 A_1 V_1 = \rho_2 A_2 V_2$$

or

$$\rho_1 A_1 V_1 - \rho_2 A_2 V_2 = 0 \tag{3.3}$$

regardless of the directions of the inflow and outflow (mass flow rate is a scalar, and it is independent of direction). This is a statement of the principle of mass conservation for a steady, one-dimensional flow, with one inlet and one outlet. For a control volume with N inlets and outlets,

$$\sum_{i=1}^{N} \rho_i A_i V_i = 0 \tag{3.4}$$

where inflows are negative and outflows are positive. This equation is called the *continuity equation* for steady one-dimensional flow through a fixed control volume.

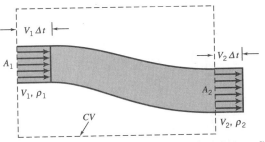

FIGURE 3.10 Steady, one-dimensional duct flow.

It is understood that the velocities are normal to the areas. When the density is constant,

$$\sum_{i=1}^{N} A_i V_i = 0$$

The dimensions of the product ρAV are given by

$$[\rho AV] = \frac{M}{L^3} L^2 \frac{L}{T} = \frac{M}{T} = \frac{mass}{unit\ time}$$

That is, ρAV is a mass flow rate, and it is measured in units of kg/s, lb_m/s or $slug/s$. It is commonly denoted by the symbol $\dot{m}$.

The dimensions of AV are

$$[AV] = L^2 \frac{L}{T} = \frac{L^3}{T} = \frac{volume}{unit\ time}$$

That is, AV is a volume flow rate, and it is measured in units of m^3/s, ft^3/s, $liter/s$, $gallon/hr$, cubic feet per minute (cfm), or m^3/s. Volume flow rate is commonly denoted by the symbol $\dot{Q}$, but this is also the symbol used to denote a heat transfer rate. To avoid confusion, we will use the symbol $\dot{q}$ to represent volume flow rate.

EXAMPLE 3.3 *Mass Conservation*

A cylinder of diameter of d and length ℓ is mounted on a "sting" support in a wind tunnel, as shown in Figure 3.11. The wind tunnel is of rectangular cross section and height $4d$. The incoming flow is steady and uniform, with constant density and a velocity V_1. Downstream of the cylinder the streamlines become parallel again, and the shape of the velocity profile is as shown. Find the magnitude of V_2 in terms of V_1.

Solution: If we draw a large control volume that encloses the cylinder but does not cut the walls of the duct (CV in the figure), we see that there is flow into the control volume over the left hand face and flow out over the right hand face. The downstream flow is not actually one-dimensional since the velocity is varying across the cross-section of the wind tunnel. Nevertheless, it can be treated as one-dimensional by considering separately the two regions where the velocity is different, and then adding the result. By applying the continuity equation in this way we obtain

$$\rho V_1(4d\ell) - \rho V_2(2d\ell) - \rho \tfrac{1}{2} V_2(2d\ell) = 0$$

so that

$$V_2 = \tfrac{4}{3} V_1$$

■

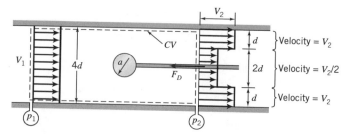

FIGURE 3.11 A cylinder mounted in a wind tunnel.

3.6 MOMENTUM EQUATION

We now consider the second basic principle of fluid motion: the momentum equation. We construct this equation for one-dimensional, steady flow by applying Newton's second law of motion to the flow through a large control volume. To begin, we need to consider the kinds of forces that can act to change the momentum of a fluid.

3.6.1 Forces

What causes the fluid to move? Fluids will begin to move when a nonzero resultant force acts on them. For example, when the pressure in one location is higher than in another location, the fluid will tend to move toward the region of lower pressure. Gravity can also cause a fluid to move: liquids flow downhill, trading their potential energy for kinetic energy of motion. Similarly, temperature differences will cause one part of a fluid to have a lower density compared to another part, and the lighter fluid will tend to rise.

There is also friction. When one layer of fluid moves with respect to an adjacent layer, a tangential viscous stress develops which can cause the fluid to speed up or slow down (see Section 1.4). Sometimes, we consider fluids where the viscosity is zero. These *inviscid* fluids do not exist in nature because all real fluids are viscous, but we can often use this approximation when the effects of viscosity are small. However, we must be careful, because neglecting viscosity can sometimes lead to spectacularly wrong answers (see, for instance, Section 7.9).

We must also include forces exerted by solid surfaces. This becomes clear when we think of a jet of water hitting a flat plate—there is a force exerted on the plate by the water, and by Newton's third law the plate exerts an equal but opposite force on the water (which acts to change the direction of motion of the fluid and therefore changes its momentum).

Other forces may also be important. For instance, if the fluid is electrically charged it can be made to move by applying a magnetic field, and Coriolis forces can be important in a rotating field (they are of crucial importance in forming our weather patterns—see Chapter 14). Mostly, we will only consider the forces due to pressure differences, viscous stress differences, and gravity.

Forces due to stress differences are called *surface* forces because they are proportional to the total surface area over which they act. For example, if a constant shear stress τ acts over an area A, the resultant shear force is equal to τA. In contrast, the acceleration due to gravity g acting on a fluid mass m introduces a force that is proportional to the mass of the fluid, mg, and it is called a *body* force.

3.6.2 Unidirectional Flow

For the case shown in Figure 3.12, the inflow and outflow are in the x-direction, and the flow is steady and one-dimensional. Since we are only concerned with the overall force acting on the fluid, a large control volume will be used. For simplicity, we choose a rectangular box. We will assume that friction is not important, and that gravity forces are negligible. The pressure outside the duct is equal to the atmospheric pressure, but over the inflow area A_1 the gauge pressure is p_1, and over the outflow area A_2 it is p_2.

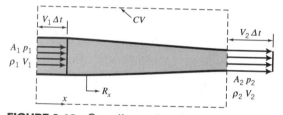

FIGURE 3.12 One-dimensional duct showing control volume (*CV*). R_x is the force exerted by the duct on the fluid, and it is equal to the force needed to hold the duct fixed.

As the fluid moves through the control volume, its velocity varies along the flow direction. Specifically, the inflow and outflow velocities, V_1 and V_2, are different. Relative motions exist, and fluid particles accelerate as they move from entry to exit. Even though the velocity field is independent of time, there is an acceleration because of spatial variations in velocity. We know from Newton's second law that a resultant force must be acting on the fluid. Where does this force come from?

1. There is a force due to pressure differences acting on the surface of the control volume. The only places where pressures other than atmospheric pressure act are the areas A_1 and A_2. The atmospheric pressure by itself does not contribute to the resultant force since it acts equally over the complete surface area of the control volume. Therefore the forces due to pressure differences are entirely due to the gauge pressures acting on areas A_1 and A_2. Since compressive pressures are taken to be positive,

   force due to pressure acting on area $A_1 = p_1 A_1$ to the right
   force due to pressure acting on area $A_2 = p_2 A_2$ to the left

2. There will also be a force exerted *by* the duct *on* the fluid (the external forces included in the momentum equation are always the forces acting *on* the fluid). This is intuitive, in that the duct will experience a force due to the action of the fluid, and therefore the duct exerts an equal but opposite force on the fluid. We will assume that the force exerted by the duct on the fluid, R_x, is positive if it acts in the positive x-direction (if the solution indicates that R_x is negative, it simply means that it acts in the negative x-direction). By Newton's third law, the force exerted *on* the duct *by* the fluid is given by $-R_x$. To hold the duct in place, there must be a restraining force acting on the duct, equal to $+R_x$. Therefore, the force exerted by a support to hold the duct in place is equal to the force acting on the fluid (neglecting the weight of the duct).

3. Forces due to friction and gravity can also be important. For the particular example considered here, they will be neglected.

The resultant force acting on the fluid, therefore, is given by the sum of the force exerted by the duct on the fluid, R_x, and the force due to pressure differences acting on the fluid, $p_1 A_1 - p_2 A_2$.

This resultant force changes the momentum of the fluid. To find this momentum change we note that

the *mass* entering during $\Delta t = \rho_1 A_1 V_1 \Delta t$

So,

$$\text{the } x\text{-}momentum \text{ entering during } \Delta t = (\rho_1 A_1 V_1 \Delta t) V_1$$

and

$$\text{the } x\text{-}momentum \text{ leaving during } \Delta t = (\rho_2 A_2 V_2 \Delta t) V_2$$

Since outflows are taken to be positive,

$$\text{the net } change \text{ of } x\text{-momentum during } \Delta t = (\rho_2 A_2 V_2^2 - \rho_1 A_1 V_1^2)\Delta t$$

and

$$\text{the net } rate \text{ change of } x\text{-momentum} = \rho_2 A_2 V_2^2 - \rho_1 A_1 V_1^2$$

This is the rate of increase in x-momentum experienced by the fluid in passing through the duct. Note that the quantity $\rho A V^2$ has the dimensions of force (MLT^{-2}) and it is measured in units of N or lb_f.

The flow is steady, so that the momentum of the fluid inside the control volume is not changing with time. Therefore the net rate of change of fluid momentum is equal to the resultant force acting on the fluid. That is,

$$R_x + p_1 A_1 - p_2 A_2 = \rho_2 A_2 V_2^2 - \rho_1 A_1 V_1^2$$

This equation is Newton's second law of motion as applied to a simple one-dimensional, steady flow, where the resultant force acting on the fluid (the left hand side) causes a change in its momentum (the right hand side). In this example, the forces all act in the x-direction, and there is no resultant force in the y-direction.

We can also use the one-dimensional continuity equation (equation 3.3 or 3.4), where $\rho_1 V_1 A_1 = \rho_2 V_2 A_2$, to eliminate V_2 so that

$$R_x = \rho_1 A_1 V_1^2 \left(\frac{\rho_1 A_1}{\rho_2 A_2} - 1 \right) + p_2 A_2 - p_1 A_1 \tag{3.5}$$

3.6.3 Bidirectional Flow

For the case shown in Figure 3.13, the inlet and outlet directions are not aligned. All other conditions are the same as in the previous example; the flow is steady, one-dimensional, and we neglect friction and gravity. In this case, however, as the fluid passes through the duct, its momentum changes in magnitude and direction. The entering momentum is all in the x-direction, whereas the exiting momentum

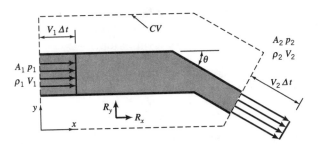

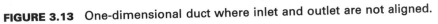

FIGURE 3.13 One-dimensional duct where inlet and outlet are not aligned.

has components in the x- and the y-directions. The force due to pressure differences also has two components, and so will the force exerted by the duct on the fluid.

In the x-direction, during the short time interval Δt, the momentum entering the control volume is $(\rho_1 V_1 A_1 \Delta t) V_1$, and the momentum leaving the control volume is $(\rho_2 V_2 A_2 \Delta t) V_2 \cos\theta$. That is,

the change of fluid x-momentum $= (\rho_2 V_2 A_2 \Delta t) V_2 \cos\theta - (\rho_1 V_1 A_1 \Delta t) V_1$

and

the rate of change of fluid x-momentum $= (\rho_2 V_2 A_2 \cos\theta - (\rho_1 V_1 A_1)) V_1$

Each term on the right hand side is of the form (mass-flow rate) × (velocity component in x-direction). Very importantly, over any inlet or outlet

> The x-component momentum is given by the mass-flow rate times the x-component of the velocity.

The force due to pressure differences in the x-direction is given by $p_1 A_1 - p_2 A_2 \cos\theta$, and so

$$R_x + p_1 A_1 - p_2 A_2 \cos\theta = \rho_2 A_2 V_2^2 \cos\theta - \rho_1 A_1 V_1^2$$

where R_x is the force exerted *by* the duct *on* the fluid in the x-direction. Since

$$\rho_1 V_1 A_1 = \rho_2 V_2 A_2$$

by mass-conservation, we can write this as

$$R_x = p_2 A_2 \cos\theta - p_1 A_1 + \rho_1 A_1 V_1^2 \left(\frac{V_2 \cos\theta}{V_1} - 1 \right)$$

We can now go through the same procedure to find the force R_y. First, we see that the momentum entering the control volume has no component in the y-direction, but the momentum leaving the control volume does. So

the rate of change of fluid y-momentum $= (\rho_2 A_2 V_2) V_2 \sin\theta$

The force due to pressure differences in the y-direction is given by $-p_2 A_2 \sin\theta$ (there is no y contribution from the force due to pressure acting on A_1 since it acts purely in the x-direction). Hence,

$$R_y - p_2 A_2 \sin\theta = \rho_2 A_2 V_2^2 \sin\theta$$

That is,

$$R_y = -p_2 A_2 \sin\theta + \rho_2 A_2 V_2^2 \sin\theta$$

where R_y is the force exerted *by* the duct *on* the fluid in the y-direction.

Finally, the magnitude of the resultant force is given by $\sqrt{R_x^2 + R_y^2}$, and it makes an angle $\arctan(R_y/R_x)$ with the positive x-direction.

3.7 VISCOUS FORCES AND ENERGY LOSSES

Fluid flows are irreversible (in a thermodynamic sense) because fluids have viscosity and whenever velocity gradients appear in the flow, there will be energy dissipation due to viscous stresses. The flow energy is conserved only if there are no velocity

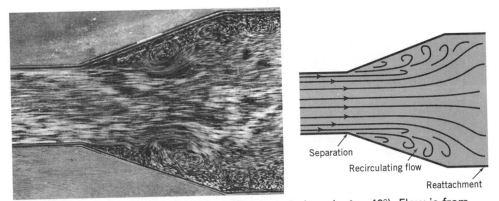

FIGURE 3.14 Separated diffuser flow (the included angle $\theta = 40°$). Flow is from left to right. From *Visualized Flow*, Japan Society of Mechanical Engineers, Pergamon Press, 1988.

gradients anywhere, or if the fluid is inviscid. These are the only conditions under which there is no energy dissipation. However in some cases, frictional effects are small compared to other effects. For instance, in the flow through a large duct where the boundary layers are very thin, viscous effects are confined to a rather small region, and the fluid friction may sometimes be neglected. This is not the case for many practical flows. In most pipe and duct flows, for example, the velocity gradients extend over the entire cross-section and frictional stresses are important everywhere.

Even if the boundary layers are thin to begin with, when the flow decelerates rapidly, or when it tries to change direction rapidly, it can *separate*. When a flow separates, it no longer follows the shape of the solid boundary. For example, in the flow shown in Figure 3.14, as the flow enters the diverging section of the duct (the "diffuser") it hardly turns at all and continues to go almost straight downstream. A region of separated flow marked by large recirculating eddies forms over the diffuser walls. In these regions, some of the fluid is actually going in a direction opposite to the bulk flow direction. Large losses can occur, as the mechanical energy of the fluid is used to drive large unsteady eddying motions, which eventually dissipate into heat. This type of flow can occur for quite small diffuser angles. To avoid the large losses associated with separation, the angle of divergence needs to be very small: the included angle θ needs to be less than about $7°$.

In contrast, when the flow is in the direction of decreasing area, as in a contraction, there is very little risk of producing large separated regions, even for very large values of θ (up to $45°$), and generally, losses are quite small.

Note that turbulent flow and separated flow are two different phenomena. In separated flow, there is backflow, so that some parts of the fluid are moving in a direction opposite to the principal flow direction, and there may be significant unsteadiness and fluctuation observed in the flow field. In turbulent flow, all the fluid is moving in the principal flow direction, although the velocity fluctuates about an average value, and the fluctuations have components in all three directions.

We see that great care needs to be taken in considering frictional effects; sometimes they are negligible, and sometimes they are not. When separation occurs, however, frictional effects and mechanical energy losses are always important. It is worth noting that sharp corners almost always produce a separated flow. The

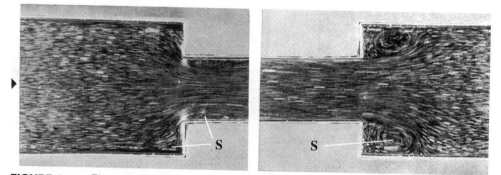

FIGURE 3.15 Flow through a sudden contraction (left), and a sudden expansion (right). Flow is from left to right. From *Visualized Flow*, Japan Society of Mechanical Engineers, Pergamon Press, 1988. *S* marks the regions of separated flow.

sudden expansion and contraction flows shown in Figure 3.15 graphically demonstrate the problem. The direction of the flow greatly influences the flow pattern and the resulting energy loss. The flow in the sudden expansion is expected to cause greater losses than the flow in the sudden contraction.

EXAMPLE 3.4 *Drag Force on a Cylinder*

Consider the flow shown in Figure 3.11. The presence of the cylinder changes the fluid pressure and momentum distributions, because the cylinder exerts a force on the fluid, $-F_D$. The force exerted by the fluid on the cylinder is equal to $+F_D$ and this force is the *drag force*. The reduced velocity region downstream of the cylinder is called the *wake*. Note that the force exerted by the strut required to hold the cylinder in place is equal to $-F_D$. In this example we wish to find F_D.

Solution: We will assume that the viscous drag forces acting on the surfaces of the control volume are negligible. When this is a reasonable approximation (see Example 3.5), we need to consider only the momentum loss in the wake and the force holding the cylinder in place. If we assume that F_D acts in the negative x-direction, as shown in the figure, then

$-F_D$ = the x-component of the force required to hold the cylinder in place

$+F_D$ = the x-component of the force of the fluid on the cylinder

$-F_D$ = the x-component of the force of the cylinder on the fluid

It is important to understand how the force required to hold the cylinder makes an appearance in the momentum equation for the fluid. In our choice of control volume, the cylinder is inside the volume, and the surface of the volume "cuts" through the sting. In effect, the sting applies a force to the fluid inside the control volume, which is equal in sign and magnitude to the force required to hold the cylinder in place.

Although the velocity varies across the tunnel at the downstream location, we will assume that the pressure is constant across the tunnel.[3] In evaluating the

---

[3] In Section 4.2.1, we show that when streamlines are parallel, only hydrostatic pressure differences can exist across streamlines.

momentum transport, we treat the downstream flow as two separate one-dimensional flows and add the result. Hence,

$$-F_D + p_1(4d\ell) - p_2(4d\ell) = \rho V_2^2(2d\ell) + \rho \left(\frac{V_2}{2}\right)^2 (2d\ell) - \rho V_1^2(4d\ell)$$

(remember that the resultant force—left hand side—is equal to the net rate of change of momentum—right hand side). Substituting $V_2 = \frac{4}{3}V_1$, as found in Example 3.3, we obtain

$$\frac{F_D}{\rho V_1^2 d\ell} = \frac{4(p_1 - p_2)}{\rho V_1^2 d\ell} - \frac{4}{9} \tag{3.6}$$

∎

EXAMPLE 3.5 *How Important Is the Wall Friction?*

We have seen that all real fluid flows have viscosity. Losses occur through viscous stresses in boundary layers and other shear layers, and in some cases, wakes are formed containing large eddies, which dissipate large amounts of energy. Yet, in many cases, the forces due to friction are small compared to the rate of change of momentum, so that frictional energy losses are small compared to the change in, for example, kinetic energy.

In Figure 3.11, we see a wake forming downstream of the cylinder, and it is clear that energy and momentum losses occur in that region. In addition, boundary layers are present on the walls of the tunnel. In writing the x-component of the momentum equation, we need to consider the wake (as we did in the previous example), but we should also consider the boundary layers on the tunnel walls. The way the flow is drawn in Figure 3.11, the boundary layers are not shown, and the wall friction is implicitly neglected. If this is a reasonable approximation, we need to consider only the momentum loss in the wake and F_D, the x-component of the force required to hold the cylinder in place.

To understand the role of the wall boundary layers, consider Figure 3.16, where the boundary layers have been shown explicitly. The boundary layers give rise to a frictional stress in the x-direction, and therefore a frictional force F_v appears in the momentum equation, where F_v represents the wall shear stress τ_w integrated over the tunnel walls. If we draw the control volume all the way to the wall, we will need to include the force F_v, and the boundary layer velocity profiles in the momentum equation, since F_v and the velocity deficits in the boundary layers contribute to the momentum transport.

However, if the control volume is drawn so that it is outside the boundary layers (CV_1 in the figure), there is no viscous friction acting on the surface of the control volume. For this control volume, $F_v = 0$ and we will obtain the earlier result given by equation 3.6. There is a catch, however. There will now be a small flow

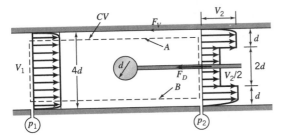

FIGURE 3.16 An alternative control volume choice.

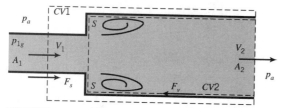

FIGURE 3.17 Alternative control volumes for the flow through a sudden expansion. The regions of separated flow are shown as recirculating eddies.

across the control volume boundaries (surfaces A and B, for example), and there will be some x-momentum transport across them. If the boundary layers grow very slowly (as they usually do), this momentum transport is small, and it can be neglected when finding F_D. It can not be neglected when trying to find F_v, as shown in Section 10.2. ∎

EXAMPLE 3.6 *Control Volume Selection*

Consider the steady flow of air through the sudden expansion shown in Figure 3.17.[4] Suppose that at the entry and exit from the control volume CV_1 (stations 1 and 2, respectively), the flow is approximately uniform so that the assumption of one-dimensional flow can be made. However, inside the control volume the flow is very complex (see, for example, Figure 3.15) and there are many losses. We can assume that at the entry to and exit from the control volume the pressures are approximately uniform across the flow, and at the exit, it is equal to atmospheric pressure. In addition, at the entrance to the sudden expansion, the flow is approximately parallel, so that the pressure is approximately equal to the flow in the recirculation zones (we will show why this is so in Section 4.2.2), and therefore the pressure acting at section s-s is uniform across the flow.

We use two different control volumes to illustrate how the control volume selection governs the information that can be obtained. First, we use a control volume that encloses the duct, and cuts through the walls of the duct at an upstream section (CV_1 in Figure 3.17). Where the control volume cuts through the walls, there will be a reaction force F_s acting on the fluid, where $-F_s$ is the force exerted by the fluid on the duct. If we use gauge pressure in the x-component momentum equation, we obtain

$$F_s + p_{1g}A_1 = -\rho V_1^2 A_1 + \rho V_2^2 A_2$$

since the only place where the pressure is not atmospheric is at station 1. There is no viscous force acting on this control volume. Outside the duct there is no fluid motion, and there are clearly no viscous stresses acting. Over the parts of the control volume surface that lie inside the duct, there are no velocity gradients, and again there are no shearing stresses acting. For this control volume, viscous forces do not play a role. By using the continuity equation ($V_1A_1 = V_2A_2$) we get

$$-F_s = p_{1g}A_1 + \rho V_1^2 A_1 \left(1 - \frac{A_1}{A_2}\right) \tag{3.7}$$

[4] This problem was adapted from an example given in *Engineering Fluid Mechanics* by Mironer, published by McGraw-Hill, 1979.

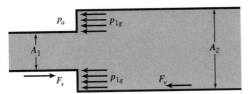

FIGURE 3.18 Free-body diagram for the sudden expansion.

If we know the geometry of the duct, we can determine F_s by measuring the gauge pressure upstream, and either the inlet or the outlet velocity.

Second, we choose a control volume that coincides with the inside surface of the sudden expansion (CV_2 in Figure 3.17). There is now no reaction force acting on the fluid inside the control volume since the control volume does not cut the walls of the duct. The pressure distribution (the normal stress distribution) along the horizontal surfaces of the control volume is unknown, but the resultant force is in the vertical direction so it does not enter the x-momentum equation. However, the viscous shearing stresses along the horizontal surfaces give rise to a horizontal frictional force on the fluid, $-F_v$, which must be included (the sign conforms to what is shown in Figure 3.17: we expect it to be a retarding force for the fluid, but its actual direction will come out of the analysis). The x-component momentum equation becomes

$$-F_v + p_1 A_1 - p_2 A_2 = -\rho V_1^2 A_1 + \rho V_2^2 A_2$$

By using the continuity equation and gauge pressures

$$F_v = p_{1g} A_2 + \rho V_1^2 A_1 \left(1 - \frac{A_1}{A_2} \right) \tag{3.8}$$

We see that F_v is positive since $A_2 > A_1$, and therefore it will act in the direction shown in the figure. If the duct is relatively short, or if the expansion ratio A_2/A_1 is large, F_v is usually small compared to the force due to pressure differences, and it can be neglected. The gauge pressure at the upstream location can then be found by measuring either V_1 or V_2.

Finally, we can draw the free-body diagram for the duct showing the forces acting in the horizontal direction (Figure 3.18). When the shear force F_v is negligible,

$$F_s = -p_{1g}(A_2 - A_1)$$

The same result can be obtained by combining equations 3.7 and 3.8 for $F_v = 0$.

■

PROBLEMS

3.1   What is meant by the term "steady, one-dimensional flow?"

3.2   Give the definitions of a streamline and a pathline. Under what condition are they the same?

3.3   Consider the velocity field given by $\mathbf{V} = x\mathbf{i} + y\mathbf{j}$. Find the general equation for the streamlines. Sketch the flowfield.

3.4   Consider the velocity field given by $\mathbf{V} = x\mathbf{i} + y\mathbf{j}$. Find the general equation for the pathlines. Compare with the results obtained in the previous problem.

3.5 Consider the velocity field given by $\mathbf{V} = x^{-1}\mathbf{i} + x^{-2}\mathbf{j}$. Find an equation for the streamline passing through the point [1, 2].

3.6 Consider the velocity field given by $\mathbf{V} = x^2\mathbf{i} - xy\mathbf{j}$. Find an equation for the streamline passing through the point [2, 1]. How long does it take a particle of fluid to move from this point to the point where $x = 4$?

3.7 Write down the principle of conservation of mass in words.

3.8 What are the dimensions of mass flow rate? What are its typical units?

3.9 Air at $60°F$ flows at atmospheric pressure through an air conditioning duct measuring 6 *in.* by 12 *in.* with a volume flow rate of 5 ft^3/s. Find:
(a) the mass flow rate,
(b) the average velocity

3.10 What are the dimensions of momentum flow rate? What are its typical units?

3.11 Water at 20°C flows through a 10 *cm* pipe with a volume flow rate of 0.5 m^3/s. Find:
(a) the mass flow rate
(b) the average velocity
(c) the momentum flow rate

3.12 The largest artery in the body is the one that supplies blood to the legs. As it comes down the trunk of the body, it splits into a Y-junction, as shown in Figure P3.12. Blood with specific gravity of 1.05 is pumped into the junction at a speed of $V_1 = 1.5$ *m/s*. The diameter of the entrance flow is $d_1 = 20$ *mm,* and for the exit flows $d_2 = 15$ *mm* and $d_3 = 12$ *mm.* If the mass flow rates at stations 2 and 3 are equal, find V_2 and V_3.

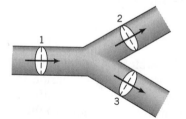

FIGURE P3.12

3.13 Air enters an air conditioning duct measuring 1 *ft* by 2 *ft* with a volume flow rate of 1000 ft^3/min. The duct supplies three classrooms through ducts that are 8 *in.* by 15 *in.* Find the average velocity through each duct.

3.14 In a hovercraft, air enters through a 1.2 m diameter fan at a rate of 135 m^3/s. It leaves through a circular exit, 2 m in diameter and 10 cm in height. Find the average velocity at the entry and the exit.

3.15 Water flows radially toward the drain in a sink, as shown in Figure P3.15. At a radius of 50 *mm,* the velocity of the water is uniform at 120 *mm/s,* and the water depth is 15 *mm.* Determine the average velocity of the water in the 30 *mm* drain pipe.

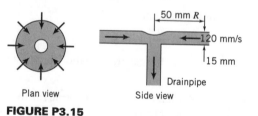

Plan view   Side view

FIGURE P3.15

3.16 A circular pool, 15 m diameter, is to be filled to a depth of 3 m. Determine the inlet flow in m^3/s and *gpm* if the pool is to be filled in 2 *hrs*. Find the number of 2 *in.* diameter hoses required if the water velocity is not to exceed 10 *ft/s*.

3.17 A tank 10 *ft* in diameter and 6 *ft* high is being filled with water at a rate of 0.3 ft^3/s. If it has a leak where it loses water at rate of 30 *gpm*, how long will it take to fill the tank if it was initially half full?

3.18 Is the tank shown in Figure P3.18 filling or emptying? At what rate is the water level rising or falling? Assume that the density is constant. All inflow and outflow velocities are steady and constant over their respective areas.

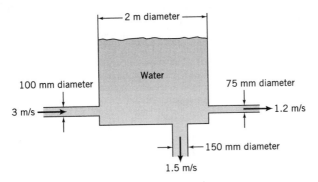

FIGURE P3.18

3.19 A gas obeying the ideal gas law flows steadily in a horizontal pipe of constant diameter between two sections, 1 and 2. If the flow is isothermal and the pressure ratio $p_2/p_1 = \frac{1}{2}$, find the velocity ratio V_2/V_1.

3.20 A gas obeying the ideal gas law enters a compressor at atmospheric pressure and 20°C, with a volume flow rate of 1 m^3/s. Find the volume flow rate leaving the compressor if the temperature and pressure at the exit are 80°C and 200 *bar*, respectively.

3.21 For the vessel shown in Figure P3.21, the flow in and out is steady, with constant density, and it may be assumed to be one-dimensional over the entry and exit planes. Neglect forces due to gravity, and assume the pressure outside the box is atmospheric. Find the mass flow rate and the volume flow rate out of area A_3, and find the x- and y-components of the force exerted on the box by the fluid, in terms of ρ, A_1, θ, and V_1.

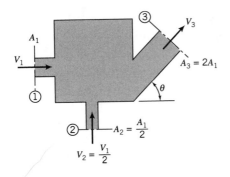

FIGURE P3.21

3.22 Find the force F required to prevent rotation of the pipe shown in Figure P3.22 about the vertical axis located at the point O. The pipe diameter is D, and the fluid density is ρ.

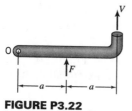

FIGURE P3.22

3.23 A porous circular cylinder, of diameter D is placed in a uniform rectangular wind tunnel section of height $4D$ and width W, as shown in Figure P3.23. The cylinder spans the width of the tunnel. An air volume flow rate of Q per unit width issues from the cylinder. The flow field is steady and has a constant density. The pressures p_1 and p_2 are uniform across the entry and exit areas, and the velocity profiles are as shown. A force F is required to hold the cylinder from moving in the x-direction. Find U_2 in terms of Q, U_1, and D. Then find F.

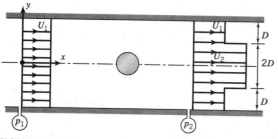

FIGURE P3.23

3.24 A jet of cross-sectional area A_1 steadily issues fluid of density ρ at a velocity V_1, into a duct of area $A_2 = 5A_1$, as shown in Figure P3.24. The jet initially has parallel streamlines. The surrounding flow in the duct has the same density ρ and a velocity $V_2 = V_1/2$. The flow mixes thoroughly, and by section B the flow is approximately uniform across the duct area. In terms of ρ and V_1, find the average velocity of flow at section B, and the pressure difference between sections A and B.

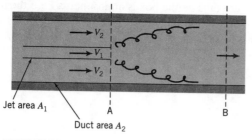

FIGURE P3.24

3.25 Water at $60°F$ enters a 6 *in.* diameter pipe at a rate of 3000 *gpm*. The pipe makes a $180°$ turn. Find the rate of change of momentum of the fluid. If the pressure in the pipe remains constant at 60 *psi,* what is the magnitude and direction of the force required to hold the pipe bend?

3.26 Air, of constant density ρ, flows through the smooth, horizontal pipe bend shown in Figure P3.26, with a constant velocity V, and exits to atmosphere. The pipe is circular and has an internal radius R. A force **F** must be applied in the horizontal plane to keep the pipe bend in place. No forces are transmitted across the flange at the inlet to the bend and the flow

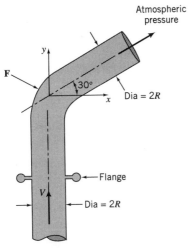

FIGURE P3.26

may be treated as one-dimensional. Find the magnitude of the force **F** in terms of ρ, R, and V. Neglect any pressure changes.

3.27 An effectively two-dimensional jet of water impinges on a wedge as shown in Figure P3.27. The wedge is supported at its apex such that the lower surface remains horizontal. If the thicknesses t_2 and t_3 are equal and $U_2 = U_3 = 2U_1$, find the angle for which the magnitude of the x- and y-components of the reaction at the apex are equal. Neglect gravitational effects. The pressure is atmospheric everywhere.

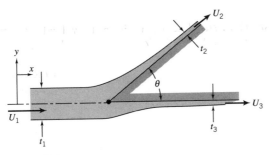

FIGURE P3.27

3.28 Air is flowing through a duct of square cross-section of height h, as shown in Figure P3.28. An orifice plate with a square hole measuring $h/2$ by $h/2$ is placed on the center-line. The

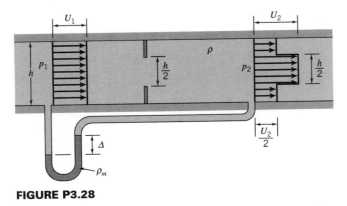

FIGURE P3.28

plate experiences a drag force F_D. The air density ρ is constant. Far upstream, the pressure and velocity are uniform and equal to p_1 and U_1, respectively. Far downstream, the pressure is p_2 and the velocity distribution is as shown. A manometer measures the pressure difference, and shows a deflection of Δ. The manometer fluid has a density ρ_m. The flow is steady, and friction can be neglected. Find the maximum velocity at the outlet, U_2, in terms of h and U_2, and the drag force F_D in terms of h, ρ, ρ_m, Δ, g, and U_1.

3.29 At approximately what included angle do you expect the flow in a diffuser to separate?

3.30 Describe the differences in the flow through a sudden contraction and the flow through a sudden expansion.

3.31 Describe the differences between turbulent flow and separated flow.

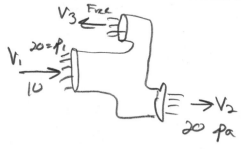

X <u>mom formula</u>

$$\text{Mom}_{leav} - \text{Mom}_{enter} = -\int p\lambda \, dA - T$$

$$\left((\rho V_2 A_2) V_2 + (\rho V_3 A_3)(-V_3) \right) - \left((\rho V_1 A_1) V_1 \right) = -T + \left(-P_a A_2 + P_a A_3 + P_1 A_1 \right)$$

Press Force acting on CS

CHAPTER

INTRODUCTION TO FLUID MOTION II

4.1 INTRODUCTION

We will now complete the introduction to the basic principles governing fluid motion by deriving and applying Bernoulli's equation and the energy equation. As in Chapter 3, these principles will be illustrated by using examples of one-dimensional, steady flows.

4.2 THE BERNOULLI EQUATION

Bernoulli's equation is obtained by applying Newton's second law along a streamline (the complete derivation is given below). It states that,

$$p_1 + \tfrac{1}{2}\rho V_1^2 + \rho g z_1 = p_2 + \tfrac{1}{2}\rho V_2^2 + \rho g z_2 = \text{constant} \qquad (4.1)$$

where z_1 and z_2 are the heights or elevations of points 1 and 2 above some horizontal reference plane, and

1. Points 1 and 2 lie on the same streamline[1];
2. The fluid has constant density;
3. The flow is steady; and
4. The flow is inviscid.

The "fluid has constant density" statement means that the expected density change in the flow field is very small. The "flow is steady" statement not only excludes unsteady flows, but also all turbulent flows. The "flow is inviscid" statement means that the Reynolds number is large and viscous friction is negligible, and that our interests lie outside boundary layers and other regions where viscous stresses are important.

Although these restrictions sound severe, Bernoulli's equation is very useful, partly because it is so simple, but mostly because it can give great insight into the trade-offs among the pressure, velocity, and height of a fluid particle. For a historical note on Daniel Bernoulli, see Section 15.6.

[1] Bernoulli's equation can be used across streamlines if the flow is *irrotational,* that is, when $\vec{\Omega} = \nabla \times \mathbf{V} = 0$. The vector $\vec{\Omega}$ is known as the vorticity—see Section 7.1.

119

We will derive Bernoulli's equation by following a fluid particle, that is, by using a Lagrangian approach.

4.2.1 Force Balance Along Streamlines

Consider a particle of fluid moving in a steady two-dimensional flow in the z-y plane of a constant density fluid (Figure 4.1). The z-direction is measured vertically up from a horizontal reference plane, so that it increases in a direction opposite to the direction of gravity, and the x-direction is into the page. The s-direction is along the streamline, and the n-direction is normal to it. Since the flow is steady, the particle will follow a streamline and its velocity at any point along the streamline is V. If we neglect friction, the forces acting in the s-direction include:

1. The component of the weight acting in the s-direction. The weight of the fluid particle is $\rho g\, dndsdx$, and its component in the s-direction is $-\rho g \sin \beta\, dndsdx$. Since $\sin \beta = \partial z/\partial s$, this force component

$$= \rho g \frac{\partial z}{\partial s}\, dndsdx$$

2. The force due to pressure acting in the s-direction

$$= \left(p - \frac{\partial p}{\partial s}\frac{ds}{2}\right) dndx - \left(p + \frac{\partial p}{\partial s}\frac{ds}{2}\right) dndx$$

$$= -\frac{\partial p}{\partial s}\, dndsdx$$

Hence the resultant force per unit volume

$$= -\frac{\partial p}{\partial s} - \rho g \frac{\partial z}{\partial s} \tag{4.2}$$

This force will accelerate the fluid particle as it moves along the streamline. In a short distance ds, the velocity changes from V to $V + \frac{\partial V}{\partial s}\, ds$, so that the rate of change of momentum of the fluid particle, per unit volume, is equal to

$$\rho \left(\frac{V + \frac{\partial V}{\partial s} ds - V}{dt}\right) = \rho V \frac{\partial V}{\partial s}$$

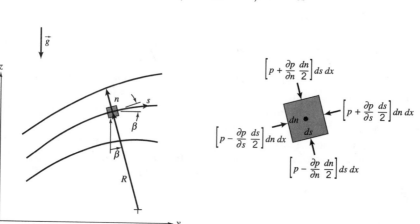

FIGURE 4.1 Fluid particle moving along a streamline.

since $V = ds/dt$. Using equation 4.2, we obtain

$$\rho V \frac{\partial V}{\partial s} = -\frac{\partial p}{\partial s} - \rho g \frac{\partial z}{\partial s} \tag{4.3}$$

This equation is called the one-dimensional Euler equation, or the Euler equation along a streamline. By multiplying through by ds, we get

$$\frac{\partial p}{\partial s} \, ds + \rho V \frac{\partial V}{\partial s} \, ds + \rho g \frac{\partial z}{\partial s} \, ds = 0$$

Now, since s is the coordinate along the streamline,

$$\frac{\partial p}{\partial s} \, ds = dp = \text{change in pressure along streamline}$$

$$\frac{\partial V}{\partial s} \, ds = dV = \text{change in velocity along streamline}$$

$$\frac{\partial z}{\partial s} \, dz = dz = \text{change in elevation along streamline}$$

so that

$$\frac{dp}{\rho} + V \, dV + g \, dz = 0 \tag{4.4}$$

No restrictions have yet been placed on the density of the fluid particle, and so this relationship holds for flows with variable density. When the density is constant, however, we can integrate this relationship along the streamline to obtain

$$\frac{p}{\rho} + \tfrac{1}{2}V^2 + gz = \text{constant}$$

which is Bernoulli's equation for steady, constant density flow, without friction, along a streamline.

Since Bernoulli's equation was derived by applying Newton's second law along a streamline, it is a form of the momentum equation. Interestingly, each term in Bernoulli's equation has the dimensions of energy, or work, per unit mass. The quantity $\tfrac{1}{2}V^2$ is the kinetic energy per unit mass, gz is the potential energy per unit mass, and p/ρ, being the integral of dp/ρ, is the work done by a unit mass of incompressible fluid against the pressure variation along the streamline. For steady, constant density flow, without friction, Bernoulli's equation indicates that the sum of pressure work, kinetic energy, and potential energy remains constant along a streamline. This connection with the energy equation is explored further in Section 4.7.

4.2.2 Force Balance Across Streamlines

Another useful result can be obtained by considering the force balance across streamlines for a steady inviscid, constant density flow. For the frictionless flow shown in Figure 4.1, the forces acting in the n-direction include:

1. The component of the weight $\rho g \, dndsdx$ acting in the n-direction, that is, $-\rho g \cos \beta \, dndsdx$. Since $\cos \beta = \partial z \, \partial n$, this component

$$= \rho g \frac{\partial z}{\partial n} \, dndsdx$$

2. The force due to pressure acting in the n-direction,

$$= \left(p - \frac{\partial p}{\partial n}\frac{dn}{2}\right)dsdx - \left(p + \frac{\partial p}{\partial n}\frac{dn}{2}\right)dsdx$$

$$= -\frac{\partial p}{\partial n}\,dndsdx$$

The resultant force per unit volume

$$= -\frac{\partial p}{\partial n} - \rho g\frac{\partial z}{\partial n} \tag{4.5}$$

This force will accelerate the fluid particle in the direction normal to the streamline. The centripetal acceleration is given by $-V^2/R$, where R is the local radius of curvature (see Figure 4.1), so that the rate of change of momentum of the fluid particle in the n-direction, per unit volume, is equal to

$$-\rho\frac{V^2}{R}$$

Using equation 4.5, we obtain

$$\boxed{\frac{\partial p}{\partial n} + \rho g\frac{\partial z}{\partial n} = \rho\frac{V^2}{R}} \tag{4.6}$$

This is the momentum equation for steady, constant density flow without friction, *across* streamlines. It is known as the Euler equation for the direction normal to a streamline.

We see that, as the streamlines become straight and the radius of curvature becomes very large, the pressure across streamlines can only vary through hydrostatic head differences. If the effects of gravity are not important, and $R \to \infty$, then equation 4.6 gives that $\partial p/\partial n = 0$, that is, the pressure is constant across streamlines. Therefore, for a steady, inviscid flow of a constant density fluid,

The pressure across straight streamlines is constant, if the effects of gravity can be neglected.

This is a very important result, and we shall make good use of it.

4.3 STAGNATION PRESSURE AND DYNAMIC PRESSURE

Bernoulli's equation leads to some interesting conclusions regarding the variation of pressure over solid bodies. Consider a steady, constant density flow impinging on a perpendicular plate (Figure 4.2). There is one streamline that divides the flow in half. Above this streamline, all the flow goes up the plate and below this streamline, all the flow goes down the plate. Along this dividing streamline, the fluid moves towards the plate. Since the flow cannot pass through the plate, the fluid must come to rest at the point where it meets the plate. In other words, it "stagnates." The fluid along the dividing, or *stagnation streamline* slows down and eventually comes to rest without deflection at the *stagnation point*.[2]

---

[2] At a stagnation point, the magnitude of the velocity is zero and its direction is indeterminate. This is an example of a singular or critical point, which is a point where streamlines can meet.

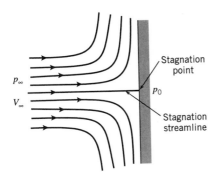

FIGURE 4.2 Stagnation point flow.

Neglecting viscous effects, Bernoulli's equation along the stagnation stream-line gives

$$p_\infty + \tfrac{1}{2}\rho V_\infty^2 + \rho g z_\infty = p_0 + \tfrac{1}{2}\rho V_0^2 + \rho g z_0$$

where the point ∞ is far upstream and point 0 is at the stagnation point (Figure 4.2). Since $z_\infty = z_0$, and $V_0 = 0$,

$$\boxed{p_\infty + \tfrac{1}{2}\rho V_\infty^2 = p_0 = \text{stagnation pressure}} \qquad (4.7)$$

The *stagnation*, or *total*, pressure, p_0, is the pressure measured at the point where the fluid comes to rest. It is the highest pressure found anywhere in the flowfield, and it occurs at the stagnation point. It is the sum of the *static* pressure (p_∞) and the *dynamic* pressure ($\tfrac{1}{2}\rho V_\infty^2$) measured far upstream. The quantity $\tfrac{1}{2}\rho V_\infty^2$ is called the dynamic pressure (first introduced in Section 1.3.6) because it arises from the motion of the fluid.

We can write equation 4.7 in words as

$$\boxed{\text{static pressure} + \text{dynamic pressure} = \text{stagnation pressure}}$$

The dynamic pressure is not really a pressure at all, it is simply a convenient name for the quantity $\tfrac{1}{2}\rho V_\infty^2$, which represents the decrease in pressure due to the increase in fluid velocity.

We can also express the pressure anywhere in the flow in the form of a nondimensional pressure coefficient C_p, where

$$C_p \equiv \frac{p - p_\infty}{\tfrac{1}{2}\rho V_\infty^2}$$

At the stagnation point $C_p = 1$, which is its maximum value. In the freestream, far from the plate, $C_p = 0$. Also, from Bernoulli's equation,

$$p - p_\infty = \tfrac{1}{2}\rho(V_\infty^2 - V^2)$$

so that

$$C_p = 1 - \left(\frac{V}{V_\infty}\right)^2, \quad \text{or} \quad \frac{V}{V_\infty} = \sqrt{1 - C_p}$$

4.4 PRESSURE–VELOCITY VARIATION

Consider again the flow shown in Figure 3.12. Since the streamlines are parallel at the entry and exit to the duct, the pressure is constant across the duct, except for hydrostatic head differences. If we ignore gravity, then the pressures over the inlet and outlet areas are constant. Along a streamline on the centerline, the Bernoulli equation and the one-dimensional continuity equation give, respectively,

$$p_1 - p_2 = \tfrac{1}{2}\rho(V_2^2 - V_1^2)$$

and

$$A_1 V_1 = A_2 V_2$$

Note that

1. With $A_2 < A_1$, $V_2 > V_1$ $\begin{bmatrix} \text{decreasing area} \\ \text{increasing velocity} \end{bmatrix}$

2. With $V_2 > V_1$, $p_2 < p_1$ $\begin{bmatrix} \text{increasing velocity} \\ \text{decreasing pressure} \end{bmatrix}$

These two observations are illustrated in Fig. 4.3. They provide an intuitive guide for analyzing fluid flows, even when the flow is not one-dimensional. For example, when fluid passes over a solid body, the pressure decreases, the flow velocity increases, and the streamlines get closer together.

EXAMPLE 4.1 An Experiment on Velocity and Pressure

An easy demonstration of the force produced by an airstream requires a piece of notebook paper and two books of about equal thickness. Place the books 4 to 5 inches apart, and cover the gap with the paper. When you blow through the passage made by the books and the paper, what do you see? Why? ∎

EXAMPLE 4.2 A Second Experiment on Velocity and Pressure

As a variation on Example 4.1, hold a piece of regular note paper by your fingertips at the corners on the short side. Then blow over the top in the direction of the long side. What do you see, and why? ∎

EXAMPLE 4.3 A Third Experiment on Velocity and Pressure

A light sphere placed in a vertical air jet becomes suspended in the jet, and it is very stable to small perturbations. You can demonstrate this using a table tennis ball and a hair dryer. Let the hair dryer blow vertically up, place the ball in the flow, and let it go. It will remain suspended by the jet of air. Push the ball down,

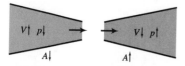

FIGURE 4.3 Pressure–velocity variation according to Bernoulli's equation.

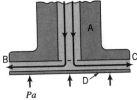

FIGURE 4.4 Attracted disk. From Martin and Connor, *Basic Physics*, 8th ed., published by Whitcombe & Tombs Pty. Ltd., Melbourne, Australia (1962), with permission.

and it springs back to its equilibrium position; push it sideways, and it rapidly returns to its original position in the center of the jet.

Why does this happen? In the vertical direction, the weight of the ball is balanced by a force due to pressure differences: the pressure over the rear half of the sphere is lower than over the front half because of losses that occur in the wake. To understand the balance of forces in the horizontal direction, remember that if the jet is approximately parallel, the pressure inside the jet in the absence of the sphere is the same as outside the jet, so that the pressure is atmospheric everywhere. Now imagine the ball placed on the edge of the jet, so that one half is exposed to the flow, and the other half is not. On the side exposed to the jet, the flow velocity increases as it passes around the sphere, and the pressure falls below atmospheric. On the side outside the jet, the pressure remains atmospheric. The differences in pressure tend to move the ball back towards the center, and therefore it is stable. ∎

EXAMPLE 4.4 *Attracted Disk*

When air is blown through a pipe, A (Figure 4.4), fitted with a flat flange, *BC*, the flat plate, *D*, is forced against the flange instead of being blown off as might be expected. The explanation is based on the relationship between pressure and velocity. The velocity in the gap between the plate and flange is initially high, but it begins to decrease as the fluid flows outward since the area through which the flow occurs increases with radius. At the same time, the pressure rises. However, at the exit (*B* and *C*), the pressure must be equal to atmospheric pressure since the streamlines at the exit are parallel. The pressure in the gap must, therefore, be everywhere less than atmospheric. Since the pressure on the outside of the plate is atmospheric, there is a resultant force holding the plate against the flange due to pressure differences. ∎

4.5 APPLICATIONS OF BERNOULLI'S EQUATION

How useful is Bernoulli's equation? How restrictive are the assumptions governing its use? Here we give some examples, and further examples will be given in later chapters.

One common use of Bernoulli's equation is as a third equation, in addition to the continuity and momentum equations, for solving flow problems. For instance, if the flow shown in Figure 3.12 was a constant density flow, we could consider a streamline along the center of the duct and apply the Bernoulli equation. Hence,

$$p_1 + \tfrac{1}{2}\rho V_1^2 = p_2 + \tfrac{1}{2}\rho V_2^2$$

since $z_1 = z_2$. Using the continuity equation, we obtain

$$p_2 = p_1 + \tfrac{1}{2}\rho V_1^2 \left(1 - \frac{A_1^2}{A_2^2}\right)$$

Then, from equation 3.5,

$$R_x = \rho A_1 V_1^2 \left(\frac{A_1}{A_2} - 1\right) + p_1(A_2 - A_1) + \tfrac{1}{2}\rho V_1^2 A_2 \left(1 - \frac{A_1^2}{A_2^2}\right)$$

Since Bernoulli's equation was used to find the pressure p_2, the force R_x can now be found if p_1, V_1, A_1, and A_2 are known.

We can go one step further and divide both sides by $\tfrac{1}{2}\rho V_1^2 A_1$, so that

$$\frac{R_x}{\tfrac{1}{2}\rho V_1^2 A_1} = 2\left(\frac{A_1}{A_2} - 1\right) + \frac{p_1}{\tfrac{1}{2}\rho V_1^2 A_1}(A_2 - A_1) + \frac{A_2}{A_1}\left(1 - \frac{A_1^2}{A_2^2}\right)$$

This step makes both sides of the equation nondimensional. The parameter on the left hand side is an example of a nondimensional force coefficient. This process of *nondimensionalization* is standard practice in fluid mechanics since it makes the answer more useful and elegant, as we shall see.

4.5.1   Pitot Tube

One of the most immediate applications of Bernoulli's equation is in the measurement of velocity with a Pitot tube. The Pitot tube (named after the French scientist and engineer Henry Pitot, 1695–1771) is perhaps the simplest and most useful fluid flow instrument ever devised (Figure 4.5). By pointing the tube directly into the flow and measuring the difference between the pressure sensed by the tube and the pressure of the surrounding air flow, the Pitot probe can provide a very accurate measure of the velocity. In fact, it is probably the most accurate method available for measuring flow velocity on a routine basis, and accuracies better than 1% are easily possible. Pitot tubes are widely used in flow measurement applications. For instance, they are standard equipment on airplanes, where a Pitot tube is combined with a static port located somewhere on the fuselage to provide the dynamic pressure of motion to the airspeed indicator in the cockpit.

To see how the Pitot tube works, consider Bernoulli's equation along the streamline that begins far upstream of the tube and comes to rest in its mouth (station 0)

$$p_\infty + \tfrac{1}{2}\rho_a V_\infty^2 + \rho_a g z_\infty = p_0 + \tfrac{1}{2}\rho_a V_0^2 + \rho_a g z_0$$

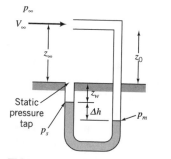

FIGURE 4.5   Pitot tube in a wind tunnel.

where ρ_a is the density of air. Since $z_\infty = z_0$, and $V_0 = 0$,

$$p_0 = p_\infty + \tfrac{1}{2}\rho_a V_\infty^2 \tag{4.8}$$

We see that the Pitot tube measures the stagnation pressure in the flow. We can also write

$$p_0 - p_\infty = \tfrac{1}{2}\rho_a V_\infty^2 \tag{4.9}$$

Therefore, to find the velocity V_∞, we need to know the density of air and the pressure difference $p_0 - p_\infty$. The density can be found from standard tables if the temperature and the atmospheric pressure are known. The pressure difference $p_0 - p_\infty$ is usually found indirectly by using a *static pressure tap* located on the wall of the wind tunnel (as shown in Figure 4.5), or on the surface of the model. If the pressure measured by a wall static pressure tap is p_s, then, by applying the hydrostatic equation three times, we find

$$
\begin{aligned}
p_s &= p_\infty + \rho_a g(z_w + z_0)\\
p_m &= p_0 + \rho_a g(z_w + z_0 + \Delta h)\\
p_m - p_s &= \rho_m g \Delta h
\end{aligned}
\tag{4.10}
$$

where ρ_m is the density of the manometer fluid. Hence,

$$p_0 - p_\infty = \rho_m g \Delta h - \rho_a g \Delta h$$

Using equation 4.9, we obtain

$$V_\infty^2 = 2g\frac{\rho_m}{\rho_a}\left(1 - \frac{\rho_a}{\rho_m}\right)\Delta h \tag{4.11}$$

Since ρ_m is always much greater than ρ_a, we have, to a very good approximation,

$$V_\infty^2 = 2g\frac{\rho_m}{\rho_a}\Delta h \tag{4.12}$$

Therefore, to determine the velocity of air using a Pitot tube, we only need a static pressure tap and a manometer. The velocity can then be found by measuring the manometer deflection, and by knowing the density of air and the density of the manometer fluid.

When using the Pitot tube, we often assume that $p_s = p_\infty$. From equation 4.10, we have (in full)

$$p_s = p_\infty\left[1 + \frac{\rho_a g(z_w + z_0)}{p_\infty}\right]$$

Since p_∞ is approximately equal to the atmospheric pressure, we see that the error in the approximation $p_s = p_\infty$ only becomes significant when $z_w + z_0$ is large enough to be some fraction of the height of the atmosphere. For example, for $z_w + z_0 = 10\ m$, the error is only about 0.1%.

The Pitot tube is sometimes combined with a static tube to make a single unit called a Pitot-static tube (Figure 4.6). A static tube is a closed tube aligned with the flow direction. At some distance from the nose, small pressure tappings are drilled in the tube wall to measure the local static pressure. In a Pitot-static tube, the two tubes are installed one inside the other. The inner tube is open at the end and measures the total pressure, while the outer tube serves as a static tube to measure the static pressure. The two tubes can be connected to the legs of a

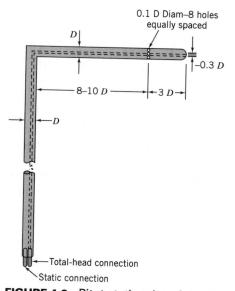

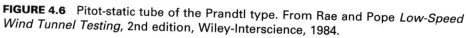

FIGURE 4.6 Pitot-static tube of the Prandtl type. From Rae and Pope *Low-Speed Wind Tunnel Testing*, 2nd edition, Wiley-Interscience, 1984.

manometer so that the manometer reads the dynamic pressure. The accuracy of a Pitot-static tube depends on its construction and installation. It is quite sensitive to misalignment with the flow direction, and to achieve good accuracy it should be calibrated against a known standard.

4.5.2 Venturi Tube and Atomizer

The relationship between pressure and velocity expressed by Bernoulli's equation is used in the design of the Venturi tube (Figure 4.7). This device is made up of a contraction followed by a diverging section, and ... ... no significant losses occur due to friction or flow separation. One application of the Venturi tube is as a flow measurement device for the steady flow of a constant-density fluid. As the fluid passes through the tube, it reaches its maximum velocity and minimum pressure at the smallest cross-sectional area. Using the one-dimensional continuity equation and Bernoulli's equation, we have

$$V_1 A_1 = V_2 A_2$$

$$\frac{p_1}{\rho} + \tfrac{1}{2}V_1^2 = \frac{p_2}{\rho} + \tfrac{1}{2}V_2^2$$

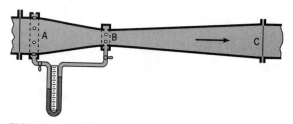

FIGURE 4.7 Venturi tube. From Martin and Connor, *Basic Physics,* 8th ed., published by Whitcombe & Tombs Pty. Ltd., Melbourne, Australia (1962), with permission.

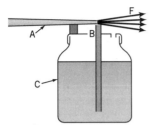

FIGURE 4.8 Atomizer. From Martin and Connor, *Basic Physics,* 8th ed., published by Whitcombe & Tombs Pty. Ltd., Melbourne, Australia (1962), with permission.

so that

$$p_1 - p_2 = \tfrac{1}{2}\rho V_1^2 \left[\left(\frac{A_1}{A_2} \right)^2 - 1 \right]$$

By measuring the pressure difference $p_1 - p_2$ and knowing the dimensions of the Venturi tube and the density of the fluid, we can find the velocity of the flow. The Venturi tube is used when an accurate velocity measurement is required in a piping system without introducing substantial losses.[3]

Carburetors also make use of the Venturi principle. Here, the low pressure at the throat of the Venturi tube is used to draw fuel into the main flow stream. If the pressure is low enough at the throat, the fuel will break up into drops and vaporize before entering the engine combustion chamber.

A further use of this concept is in an atomizer (Figure 4.8). A stream of air blown through the pipe *A* passes directly above the open end of the tube *B,* the other end of which dips into a liquid in the container *C,* which is open to the atmosphere. The tube is designed so that the airstream diverges after passing *B.* Far from the tube exit, at *F,* the pressure must be atmospheric, so that the pressure at *B* must be below atmospheric (since the flow must slow down as it goes from *B* to *F*). Therefore, the pressure of the atmosphere acting on the surface of the liquid in *C* forces it up to *B* where it is entrained in the airstream and carried forward as a spray. Paint sprayers, perfume sprayers, and snowmakers at ski resorts all operate in this way.

4.5.3 Siphon

A siphon can be used to drain liquid from an open tank. It consists of a tube with one end immersed in the liquid and the other end kept outside the liquid and below the level of the free surface (see Figure 4.9). If there are no losses, and the flow is steady, we can use Bernoulli's equation to find the speed of the jet issuing from the open end of the tube, and the minimum pressure in the tube.

When the tank is large, the level of the water falls very slowly and we can assume that the velocity of the free surface is almost zero. This is called the *quasi-steady* assumption, where the flow is not really steady but it is steady enough to apply Bernoulli's equation, for example. To draw a streamline, we should remember that a streamline is a line which is tangential to the instantaneous velocity vector. If we took a short-exposure photograph of some marked fluid particles, we would

---

[3] In fully-developed pipe flow, where the flow is not one-dimensional, a kinetic energy correction factor is required—see Section 9.8.1.

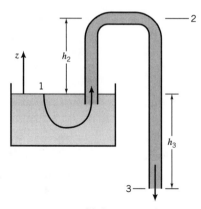

FIGURE 4.9 Siphon, used to transfer liquids from a higher to a lower level.

see that the velocity vectors throughout most of the tank are directed towards the entrance of the tube. So we can start the streamline at the bottom of the tank, or halfway up, or at the surface of the water. However to make the streamline useful, it needs to be chosen so that it connects a point where we have information with a point where we need information. At the surface we know that $V_1 = 0$, and that the pressure is atmospheric. Since two parameters are known at the surface, this is a good starting place for the streamline. To find the conditions at the exit of the tube, we draw the streamline to connect to a point in the exit plane. Then

$$\frac{p_a}{\rho} + 0 + gz_3 = \frac{p_3}{\rho} + \tfrac{1}{2}V_3^2 + 0$$

What about p_3? Close to the exit of the jet the flow is parallel and the pressure surrounding the jet is atmospheric. In a steady, constant density flow, the pressure can only vary across straight streamlines due to hydrodynamic pressure gradients (see Section 4.2.2). There can be no hydrostatic head differences across the jet since it is in free fall, so that with $V_1 = 0$, and $p_3 = p_1 = p_a$,

$$V_3 = \sqrt{2gz_3}$$

The cross-sectional area of the tube does not vary and, therefore, the velocity in all parts of the tube is also equal to V_3. Since the velocity in the tube is constant, the minimum pressure will occur at the maximum height of the tube, that is, at point 2. Here,

$$\frac{p_1}{\rho} + \tfrac{1}{2}V_1^2 + 0 = \frac{p_2}{\rho} + \tfrac{1}{2}V_2^2 + gz_2$$

With $V_1 = 0, p_1 = p_a$, and $V_2 = V_3 = \sqrt{2gz_3}$

$$\frac{p_a}{\rho} = \frac{p_2}{\rho} + gz_3 + gz_2$$

That is,

$$\frac{p_2}{\rho} = \frac{p_a}{\rho} - g(z_3 + z_2)$$

We see that the pressure at point 2 is below atmospheric. If the elevation of point 2 is high enough, the pressure can become equal to the vapor pressure of the liquid.

The vapor pressure is the pressure at which a liquid will boil. If the liquid pressure is greater than the vapor pressure, the only exchange between the liquid and vapor phases is evaporation at a free surface. If the liquid pressure falls below the vapor pressure, vapor bubbles begin to appear in the liquid. For water at 68°F the vapor pressure is 49 lb_f/ft^2, which is equal to 0.0232 *atm*. Since 1 atmosphere = 33.9 *ft*, we find that if

$$z_2 + z_3 > (1 - 0.0232) \times 33.9 \ ft = 33.1 \ ft$$

the water will boil and a vapor lock can form.

Since the pressure decreases with altitude, the air pressure at higher elevations comes closer to the vapor pressure of water, and water will boil at a lower temperature as the altitude increases. At an altitude of 10,000 *ft*, for example, where the atmospheric pressure is 10.1 *psi*, water boils at 193°F, not 212°F.

A liquid can also boil if its velocity is high enough so that the pressure in the liquid drops below the vapor pressure. The formation of vapor bubbles is then called *cavitation,* and it can cause severe erosion on marine propellers, where low pressure regions are found near the tips of the blades (Figure 8.1). Just below the water surface, the pressure equals the vapor pressure at a speed of about 50 *ft/s* (= 34 *mph*). Cavitation can therefore be a problem at relatively modest speeds, although in practice it does not usually occur until the pressure is well below the vapor pressure. At greater depths, the pressure in the surrounding fluid increases, and the onset of cavitation is delayed to higher velocities.

4.6 BERNOULLI'S EQUATION AND DRAINING TANKS

Consider water draining out of a large tank (Figure 4.10). The tank is large enough so that we can use the quasi-steady assumption. Along a streamline connecting a point at the surface with a point in the exit plane of the jet, we have

$$\frac{p_1}{\rho} + 0 + gH = \frac{p_2}{\rho} + \tfrac{1}{2}V_2^2 + 0$$

If the streamlines at the exit of the jet are parallel, $p_2 = p_a$. Also, the jet is in free fall, so that hydrostatic head differences across the jet are zero. With $V_1 = 0$, and $p_2 = p_1 = p_a$,

$$V_2 = \sqrt{2gH}$$

This is known as Torricelli's formula.

What is the discharge? The volume discharge $\dot{q}$ is the volume outflow per unit time, that is, the volume flow rate out of the tank. In a short time Δt, the

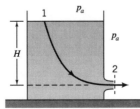

FIGURE 4.10 Water draining out of a large tank.

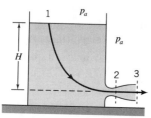

FIGURE 4.11 Tank exit with an expanding drain.

volume of fluid leaving the tank is $A_2 V_2 \Delta t$, where A_2 is the cross-sectional area of the jet. So the volume outflow per unit time is simply $A_2 V_2$. Similarly, the mass flow discharge, or the mass flow rate leaving the tank is given by $\rho A_2 V_2$. Hence, for the flow leaving the tank,

$$\dot{q} = \text{volume discharge rate} = A_2 V_2 = A_2 \sqrt{2gH}$$
$$\dot{m} = \text{mass discharge rate} = \rho A_2 V_2 = \rho A_2 \sqrt{2gH}$$

What if the orifice was modified to have a diverging section, so that the area increases from A_2 to A_3 (Figure 4.11)? If the streamlines are parallel at the exit, the pressure at the exit, p_3, is atmospheric. If there are no losses, we see that the exit velocity is still $\sqrt{2gH}$. The volume discharge, however, has increased to $A_3 \sqrt{2gH}$ so that it is greater than before by the ratio A_3/A_2. Provided there are no losses, the exit velocity is independent of the exit area but the discharge increases as the exit area increases. What has happened at the point A_2? Because of continuity, the velocity at A_2 is larger than the velocity at A_3 by the ratio A_3/A_2. Since the pressure at A_3 is atmospheric, the pressure at A_2 is below atmospheric, and the converging-diverging nozzle acts similar to a Venturi tube (see Section 4.5.2). If the pressure at station 2 drops below the vapor pressure, cavitation can occur.

EXAMPLE 4.5 *The Flow in a Jet*

Consider a tank draining through a small orifice, where the orifice outlet points up at an angle θ, as in Figure 4.12. The magnitude of the exit velocity is still $V_e = \sqrt{2gH}$. As the jet issues into the atmosphere, the vertical component of the velocity (w) decreases under the action of gravity. The vertical velocity goes to zero at the top of the jet trajectory and then becomes negative. The horizontal component of the fluid velocity (u) remains constant throughout the trajectory when air friction is neglected, and the only force acting is due to gravity. Find the maximum height to which the jet rises.

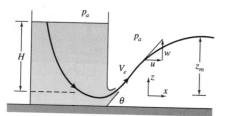

FIGURE 4.12 Tank draining with exit pointing up at angle θ.

Solution: Consider a streamline that starts at the exit and follows the path of the jet. The pressure everywhere outside the jet is atmospheric, and since there are no pressure gradients across the jet (it is in free fall), the pressure inside the jet is also atmospheric. If there are no losses,

$$\tfrac{1}{2}V_e^2 = \tfrac{1}{2}V^2 + gz = \text{constant}$$

We can write $V^2 = u^2 + w^2$, where u and w are the horizontal and vertical components of V. Similarly, for the exit velocity, $V_e^2 = u_e^2 + w_e^2$. Since the horizontal component of the fluid velocity remains constant (there are no forces acting on the fluid in this direction), $u_e = u = \text{constant}$. Bernoulli's equation reduces to

$$\tfrac{1}{2}w_e^2 = \tfrac{1}{2}w^2 + gz \tag{4.13}$$

The highest point of the trajectory is given by point where $z = z_m$ and $w = 0$. That is,

$$gz_m = \tfrac{1}{2}w_e^2$$

We can check this answer by taking limits. For $\theta = 90°$, $z_m = H$, and for $\theta = 0°$, $z_m = 0$, as expected. Now, $w_e = V_e \sin\theta = \sqrt{2gH}\sin\theta$, so that

$$gz_m = \tfrac{1}{2}(2gH\sin^2\theta)$$

and

$$z_m = H\sin^2\theta$$

In addition, by using equation 4.13, it can be shown that the path of a fluid particle is parabolic (following a fluid particle, we can use $u = dx/dt$, and $w = dz/dt$). ∎

EXAMPLE 4.6 *Forces Exerted by an Exiting Jet*

A tank sits on a balance scale to measure the vertical and horizontal forces F_z and F_x. The tank has an opening near the bottom that points up at an angle θ to the horizontal (Figure 4.13). The water level is kept constant so that the problem is steady. We will assume that there are no losses, and that the fluid is of constant density ρ.

(a) What is F_z?
(b) What is F_x?
(c) Find the weight of the water in flight.

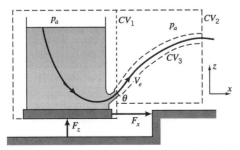

FIGURE 4.13 Reactive forces acting when a tank drains.

Solution: For part (a), we use the control volume labeled CV_1. The momentum equation in the z-direction gives

$$-\text{(weight of water in tank)} - \text{(weight of tank)} + F_z = \rho V_e A_e (V_e \sin \theta)$$

where V_e is the velocity and A_e is the cross-sectional area of the jet at the exit. Note that the z-momentum term is given by the product of the mass flow rate $\rho V_e A_e$ and the z-component of the velocity $V_e \sin \theta$. We see that F_z has two parts: a static part due to the combined weight of the water and tank, and a dynamic part that depends on the rate of outflux of z-momentum.

For part (b), to find the horizontal force F_x we can use either CV_1 or CV_2, since the horizontal component of the momentum does not change in the x-direction (no forces act in that direction). So, for CV_1 or CV_2

$$F_x = \text{net outflow of } x\text{-momentum}$$

so

$$F_x = \rho V_e A_e (V_e \cos \theta)$$

The x-momentum term is given by the product of the mass flow rate $\rho V_e A_e$ and the x-component of the velocity $V_e \cos \theta$.

For part (c), we use the control volumes labeled CV_2 and CV_3. For the volume CV_2, the momentum equation in the z-direction is

$$-\text{(weight of water in tank)} - \text{(weight of tank)} - \text{(weight of water in flight)} + F_z = 0$$

For control volume CV_3, the momentum equation in the z-direction becomes

$$-\text{(weight of water in flight} = \text{net outflow of } z\text{-momentum}$$
$$= -\rho(V_e \sin \theta)V_e A_e$$

which could also have been deduced from the results obtained using control volumes CV_1 and CV_2.   ∎

EXAMPLE 4.7 *Forces Exerted by an Entering Jet*

Just to complicate matters more, consider what happens if the jet from the previous examples falls into a second tank placed on another scale (Figure 4.14).
(a) What is F_z'?
(b) What is F_z''?

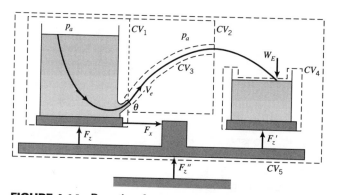

FIGURE 4.14 Reactive forces acting when a tank is filling.

Solution: For part (a), we use the control volume labeled CV_4. The momentum equation in the z-direction becomes

$$-(\text{weight of water in tank 2}) - (\text{weight of tank 2}) + F_z' = \text{net outflow of } z\text{-momentum}$$
$$= \rho V_e A_e (w_E)$$

where w_E is the vertical component of the velocity of the jet at the point where it enters the water contained in the second tank (the z-momentum term is given by the product of the mass flow rate $\rho V_e A_e$ and the z-component of the velocity of the jet).

We can use Bernoulli's equation to find w_E. Between the exit from the first tank and the entry to the second one,

$$\frac{p_e}{\rho} + \tfrac{1}{2}(u_e^2 + w_e^2) + gz_e = \frac{p_E}{\rho} + \tfrac{1}{2}(u_E^2 + w_E^2) + gz_E$$

Since $p_e = p_E$, $u_e = u_E$, and $w_e = V_e \sin \theta = \sqrt{2gH} \sin \theta$, we need only to know the distance $z_e - z_E$ to find w_E.

For part (b), we consider the z-momentum equation for control volume CV_5

$$-(\text{weight of entire system}) + F_z'' = \text{net outflow of } z\text{-momentum} = 0$$

and so F_z'' is constant. As F_z decreases, F_z' increases so that F_z'' remains constant. ∎

EXAMPLE 4.8 *Jet on a Cart*

A pipe of cross-sectional area A is connected by a flange to a large, pressurized tank, which supplies air of constant density ρ to a jet of cross-sectional area $\frac{1}{4}A$, as shown in Figure 4.15. The tank sits on a cart equipped with wheels so that it can roll freely, and the jet exits to atmosphere with a velocity V. Find:
(a) The gauge pressure p_g in the pipe at the flange;
(b) The force holding the pipe to the tank at the flange; and
(c) The tension in the string holding the cart.

Solution: For part (a), we will assume that the flow is quasi-steady (it is a large tank), and that there are no losses downstream of the flange[4] so that we can apply

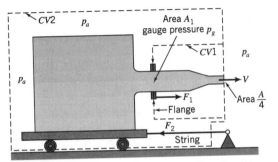

FIGURE 4.15 Jet on a cart, held in place by a string.

---

[4] When there are no losses in a nozzle, it is sometimes called a *smooth* nozzle. The word "smooth" is often used to indicate that there are no losses in a flow. This is not true for a smooth pipe or duct, where losses can never be neglected.

Bernoulli's equation between a point in the pipe at the location of the flange, where the velocity is V_1, and a point in the exit plane of the jet, where the pressure is atmospheric. Hence,

$$p_g = \tfrac{1}{2}\rho V_1^2 = \tfrac{1}{2}\rho V^2$$

From continuity we have

$$V_1 A = V\frac{A}{4}$$

so that

$$p_g = \tfrac{1}{2}\rho\left(V^2 - \frac{V^2}{16}\right)$$
$$p_g = \tfrac{15}{32}\rho V^2$$

For part (b), we use control volume $CV1$ and apply the x-momentum equation. Let $+F_1$ be the x-component of the force holding the pipe to the flange, so that $-F_1$ is the x-component of the force by the fluid on the pipe, and $+F_1$ is the x-component of the force exerted by the pipe on the fluid. Therefore,

$$F_1 + p_g A = -\rho V_1^2 A + \rho V^2\frac{A}{4} = -\rho\frac{V^2}{16}A + \rho V^2\frac{A}{4},$$

so that

$$F_1 = -p_g A + \tfrac{3}{16}\rho V^2 A = -\tfrac{15}{32}\rho V^2 A + \tfrac{3}{16}\rho V^2 A$$

and therefore

$$F_1 = -\tfrac{9}{32}\rho V^2 A$$

For part (c), we use control volume $CV2$ and again use the x-momentum equation. Let $+F_2$ be the x-component of the force holding the cart, so that $-F_2$ is the x-component of the force by the fluid on the cart, and $+F_2$ is the x-component of the force exerted by the cart on the fluid. Therefore

$$F_2 = +\rho V^2\frac{A}{4}$$

∎

4.7 *ENERGY EQUATION

Here, we use a control volume formulation to derive the energy equation for one-dimensional, steady flow. We will also explore its relationship to Bernoulli's equation. To do so, we need to use the first law of thermodynamics. We assume some previous familiarity with the study of thermodynamics, so the next section simply summarizes the basic concepts without a deep discussion of the underlying principles.

4.7.1 First Law of Thermodynamics

The first law of thermodynamics states:

> For a closed system, the sum of the work and heat interactions is equal to the total change in energy of the system.

In terms of fluid mechanics, a *closed system* is a fixed mass of fluid, with no inflow or outflow. That is,

$$\Sigma Q + \Sigma W = \Delta E = \Delta(\hat{U} + PE + KE) \tag{4.14}$$

By "work and heat interactions" we mean the work W done on or by the system, and the heat Q transferred to or from the system. Work and heat transferred *from* the surroundings *to* the system are taken to be positive.[5] The "total change in energy," ΔE, is made up of three components: the potential energy PE, the kinetic energy KE, and the *internal energy* $\hat{U}$. The internal energy is the energy stored in the fluid due to molecular activity and molecular bonding forces. We have used the symbol $\hat{U}$ for the internal energy, rather than the more common symbol U to avoid possible confusion with the x-component of the velocity. We will reserve the symbol V for the magnitude of the velocity vector.

As an example, consider the first law applied to a ball rolling up and down the inside of a bowl (Figure 4.16). The ball and the bowl make up a system. If we assume that there are no work and heat interactions (the system is isolated from its surroundings), then as the ball rolls down it loses potential energy while gaining kinetic energy. At the same time, friction raises the temperatures of the ball and the bowl, so that the internal energy of the system increases. The first law states that the change in total energy is zero, so that the sum of the potential, kinetic, and internal energy remains constant. The total energy of the system can only be changed if there is a transfer of work or heat. For this to happen, the system can no longer be isolated from its surroundings. Cooling the system, so that heat is removed from the system, is a negative heat transfer. Lifting the ball to a higher elevation, so that work is done on the system, is a positive work transfer.

Work can also be done by compression. If we think of a gas contained in a cylinder as it is compressed by a piston, work is being done in changing the volume of the gas. As the piston moves from position x_1 to x_2, the work done is

$$W = -\int_1^2 pA\,dx$$

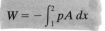

FIGURE 4.16 A simple thermodynamic system.

[5] This sign convention seems intuitively attractive, but it is not universally adopted. When consulting other textbooks it is important to check what sign convention is used to avoid possible confusion.

which is positive since the work is done *on* the system. If the mass of the gas inside the cylinder is m, with a density ρ and volume $\forall$, then

$$\forall = \frac{m}{\rho} = mv$$

where $v = 1/\rho$ is the *specific volume*. Since $d\forall = A\,dx$, and the mass is constant, $d\forall = m\,dv$, and we have

$$W = -\int_1^2 p\,d\forall$$

$$= m - \int_1^2 p\,dv$$

Therefore, the work done is proportional to the area under the curve from points 1 to 2 on a p-v diagram (Figure 4.17).

The heat input Q is related to temperature changes. The change in temperature for a given heat interaction depends on material properties such as its specific heats. For the SI system of units:

> Specific heat is defined as the amount of heat required to raise the temperature of 1 kg of substance by 1°K.

The amount of heat, δQ, required to produce a small change in temperature, dT, is given by

$$\delta Q = mC\,dT$$

where C is the specific heat. A different amount of heat δQ is usually required if the process occurs at a constant volume, $(\delta Q)_v$, or at a constant pressure, $(\delta Q)_p$. The specific heat at constant volume C_v is defined as

$$C_v = \frac{1}{m}\frac{(\delta Q)_v}{dT}, \quad \text{or} \quad (\delta Q)_v = mC_v\,dT \tag{4.15}$$

The units of C_v are $J/(kg\,K)$, that is, $N \cdot m/(kg\,K)$, or $ft \cdot lb_f/(slug\,R)$. For a constant C_v, we can integrate equation 4.15 and get

$$Q = mC_v\Delta T \quad \text{(constant volume)} \tag{4.16}$$

Similarly, for a process at constant pressure

$$C_p = \frac{1}{m}\frac{(\delta Q)_p}{dT}, \quad \text{or} \quad (\delta Q)_p = mC_p\,dT \tag{4.17}$$

and if C_p is constant,

$$Q = mC_p\Delta T \quad \text{(constant pressure)} \tag{4.18}$$

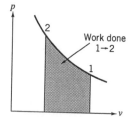

FIGURE 4.17 Work done by compression.

Note that if the mass of fluid is thermally insulated from its surroundings so that all heat interactions are zero ($Q = 0$), the change of state of the fluid is called "adiabatic."

For a fluid in motion, we need to be more precise in defining which part of the heat interaction is associated with the specific heats. In particular, we define:

$$C_v = \left(\frac{\partial \hat{u}}{\partial T}\right)_v, \quad \text{and} \quad C_p = \left(\frac{\partial h}{\partial T}\right)_p \qquad (4.19)$$

where $\hat{u}$ is the internal energy per unit mass ($= \hat{U}/\rho$), and h is the enthalpy per unit mass, defined by

$$h = \hat{u} + \frac{p}{\rho}, \quad \text{or} \quad h = \hat{u} + pv \qquad (4.20)$$

4.7.2 One-Dimensional Flow

We now apply the first law to an open system, where fluid is flowing in and out of a control volume. We are interested in the rates of change in total energy, work done on the fluid, and heat transferred from the surroundings.

Consider a streamtube in steady flow, with inflow over area A_1 and outflow over area A_2 (Figure 4.18). We will first examine the rate of change of total energy. The total energy is the sum of the internal energy, and the kinetic and potential energies. During a short time interval Δt,

the *total energy* entering during $\Delta t = (\hat{u}_1 + \frac{1}{2}V_1^2 + gz_1)\rho_1 A_1 V_1 \Delta t$

and

the *total energy* leaving during $\Delta t = (\hat{u}_2 + \frac{1}{2}V_2^2 + gz_2)\rho_2 A_2 V_2 \Delta t$

where $\hat{u}$ is the internal energy per unit mass, $\frac{1}{2}V^2$ is the kinetic energy per unit mass, and gz is the potential energy per unit mass. Hence

the net *change* of total energy during $\Delta t = \Delta E = \dot{m}[(\hat{u}_2 + \frac{1}{2}V_2^2 + gz_2) - (\hat{u}_1 + \frac{1}{2}V_1^2 + gz_1)]\,\Delta t$

and

the net *rate* of change of total energy $= \Delta \dot{E} = \dot{m}[(\hat{u}_2 + \frac{1}{2}V_2^2 + gz_2) - (\hat{u}_1 + \frac{1}{2}V_1^2 + gz_1)]$

where we have used the fact that mass is conserved so that

$$\dot{m} = \rho_1 A_1 V_1 = \rho_2 A_2 V_2$$

The quantity $\Delta \dot{E}$ is the rate of increase in total energy experienced by the fluid in passing through the duct. Note that energy has the dimensions of force times a distance (ML^2T^{-2}) and it is measured in terms of joules or $ft \cdot lb_f$.

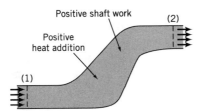

FIGURE 4.18 Streamtube in steady flow.

The flow is steady, so that the energy inside the control volume is not changing with time. Writing the first law of thermodynamics in terms of rates of change,

$$\Delta \dot{E} = \dot{Q} + \dot{W} \tag{4.21}$$

we obtain

$$\dot{m}[(\hat{u}_2 + \tfrac{1}{2}V_2^2 + gz_2) - (\hat{u}_1 + \tfrac{1}{2}V_1^2 + gz_1)] = \dot{Q} + \dot{W} \tag{4.22}$$

We will now examine the rate of change of work done on the fluid. The work term can be split into three parts

$$\dot{W} = \dot{W}_{pressure} + \dot{W}_{viscous} + \dot{W}_{shaft}$$

where $\dot{W}_{pressure}$ is the work done on surfaces by forces due to pressure, $\dot{W}_{viscous}$ is the shear work done on surfaces by viscous stresses, and $\dot{W}_{shaft}$ is the work done by a machine on the system (the machine can be a pump, fan, piston, and so forth).

We treat the pressure work term separately. The work done on the fluid during Δt at station 1 by pressure is given by

$$p_1 A_1 \, ds = p_1 A_1 V_1 \, \Delta t = \frac{p_1}{\rho_1} \dot{m} \, \Delta t$$

Similarly, the work done on the fluid during Δt at station 2 by pressure is given by

$$-\frac{p_2}{\rho_2} \dot{m} \, \Delta t$$

and so

$$\dot{W}_{pressure} = \left(\frac{p_1}{\rho_1} - \frac{p_2}{\rho_2} \right) \dot{m} \, \Delta t$$

Therefore

$$\frac{p_1}{\rho_1} + \hat{u}_1 + \tfrac{1}{2}V_1^2 + gz_1 = \frac{p_2}{\rho_2} + \hat{u}_2 + \tfrac{1}{2}V_1^2 + gz_2 - \frac{\dot{Q} + \dot{W}'}{\dot{m}} \tag{4.23}$$

where

$$\dot{W}' = \dot{W}_{viscous} + \dot{W}_{shaft}$$

so that $\dot{W}'$ is the net amount of work done on the fluid exclusive of the pressure work. For example, if a pump was put inside the streamtube, $\dot{W}'$ would be equal to the pump energy input rate. When viscous forces are present, $\dot{W}'$ is equal to the rate of work done on the fluid by viscous forces.

For adiabatic flows, the energy equation is often written in the form

$$\frac{p_1}{\rho_1} + \tfrac{1}{2}V_1^2 + gz_1 = \frac{p_2}{\rho_2} + \tfrac{1}{2}V_1^2 + gz_2 + gh_{\ell T} - gH \tag{4.24}$$

The change in internal energy is represented as $gh_{\ell T}$, and $h_{\ell T}$ is called the *total head loss* since it represents the irreversible conversion of mechanical energy to heat (see Section 9.8.1). The work per unit mass is represented by gH, and H is called the *head*. It represents the work done on the fluid by a device such as a pump or turbine (see Section 13.2). The quantities $h_{\ell T}$ and H both have the dimensions of length.

By introducing the enthalpy h $(= \hat{u} + p/\rho)$, the energy equation can also be written as:

$$h_1 + \tfrac{1}{2}V_1^2 + gz_1 = h_2 + \tfrac{1}{2}V_1^2 + gz_2 - \frac{\dot{Q} + \dot{W}'}{\dot{m}}$$

(4.25)

Equations 4.23 and 4.25 are both called the one-dimensional energy equation for steady flow.

4.7.3 Relation to Bernoulli's Equation

The one-dimensional energy equation is particularly useful when there is no heat interaction (adiabatic flow) and no shaft and viscous work interactions. Under these conditions

$$\frac{p_1}{\rho_1} + \hat{u}_1 + \tfrac{1}{2}V_1^2 + gz_1 = \frac{p_2}{\rho_2} + \hat{u}_2 + \tfrac{1}{2}V_1^2 + gz_2$$

(4.26)

The energy equation for frictionless, adiabatic flow is therefore very similar to the Bernoulli equation (equation 4.1), except that the energy equation takes account of the changes in density and internal energy. For constant density, frictionless flow, it turns out that the internal energy $\hat{u}$ will remain constant, and then the one-dimensional energy equation reduces to Bernoulli's equation, even though it was derived from entirely different principles.

As we pointed out earlier, the terms in Bernoulli's equation all have the dimensions of energy per unit mass, and it can be interpreted as expressing the conservation of mechanical energy. The energy equation, however, is more general. In particular, if there is friction, mechanical energy will not be conserved along streamlines and Bernoulli's equation cannot be used (see Section 9.8). Similarly, if there is work done on the fluid by a pump, for instance, we must use the energy equation with terms that represent the shaft work.

Note also that for a one-dimensional steady flow, with no losses, the Bernoulli equation can be written as

$$\frac{p}{\rho} + \tfrac{1}{2}V^2 + gz = B$$

where B is sometimes called the Bernoulli constant. By comparing this equation with equation 4.23, we see that when losses occur, the Bernoulli constant will decrease, and the amount it decreases may be identified with the amount of mechanical energy dissipated. Alternatively, adding heat to the fluid or passing it through a pump will increase the Bernoulli constant (both processes add energy to the flow).

EXAMPLE 4.9 *Power Required to Drive a Pump*

A pump delivers 20 l/s (liters per second) of water at 5°C, increasing the pressure from 1.5 *atm* to 4.0 *atm* (see Figure 4.19). The inlet diameter is 10 *cm*, and the outlet diameter is 2.5 *cm*. If there is no heat transfer to the fluid, and there is no work done by viscous forces, find the power required to drive the pump. The inlet and outlet are at the same height, and the change in internal energy can be ignored. The flow can be assumed to be one-dimensional.

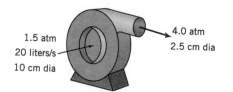

1.5 atm
20 liters/s
10 cm dia

4.0 atm
2.5 cm dia

FIGURE 4.19   Control volume for the pump.

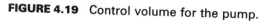

Solution: Equation 4.23 gives the one-dimensional energy equation as applied to the control volume shown in the figure. With $\dot{Q} = 0$, and $\dot{W}' = \dot{W}_{shaft}$,

$$\frac{p_1}{\rho_1} - \frac{p_2}{\rho_2} + \tfrac{1}{2}V_1^2 - \tfrac{1}{2}V_1^2 = \frac{\dot{W}_{shaft}}{\dot{m}}$$

To find $\dot{W}_{shaft}$, the rate at which the pump is delivering shaft work to the fluid, we need to find $\dot{m}$, V_1, and V_2. Now,

$$\dot{m} = \rho\dot{q} = 1000\ kg/m^3 \times 20\ l/s \div 10^3\ l/m^3 = 20\ kg/s$$

Also,

$$V_1 = \frac{\dot{q}}{A_1} = \frac{20\ l/s}{10^3\ l/m^3} \times \frac{1}{\frac{\pi}{4}(0.1)^2\ m^2} = 2.55\ m/s$$

and

$$V_2 = \frac{\dot{q}}{A_2} = \frac{20\ l/s}{10^3\ l/m^3} \times \frac{1}{\frac{\pi}{4}(0.025)^2\ m^2} = 40.7\ m/s$$

With $p_1 = 101,325\ Pa$ and $p_2 = 405,300\ Pa$, we obtain

$$\dot{W}_{shaft} = 20\ kg/s \left[\frac{(101,325 - 405,300)\ Pa}{1000\ kg/m^3} + \tfrac{1}{2}(40.7^2 - 2.55^2)\ m/s^2 \right]$$

$$= 10,420\ watt$$

This is equivalent to $\frac{10,420}{746}\ hp = 14.0\ hp$. ∎

EXAMPLE 4.10   *Enthalpy Change Produced by Pump*

In the previous example, what is the change in enthalpy?

Solution: We have

$$h_2 - h_1 = \left(\hat{u}_2 + \frac{p_2}{\rho_2} \right) - \left(\hat{u}_1 + \frac{p_1}{\rho_1} \right)$$

$$= \frac{p_2 - p_1}{\rho} = \frac{(405,300 - 101,325)\ Pa}{1000\ kg/m^3} = 304\ m^2/s^2$$

∎

PROBLEMS

4.1   Write down Bernoulli's equation. Under what conditions does this relationship hold?

4.2   Explain the terms "total pressure" and "dynamic pressure." What does a Pitot tube measure? What does a Pitot-static tube measure?

4.3   Draw a sketch illustrating a simple stagnation point in a uniform flow. What is the stagnation pressure? Why do you think it is sometimes called the "total" pressure?

4.4   For the flows shown in Figure P4.4, give reasons why Bernoulli's equation can or cannot be used between points:

(a) 1 and 2   (d) 7 and 8
(b) 3 and 4   (e) 8 and 9
(c) 5 and 6   (f) 9 and 10

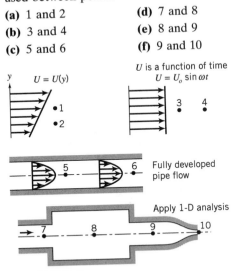

FIGURE P4.4

4.5   For the duct flow shown in Figure P4.5, the pressure at the exit (point 4) is atmospheric, the density ρ is constant, and the duct has a constant width w.

(a) Sketch the flow pattern.
(b) Can you use Bernoulli's equation between points 2 and 4? Why?
(c) Are the pressures at points 2 and 3 equal? Why?
(d) Find the gauge pressure at point 2 in terms of ρ and V_1.
(e) Find the gauge pressure at point 1 in terms of ρ and V_1.

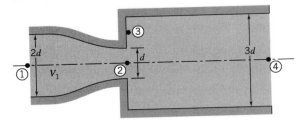

FIGURE P4.5

4.6   Fluid passes through a fan placed in a duct of constant area, as shown in Figure P4.6. The density is constant.

(a) Is the volume flow rate at station 1 equal to that at station 2? Why?
(b) Can Bernoulli's equation be applied between stations 1 and 2? Why?

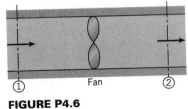

FIGURE P4.6

4.7   An airplane is moving at a speed of 250 *mph* at an elevation of 12,000 *ft*. By using Bernoulli's equation, find the pressure at the stagnation point, and at a point on the upper surface of the wing where the local velocity is 350 *mph*. Assume a Standard Atmosphere (Table A–C.6).

4.8   An airplane is moving through still air at 60 *m/s*. At some point on the wing, the air pressure is -1200 *N/m²* gauge. If the density of air is 0.8 *kg/m³*, find the velocity of the flow at this point. Carefully list the assumptions you have made in your analysis. Express your answer in terms of the nondimensional pressure coefficient C_p.

4.9   Water flows steadily at a rate of 0.6 *ft³/s* through a horizontal cone-shaped contraction. The diameter of the contraction decreases from 4.0 *in.* to 3.0 *in.* over a length of 1.2 *ft*. Assuming that conditions are uniform over any cross section, find the rate of change of pressure with distance in the direction of the flow at the section 0.6 *ft* from the end of the contraction.

4.10   Consider the steady, smooth flow of air through the circular duct shown in Figure P4.10. The air has constant density ρ, and the duct exits to atmospheric pressure.
(a) Find p_1, the gauge pressure at station 1, in terms of ρ and V_1.
(b) Find the direction of the force **F** acting to hold the duct in place. Assume one-dimensional flow.

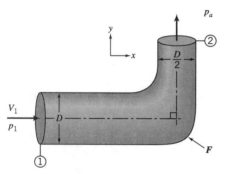

FIGURE P4.10

4.11   An incompressible, one-dimensional air flow of density ρ exits steadily into the ambient atmosphere from the contraction shown in Figure P4.11. The contraction is bolted onto a constant area duct at station 1, and the area ratio $A_1/A_2 = 4$. If the total force in the bolts is F_x, find $F_x/\rho U_1^2 A_1$. Show all your work, and state your assumptions clearly. Would your analysis hold if the flow direction was reversed?

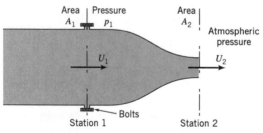

FIGURE P4.11

4.12   Neglecting friction, find the axial force produced at the flange when water discharges at 200 *gpm* into atmospheric pressure from the circular nozzle shown in Figure P4.12.

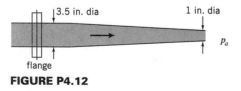

FIGURE P4.12

4.13 Air of constant density ρ flows steadily through the smooth circular pipe of radius R with a velocity V, as shown in Figure P4.13. Find the force **F** transmitted across the flange in terms of ρ, R, and V. Ignore gravity. For this problem, the nozzle is not attached to the exit.

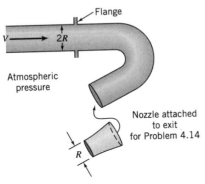

FIGURE P4.13

4.14 A nozzle with an exit radius of $R/2$ is attached to the exit of the pipe bend described in the previous problem. Find the new force **F'** transmitted across the flange in terms of ρ, R, and V. Neglect losses.

4.15 Air of constant density ρ flows steadily through the horizontal pipe system shown in Figure P4.15. The air flow exits to atmosphere through a $4:1$ contraction. Assume one-dimensional flow, neglect losses, and the weight of the piping.

(a) Find the gauge pressure measured by the gauge near station 1 in terms of ρ and U_1.

(b) Find the resultant force acting in the flange bolts at station 1.

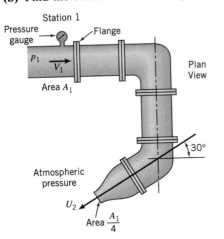

FIGURE P4.15

4.16 An incompressible fluid of density ρ flows smoothly and steadily from left to right through the nozzle shown in Figure P4.16(a). Determine the magnitude and the direction of the force

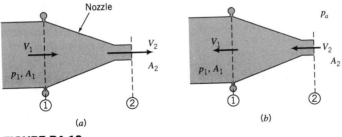

(a)   (b)

FIGURE P4.16

exerted by the fluid on the nozzle in terms of the density ρ, the inlet velocity V_1, and the inlet area A_1, given that the pressures and velocities at stations 1 and 2 are uniform across the areas A_1 and A_2, and that $A_1/A_2 = 4$. The pressure outside the nozzle is equal to atmospheric pressure, and the flow may be assumed to be one-dimensional.

4.17 Consider the previous problem in the case where the flow is from right to left [Figure P4.16(b)]. Is anything different? What additional information (beyond the density ρ, the inlet velocity V_1, and the inlet area A_1) would you require to determine the force exerted by the fluid on the nozzle with this new flow direction?

4.18 Water in an open cylindrical tank 10 *ft* in diameter and 6 *ft* deep drains into the atmosphere through a 2 *in.* diameter nozzle in the bottom of the tank. Neglecting friction and the unsteadiness of the flow, find the volume of water discharged in 20 *s*.

4.19 Water flows into a large circular tank at a rate of $\dot{q}$ *m/s*, as shown in Figure P4.19. Water also leaves through a smooth circular nozzle of diameter d near the base of the tank. Determine the height h for the flow to be independent of time.

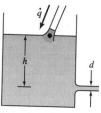

FIGURE P4.19

4.20 Given the siphon arrangement shown in Figure P4.20, find the exit velocity of the siphon assuming no losses. What limit exist on the maximum value of L if the siphon is to continue to function?

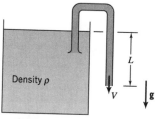

FIGURE P4.20

4.21 Water exits from a reservoir as shown in Figure P4.21. As H increases, the exit velocity increases until a critical elevation is reached and cavitation occurs. Find this value of H. Assume uniform flow and no losses, and that the vapor pressure is 0.25 *psia*.

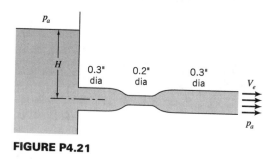

FIGURE P4.21

4.22   A tube of constant area A_t is used as a siphon, and it draws fluid smoothly from an infinitely large reservoir, as shown in Figure P4.22. The fluid exits with velocity V_e at an angle θ to the horizontal, at a distance H_1 below the surface of the reservoir.

(a) Explain what the practical restrictions are on the maximum height of the siphon tube, H_2.

(b) Express the ratio of the cross-sectional area of the jet to tube cross-sectional area as a function of H_1 and height y.

(c) Find H_3, the maximum height reached by the jet, as a function of H_1 and θ.

(d) Find the volume of water contained in the jet between stations A and B.

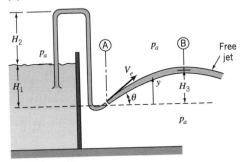

FIGURE P4.22

4.23   Water leaves a small hole in the vertical side of a bucket in a continuous jet.

(a) If the head of water above the hole is 1.25 m, what is the jet velocity at exit?

(b) If the jet strikes the ground at a point situated 2.21 m horizontally from the hole, and the ground is 1 m below the hole, recalculate the jet velocity at exit. Why might this result be different from that obtained in part **(a)**?

4.24   Water flows from a large open reservoir and discharges through a circular, horizontal pipe fitted with a nozzle into air at atmospheric pressure, as shown in Figure P4.24. Subsequently, it strikes the ground a distance x upstream of the nozzle. Neglecting losses, find:

(a) the velocity at the nozzle exit

(b) the velocity and pressure in the pipe near the nozzle

(c) the distance x.

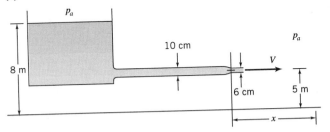

FIGURE P4.24

4.25   A jet issues smoothly from a hole in a tank, as shown in Figure P4.25. Express the maximum height of the jet as a function of the angle θ and the depth of the tank H (which is held

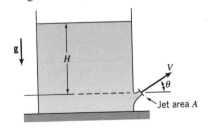

FIGURE P4.25

constant). What is the horizontal component of the force exerted on the tank by the jet? Ignore losses.

4.26   A jet of constant density fluid rises into the atmosphere, without loss, from the bottom of a large tank, as shown in Figure P4.26.

(a) Find the exit velocity V in terms of g and H. State all assumptions clearly.

(b) Find the horizontal component of the force the jet exerts on the tank in terms of ρ, V_1, θ, and the exit area A.

(c) As the jet rises, its vertical velocity decreases but its horizontal velocity remains constant. Find the maximum height to which the jet rises in terms of H and θ.

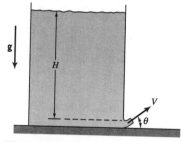

FIGURE P4.26

4.27   A jet of constant density fluid rises into the atmosphere, without loss, from a large tank at a constant angle θ, as shown in Figure P4.27. The exit is located very near the bottom, and the change in water depth H with time may be neglected. (a) What is the maximum height, relative to the tank bottom, to which the jet rises? (b) What is the vertical force F required to support the tank? Neglect the weight of the tank itself.

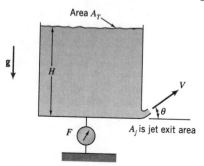

FIGURE P4.27

4.28   A large tank smoothly issues a jet of water from an orifice at a depth H below the surface, as shown in Figure P4.28. The orifice has an exit area A_2 and it points upwards at an angle of θ to the horizontal. Assuming that $A_1 \gg A_2$ and that there are no losses, (a) find the

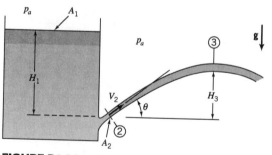

FIGURE P4.28

maximum height to which the jet rises ($= H_3$) in terms of H_1 and θ. (b) Find the volume of water in flight between the jet exit and the point of maximum jet height in terms of A_2, H_1, and θ.

4.29   Water of density ρ issues smoothly from a spigot of circular cross-section into the atmosphere, as shown in Figure P4.29. At the flange, the velocity is V. The diameter decreases from D at the flange to $D/4$ at the exit. Ignoring losses, (a) find the gauge pressure at the flange in terms of ρ and V, and (b) find the magnitude and direction of the force exerted by the water on the spigot in terms of ρ, D, and V (ignore the weight of water contained in the spigot).

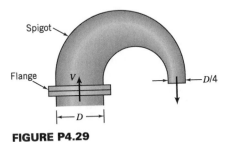

FIGURE P4.29

4.30   For incompressible fluid flowing in a duct, show that the decrease in total pressure ($=$ static pressure $+$ dynamic pressure) after a sudden enlargement is

$$\frac{1}{2}\rho U_1^2 \left(1 - \frac{U_2}{U_1}\right)^2$$

where U_1 is the velocity upstream, and U_2 the velocity well downstream of the sudden change in the cross-section. Note that at the point where the fluid enters the enlargement, the streamlines are nearly parallel. Explain why the velocity U_2 must be taken to be well downstream of the sudden enlargement and support your explanation with a sketch of the flow pattern. Give explanations for any assumptions you have made regarding the pressure distributions.

4.31   Consider the steady flow of a fluid of constant density in a duct of constant width w. The fluid flows smoothly up a bump, as shown in Figure P4.31, and then separates such that the streamlines are initially straight and parallel. Determine the magnitude and the direction of the horizontal component of the force exerted by the fluid on the bump in terms of the density, the inlet velocity V_1, the inlet height H_1, and the width w, given that the pressures and velocities at stations 1 and 2 are uniform across the heights H_1 and H_2, and that $H_1/H_2 = 2$.

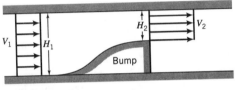

FIGURE P4.31

4.32   A fluid of density ρ_1 flows through a circular nozzle as shown in Figure P4.32. The pressure difference between sections 1 and 2 is measured using a manometer filled with a liquid of density ρ_2. Find the pressure difference $p_1 - p_2$ and the area ratio A_1/A_2 in terms of $z_1 - z_2$, D, ρ_1, ρ_2, and V_1.

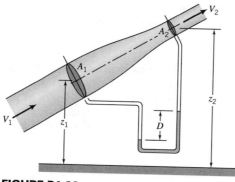

FIGURE P4.32

4.33 Air of constant density ρ_a flows steadily through a circular pipe of diameter D, which is downstream of a frictionless nozzle of diameter d as shown in Figure P4.33. Assume one-dimensional flow. If the manometer reads a deflection of h, find the velocity V at the nozzle exit in terms of h, D, d, ρ_a, and the density of the manometer fluid ρ_m.

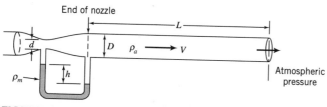

FIGURE P4.33

4.34 A fluid flows steadily from left to right through the duct shown in Figure P4.34. Another fluid of a different density enters from a second duct at right angles to the first. The two fluids mix together and at station 3 the resultant fluid has a uniform composition. This mixed fluid then exits to atmosphere through a smooth contraction without further change in density. At stations 1, 2, 3, and 4, the flow properties and parameters are constant over their respective areas. Find the pressure p_1 in terms of ρ_1, A_1, and U_1. Ignore the effect of gravity. Assume $U_2 = 2U_1$, $\rho_2 = 3\rho_1$, $\rho_3 = 2\rho_1$, $\rho_3 = \rho_4$, $A_2 = A_1/4$, $A_1 = A_3$, $A_4 = A_1/4$.

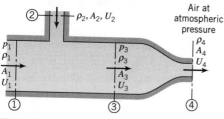

FIGURE P4.34

4.35 Two gases of density ρ_1 and ρ_3 are being mixed in the device shown in Figure P4.35. The device is two-dimensional, of height d and constant width. Gas of density ρ_1 and velocity U_1 enters from the left, and it is mixed with gas of density ρ_3 and velocity U_3 entering through two ducts of size $d/4$. The mixing is complete at station 2, so that both the velocity U_3 and density ρ_3 are uniform across the duct. The mixture then accelerates smoothly through a contraction to exit at atmospheric pressure with a velocity U_4 and an unchanged density ($\rho_4 = \rho_2$). You are given that $\rho_3 = 2\rho_1$, and $U_3 = 3U_1 = U_4$. Find the ratio ρ_2/ρ_1, and the pressure coefficient C_p, where

$$C_p = \frac{p_1 - p_2}{\frac{1}{2}\rho_1 U_1^2}$$

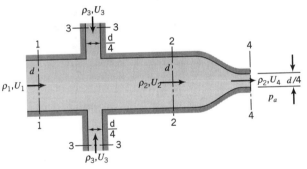

FIGURE P4.35

4.36 Water issues steadily from the smooth, circular funnel shown in Figure P4.36 under the action of gravity.

(a) Find the area ratio A_3/A_2 in terms of h_1 and h_2.

(b) By using the momentum equation, find the volume of fluid contained in the jet between stations 2 and 3 in terms of h_1, h_2, and A_2. Assume that $A_1 \gg A_2$.

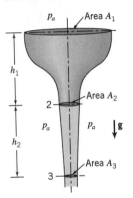

FIGURE P4.36

4.37 A circular hovercraft, of weight Mg, hovers a distance h above the ground, as shown in Figure P4.37. Far from the inlet the air is at atmospheric pressure and may be considered stationary. The density remains constant throughout. The expansion downstream of the fan occurs without loss, and the exit streamlines are parallel to the ground. Find h in terms of the inlet velocity V_i, the diameter of the fan d, the diameter of the exit plane D, the density ρ, and the weight Mg. Assume one-dimensional flow over the entry and exit areas.

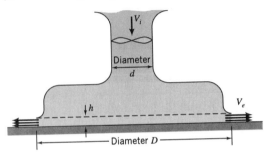

FIGURE P4.37

4.38 An axisymmetric body of cross-sectional area a moves steadily down a tube of cross-sectional area A, which is filled with a fluid of constant density ρ, as shown in Figure P4.38. The flow over the main part of the body is streamlined, but the flow separates over the rear part of the body such that the pressure over the section x-x immediately downstream of the base is uniform. Neglecting viscous shear forces at wall, show that the velocity of the body is given by

$$V = \frac{A-a}{a}\sqrt{\frac{2F}{\rho A}}$$

where F is the force necessary to keep the body moving at a constant speed.

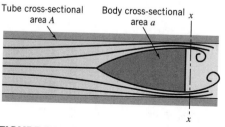

Tube cross-sectional area A   **Body cross-sectional area a**

FIGURE P4.38

4.39 Air enters a duct at a speed of 100 m/s and leaves it at 200 m/s. If no heat is added to and no work is done by the air, what is the change in temperature of the air as it passes through the duct?

4.40 Air enters a machine at 373°K with a speed of 200 m/s and leaves it at 293°C. If the flow is adiabatic, and the work output by the machine is 10^5 $N \cdot m/kg$, what is the exit air speed? What is the exit speed when the machine is delivering no work?

4.41 Two jets of air with the same mass flow rate mix thoroughly before entering a large closed tank. One jet is at 400°K with a speed of 100 m/s and the other is at 200°K with a speed of 300 m/s. If no heat is added to the air, and there is no work done, what is the temperature of the air in the tank?

4.42 A basketball is pumped up isothermally using air that is originally at 20°C and 1.05×10^5 Pa. As a result, the air is compressed to 20% of its original volume. If the mass of air is 0.1 kg, find:
(a) the final pressure
(b) the work required
(c) the changes in internal energy and enthalpy

CHAPTER 5

EQUATIONS OF MOTION IN INTEGRAL FORM

In Chapters 3 and 4, we derived the continuity, momentum, and energy equations for one-dimensional, steady flows. We now extend this work to include two- and three-dimensional, unsteady flows, for fixed control volumes. We first introduce the concept of flux, which simplifies the derivations considerably.

5.1 FLUX

When a fluid flows across the surface area of a control volume, it carries with it many properties. For example, if the fluid has a certain temperature, it carries this temperature with it across the surface into the control volume. If it has a particular density, it carries this density with it. The fluid can also carry momentum and energy. This "transport" of fluid properties by the flow across a surface is embodied in the concept of *flux*.

The flux of (something) is the amount of that (something) being transported across a surface per unit time.

Consider *volume flux*, that is, the volume of all fluid particles going through an area dA in time Δt [Figure 5.1(a)]. For a three-dimensional, time-dependent flow, the velocity and density vary in space and time. If we mark a number of fluid particles and visualize their motion over a short time Δt, we can identify the fluid particles that pass through dA during this time interval [Figure 5.1(b)].

If the area dA is small enough, the distributions of ρ and $\mathbf{V}$ can be approximated by their average values over the area. We can now find the volume which contains all the particles that will pass through dA in time Δt. From Figure 5.1(b), this volume is given by $(V \Delta t \cos \theta)\, dA = (\mathbf{n} \cdot \mathbf{V} \Delta t)\, dA$ where $\mathbf{n}$ is the unit normal which defines the orientation of the surface dA [Figure 5.1(c)]. That is,

$$\left\{ \begin{array}{c} \text{amount of volume per unit time} \\ \text{transported across an area} \\ dA \text{ with a direction } \mathbf{n} \end{array} \right\} = \left\{ \begin{array}{c} \text{total} \\ \text{volume} \\ \text{flux} \end{array} \right\} = (\mathbf{n} \cdot \mathbf{V})\, dA$$

where $\mathbf{n} \cdot \mathbf{V}\, dA$ is the volume flux (dimensions of $L^3 T^{-1}$ = volume flow rate), and $\mathbf{n} \cdot \mathbf{V}$ represents the volume flux per unit area (dimensions of LT^{-1}).

Once we have written the volume flux we can easily write down other fluxes, such as

$$\text{mass flux} = \mathbf{n} \cdot \rho \mathbf{V}\, dA$$
$$\text{momentum flux} = (\mathbf{n} \cdot \rho \mathbf{V})\mathbf{V}\, dA$$
$$\text{kinetic energy flux} = \tfrac{1}{2}(\mathbf{n} \cdot \rho \mathbf{V})V^2\, dA$$

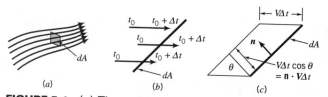

FIGURE 5.1 (a) Three-dimensional, unsteady flowfield with $\mathbf{V}(x, y, z, t)$ and $\rho(x, y, z, t)$. (b) Edge-on view of element of control surface of area dA. (c) Volume swept through dA in time Δt.

The dimensions of mass flux are MT^{-1} (= mass flow rate), for momentum flux they are MLT^{-2} (= force), and for kinetic energy flux they are ML^2T^{-3} (= force × velocity = power).

EXAMPLE 5.1 *Flux*

A uniform flow of air with a velocity of 10 *m/s* and density 1.2 *kg/m³* passes at an angle of 45° through an area of 0.1 *m²* (see Figure 5.2). Find the volume flux, the mass flux, the x-component of the momentum flux, and the kinetic energy flux passing through the area.

Solution: We choose a coordinate system where the normal to the area lies in the x-direction, so that $\mathbf{n} = \mathbf{i}$. The velocity vector in Cartesian coordinates is given by

$$\mathbf{V} = u\mathbf{i} + v\mathbf{j} + w\mathbf{k}$$

where u, v, and w are the velocity components in the x, y, and z directions, respectively. In this particular case,

$$\mathbf{V} = -\frac{10}{\sqrt{2}}\mathbf{i} + \frac{10}{\sqrt{2}}\mathbf{j} \ m/s = 7.07\,(-\mathbf{i} + \mathbf{j}) \ m/s$$

The flow is independent of time (steady), and two-dimensional (it depends on only two space coordinates).
Therefore

$$
\begin{aligned}
\text{volume flux} &= \int \mathbf{n} \cdot \mathbf{V} \, dA \\
&= \mathbf{i} \cdot (7.07 \ m/s)(-\mathbf{i} + \mathbf{j})\, A \\
&= -7.07 \ m/s \times 0.1 \ m^2 \\
&= -0.707 \ m^3/s
\end{aligned}
$$

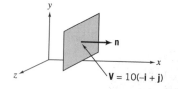

FIGURE 5.2 Notation for Example 5.1.

The volume flux is negative because it is an influx. Also,

$$\text{mass flux} = \int \mathbf{n} \cdot \rho \mathbf{V} \, dA$$
$$= \rho \times \text{volume flux}$$
$$= -1.2 \, kg/m^3 \times 7.07 \, m^3/s$$
$$= -8.50 \, kg/s$$

$$x\text{-momentum flux} = \int \mathbf{i} \cdot (\mathbf{n} \cdot \rho \mathbf{V}) \mathbf{V} \, dA$$
$$= \int (\mathbf{n} \cdot \rho \mathbf{V})(\mathbf{i} \cdot \mathbf{V}) \, dA$$
$$= \text{mass flux} \times \mathbf{i} \cdot \mathbf{V}$$
$$= 8.50 \, kg/s \times 7.07 \, m/s$$
$$= 60 \, kg \, m/s^2 = 60 \, N$$

The momentum flux is positive because it is an influx of negative momentum. Finally,

$$\text{kinetic energy flux} = \int \tfrac{1}{2}(\mathbf{n} \cdot \rho \mathbf{V})V^2 \, dA$$
$$= \tfrac{1}{2} \times \text{mass flux} \times V^2$$
$$= -\tfrac{1}{2} \times 8.50 \, kg/s \times 10 \, m^2/s^2$$
$$= -425 \, kg \cdot m^2/s^3 = -425 \, watt$$ ∎

EXAMPLE 5.2 *Flux of Mass and Momentum*

Water drains from a large tank through a square outlet measuring $b \times b$, as shown in Figure 5.3. The water level in the tank is kept constant by an external supply pipe so that the flow is steady. The velocity varies across the outlet area in a vertical direction but it is constant across its width (into the page). Find the flux of volume, mass, and momentum leaving the tank.

Solution: If there are no losses, we can apply Bernoulli's equation between a point on the surface of the water and a point in the exit plane at a distance $H + x$ below the surface of the water. Since the pressures at these points are equal to atmospheric pressure, and the velocity at the surface is zero, we find from Torricelli's formula

$$V(x) = \sqrt{2g(H + x)}$$

To find the total volume flux leaving the tank, we evaluate the integral

$$\text{volume flux} = \int \mathbf{n}_e \cdot \mathbf{V} \, dA_e$$

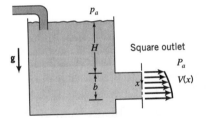

FIGURE 5.3 Tank draining through a square hole.

where the subscript e denotes the exit. In this case, $\mathbf{n}_e = \mathbf{i}$ and $\mathbf{V} = V\mathbf{i}$, so that

$$\text{volume flux} = \int \mathbf{i} \cdot V\mathbf{i} \, dA_e$$

$$= \int V \, dA_e$$

$$= \int_0^b \sqrt{2g(H + x)} \, b \, dx$$

$$\Rightarrow \text{volume flux} = \tfrac{2}{3}\sqrt{2g}[(H + b)^{3/2} - H^{3/2}]$$

$$\Rightarrow \text{mass flux} = \tfrac{2}{3}\rho\sqrt{2g}[(H + b)^{3/2} - H^{3/2}]$$

The momentum flux is given by

$$\text{momentum flux} = \int (\mathbf{n}_e \cdot \rho\mathbf{V})\mathbf{V} \, dA_e$$

$$= \int (\mathbf{i} \cdot \rho V\mathbf{i})V\mathbf{i} \, dA_e$$

$$= \int \rho V^2 \, dA_e$$

$$= \int_0^b 2\rho g(H + x) \, b \, dx$$

$$= 2\rho g(Hb + \tfrac{1}{2}b^2) \qquad \blacksquare$$

5.2 CONTINUITY EQUATION

Consider mass conservation for the fixed control volume CV shown in Figure 5.4. At any instant of time, a given mass of fluid occupies the space defined by CV. The total mass of this fluid (the "system" in the language of thermodynamics—see Section 4.7) remains constant by definition. To apply the conservation of mass to the fixed control volume, we need to consider the instantaneous mass flux through its surface, and the corresponding rate of change of mass contained inside. In Section 3.5, we found that when the flow is steady, the mass flow rates in and out of the control volume must be equal so that the mass inside the control volume remains constant. When the flow is unsteady, however, the outflux and influx of mass are not equal, and the total mass contained inside the control volume varies with time. That is

1. A fluid element of volume $d\forall$ has a mass $\rho \, d\forall$. Therefore the mass of fluid inside the control volume at any time is $\int \rho \, d\forall$. Hence,

$$\text{rate of change of mass in CV} = \frac{\partial}{\partial t}\int \rho \, d\forall \qquad (5.1)$$

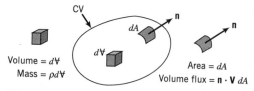

Volume = $d\forall$
Mass = $\rho d\forall$

Area = dA
Volume flux = $\mathbf{n} \cdot \mathbf{V} \, dA$

FIGURE 5.4 Fixed control volume for derivation of the integral form of the continuity equation.

This rate of change of mass will be negative if the mass inside the control volume is decreasing with time (that is, when the outflux exceeds the influx). The partial derivative with respect to time is used to emphasize that, since the volume is fixed in shape and location, the integral is only a function of time.

2. For a small element of surface area dA, the mass outflux through dA per unit time $= \rho \mathbf{V} \cdot \mathbf{n} \, dA$. Therefore

$$\text{total mass flux out of CV} = \int \mathbf{n} \cdot \rho \mathbf{V} \, dA \tag{5.2}$$

The integrand will be positive when the direction of the flow is in the same direction as the outward-facing unit normal vector $\mathbf{n}$ (an outflux), and negative when the direction of the flow is opposite to that of $\mathbf{n}$ (an influx). From equations 5.1 and 5.2, the conservation of mass requires that

$$\frac{\partial}{\partial t} \int \rho \, d\forall + \int \mathbf{n} \cdot \rho \mathbf{V} \, dA = 0 \tag{5.3}$$

This is the integral form of the continuity equation for a fixed control volume, in a three-dimensional, time-dependent flow. In words

$$\left\{ \begin{array}{c} \text{rate of increase of mass} \\ \text{inside control volume} \end{array} \right\} + \left\{ \begin{array}{c} \text{net mass flux} \\ \text{out of control volume} \end{array} \right\} = \{0\}$$

When the mass inside the control volume is fixed,

$$\int \mathbf{n} \cdot \rho \mathbf{V} \, dA = 0 \qquad \text{fixed mass} \tag{5.4}$$

When the flow is steady, the flow properties do not depend on time, and since the control volume is fixed,

$$\int \mathbf{n} \cdot \rho \mathbf{V} \, dA = 0 \qquad \text{steady flow} \tag{5.5}$$

Equations 5.4 and 5.5 are identical, although one applies when the mass inside the control volume is fixed, and the other when the flow is steady. These conditions have somewhat different implications, depending on the flow. When the flow is steady, for example, the inlet and outlet mass flow rates, and the mass inside the control volume, do not change with time. When the only restriction is that the mass inside the control volume is constant in time, however, it is possible for the inlet and outlet mass flow rates to be unsteady, as long as they are equal.

Finally, for steady or unsteady constant density flow,

$$\int \mathbf{n} \cdot \mathbf{V} \, dA = 0 \qquad \text{constant density flow} \tag{5.6}$$

EXAMPLE 5.3   *Steady, One-Dimensional Flow*

Flow in a diverging duct is illustrated in Figure 5.5. Apply the continuity equation for steady flow through the control volume.

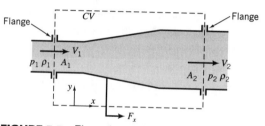

FIGURE 5.5 Flow through a diverging duct.

Solution: For the large control volume shown, there is inflow over area A_1 and outflow over A_2. If the flow is steady, then

$$\int \mathbf{n} \cdot \rho \mathbf{V}\, dA = \int_{A_1} \mathbf{n_1} \cdot \rho_1 \mathbf{V_1}\, dA_1 + \int_{A_2} \mathbf{n_2} \cdot \rho_2 \mathbf{V_2}\, dA_2 = 0$$

If the velocities V_1 and V_2 are normal to the inflow and outflow areas, then $\mathbf{n_1} \cdot \rho_1 \mathbf{V_1} = -\rho_1 V_1$, and $\mathbf{n_2} \cdot \rho_2 \mathbf{V_2} = \rho_2 V_2$. Hence,

$$-\int_{A_1} \rho_1 V_1\, dA + \int_{A_2} \rho_2 V_2\, dA = 0$$

If the densities and velocities are uniform over their respective areas, we have

$$-\rho_1 V_1 A_1 + \rho_2 V_2 A_2 = 0$$

and we recover the result for a one-dimensional steady flow first given in Section 3.5.

∎

EXAMPLE 5.4 *Steady, Two-Dimensional Flow*

In many cases the velocity is not uniform over the inlet and outlet areas and the one-dimensional assumption cannot be made. However, the streamlines of the flow entering and leaving the control volume are often parallel, and then it may be possible to assume that the pressure is uniform over the inlet and outlet areas. (Remember: when gravity is not important, or when hydrostatic pressure differences are negligibly small, the pressure is constant across parallel streamlines—Section 4.2.2.)

Consider the steady duct flow shown in Figure 5.6. The duct has a constant width W, and the inflow and outlet velocities vary in the x-direction, *and* in the y-direction. The flow is two-dimensional since the velocity distributions depend on two space variables (x and y). For this particular problem, the velocity distribution over the inlet is parabolic (this is an exact result for laminar flow in a long duct—see Section 9.5.1), and as the area expands, the profile becomes less "full" at the exit.

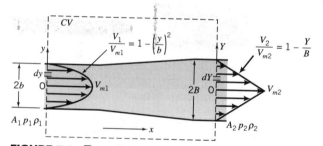

FIGURE 5.6 Two-dimensional duct showing control volume.

The flow is steady. The pressure outside the duct is atmospheric everywhere, and over the inlet and outlet areas the gauge pressures are p_1 and p_2, and the densities are ρ_1 and ρ_2, respectively. The pressures and densities are uniform over A_1 and A_2. Find the velocity ratio V_{m2}/V_{m1}.

Solution: Consider the conservation of mass for steady flow.

$$\int \mathbf{n} \cdot \rho \mathbf{V}\, dA = 0$$

The integral is over the entire surface of the control volume. Areas A_1 and A_2 are the only places where mass is entering or leaving the control volume, and so

$$\int \mathbf{n} \cdot \rho \mathbf{V}\, dA = 0 = \int \mathbf{n_1} \cdot \rho_1 \mathbf{V_1}\, dA_1 + \int \mathbf{n_2} \cdot \rho_2 \mathbf{V_2}\, dA_2$$

From Figure 5.5, we have $\mathbf{V_1} = V_1\mathbf{i}$, $\mathbf{V_2} = V_2\mathbf{i}$, $\mathbf{n_1} = -\mathbf{i}$, and $\mathbf{n_2} = \mathbf{i}$. Therefore,

$$\int \mathbf{n} \cdot \rho \mathbf{V}\, dA = 0 = \int -\mathbf{i} \cdot \rho_1 V_1 \mathbf{i}\, dA_1 + \int \mathbf{i} \cdot \rho_2 V_2 \mathbf{i}\, dA_2$$

$$= \int (-\rho_1 V_1)\, dA_1 + \int (+\rho_2 V_2)\, dA_2$$

so that

$$\int \rho_1 V_1\, dA_1 = \int \rho_2 V_2\, dA_2 \tag{5.7}$$

Two separate coordinate systems are used. For the inlet area the origin of the y-axis is located at the center of area A_1, which has a height $2b$, and for the outlet area the origin of the Y-axis is located at the center of area A_2, which has a height $2B$. For dA_1 we take a thin strip of height dy and width W, so that the mass flow rate entering over area A_1 is given by

$$\int \rho_1 V_1\, dA_1 = \int_{-b}^{+b} (\rho_1 V_1) W\, dy = 2 \int_0^{+b} (\rho_1 V_1) W\, dy$$

because the velocity distribution is symmetric. Similarly, the mass flow rate leaving over area A_2 is given by

$$\int \rho_2 V_2\, dA_2 = 2 \int_0^{+B} (\rho_2 V_2) W\, dY$$

where we have used symmetry to avoid having to write separate integrals for the limits $-B$ to 0, and 0 to $+B$. This would be necessary because the two halves of the triangular distribution are described by different equations

$$1 - \frac{Y}{B} \quad \text{and} \quad 1 + \frac{Y}{B}$$

for the top and bottom halves of A_2, respectively. Substituting these results in equation 5.7, and using the fact that the densities are uniform over the inlet and outlet areas,

$$\rho_1 \int_0^b V_{m1} \left(1 - \left(\frac{y}{b}\right)^2\right) dy = \rho_2 \int_0^B V_{m2} \left(1 - \frac{Y}{B}\right) dY$$

$$\rho_1 V_{m1} \left[y - \frac{y^3}{3b^2}\right]_0^b = \rho_2 V_{m2} \left[Y - \frac{Y^2}{2B}\right]_0^B$$

$$\rho_1 V_{m1} \frac{2b}{3} = \rho_2 V_{m2} \frac{B}{2}$$

Finally,

$$\frac{V_{m2}}{V_{m1}} = \frac{4b}{3B}\frac{\rho_1}{\rho_2}$$

∎

EXAMPLE 5.5 *Squeezing a Liquid Film*

In this unsteady flow problem, a fluid of density ρ held between two plates is undergoing a simple strain rate as the plates move toward each other (Figure 5.7).[1] The top plate is moving toward the bottom plate at a velocity $V_p(t)$, and as it moves, oil is squeezed out from between the plates. The plates are long and parallel, of width W. Find the velocity u as a function of distance along the gap, x, at any instant of time. The flow is one-dimensional.

Solution: We will use two different, fixed control volumes. First, we use a control volume of length dx, located a distance x from the centerplane (*CV*1 in Figure 5.7). To apply the continuity equation, we start by considering the unsteady term, that is,

$$\frac{\partial}{\partial t}\int \rho\, d\forall$$

where $\forall$ is the volume of $CV1$. The integral represents the mass contained in the control volume, which is equal to $\rho bW\, dx$. Hence,

$$\frac{\partial}{\partial t}\int \rho\, d\forall = \frac{\partial}{\partial t}(\rho bW\, dx) = \rho W\frac{\partial b}{\partial t}\, dx$$

Now, $\partial b/\partial t = db/dt = -V_p$, since b is only a function of time, and the negative sign accounts for the fact that the gap decreases with time. So,

$$\frac{\partial}{\partial t}\int \rho\, d\forall = -\rho WV_p\, dx \qquad (5.8)$$

Next, we consider the mass flux term

$$\int \mathbf{n}\cdot\rho\mathbf{V}\, dA$$

Here, $\mathbf{V} = u\mathbf{i} + v\mathbf{j} + w\mathbf{k}$, and A is the surface area of $CV1$. On the left face of the control volume, $\mathbf{n} = -\mathbf{i}$ and $\mathbf{V} = u\mathbf{i}$. On the right hand face, $\mathbf{n} = \mathbf{i}$ and $\mathbf{V} = (u + du)\mathbf{i}$, where du can be found using a Taylor series expansion

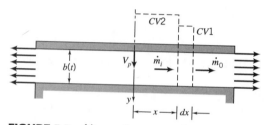

FIGURE 5.7 Alternative control volumes for an unsteady flow.

---

[1] This problem and the next one were adapted from examples given in the book *Engineering Fluid Mechanics* by Alan Mironer, published by McGraw-Hill, 1979.

$(= (\partial u/\partial x)dx)$. Hence,

$$\int \mathbf{n} \cdot \rho \mathbf{V} \, dA = -\rho b W u + \rho b W \left(u + \frac{\partial u}{\partial x} \, dx \right) = \rho b W \frac{\partial u}{\partial x} \, dx \tag{5.9}$$

Adding the unsteady term and the mass flux term in the continuity equation (equations 5.8 and 5.9), we obtain

$$V_p + b \frac{\partial u}{\partial x} = 0$$

The most general solution to this partial differential equation is

$$u = \frac{V_p}{b} x + f(t)$$

The unknown function $f(t)$ can be found using the boundary conditions on the centerplane where $x = 0$ and $u = 0$. Hence $f(t) = 0$, and

$$u = \frac{x}{b} V_p \tag{5.10}$$

Second, we use a control volume of length x, with its left face located at the centerplane where $x = 0$ ($CV2$ in Figure 5.7). For this control volume, we have

$$\frac{\partial}{\partial t} \int \rho \, d\forall = \rho W \frac{\partial b}{\partial t} x = -\rho V_p W_x$$

There is no mass flux over the left face of this control volume, and there is only an outflux over the right face, given by $\rho u b W$. The continuity equation gives

$$-\rho V_p W x + \rho u b W = 0$$

That is,

$$u = \frac{x}{b} V_p$$

as before (equation 5.10), but in a much more direct way. Choosing the control volume wisely always simplifies the problem. ■

EXAMPLE 5.6 *Moving Piston in a Cylinder*

A leakproof piston moves with velocity V into a cylinder filled with liquid of density ρ (Figure 5.8). The cylinder has a cross-sectional area A_c, and the spout has an exit cross-sectional area A_s. Find U, the velocity at the exit from the spout, at any instant of time. The flow is one-dimensional.

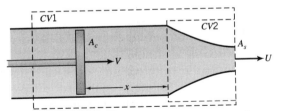

FIGURE 5.8 Alternative control volumes for an unsteady syringe flow.

Solution: We will again use two different control volumes. First, we use a control volume containing the piston, cylinder, and spout (CV1 in Figure 5.8). We begin with the unsteady term in the continuity equation. The mass contained in the control volume is equal to $(\rho A_c x + m_s)$, where m_s is the mass contained in the spout. Then,

$$\frac{\partial}{\partial t}\int \rho \, d\forall = \frac{\partial}{\partial t}(\rho A_c x + m_s) = \rho A_c \frac{\partial x}{\partial t} + \frac{\partial m_s}{\partial t}$$

The spout always contains the same amount of mass (even though there is flow through it), so that $\partial m_s/\partial t = 0$. Also, $\partial x/\partial t = dx/dt = -V$, since x is only a function of time, and the negative sign accounts for the fact that the length of the volume of fluid in the barrel of the piston decreases with time. So,

$$\frac{\partial}{\partial t}\int \rho \, d\forall = -\rho A_c V$$

The mass flux term is

$$\int \mathbf{n} \cdot \rho \mathbf{V} \, dA = \rho U A_s$$

since the only place where mass is entering or leaving the control volume is over the spout exit area. Adding the unsteady term and the mass flux term in the continuity equation, we obtain the result

$$U = \frac{A_c}{A_s} V \tag{5.11}$$

Second, we select a control volume not containing the piston or cylinder but only the spout (CV2 in Figure 5.8). Since the spout is always full of fluid and the mass of fluid inside this control volume is therefore not changing with time, the flow is steady for this choice of control volume, so that

$$\frac{\partial}{\partial t}\int \rho \, d\forall = 0$$

We see that there is a mass influx over the left face of the control volume, and there is an outflux over the right face, so that

$$\int \mathbf{n} \cdot \rho \mathbf{V} \, dA = \int -\mathbf{i} \cdot \rho V \mathbf{i} \, dA_c + \int \mathbf{i} \cdot \rho U \mathbf{i} \, dA_s = -\rho V A_c + \rho U A_s = 0$$

Hence,

$$U = \frac{A_c}{A_s} V$$

as before (equation 5.11). ■

5.3 MOMENTUM EQUATION

To find the integral form of the three-dimensional, time-dependent momentum equation, we will use a fixed control volume similar to that used in the derivation of the continuity equation (see Figure 5.9). The momentum of the particular mass of fluid occupying the control volume at any instant (the "system") will change

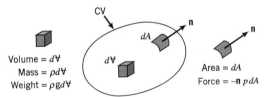

FIGURE 5.9 Fixed control volume for derivation of the integral form of the momentum equation.

under the action of a resultant force according to Newton's second law. To express the rate of change of momentum of this mass in terms of a fixed control volume, we need to consider the instantaneous momentum flux through its surface and the rate of change of momentum of the fluid contained inside.

We saw in Chapter 3 that the momentum of the fluid contained in a fixed control volume can change by a nonzero momentum flux through the surface of the control volume. If the momentum influx is smaller (say) than the momentum outflux, there is a net positive momentum outflux that will tend to decrease the momentum of the fluid in the control volume.

The momentum of the fluid contained in the control volume at any instant will also change if its density or velocity is changing with time.

These mechanisms are similar to the ones governing the conservation of mass, where a net outflux of mass leads to an unsteady variation in the amount of mass contained in the control volume. In other words, the sum of the rate of change of mass in the control volume and the net outflux of mass from the control volume must be zero, since mass must be conserved. For the momentum, however, the sum of the rate of change of momentum in the control volume and the net outflux of momentum is not necessarily zero: momentum is not conserved if there is a resultant force acting on the fluid. That is,

$$\left\{ \begin{array}{c} \text{rate of change of} \\ \text{momentum inside} \\ \text{control volume} \end{array} \right\} + \left\{ \begin{array}{c} \text{net outflux of} \\ \text{momentum from} \\ \text{control volume} \end{array} \right\} = \left\{ \begin{array}{c} \text{resultant force} \\ \text{acting on fluid} \\ \text{in control volume} \end{array} \right\}$$

The resultant force is equal to the sum of an unsteady term and a flux term. We will now consider each term in this momentum balance, starting with the unsteady term (the first one on the left).

5.3.1 Unsteady Term

For the rate of change of momentum inside the control volume, consider an element of volume $d\forall$. The mass of this volume is $\rho\, d\forall$, and its momentum is $\rho\mathbf{V}\, d\forall$. The total momentum contained in the control volume is found by integration, and its rate of change with time is found by differentiating with respect to time. That is

$$\left\{ \begin{array}{c} \text{rate of change of momentum} \\ \text{inside control volume} \end{array} \right\} = \frac{\partial}{\partial t} \int \rho\mathbf{V}\, d\forall \qquad (5.12)$$

This quantity is positive if the momentum inside the control volume is increasing with time. The partial derivative with respect to time is used to emphasize that, since the volume is fixed in shape and location, the integral (but not the integrand) depends only on time.

5.3.2 Flux Term

For an element of surface area dA, we have a volume flux $\mathbf{n} \cdot \mathbf{V} dA$ (see Section 5.1). The mass flux is therefore given by $\mathbf{n} \cdot \rho \mathbf{V}\, dA$, and the momentum flux is given by $(\mathbf{n} \cdot \rho \mathbf{V})\mathbf{V}\, dA$. The flux is positive if the velocity is directed along $\mathbf{n}$, that is, when it is an outflux. That is

$$\left\{ \begin{matrix} \text{net outflux of momentum} \\ \text{from control volume} \end{matrix} \right\} = \int (\mathbf{n} \cdot \rho \mathbf{V})\mathbf{V}\, dA \qquad (5.13)$$

5.3.3 Resultant Force

We have surface forces, body forces, and forces due to external surfaces. They act on the fluid mass that is coincident with the control volume at a particular instant of time.

1. Surface forces include viscous forces acting over the surface of the control volume, and forces due to pressure differences acting normal to the surface. For now, we will write the force due to viscous friction simply as $\mathbf{F_v}$. With respect to the force due to pressure differences, consider an element of surface area dA. The force due to pressure acts with a magnitude $p\, dA$. The direction of the force is normal to the surface, and by convention pressure forces are positive if they are compressive, so that the vector force due to pressure acting on dA is $-\mathbf{n}p\, dA$. That is

$$\left\{ \begin{matrix} \text{resultant force due to pressure} \\ \text{differences acting on the fluid} \\ \text{contained in the control volume} \end{matrix} \right\} = \int -\mathbf{n}p\, dA \qquad (5.14)$$

2. Body forces include gravitational, magnetic, and electrical forces acting over all the fluid contained in the control volume. The only body force considered here is the force due to gravity. An element of volume $d\forall$ has a mass $\rho\, d\forall$, and the vector force due to gravity acting on this mass is $\rho\mathbf{g}\, d\forall$. That is

$$\left\{ \begin{matrix} \text{resultant force due to gravity} \\ \text{acting on the fluid contained} \\ \text{in the control volume} \end{matrix} \right\} = \int \rho\mathbf{g}\, d\forall = \mathbf{g} \int \rho\, d\forall = m\mathbf{g} \qquad (5.15)$$

   where m is the total mass of fluid contained in the control volume.

3. The forces due to external surfaces, $\mathbf{F_{ext}}$, are the forces applied to the fluid by the walls of a duct, the surfaces of a deflector, or the forces acting in the solid cut by the control volume. An example of the latter is when a control volume is drawn to cut through the solid walls of a duct; we must include the forces exerted by the walls in the force balance on the fluid. (Remember: when a fluid exerts a force on a solid surface, an equal but opposite force acts on the fluid.)

By combining the terms given in equations 5.13 to 5.15, and including the viscous forces $\mathbf{F_v}$ and the forces due to external surfaces $\mathbf{F_{ext}}$, we obtain the integral form of the momentum equation for a fixed control volume.

$$\boxed{\mathbf{F_v} + \mathbf{F_{ext}} - \int \mathbf{n}p\, dA + \int \rho\mathbf{g}\, d\forall = \frac{\partial}{\partial t} \int \rho\mathbf{V}\, d\forall + \int (\mathbf{n} \cdot \rho\mathbf{V})\mathbf{V}\, dA} \qquad (5.16)$$

This is a vector equation, so that in Cartesian coordinates it has components in the x-, y- and z-directions.

EXAMPLE 5.7 *Steady, One-Dimensional Flow*

In Section 5.2, we considered the continuity equation applied to a simple diverging duct (Figure 5.5). We now find the x-component of the force exerted by the duct on the fluid. The flow is taken to be inviscid, steady, and horizontal.

Solution: Consider the x-component of the momentum equation. This equation is found by taking the dot product of equation 5.16 with the unit vector in the x-direction, **i.** That is

$$\mathbf{i} \cdot \mathbf{F}_{ext} - \mathbf{i} \cdot \int \mathbf{n}p \, dA + \mathbf{i} \cdot \int \rho \mathbf{g} \, d\forall = \mathbf{i} \cdot \int (\mathbf{n} \cdot \rho \mathbf{V}) \mathbf{V} \, dA$$

so that

$$F_x - \int \mathbf{i} \cdot \mathbf{n}p \, dA + 0 = \int (\mathbf{n} \cdot \rho \mathbf{V}) \, \mathbf{i} \cdot \mathbf{V} \, dA$$

where F_x is the x-component of the force exerted by the duct on the fluid, and it was taken to be positive in the x-direction (the actual direction will come out as part of the solution—if we find that F_x is negative, this means it actually points in the negative x-direction).

Remember: on the right hand side the dot product with the unit vector goes with the second **V**, not the first—the first **V** is contained in the mass flux term, and it already forms a dot product with the unit normal vector **n,** so that the mass flux term is a scalar.

We now evaluate the integrals over A_1 and A_2.

(a)
$$-\int \mathbf{i} \cdot \mathbf{n}p \, dA = -\int -p_1 \, dA_1 - \int + p_2 \, dA_2$$
$$= \int p_1 \, dA_1 - \int p_2 \, dA_2$$

(b)
$$\int (\mathbf{n} \cdot \rho \mathbf{V}) \mathbf{i} \cdot \mathbf{V} \, dA = \int (-\rho_1 V_1)(+V_1) \, dA_1 + \int (\rho_2 V_2)(+V_2) \, dA_2$$
$$= -\int \rho V_1^2 \, dA_1 + \int \rho V_2^2 \, dA_2$$

Therefore,

$$F_x = -\int p_1 \, dA_1 + \int p_2 \, dA_2 - \int \rho_1 V_1^2 \, dA_1 + \int \rho_2 V_2^2 \, dA_2$$

For a one-dimensional flow, this simplifies to

$$F_x = p_2 A_2 - p_1 A_1 + \rho_2 V_2^2 A_2 - \rho_1 V_1^2 A_1 \qquad \blacksquare$$

EXAMPLE 5.8 *Steady, Two-Dimensional Flow*

Consider the forces acting on the fluid in the steady, two-dimensional flow of width W shown in Figure 5.6.

Solution: If we ignore gravity and friction, the only forces acting on the fluid will be forces due to pressure differences, and the force exerted by the duct on the fluid. We begin with the x-component of the momentum equation (equation 5.16).

That is

$$F_x + p_1A_1 - p_2A_2 = \int (\mathbf{n} \cdot \rho \mathbf{V})\mathbf{i} \cdot \mathbf{V} \, dA$$
$$= -\int \rho V_1^2 \, dA_1 + \int \rho V_2^2 \, dA_2$$
$$= 2W \int_0^{+B} \rho V_2^2 \, dY - 2W \int_0^{+b} \rho V_1^2 \, dy$$

which can be used to find F_x after substituting for V_1 and V_2 in terms of their variations with y and Y, respectively, and integrating.

What about the y-direction? We see that the momentum only changes in the x-direction, and that the forces due to pressure differences only act in the x-direction. So the y-component of the force exerted by the duct on the fluid must be zero. ∎

EXAMPLE 5.9   *Lift and Drag on an Airfoil*

Consider an airfoil placed in a wind tunnel of constant cross-sectional area (height h and width W). The flow is steady and of constant density, and the airfoil develops a lift force and a drag force (see Figure 5.10). The lift force F_L is defined as the force on a body normal to the direction of the incoming flow, and the drag force F_D is defined as the force in the direction of the incoming flow. The incoming flow is uniform across the wind tunnel with a velocity magnitude V_1, but downstream the velocity varies across the tunnel in the y-direction. The velocity in the wake of the airfoil is less than V_1, and therefore, by conservation of mass, the velocity outside the wake must be greater than V_1. There is also a pressure difference, so that p_2 is less than p_1, but since the streamlines at stations 1 and 2 are parallel, the pressures are constant across the wind tunnel (we will ignore gravity). How can we find the lift and drag forces?

Solution: We start with the x-momentum balance for the control volume shown in Figure 5.10

$$-F_D - F_v - \int \mathbf{i} \cdot \mathbf{n} p \, dA + 0 = \int (\mathbf{n} \cdot \rho \mathbf{V})\,\mathbf{i} \cdot \mathbf{V} \, dA$$

where $-F_D$ is the force exerted by the airfoil on the fluid. We will neglect F_v, the viscous force exerted by the walls of the tunnel on the fluid. Hence,

$$-F_D + (p_1 - p_2)hW = -\rho V_1^2 hW + 2 \int_0^{h/2} \rho V_2^2 W \, dy$$

For the velocity distribution shown in the figure,

$$F_D = (p_1 - p_2)hW + \rho V_1^2 hW - \tfrac{1}{3}\rho V_{2m}^2 Wh$$

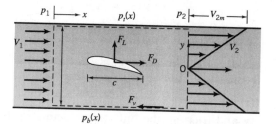

FIGURE 5.10   Lift and drag on an airfoil in a wind tunnel.

We can show that $V_{2m} = 2V_1$ by using the continuity equation. In addition, we can make the drag force nondimensional by dividing through by $(\frac{1}{2}\rho V_1^2 Wh)$ (we recognize this as the upstream dynamic pressure multiplied by the cross-sectional area of the tunnel). Therefore

$$\frac{F_D}{\frac{1}{2}\rho V_1^2 Wh} = \frac{(p_1 - p_2)}{\frac{1}{2}\rho V_1^2} - \frac{2}{3}$$

Non-dimensionalizing has cleaned up the final expression, and it has also revealed the presence of a pressure coefficient on the right hand side, similar to that introduced in Section 4.3, as well as a new nondimensional parameter called a *drag coefficient, C_D',* on the left hand side, defined by

$$C_D' = \frac{F_D}{\frac{1}{2}\rho V_1^2 Wh}$$

In the usual form of the drag coefficient, the plan area of the wing is used instead of the cross-sectional area of the tunnel. That is,

$$C_D \equiv \frac{F_D}{\frac{1}{2}\rho V_1^2 Wc} \tag{5.17}$$

where c is the chord length of the airfoil (the distance between its leading and trailing edges).

For the y-momentum balance

$$-F_L - \int \mathbf{j} \cdot \mathbf{n} p \, dA + 0 = \int (\mathbf{n} \cdot \rho \mathbf{V}) \mathbf{j} \cdot \mathbf{V} \, dA$$

where $-F_L$ is the force exerted by the airfoil on the fluid. Since there is no flow in the y-direction

$$-F_L + \int_b p_b W \, dx - \int_t p_t W \, dx = 0$$

so that

$$F_L = W \left(\int_b p_b \, dx - \int_t p_t \, dx \right)$$

where p_b and p_t are the pressure distributions over the bottom and top faces of the control volume. Therefore, the lift force can be found by measuring the pressure distributions on the upper and lower tunnel walls. When we divide through by $(\frac{1}{2}\rho V_1^2 Wh)$, we obtain

$$\frac{F_L}{\frac{1}{2}\rho V_1^2 Wh} = \frac{1}{\frac{1}{2}\rho V_1^2 h} \left(\int_b p_b \, dx - \int_t p_t \, dx \right)$$

We now have a nondimensional *lift coefficient C_L'* on the left hand side. In its more usual form, it is defined using the plan area of the wing so that

$$C_L \equiv \frac{F_L}{\frac{1}{2}\rho V_1^2 Wc} \tag{5.18}$$

∎

EXAMPLE 5.10 *Unsteady Flow and Moving Control Volumes*

A steady jet of water having a velocity V_j hits a deflector, which is moving to the right at a constant velocity V_d (see Figure 5.11). The deflector turns the flow through

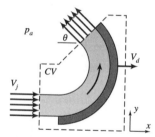

FIGURE 5.11   Moving deflector: unsteady flow.

an angle $\pi - \theta$. Find **F**, the force vector exerted by the fluid on the deflector. We will assume that the effects of gravity and friction can be neglected.

Solution: The first thing to note is that the problem is unsteady for a stationary observer. This complicates the analysis considerably since we would need to use the unsteady form of the continuity and momentum equations, and Bernoulli's equation cannot be used. If the observer moves with the deflector, however, the problem becomes steady and these complications are avoided. A velocity transformation involving the superposition of a constant translating velocity (but not an angular velocity) has no effect on the forces acting on a system. That is, the forces are the same whether a motion is viewed in a stationary coordinate system or one that is moving with a constant velocity. The control volume for this steady flow is shown in Figure 5.12.

Along the surface of the jet, we see that the pressure is constant and equal to atmospheric pressure. The pressure is also constant across the jet over the inlet and outlet areas, since the streamlines are parallel and the jet is in free fall. Since the flow in this framework is steady and there is no friction, Bernoulli's equation applied along any streamline that starts at the inlet and finishes at the outlet indicates that the magnitude of the inlet and outlet velocities are equal. The deflector changes the velocity direction, but not its magnitude. Then, from the continuity equation, we know that the cross-sectional area of the jet, A, must remain constant.

The momentum equation gives

$$-\mathbf{F} - \int p\mathbf{n}\, dA = \int (\mathbf{n} \cdot \rho\mathbf{V})\mathbf{V}\, dA$$

where $+\mathbf{F}$ is the force exerted by the fluid on the deflector, and $-\mathbf{F}$ is the force exerted on the fluid by the deflector. The pressure is constant everywhere, so

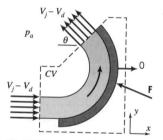

FIGURE 5.12   Moving deflector: steady flow.

that $\int p\mathbf{n}\,dA = 0$ for the control volume shown. Over the inlet, the velocity is $(V_j - V_d)\,\mathbf{i}$, and $\mathbf{n} = -\mathbf{i}$. Over the outlet the velocity magnitude is the same, but its direction is different, so that the outlet velocity is $(V_j - V_d)\,(-\cos\theta\mathbf{i} + \sin\theta\mathbf{j})$, and $\mathbf{n} = -\cos\theta\mathbf{i} + \sin\theta\mathbf{j}$. Hence,

$$-\mathbf{F} = -\rho(V_j - V_d)^2\,\mathbf{i}A + \rho(V_j - V_d)^2\,(-\cos\theta\mathbf{i} + \sin\theta\mathbf{j})A$$
$$= -\rho A(V_j - V_d)^2((1 + \cos\theta)\mathbf{i} - \sin\theta\mathbf{j})$$

and so

$$\mathbf{F} = \rho A(V_j - V_d)^2((1 + \cos\theta)\mathbf{i} - \sin\theta\mathbf{j}$$

We can check this answer by taking the limit when $V_j = V_d$. At this point, $\mathbf{F} = 0$, as expected. ∎

5.4 REYNOLDS' TRANSPORT THEOREM

It is now possible to express the conservation concepts embodied in the integral forms of the continuity and the momentum equations in a more general way. First, we note that the continuity and momentum equations for a fixed control volume (equations 5.3 and 5.16) are given by

$$\frac{\partial}{\partial t}\int \rho\,d\forall + \int \mathbf{n}\cdot\rho\mathbf{V}\,dA = 0 \tag{5.19}$$

and

$$\frac{\partial}{\partial t}\int \rho\mathbf{V}\,d\forall + \int (\mathbf{n}\cdot\rho\mathbf{V})\mathbf{V}\,dA = \mathbf{F}_{ext} + \mathbf{F}_v - \int \mathbf{n}p\,dA + \int \rho\mathbf{g}\,d\forall \tag{5.20}$$

The left hand sides of these equations resemble each other closely. For the continuity equation, the terms on the left hand side represent the total rate of change of mass of a fixed mass of fluid due to the rate of change of mass contained in the control volume plus the outflux of mass through the surface of the control volume. Similarly, for the momentum equation, the terms on the left hand side represent the total rate of change in the momentum of a fixed mass of fluid due to the rate of change of momentum of the fluid contained in the control volume plus the outflux of momentum through the surface of the control volume.

The left hand sides of equations 5.19 and 5.20, therefore, describe the total rate of change in a property of the fluid (mass, momentum) contained in a given volume, accounting for the flow in and out of the control volume. In other words, they provide a link between the total rate of change of a property of a fixed *mass* of fluid and the rate of change of that property in a fixed *volume* of fluid. For instance, for a fixed mass of fluid, the rate of change of the property called mass is obviously zero. For a fixed volume of fluid, the same physical observation is expressed as a balance between the rate of change of mass in the control volume and the mass transported in and out over the surface of the control volume.

In more general terms, we can define an *extensive* property of the fluid B, which could be mass, momentum, energy, and so on. An *intensive* property b can also be defined, which is simply the property B per unit mass, so that

$$B = mb$$

where m is the mass of fluid being considered. The value of B is directly proportional to the amount of mass being considered, whereas the value of b is independent of the amount of mass. It follows that

$$B_{sys} = \int_{CV} \rho b \, d\mathsf{V}$$

where the subscript *sys* denotes a *system*, that is, a fixed mass of fluid.

The question is, therefore, what is the rate of change of B_{sys} (following a fixed mass of fluid) in terms of the rate of change of B_{CV}, which is the amount of B contained in the control volume at any time. The Reynolds transport theorem for a fixed control volume states that

$$\frac{DB_{sys}}{dt} = \frac{\partial B_{CV}}{\partial t} + \int (\mathbf{n} \cdot \rho \mathbf{V}) b \, dA$$

and therefore

$$\frac{DB_{sys}}{dt} = \frac{\partial}{\partial t} \int \rho b \, d\mathsf{V} + \int (\mathbf{n} \cdot \rho \mathbf{V}) b \, dA \tag{5.21}$$

The rate of change of B (following a mass of fluid) is equal to the rate of change of B contained in the control volume plus the net outflux of B through the surface of the control volume. Equation 5.21 provides a link between control volume and system concepts. It serves the same function for a large control volume as the total derivative serves for a small control volume (see Section 6.1). We have not provided a formal derivation here,[2] but the link with the continuity and momentum equations is clear. For example, when the property B is mass,

$$B = m, \quad \text{and} \quad b = \frac{dB}{dm} = 1$$

and we obtain the continuity equation, equation 5.19. When the property B is linear momentum,

$$\mathbf{B} = m\mathbf{V}, \quad \text{and} \quad b = \frac{d\mathbf{B}}{dm} = \mathbf{V}$$

and we obtain the left hand side of the momentum equation, equation 5.20.

The power of the Reynolds' transport theorem lies in the fact that the property B can be anything that is *transported* by the fluid: mass, linear momentum, angular momentum, kinetic energy, and so forth.

5.5 *ENERGY EQUATION

We can now use the Reynolds transport theorem to derive the three-dimensional, unsteady form of the energy equation for a fixed control volume. In this case, the extensive property is total energy E. The first law of thermodynamics states that

[2] Formal derivations can be found in White *Fluid Mechanics*, McGraw-Hill 1986; Shapiro *Elements of Gasdynamics*, Wiley 1962; Munson, Young & Okiishi *Fundamentals of Fluid Mechanics*, Wiley 1998; and Potter & Wiggert *Mechanics of Fluids*, Prentice Hall 1997.

for a fixed mass of fluid

$$\left\{\begin{array}{c} \text{rate of change of} \\ \text{total energy of} \\ \text{a system} \end{array}\right\} = \left\{\begin{array}{c} \text{net rate of energy} \\ \text{addition by heat} \\ \text{transfer to fluid} \end{array}\right\} + \left\{\begin{array}{c} \text{net rate of energy} \\ \text{addition by work} \\ \text{done on fluid} \end{array}\right\}$$

That is,

$$\frac{DE_{sys}}{Dt} = \dot{Q}_{sys} + \dot{W}_{sys}$$

For a fixed control volume, with $B = E = me$ and $b = dB/dm = e$, the Reynolds transport theorem gives

$$\dot{Q} + \dot{W} = \frac{\partial}{\partial t} \int \rho e \, d\forall + \int (\mathbf{n} \cdot \rho \mathbf{V}) e \, dA \tag{5.22}$$

where $\dot{Q}$ and $\dot{W}$ are the rates of heat and work energies transferred to the fluid contained in the control volume at time t.

As in Section 4.7.2, the work done on the fluid can be split three ways

$$\dot{W} = \dot{W}_{pressure} + \dot{W}_{viscous} + \dot{W}_{shaft}$$

where $\dot{W}_{pressure}$ is the work done on the control volume surface by forces due to pressure, $\dot{W}_{viscous}$ is the shear work done on the control volume surface by viscous stresses, and $\dot{W}_{shaft}$ is the shaft work done by a machine on the fluid inside the control volume (a pump, fan, piston, and so forth), all per unit time.

For a surface element dA, the rate of doing work by pressure forces is given by the force due to pressure times the velocity component normal to the surface into the control volume. That is

$$d\dot{W}_{pressure} = -p(\mathbf{n} \cdot \mathbf{V}) \, dA$$

and the total pressure work is given by

$$\dot{W}_{pressure} = -\int p(\mathbf{n} \cdot \mathbf{V}) \, dA$$

Similarly, the rate of doing work by shear forces is given by the force due to shear times the velocity component normal to the surface into the control volume. That is

$$d\dot{W}_{viscous} = -(\tau \cdot \mathbf{V}) \, dA$$

and the total shear work is given by

$$\dot{W}_{viscous} = -\int (\tau \cdot \mathbf{V}) \, dA$$

By a suitable choice of control volume, the shear work can often be made zero. For example, at a solid surface, the no-slip condition makes the velocity at the wall zero, so that the rate of work by viscous forces is also zero. At an inlet or outlet, the flow is usually aligned with the unit normal vector $\mathbf{n}$, and the shear work is again zero. The shear work is rarely important for large control volumes, and we will neglect it from now on. The work term then reduces to

$$\dot{W} = \dot{W}_{shaft} - \int p(\mathbf{n} \cdot \mathbf{V}) \, dA$$

Hence, the integral form of the energy equation for a fixed control volume is given by

$$\dot{Q} + \dot{W}_{shaft} = \frac{\partial}{\partial t} \int \rho(\hat{u} + \tfrac{1}{2}V^2 + gz)\, d\forall$$

$$+ \int (\mathbf{n} \cdot \rho\mathbf{V})\left(\hat{u} + \frac{p}{\rho} + \tfrac{1}{2}V^2 + gz\right) dA \tag{5.23}$$

where the total specific energy $e = \hat{u} + \tfrac{1}{2}V^2 + gz$. The pressure work term has been combined with the energy flux term. This is common practice.

EXAMPLE 5.11 *Energy Equation Applied to Duct Flow*

An incompressible fluid flows steadily through a duct, as shown in Figure 5.13. The duct has a width W. The velocity at entrance to the duct is uniform and equal to V_0, while the velocity at the exit varies linearly to its maximum value, which is equal to V_0. Find the net enthalpy transport out of the duct in terms of ρ, V_0, $\dot{Q}$, W, and δ_1, given that the heat transfer rate is $\dot{Q}$.

Solution: Since the flow is steady and there is no shaft work (we will ignore the work done by viscous forces), the energy equation (equation 5.23) becomes

$$\dot{Q} = \int (\mathbf{n} \cdot \rho\mathbf{V})\left(\hat{u} + \frac{p}{\rho} + \tfrac{1}{2}V^2 + gz\right) dA$$

Ignoring potential energy changes, we find that

$$\int (\mathbf{n} \cdot \rho\mathbf{V})h\, dA = -\int (\mathbf{n} \cdot \rho\mathbf{V})V^2\, dA + \dot{Q}$$

where the enthalpy $h = \hat{u} + p/\rho$. The left hand side represents the net enthalpy transport out of the duct, so we need to evaluate the right hand side, which represents the net transport of kinetic energy out of the duct plus the net rate of heat transfer in. For the control volume shown, over the inflow area

$$\int (\mathbf{n} \cdot \rho\mathbf{V})\tfrac{1}{2}V^2\, dA = (-\rho V_0)V_0^2\, W\delta_1$$

$$= -\tfrac{1}{2}\rho V_0^3\, W\delta_1$$

Over the outflow area

$$\int \left[\mathbf{n} \cdot \rho\mathbf{V})\tfrac{1}{2}V^2\, dA = \int_0^{h_2}\left(-\rho V_0\left(\frac{y}{\delta_2}\right)\right)\tfrac{1}{2}V_0^2\left(\frac{y}{\delta_1}\right)^2 W\, dy\right.$$

$$= \tfrac{1}{2}\rho V_0^3\, W \int_0^{h_2}\left(\frac{y}{\delta_2}\right)^3 dy$$

$$= \tfrac{1}{8}\rho V_0^3\, W\delta_2$$

We can relate δ_1 to δ_2 by using the continuity equation, where, for steady flow

$$\int \mathbf{n} \cdot \rho\mathbf{V}\, dA = 0$$

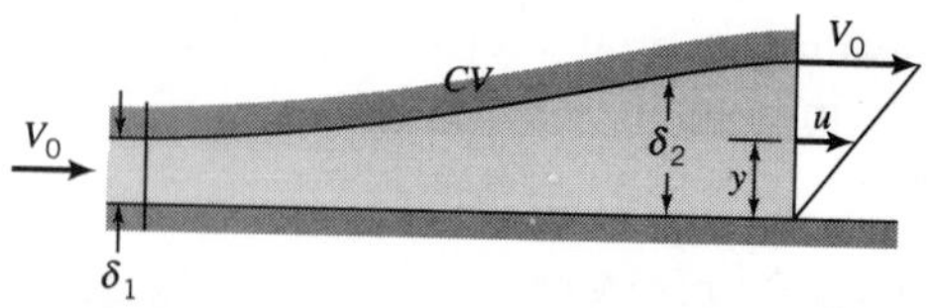

FIGURE 5.13 Control volume for Example 5.11.

(equation 5.5). That is,

$$-\rho V_0 W \delta_1 + \int_0^{h_2} \rho V_0 \left(\frac{y}{\delta_2}\right) W \, dy = 0$$

so that

$$\delta_1 = \tfrac{1}{2} \delta_2$$

Finally,

$$\text{enthalpy flux} = \int (\mathbf{n} \cdot \rho \mathbf{V})h \, dA = -\tfrac{1}{2}\rho V_0^3 W \delta_1 + \tfrac{1}{8}\rho V_0^3 W \delta_2 + \dot{Q}$$
$$= -\tfrac{1}{2}\rho V_0^3 W \delta_1 + \tfrac{1}{4}\rho V_0^3 W \delta_1 + \dot{Q}$$
$$= -\tfrac{1}{4}\rho V_0^3 W \delta_1 + \dot{Q}$$

When the enthalpy flux is negative, it represents an influx. ∎

PROBLEMS

5.1 For unsteady flow, write down the integral form of the continuity equation. Explain each term in words.

5.2 For unsteady inviscid flow, write down the integral form of the momentum equation. Explain each term in words.

5.3 The velocity distribution of a two-dimensional flow between parallel plates a distance $2h$ apart is given by $V = V_m(1 - y^2/h^2)$, where V_m is the maximum velocity, as shown in Figure P5.3. Find the average velocity and the mass flow rate.

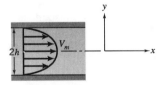

FIGURE P5.3

5.4 Two rectangular air-conditioning ducts, of constant width W (into the page), meet at right angles as shown in Figure P5.4. The flow is steady and the density is constant. All velocities are normal to the exit and entrance areas. The velocity profiles at stations 1 and 2 are parabolic. Find the mass flow rate at station 3. Is it in or out?

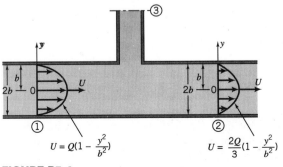

FIGURE P5.4

5.5 In the rectangular duct shown in Figure P5.5, two parallel streams of a constant density gas enter on the left with constant velocities U_1 and U_2. After mixing, the gas exits on the right with a parabolic profile and a maximum value of U_3. Find U_3 in terms of U_1 and U_2.

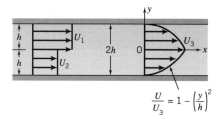

$$\frac{U}{U_3} = 1 - \left(\frac{y}{h}\right)^2$$

FIGURE P5.5

5.6 Figure P5.6 shows a sketch of a simple carburetor. The air enters from the left with a uniform velocity profile, and flows through a contraction where fuel-rich air of the same density enters at a rate of $\dot{q}\, m^3/s$. The mixture then exits on the right with a triangular velocity profile as shown. The flow is steady, and the cross-sectional areas at stations 1 and 2 are rectangular of width W. Assume that all velocities are normal to their respective areas. Find $\dot{q}$ when $U_2 = 2U_1$.

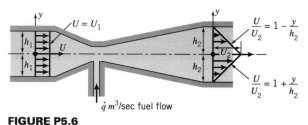

$\dot{q}\, m^3/sec$ fuel flow

FIGURE P5.6

5.7 A two-dimensional duct of constant width carries a steady flow of constant density air as shown in Figure P5.7. The incoming velocity profile is of constant magnitude U_0 and the outgoing velocity profile is parabolic according to $U = U_m(1 - (2y/h)^2)$. (a) Find U_m as a function of U_0, b, and h. (b) Will U_m increase or decrease if a heater is inserted into the duct as shown?

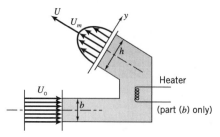

FIGURE P5.7

5.8 A cylinder of mass M per unit length falls down a channel of width D, as shown in Figure P5.8. The cylinder has reached its terminal velocity V_t, and the flow is steady when viewed by an observer traveling with the cylinder. That same observer sees a sinusoidally shaped velocity distribution in the wake, with an amplitude of U_0, as shown. The pressure everywhere is uniform. Find U_0 in terms of V_t, and express M in terms of V_t, g, D, and the fluid density ρ.

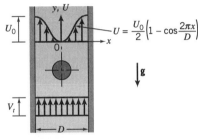

FIGURE P5.8

5.9 A constant density fluid flows steadily through a duct of width W, as shown in Figure P5.9. Find U_m in terms of U_1 and U_2. What is the momentum flux passing through the exit of the duct?

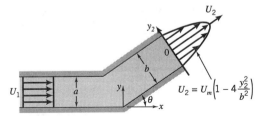

FIGURE P5.9

5.10 For the rectangular duct of width W shown in Figure P5.10, the flow is steady and the density is constant. Find U_m in terms of U_1, a, and b, and find the y-component of the momentum flux leaving the duct.

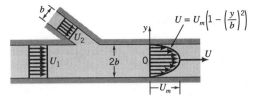

FIGURE P5.10

5.11 For the constant density flow in the rectangular duct of width W shown in Figure P5.11, find U_m in terms of a, b, and U_1, and the (vector) momentum flux leaving the duct.

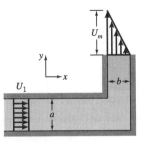

FIGURE P5.11

5.12 A fluid enters the rectangular duct shown in Figure P5.12 with constant velocity U_1 and density ρ_1. At the exit plane, the fluid velocity is U_2 and the density has a profile described by a square root relationship, with a maximum value of ρ_2. Find ρ_2 in terms of ρ_1 when $U_1 = U_2$. Also find the vector momentum flux leaving the duct.

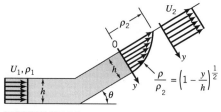

FIGURE P5.12

5.13 A fluid flows through a two-dimensional duct of width W, as shown in Figure P5.13. At the entrance to the box (face 1), the flow is one-dimensional and the fluid has density ρ_1 and velocity U_1. At the exit (face 2), the fluid has a uniform density ρ_2 but the velocity varies across the duct as shown. Find the rate at which the average density inside the box is changing with time.

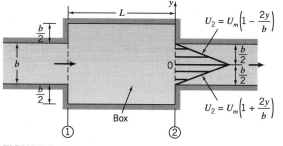

FIGURE P5.13

5.14 The wake of a body is approximated by a linear profile as sketched in Figure P5.14. The flow is incompressible, steady, and two-dimensional. Outside the wake region the flow is inviscid and the velocity is U_2. The upstream velocity, U_1, is uniform.

(a) Find U_2/U_1 as a function of δ/H.

(b) Find the pressure coefficient $(p_2 - p_1)/(\frac{1}{2}\rho U_1^2)$ as a function of δ/H.

(c) Find the nondimensional drag (per unit body span) coefficient $D/(\frac{1}{2}\rho U_1^2 H)$ as a function of δ/H.

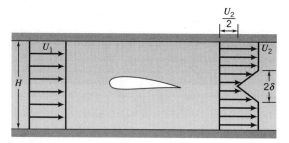

FIGURE P5.14

5.15 Air of constant density flows steadily through the rectangular duct shown in Figure P5.15. The duct has a constant width W into the page. At the entrance, the velocity is constant across the area and equal to U_{av}. The velocity at the exit has a parabolic distribution across the duct, with a maximum value U_m, and it is constant in the direction into the page. The pressure is constant everywhere. By using the continuity equation, find the ratio b/a such that $U_m = U_{av}$. Then find the force **F** exerted by the flow on the duct, assuming that the wall friction is negligible.

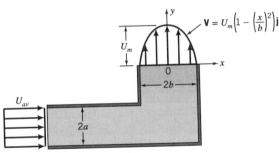

FIGURE P5.15

5.16 A cylinder is held in a two-dimensional duct of constant width W and height D. As a result, the downstream velocity distribution becomes as shown in Figure P5.16. The density ρ is constant, and the flow is steady. Find U_2 in terms of U_1, and find the force exerted by the fluid on the cylinder in terms of ρ, U_1, W, and D. Neglect frictional effects and assume that $p_1 = p_2$.

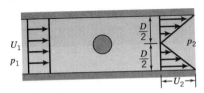

FIGURE P5.16

5.17 A propeller is placed in a constant area circular duct of diameter D, as shown in Figure P5.17. The flow is steady and the fluid has a constant density ρ. The pressure p_1 and p_2 are uniform across the entry and exit areas, and the velocity profiles are as shown. Find the thrust T produced by the propeller on the fluid in terms of U_1, U_m, ρ and D, and p_1 and p_2.

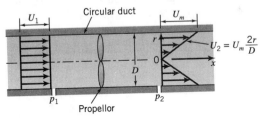

FIGURE P5.17

5.18 Air of constant density enters a rectangular duct of width W and height D_1 with a uniform velocity V_1. The top wall diverges (as shown in Figure P5.18) so that the pressure remains uniform everywhere. Downstream, on the top and bottom wall, the velocity profiles are

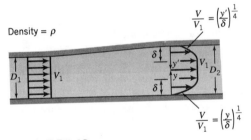

FIGURE P5.18

given by $V/V_1 = (y/\delta)^{1/4}$. Given that $D_2 = 1.1D_1$, find δ in terms of D_1, and find the magnitude and direction of the force F exerted by the air on the duct in terms of a nondimensional force coefficient $F/\rho V_1^2 D_1 W$ (ignore viscous stresses).

5.19   A fluid of constant density ρ enters a duct of width W and height h_1, with a parabolic velocity profile with a maximum value of V_1, as shown in Figure P5.19. At the exit plane, the duct has height h_2 and the flow has a parabolic velocity profile with a maximum value of V_2. The pressures at the entry and exit stations are p_1 and p_2, respectively, and they are uniform across the duct.

(a) Find V_2 in terms of V_1, h_1, and h_2.

(b) Find the magnitude and direction of the horizontal force F exerted by the fluid on the step in terms of ρ, V_1, W, p_1 and p_2, h_1 and h_2. Ignore friction. Note that at the point where the flow separates off the step, the flow streamlines can be assumed to be parallel: this observation provides information about the pressure on the vertical face of the step.

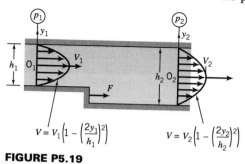

FIGURE P5.19

5.20   Two fences are placed inside a horizontal duct of height H and width W, as shown in Figure P5.20. Constant density air flows steadily from left to right, and the velocity upstream of the fences (station 1) is constant across the area and equal to U_{av}. The velocity downstream of the fences (station 2) has a parabolic distribution across the duct with a maximum value U_m. The pressures at stations 1 and 2 are p_1 and p_2, respectively, and they are constant across the duct area.

(a) Find the ratio U_m/U_{av}.

(b) Find the force F acting on the fences, assuming that the wall friction is negligible. Express the result in terms of the nondimensional drag coefficient C_D, and the nondimensional pressure coefficient C_p, where: $C_D = F/(\frac{1}{2}\rho U_{av}^2 HW)$, and $C_p = (p_1 - p_2)/(\frac{1}{2}\rho U_{av}^2)$.

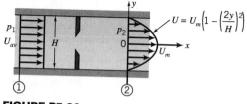

FIGURE P5.20

5.21   Air is flowing steadily into a rectangular duct of constant width W, as shown in Figure P5.21. There is inflow over area A_1 ($= 3hW$) and outflow over area A_3 ($= 2hW$). The minimum area is A_2 ($= hW$), and the density of air ρ is constant everywhere. Across A_1 and A_2 the velocities are constant and equal to U_1 and U_2, respectively, as shown. Across A_3 the velocity distribution is parabolic with a maximum value of U_m. The pressures are constant over each area: over area A the gauge pressure is p_1, over area A_2 the gauge pressure is p_2, and over A_3 the pressure is atmospheric. There are no losses.

(a) Find the ratios U_2/U_1, and U_m/U_1.

(b) Find the pressure difference $p_1 - p_2$ in terms of ρ and U_1.

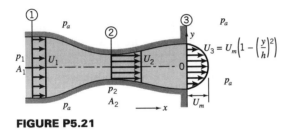

FIGURE P5.21

(c) Find the force F acting on the fluid (magnitude and sign) between stations 1 and 3, in terms of ρ, U_1, and p_1.

5.22 The entrance region of a parallel, rectangular duct flow is shown in Figure P5.22. The duct has a width W and a height D. The fluid density is constant, and the flow is steady. The velocity variation in the boundary layer of thickness δ at station 2 is assumed to be linear, and the pressure at any cross-section is uniform. Ignore the flow over the side walls of the duct (that is, $W \gg D$).

(a) Using the continuity equation, show that $U_1/U_2 = 1 - \delta/D$.

(b) Find the pressure coefficient $C_p = (p_1 - p_2)/(\frac{1}{2}\rho U_1^2)$.

(c) Show that

$$\frac{-F_v}{\frac{1}{2}\rho U_1^2 WD} = \frac{U_2^2}{U_1^2}\left(1 - \frac{8}{3}\frac{\delta}{D}\right) - 1$$

where F_v is the total viscous force acting on the walls of the duct.

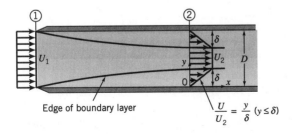

FIGURE P5.22

5.23 A fluid of constant density flows over a flat plate of length L and width W. At the leading edge of the plate the velocity is uniform and equal to U_0. A boundary layer forms on the plate so that at the trailing edge the velocity profile is parabolic, as shown in Figure P5.23.) Find:

(a) The volume flow rate $\dot{q}$ leaving the top surface of the control volume where $y = \delta$.

(b) The x-component of the momentum flux leaving the control volume through the same surface.

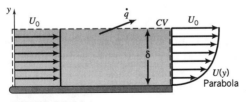

FIGURE P5.23

5.24 A cylinder of width W is placed near a plane wall, as shown in Figure P5.24. The incoming flow has a uniform velocity U_∞, and the downstream flow has a linear velocity profile $U = U_\infty y/H$. Assuming steady, constant density, constant pressure flow, find:

(a) The average velocity in the y-direction over the horizontal plane located at $y = H$.

(b) The force exerted on the cylinder by the fluid. Neglect viscous effects.

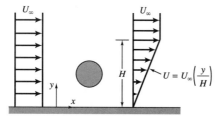

FIGURE P5.24

5.25 A model of a two-dimensional semicircular hut was put in a wind tunnel, and the downstream velocity profile was found to be as shown in Figure P5.25. Here, U_∞ is the freestream velocity, ρ is the air density, and D is the hut diameter. Assume that viscous effects and pressure variations can be neglected.

(a) Draw the flow pattern over the hut (remember that continuity must be satisfied).

(b) Find the average velocity in the y-direction over the horizontal plane located at $y = D$.

(c) Calculate the nondimensional force coefficient C_D, where $C_D = F/(\frac{1}{2}\rho U_\infty^2 D)$, and F is the force acting on the hut per unit span.

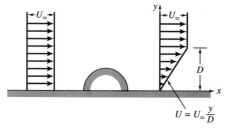

FIGURE P5.25

5.26 A horizontal jet of water, of constant velocity U_j, strikes a deflector such that the jet direction is smoothly changed, as shown in Figure P5.26. Find the ratio of the vertical and horizontal components of the force required to hold the deflector stationary in terms of the angle θ. Neglect gravitational effects. Does this force ratio change when the deflector moves to the right with velocity U_b?

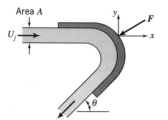

FIGURE P5.26

5.27 A horizontal jet of air of width W strikes a stationary scoop as indicated in Figure P5.27. The jet velocity is V and the jet height is D. If the jet height remains constant as the air flows over the plate surface, find:

(a) The force **F** required to hold the plate stationary.

(b) The change in **F** when the plate moves to the right at a constant speed $V/2$.

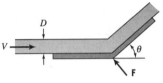

FIGURE P5.27

5.28 Water from a stationary nozzle impinges on a moving vane with a turning angle of $\theta = 60°$, as shown in Figure P5.28. The vane moves at a constant speed $U = 10$ *m/s,* and the jet exit velocity is $V = 30$ *m/s.* The nozzle has an exit area of 0.005 m^2. Find the force required to keep U constant.

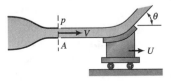

FIGURE P5.28

5.29 A snowplow mounted on a truck clears a path 12 *ft* wide through heavy wet snow, as shown in Figure P5.29. The snow is 8 *in.* deep and its density is 10 lb_m/ft^3. The truck travels at 20 *mph,* and the plow is set at an angle of 45° from the direction of travel. The snow is therefore discharged from the plow at an angle of 45° from the direction of travel and 45° above the horizontal, as shown in Figure P5.29. Find the force required to push the plow.[3]

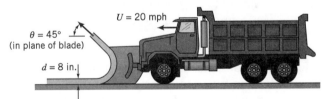

FIGURE P5.29

5.30 A pump submerged in water contained in a cart ejects the water into the atmosphere, as shown in Figure P5.30. The flow area leaving the ejector is 0.01 m^2.

(a) If the ejector flow velocity is 3 *m/s,* the flow will be returned to the cart. Find F, the force necessary to restrain the cart.

(b) If the ejector flow velocity is 4 *m/s,* the flow will just clear the sides of the cart which are at the same height as the ejector. Find F, the force necessary to restrain the cart.

(c) Find L, the distance between the ejector exit and the side of the cart for the flow in part **(b)**, where the exit of the jet and the sides of the cart are at the same height.

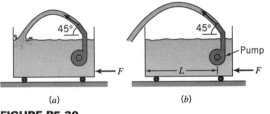

(a)   (b)

FIGURE P5.30

---

[3] Adapted from Fox & McDonald, *Introduction to Fluid Mechanics,* 4th edition, published by John Wiley & Sons, 1992.

CHAPTER 6

DIFFERENTIAL EQUATIONS OF MOTION

In previous chapters, we have mostly used large control volumes to find the overall balances of mass and momentum, without describing the specific flow behavior inside the control volume. For some purposes, this approach can provide very useful information, as in finding the flow rate through the system or the external forces acting on the fluid. For many other purposes, we need to have a more detailed knowledge of the flow behavior. For instance, the performance of an airplane depends critically on its shape. For an airplane in cruise, the lift is primarily generated by the pressure differences between the bottom and top surfaces of the wing. A large part of the drag, on the other hand, is due to the viscous friction between the fluid and the surface of the airplane. Both lift and drag are strongly dependent on the shapes of the wing and the fuselage. A large control volume cannot give very specific insight into how these shapes can be designed. We need to consider the local flow behavior, and this can only be obtained by using very small control volumes or fluid elements. This approach leads to a set of differential equations describing the detailed motion of the fluid.

We can obtain these equations directly from their integral forms. This is shown in Section A–A.10. Here, we choose to derive them using small, fixed control volumes, which means we adopt an Eulerian description of the flowfield. To begin this analysis, we need to express the rate of change of velocity, density, and pressure of a fluid particle in the Eulerian system.

6.1 RATE OF CHANGE FOLLOWING A FLUID PARTICLE

How are we to describe the displacement, velocity, and acceleration of a fluid? As indicated in Section 3.2, there are two possibilities: the Lagrangian system and the Eulerian system. In the Lagrangian system, we follow fluid particles, which are of fixed mass, whereas in the Eulerian system, we look for a "field" description, which gives all the details of the entire flow field at any position and time, and we do not follow individual fluid particles.

To apply Newton's laws of mechanics, as well as other principles such as the conservation of mass, it is necessary to refer to fixed quantities of matter, that is, fluid particles. A Lagrangian system would therefore seem more suitable than an Eulerian system for deriving the equations of motion of a fluid. On the other hand, a full description of the flow field in a Lagrangian system requires that a very large number of fluid particles be followed, and it is usually easier to adopt an Eulerian point of view.

To make the Eulerian representation useful, we need to find a way to express the rate of change of properties of a fluid particle of fixed mass in Eulerian coordi-

182

nates. The *total derivative* provides us with the required link between the Lagrangian and Eulerian descriptions for small, fixed control volumes, in the same way as the Reynolds transport theorem (Section 5.4) provides a link between the two descriptions for large, fixed control volumes.

In the Eulerian system, the density and velocity are functions of four independent variables: x, y, z, and t. In Cartesian coordinates,

$$\rho = \rho(x, y, z, t)$$

and

$$\mathbf{V} = u\mathbf{i} + v\mathbf{j} + w\mathbf{k}$$

where $u(x, y, z, t)$, $v(x, y, z, t)$, and $w(x, y, z, t)$ are the velocity components in the x-, y-, and z-directions.

Consider a fluid particle as it moves a distance Δx, Δy, and Δz in space during a time Δt. The velocity changes by an amount $\Delta \mathbf{V}$, and by the chain rule of differentiation

$$\Delta \mathbf{V} = \frac{\partial \mathbf{V}}{\partial t} \Delta t + \frac{\partial \mathbf{V}}{\partial x} \Delta x + \frac{\partial \mathbf{V}}{\partial y} \Delta y + \frac{\partial \mathbf{V}}{\partial z} \Delta z$$

Dividing through by Δt:

$$\frac{\Delta \mathbf{V}}{\Delta t} = \frac{\partial \mathbf{V}}{\partial t} + \frac{\partial \mathbf{V}}{\partial x} \frac{\Delta x}{\Delta t} + \frac{\partial \mathbf{V}}{\partial y} \frac{\Delta y}{\Delta t} + \frac{\partial \mathbf{V}}{\partial z} \frac{\Delta z}{\Delta t}$$

and as Δt approaches small values,

$$\left(\frac{d\mathbf{V}}{dt} \right)_{sys} = \frac{\partial \mathbf{V}}{\partial t} + \left(\frac{dx}{dt} \right)_{sys} \frac{\partial \mathbf{V}}{\partial x} + \left(\frac{dy}{dt} \right)_{sys} \frac{\partial \mathbf{V}}{\partial y} + \left(\frac{dz}{dt} \right)_{sys} \frac{\partial \mathbf{V}}{\partial z}$$

where we have identified the derivatives pertaining to the properties of the fluid particle (the "system"). We use the following short hand rotation to represent the time rate of change of a system:

$$\frac{D}{Dt} \equiv \left(\frac{d}{dt} \right)_{sys}$$

so that:

$$\frac{D\mathbf{V}}{Dt} = \frac{\partial \mathbf{V}}{\partial t} + u \frac{\partial \mathbf{V}}{\partial x} + v \frac{\partial \mathbf{V}}{\partial y} + w \frac{\partial \mathbf{V}}{\partial z} \tag{6.1}$$

since $u = dx/dt$, and so on. The special symbol D/Dt denotes the *total derivative*, that is, the derivative of a system quantity that depends on three space variables as well as time. In Cartesian coordinates,

$$\frac{D}{Dt} = \frac{\partial}{\partial t} + u \frac{\partial}{\partial x} + v \frac{\partial}{\partial y} + w \frac{\partial}{\partial z} \tag{6.2}$$

The first term in the definition of the total derivative represents the variation with time at a given point. It is called the *local* part of the total derivative, and it is the only source of variation in a field that is uniform in space. When the flow is unsteady, the local part is nonzero. In other words, if the partial derivative of the velocity, pressure, or density with respect to time is not zero, the flowfield is unsteady.

The last three terms in the definition of the total derivative represent the variation that occurs in space, and it is called the *convective* part of the total

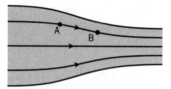

FIGURE 6.1 Steady flow through a converging duct.

derivative. In a nonuniform flow field, the velocity can vary in all three directions. As a fluid particle passes through a given point, its velocity changes in response to the spatial variation of velocity. The rate at which it is changing its velocity will depend on the spatial gradients of the velocity, and the speed at which it is moving from one point to another.

Consider steady, constant density flow through a converging duct (Figure 6.1). The velocity at point A does not vary with time, but the velocity at point B is greater than at point A because the cross-sectional area is smaller. Therefore a fluid particle experiences an acceleration as it moves from A to B: this part of the acceleration is described by the convective acceleration. If the mass flow rate through the duct was also unsteady, the velocity at each point would change with time and position, and there would be an additional contribution from the local acceleration.

So the total derivative provides information on the rate at which the properties of a fluid particle change when the flow field is expressed in an Eulerian system. It is, in fact, the rate of change *following* a fluid particle, and therefore it is also called the *particle derivative*, the *substantial derivative*, or the *material derivative*. It can operate on a vector or a scalar, and it is used to find the rate of change of velocity, density, pressure, or energy.

The symbol D/Dt denotes an "operator" which, when acting on the velocity, gives the acceleration in an Eulerian system. It is sometimes written as

$$\frac{D}{Dt} = \frac{\partial}{\partial t} + \mathbf{V} \cdot \nabla$$

It is incorrect to think of $\mathbf{V} \cdot \nabla$ as the scalar product of two vectors. It is best thought of as another operator such that, in Cartesian coordinates,

$$\mathbf{V} \cdot \nabla = u \frac{\partial}{\partial x} + v \frac{\partial}{\partial y} + w \frac{\partial}{\partial z}$$

Scalar products are commutative, which means that the order of multiplication does not change the result. In contrast, $\mathbf{V} \cdot \nabla$ and $\nabla \cdot \mathbf{V}$ are not the same at all: one is an operator and the other is an operation (see Section A-A.9).

6.1.1 Acceleration in Cartesian Coordinates

Equation 6.1 expresses the *total acceleration* following a fluid particle in Cartesian coordinates. As we saw in the previous section, the first term on the right hand side is the local acceleration, that is, the rate of change of velocity of a fluid particle at a given point due to unsteady effects. The last three terms on the right hand side together make up the convective acceleration, which is the rate of change of velocity due to the change of position of fluid particles. Therefore the acceleration given in equation 6.1 contains time and space derivatives of the vector velocity $\mathbf{V}$.

In a Cartesian system,

$$\mathbf{V} = u\mathbf{i} + v\mathbf{j} + w\mathbf{k}$$

so that

$$\frac{\partial \mathbf{V}}{\partial t} = u\frac{\partial \mathbf{i}}{\partial t} + \mathbf{i}\frac{\partial u}{\partial t} + v\frac{\partial \mathbf{j}}{\partial t} + \mathbf{j}\frac{\partial v}{\partial t} + w\frac{\partial \mathbf{k}}{\partial t} + \mathbf{k}\frac{\partial w}{\partial t}$$

and

$$\frac{\partial \mathbf{V}}{\partial x} = u\frac{\partial \mathbf{i}}{\partial x} + \mathbf{i}\frac{\partial u}{\partial x} + v\frac{\partial \mathbf{j}}{\partial x} + \mathbf{j}\frac{\partial v}{\partial x} + w\frac{\partial \mathbf{k}}{\partial x} + \mathbf{k}\frac{\partial w}{\partial x}$$

and so on.

The unit vectors $\mathbf{i}, \mathbf{j},$ and $\mathbf{k}$ have a constant magnitude and direction, and their derivatives are zero. That is

$$\frac{\partial \mathbf{V}}{\partial t} = \mathbf{i}\frac{\partial u}{\partial t} + \mathbf{j}\frac{\partial v}{\partial t} + \mathbf{k}\frac{\partial w}{\partial t}$$

and

$$\frac{\partial \mathbf{V}}{\partial x} = \mathbf{i}\frac{\partial u}{\partial x} + \mathbf{j}\frac{\partial v}{\partial x} + \mathbf{k}\frac{\partial w}{\partial x}$$

and so on.

Collecting all the terms, we obtain

$$\frac{D\mathbf{V}}{Dt} = \left(\frac{\partial u}{\partial t} + u\frac{\partial u}{\partial x} + v\frac{\partial u}{\partial y} + w\frac{\partial u}{\partial z}\right)\mathbf{i}$$

$$+ \left(\frac{\partial v}{\partial t} + u\frac{\partial v}{\partial x} + v\frac{\partial v}{\partial y} + w\frac{\partial v}{\partial z}\right)\mathbf{j}$$

$$+ \left(\frac{\partial w}{\partial t} + u\frac{\partial w}{\partial x} + v\frac{\partial w}{\partial y} + w\frac{\partial w}{\partial z}\right)\mathbf{k} \tag{6.3}$$

which reveals the true complexity of the acceleration in an Eulerian description.

6.1.2 Acceleration in Cylindrical Coordinates

The form of the operator D/Dt depends on the coordinate system. In a cylindrical coordinate system

$$\mathbf{V}(r, \theta, z, t) = u_r\mathbf{e}_r + u_\theta\mathbf{e}_\theta + u_z\mathbf{e}_z$$

where u_r, u_θ, and u_z are the velocity components in the $\mathbf{e}_r$, $\mathbf{e}_\theta$, and $\mathbf{e}_z$ directions (Figure 6.2). Given that

$$u_r = \frac{dr}{dt}, \quad u_\theta = r\frac{d\theta}{dt}, \quad \text{and} \quad u_z = \frac{dz}{dt}$$

we find that

$$\frac{D\mathbf{V}}{Dt} = \frac{\partial \mathbf{V}}{\partial t} + u_r\frac{\partial \mathbf{V}}{\partial r} + \frac{u_\theta}{r}\frac{\partial \mathbf{V}}{\partial \theta} + u_z\frac{\partial \mathbf{V}}{\partial z} \tag{6.4}$$

In finding the derivative $\partial \mathbf{V}/\partial \theta$, we need to remember that $\mathbf{e}_r$ and $\mathbf{e}_\theta$ are not constant vectors. They depend on θ, and their derivatives with respect to θ do not vanish.

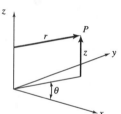

FIGURE 6.2 Cylindrical coordinate system.

That is:

$$\frac{\partial \mathbf{V}}{\partial \theta} = \frac{\partial}{\partial \theta}(u_r \mathbf{e_r} + u_\theta \mathbf{e_\theta} + u_z \mathbf{e_z}) = \mathbf{e_r}\frac{\partial u_r}{\partial \theta} + u_r\frac{\partial \mathbf{e_r}}{\partial \theta} + \mathbf{e_\theta}\frac{\partial u_\theta}{\partial \theta} + u_\theta\frac{\partial \mathbf{e_\theta}}{\partial \theta} + \mathbf{e_z}\frac{\partial u_z}{\partial \theta}$$

When the coordinate θ changes by a small amount $d\theta$, the unit vectors $\mathbf{e_r}$ and $\mathbf{e_\theta}$ change by an amount $d\mathbf{e_r}$ and $d\mathbf{e_\theta}$. From Figure 6.3, we see that the direction of $d\mathbf{e_\theta}$ is opposite to the direction of $\mathbf{e_r}$. Since the triangles abc and lmn are similar, the magnitude of $d\mathbf{e_\theta}$ is $d\theta$, and therefore

$$d\mathbf{e_\theta} = -\mathbf{e_r}\,d\theta$$

Similarly,

$$d\mathbf{e_r} = \mathbf{e_\theta}\,d\theta$$

and so we have

$$\frac{\partial \mathbf{e_r}}{\partial \theta} = \mathbf{e_\theta}$$

and

$$\frac{\partial \mathbf{e_\theta}}{\partial \theta} = -\mathbf{e_r}$$

It follows that the acceleration in cylindrical coordinates, written out in full, is given by

$$\frac{D\mathbf{V}}{Dt} = \left(\frac{\partial u_r}{\partial t} + u_r\frac{\partial u_r}{\partial r} + \frac{u_\theta}{r}\frac{\partial u_r}{\partial \theta} + u_z\frac{\partial u_r}{\partial z} - \frac{u_\theta^2}{r}\right)\mathbf{e_r}$$

$$+ \left(\frac{\partial u_\theta}{\partial t} + u_r\frac{\partial u_\theta}{\partial r} + \frac{u_\theta}{r}\frac{\partial u_\theta}{\partial \theta} + u_z\frac{\partial u_\theta}{\partial z} + \frac{u_r u_\theta}{r}\right)\mathbf{e_\theta}$$

$$+ \left(\frac{\partial u_z}{\partial t} + u_r\frac{\partial u_z}{\partial r} + \frac{u_\theta}{r}\frac{\partial u_z}{\partial \theta} + u_z\frac{\partial u_z}{\partial z}\right)\mathbf{e_z} \tag{6.5}$$

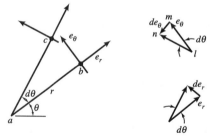

FIGURE 6.3 Rate of change of unit vectors in a cylindrical coordinate system.

EXAMPLE 6.1 *Rate of Change of Temperature Following a Fluid Particle*

Given the Eulerian, Cartesian velocity field $\mathbf{V} = 3\mathbf{i} + 2x\mathbf{j}$, and a temperature field described by $T = 4y^2$, find the rate of change of temperature following a fluid particle.

Solution: We need to find the total derivative of the temperature, that is,

$$\frac{DT}{Dt} = \frac{\partial T}{\partial t} + u\frac{\partial T}{\partial x} + v\frac{\partial T}{\partial y} + w\frac{\partial T}{\partial z}$$

For these particular velocity and temperature fields,

$$\frac{\partial T}{\partial t} = 0$$

$$u\frac{\partial T}{\partial x} = 3\frac{\partial 4y^2}{\partial x} = 0$$

$$v\frac{\partial T}{\partial y} = 2x\frac{\partial 4y^2}{\partial y} = 16xy$$

Therefore,

$$\frac{DT}{Dt} = 16xy \qquad\blacksquare$$

EXAMPLE 6.2 *Acceleration of a Fluid Particle*

Given the Eulerian, Cartesian velocity field $\mathbf{V} = 2t\mathbf{i} + xz\mathbf{j} - t^2y\mathbf{k}$, find the acceleration following a fluid particle.

Solution: We need to find the total derivative of the velocity, that is,

$$\frac{D\mathbf{V}}{Dt} = \frac{\partial \mathbf{V}}{\partial t} + u\frac{\partial \mathbf{V}}{\partial x} + v\frac{\partial \mathbf{V}}{\partial y} + w\frac{\partial \mathbf{V}}{\partial z}$$

For this particular velocity field,

$$\frac{\partial \mathbf{V}}{\partial t} = \frac{\partial 2t}{\partial t}\mathbf{i} + \frac{\partial xz}{\partial t}\mathbf{j} - \frac{\partial t^2y}{\partial t}\mathbf{k} = 2\mathbf{i} + 0 - 2ty\mathbf{k}$$

$$u\frac{\partial \mathbf{V}}{\partial x} = 2t\left[\frac{\partial 2t}{\partial x}\mathbf{i} + \frac{\partial xz}{\partial x}\mathbf{j} - \frac{\partial t^2y}{\partial x}\mathbf{k}\right] = 2t[0 + z\mathbf{j} - 0] = 2tz\mathbf{j}$$

$$v\frac{\partial \mathbf{V}}{\partial y} = xz\left[\frac{\partial 2t}{\partial y}\mathbf{i} + \frac{\partial xz}{\partial y}\mathbf{j} - \frac{\partial t^2y}{\partial y}\mathbf{k}\right] = xz[0 + 0 - t^2\mathbf{k}] = -xzt^2\mathbf{k}$$

$$w\frac{\partial \mathbf{V}}{\partial z} = -t^2y\left[\frac{\partial 2t}{\partial z}\mathbf{i} + \frac{\partial xz}{\partial z}\mathbf{j} - \frac{\partial t^2y}{\partial z}\mathbf{k}\right] = -t^2y[0 + x\mathbf{j} - 0] = -xyt^2\mathbf{j}$$

Finally,

$$\frac{D\mathbf{V}}{Dt} = 2\mathbf{i} - 2ty\mathbf{k} + 2tz\mathbf{j} - xzt^2\mathbf{k} - xyt^2\mathbf{j} = 2\mathbf{i} + (2tz - xyt^2)\mathbf{j} - (2ty + xzt^2)\mathbf{k} \qquad\blacksquare$$

6.2 CONTINUITY EQUATION

We are now ready to derive the continuity equation in differential form for an Eulerian system, in Cartesian coordinates. Consider an elemental volume $dx\,dy\,dz$

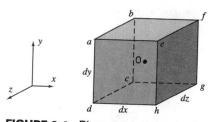

FIGURE 6.4 Elemental control volume for derivation of the differential form of the continuity equation.

(Figure 6.4). At the point 0, in the middle of the box, at time t,

$$u = u_0, \ v = v_0, \ w = w_0, \text{ and } \rho = \rho_0$$

First, we find the net mass outflow per unit time. By using a Taylor series expansion, and dropping higher-order terms:

On face $abcd$

$$u = u_0 - \left(\frac{\partial u}{\partial x}\right)_0 \frac{dx}{2}$$

and

$$\rho = \rho_0 - \left(\frac{\partial \rho}{\partial x}\right)_0 \frac{dx}{2}$$

On face $efgh$

$$u = u_0 + \left(\frac{\partial u}{\partial x}\right)_0 \frac{dx}{2}$$

and

$$\rho = \rho_0 + \left(\frac{\partial \rho}{\partial x}\right)_0 \frac{dx}{2}$$

We can find the mass flux through the box by finding the mass flux through each of its six faces and summing the result. We start with faces $abcd$ and $efgh$

mass inflow through $abcd$ in time $dt = (\rho u \, dA \, dt)_{abcd}$

$$= \left[\rho_0 - \left(\frac{\partial \rho}{\partial x}\right)_0 \frac{dx}{2}\right]\left[u_0 - \left(\frac{\partial u}{\partial x}\right)_0 \frac{dx}{2}\right] dy \, dz \, dt$$

mass outflow through $efgh$ in time $dt = (\rho u \, dA \, dt)_{efgh}$

$$= \left[\rho_0 + \left(\frac{\partial \rho}{\partial x}\right)_0 \frac{dx}{2}\right]\left[u_0 + \left(\frac{\partial u}{\partial x}\right)_0 \frac{dx}{2}\right] dy \, dz \, dt$$

net outflow through $abcd$ and $efgh = \left[u_0 \left(\frac{\partial \rho}{\partial x}\right)_0 + \rho_0 \left(\frac{\partial u}{\partial x}\right)_0\right] dx \, dy \, dz \, dt$

$$= \left[\frac{\partial(\rho u)}{\partial x}\right]_0 dx \, dy \, dz \, dt$$

Similar expressions can be derived for the other faces

net outflow through $cdhg$ and $abfe = \left[\frac{\partial(\rho v)}{\partial y}\right]_0 dx \, dy \, dz \, dt$

net outflow through $cbfg$ and $aehd = \left[\frac{\partial(\rho w)}{\partial z}\right]_0 dx \, dy \, dz \, dt$

By adding up the contributions over all six faces, we have

$$\text{total net mass outflow in time } dt = \left[\frac{\partial(\rho u)}{\partial x} + \frac{\partial(\rho v)}{\partial y} + \frac{\partial(\rho w)}{\partial z} \right] dx\, dy\, dz\, dt$$

(we have dropped the subscript because the result should not depend on the particular point that was considered). This must be equal to the decrease in mass contained in this volume during the same time interval (mass must be conserved), so that

$$\left[\frac{\partial(\rho u)}{\partial x} + \frac{\partial(\rho v)}{\partial y} + \frac{\partial(\rho w)}{\partial z} \right] dx\, dy\, dz\, dt = -dx\, dy\, dz\, \frac{\partial \rho}{\partial t}\, dt$$

That is,

$$\frac{\partial(\rho u)}{\partial x} + \frac{\partial(\rho v)}{\partial y} + \frac{\partial(\rho w)}{\partial z} = -\frac{\partial \rho}{\partial t} \tag{6.6}$$

In terms of the divergence operator (see Section A.5), we have

$$\nabla \cdot \rho \mathbf{V} = -\frac{\partial \rho}{\partial t} \tag{6.7}$$

Alternatively, Equation 6.6 can be expanded to give

$$u\frac{\partial \rho}{\partial x} + \rho\frac{\partial u}{\partial x} + v\frac{\partial \rho}{\partial y} + \rho\frac{\partial v}{\partial y} + w\frac{\partial \rho}{\partial z} + \rho\frac{\partial w}{\partial z} = -\frac{\partial \rho}{\partial t}$$

Hence,

$$\frac{\partial u}{\partial x} + \frac{\partial v}{\partial y} + \frac{\partial w}{\partial z} = -\frac{1}{\rho}\frac{D\rho}{Dt}$$

or

$$\nabla \cdot \mathbf{V} = -\frac{1}{\rho}\frac{D\rho}{Dt} \tag{6.8}$$

Equations 6.7 and 6.8 are two forms of the continuity equation in differential form. Since they are written in vector form, they are independent of the coordinate system. By definition, an incompressible fluid is a fluid where $D\rho/Dt = 0$. From equation 6.8, we see that this requires $\nabla \cdot \mathbf{V} = 0$. A fluid of constant density is obviously incompressible, but it is possible to have an incompressible fluid with variable density, as long as $D\rho/Dt = 0$ (as in a stratified fluid, for example).

6.2.1 Particular Forms

In Cartesian coordinates,

$$\frac{\partial(\rho u)}{\partial x} + \frac{\partial(\rho v)}{\partial y} + \frac{\partial(\rho w)}{\partial z} = -\frac{\partial \rho}{\partial t} \tag{6.9}$$

In cylindrical coordinates,

$$\frac{1}{r}\frac{\partial(r\rho u_r)}{\partial r} + \frac{1}{r}\frac{\partial(\rho u_\theta)}{\partial \theta} + \frac{\partial(\rho u_z)}{\partial z} = -\frac{\partial \rho}{\partial t} \tag{6.10}$$

EXAMPLE 6.3   *Incompressible Flow*

Determine if the Eulerian velocity field $\mathbf{V} = 2x\mathbf{i} + t^2\mathbf{j}$ is incompressible.

Solution: A velocity field is incompressible if $\nabla \cdot \mathbf{V} = 0$. For the Cartesian velocity field given here,

$$\nabla \cdot \mathbf{V} = \frac{\partial u}{\partial x} + \frac{\partial v}{\partial y} + \frac{\partial w}{\partial z} = \frac{\partial 2x}{\partial x} + \frac{\partial t^2}{\partial y} = 2$$

and therefore this flow field is compressible.   ∎

6.3   MOMENTUM EQUATION

Here we derive the momentum equation in differential form. When the fluid is inviscid, the equation that results is called the Euler equation, or the momentum equation for inviscid flow. When the fluid is viscous, the equation is known as the Navier–Stokes equation. We begin with inviscid flows.

Consider a fixed control volume of infinitesimal size (Figure 6.5). There is flow through the six faces of the volume, and surface forces and body forces act on the fluid particle that occupies the volume at a particular instant in time. The only surface force taken into account is that due to pressure differences, and the only body force considered is that due to gravity. The volume element $dxdydz$ is similar to the one used to derive the continuity equation except that only one face is shown, and this face has an arbitrary orientation with respect to the gravitational vector $\mathbf{g}$ (that is, $\mathbf{g}$ may have components in or out of the page, as well as being at an angle to the x- and y-axes). The point 0 is at the center of the control volume.

The resultant force in the x-direction, F_x, has two contributions: the force due to pressure differences acting on the two faces of area $dy\,dz$, and the x-component of the force due to the weight of the fluid contained in the volume $dx\,dy\,dz$. Using a Taylor series expansion about the center of the volume, and neglecting second-order terms, we have

$$F_x = \left[p_0 - \frac{\partial p}{\partial x}\bigg|_0 \frac{dx}{2} \right] dy\,dz - \left[p_0 + \frac{\partial p}{\partial x}\bigg|_0 \frac{dx}{2} \right] dy\,dz + \rho_0\, \mathbf{g} \cdot \mathbf{i}\, dx\,dy\,dz$$

That is,

$$F_x = - \frac{\partial p}{\partial x}\bigg|_0 dx\,dy\,dz + \rho_0\, \mathbf{g}_x\, dx\,dy\,dz$$

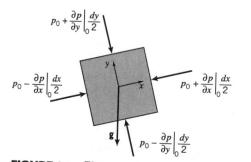

FIGURE 6.5   Elemental control volume for derivation of the differential form of the momentum equation.

where g_x is the x-component of the gravitational vector **g.** Similarly for the y- and z-directions:

$$F_y = -\frac{\partial p}{\partial y}\bigg|_0 dx\,dy\,dz + \rho_0\,\mathbf{g}_y\,dx\,dy\,dz$$

$$F_z = -\frac{\partial p}{\partial z}\bigg|_0 dx\,dy\,dz + \rho_0\,\mathbf{g}_z\,dx\,dy\,dz$$

so that

$$\mathbf{F} = F_x\mathbf{i} + F_y\mathbf{j} + F_z\mathbf{k}$$

$$= -\left(\mathbf{i}\frac{\partial p}{\partial x} + \mathbf{j}\frac{\partial p}{\partial y} + \mathbf{k}\frac{\partial p}{\partial z}\right) + \rho g_x\mathbf{i} + \rho g_y\mathbf{j} + \rho g_z\mathbf{k}$$

where the subscript has been dropped since the result should be independent of the particular location chosen for the derivation. In terms of the gradient operator (see Section A–A.4), we have

$$\mathbf{F} = -\nabla p + \rho\mathbf{g}$$

By Newton's second law of motion, **F** is equal to the rate of change of momentum following a fluid particle. For a velocity field in an Eulerian system, the acceleration following a fluid particle is given by $D\mathbf{V}/Dt$, and

$$\left\{\begin{array}{c}\text{the rate of change of momentum}\\ \text{following a fluid particle}\end{array}\right\} = \rho\,dx\,dy\,dz\,\frac{D\mathbf{V}}{Dt}$$

Therefore

$$\rho\frac{D\mathbf{V}}{Dt} = -\nabla p + \rho\mathbf{g} \tag{6.11}$$

This is the differential form of the linear momentum equation for an inviscid fluid, in vector form. It is called the Euler equation, and it holds for compressible and incompressible flows (for a historical note on Leonhard Euler, see Section 15.7). In words, we can write the Euler equation as

$$\underbrace{\rho\frac{D\mathbf{V}}{Dt}}_{\substack{\text{inertia force}\\ \text{(mass} \times \text{acceleration)}}} = \underbrace{-\nabla p}_{\substack{\text{surface force due to}\\ \text{pressure differences}}} + \underbrace{\rho\mathbf{g}}_{\substack{\text{body force due to}\\ \text{gravitational attraction}}}$$

All forces are per unit volume.

6.3.1 Euler Equation in Cartesian Coordinates

To take the next step, we need to introduce the concept of a potential function. We will simply define a function ϕ_g such that

$$\mathbf{g} = -g\nabla\phi_g$$

The vector $\nabla\phi_g$ is a unit vector that points in the direction opposite to the vector **g.** The scalar parameter ϕ_g is called a *potential function,* and here it represents altitude or elevation. The term "potential" is clearly connected with the idea of potential energy, which only depends on altitude. If a body moves from point a to point b, its potential energy will be unchanged as long as the two points have the same elevation. Gravity is called a "conservative" force field since the potential energy of a body depends only on its elevation and not on the particular path used to get from a to b. The quantity that measures the change in potential energy is the elevation, which is also the potential function associated with the conservative force field due to gravity, ϕ_g.

The Euler equation may then be written as

$$\frac{D\mathbf{V}}{Dt} = -\frac{1}{\rho}\nabla p + \mathbf{g} = -\frac{1}{\rho}\nabla p - g\nabla \phi_g$$

Therefore, in Cartesian coordinates, we have

$$\frac{\partial u}{\partial t} + u\frac{\partial u}{\partial x} + v\frac{\partial u}{\partial y} + w\frac{\partial u}{\partial z} = -\frac{1}{\rho}\frac{\partial p}{\partial x} - g\frac{\partial \phi_g}{\partial x} \tag{6.12}$$

$$\frac{\partial v}{\partial t} + u\frac{\partial v}{\partial x} + v\frac{\partial v}{\partial y} + w\frac{\partial v}{\partial z} = -\frac{1}{\rho}\frac{\partial p}{\partial y} - g\frac{\partial \phi_g}{\partial y} \tag{6.13}$$

$$\frac{\partial w}{\partial t} + u\frac{\partial w}{\partial x} + v\frac{\partial w}{\partial y} + w\frac{\partial w}{\partial z} = -\frac{1}{\rho}\frac{\partial p}{\partial z} - g\frac{\partial \phi_g}{\partial z} \tag{6.14}$$

6.3.2   Euler Equation in Cylindrical Coordinates

In cylindrical coordinates, we have

$$\frac{\partial u_r}{\partial t} + u_r\frac{\partial u_r}{\partial r} + \frac{u_\theta}{r}\frac{\partial u_r}{\partial \theta} + u_z\frac{\partial u_r}{\partial z} - \frac{u_\theta^2}{r} = -\frac{1}{\rho}\frac{\partial p}{\partial r} - g\frac{\partial \phi_g}{\partial r} \tag{6.15}$$

$$\frac{\partial u_\theta}{\partial t} + u_r\frac{\partial u_\theta}{\partial r} + \frac{u_\theta}{r}\frac{\partial u_\theta}{\partial \theta} + u_z\frac{\partial u_\theta}{\partial z} + \frac{u_r u_\theta}{r} = -\frac{1}{\rho r}\frac{\partial p}{\partial \theta} - g\frac{1}{r}\frac{\partial \phi_g}{\partial \theta} \tag{6.16}$$

$$\frac{\partial u_z}{\partial t} + u_r\frac{\partial u_z}{\partial r} + \frac{u_\theta}{r}\frac{\partial u_z}{\partial \theta} + u_z\frac{\partial u_z}{\partial z} = -\frac{1}{\rho}\frac{\partial p}{\partial z} - g\frac{\partial \phi_g}{\partial z} \tag{6.17}$$

6.3.3   Navier–Stokes Equations

We now consider the effects of viscosity. In Section 1.4.1, we indicated that the viscous stress for a Newtonian fluid is given by the coefficient of viscosity times the velocity gradient. We need to add here that this is strictly true only for incompressible flow, and the following derivation will also be restricted to incompressible flow.

When the only velocity gradient that acts is the gradient of the x-component of velocity in the y-direction (as in Figure 6.6), the principal shear stress is $\tau_{yx} = \mu(\partial u/\partial y)$, where the subscript yx denotes a stress that acts in the x-direction and

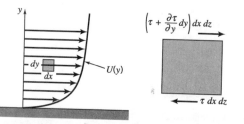

FIGURE 6.6 Viscous flow showing an element in shear. Left: the velocity $U = U(y)$, so that $\tau = \tau_{yx}$. Right: notation for fluid element.

is associated with a velocity gradient in the y-direction. When the viscosity is constant, the resultant force acting on the element shown in Figure 6.6 due to this viscous stress is given by

$$\left(\tau_{yx} + \frac{\partial \tau_{yx}}{\partial y} dy\right) dx\, dz - \tau_{yx}\, dx\, dz = \frac{\partial \tau_{yx}}{\partial y} dx\, dy\, dz = \mu \frac{\partial^2 u}{\partial y^2} dx\, dy\, dz$$

That is, for the case where $(\partial u / \partial y)$ is the only velocity gradient and μ is constant, the resultant force due to viscous friction in the x-direction, per unit volume, is given by[1]

$$\frac{\partial \tau_{yx}}{\partial y} = \mu \frac{\partial^2 u}{\partial y^2}$$

We see that a resultant force will occur only if the viscous stress varies in the flow, that is, when gradients of the stress τ_{yx} are present. If the shear stress is uniform throughout the flow, fluid particles will distort but there will be no resultant force due to viscous stresses. In other words, the viscous stress will not contribute to the acceleration of fluid particles.

Normal stresses due to extensional strain rates also lead to viscous stresses (see Section 1.4). A similar analysis to that given previously shows that for a flow where $(\partial u / \partial x)$ is the only velocity gradient and μ is constant, the resultant force due to viscous friction in the x-direction, per unit volume, is given by

$$\frac{\partial \tau_{xx}}{\partial x} = \mu \frac{\partial^2 u}{\partial x^2}$$

In the general case, where velocity gradients act in all directions, the x-component of the viscous force per unit volume in Cartesian coordinates becomes

$$\mu \left(\frac{\partial \tau_{xx}}{\partial x} + \frac{\partial \tau_{yx}}{\partial y} + \frac{\partial \tau_{zx}}{\partial z}\right)$$

By expressing the stresses in terms of the velocity gradients, the x-component of the viscous force becomes

$$\mu \left(\frac{\partial^2 u}{\partial x^2} + \frac{\partial^2 u}{\partial y^2} + \frac{\partial^2 u}{\partial z^2}\right) = \mu \nabla^2 u$$

where ∇^2 is the Laplacian operator (see Section A.6). In vector notation, therefore, the total viscous force per unit volume is given by $\mu \nabla^2 \mathbf{V}$. In Cartesian coordinates,

$$\nabla^2 \mathbf{V} = \frac{\partial^2 \mathbf{V}}{\partial x^2} + \frac{\partial^2 \mathbf{V}}{\partial y^2} + \frac{\partial^2 \mathbf{V}}{\partial z^2}$$

This viscous force per unit volume can be added to the Euler equation, and we obtain

$$\rho \frac{D\mathbf{V}}{Dt} = -\nabla p + \rho \mathbf{g} + \mu \nabla^2 \mathbf{V} \tag{6.18}$$

This equation is known as the Navier–Stokes equation. It is the momentum equation for the flow of a viscous fluid. It was derived independently by Navier and Stokes in the early nineteenth century (for some historical notes see Sections 15.10 and

---

[1] The dimensions of $\partial \tau_{yx} / \partial y$ are stress/length = force/area × length = force/volume.

15.14). In the form written here it only applies to incompressible, constant viscosity flow.[2]

The Euler and Navier–Stokes equations are nonlinear, partial differential equations, and no general solutions exist. The main source of difficulty in both equations is the acceleration term, since it involves products of velocity. That is, it is nonlinear (note, for example, that $u\partial u/\partial x$ can be written as $\frac{1}{2}\partial u^2/\partial x$). Analytical solutions exist only under particular conditions, such as incompressible, irrotational flow (Chapter 7), or fully developed flows where the acceleration term is zero (Chapter 9). Numerical techniques must be used for other cases, and it is now routinely possible, for instance, to solve the Euler equation for an entire airplane at transonic Mach numbers, as long as it is not maneuvering too quickly. When viscous effects are included, however, numerical solutions require a great deal more computer speed and memory, and full Navier–Stokes solutions are currently possible only at low Reynolds numbers, that is, at Reynolds numbers not much greater than the value where transition to turbulence occurs.

6.3.4 Boundary Conditions

The differential equations of motion are complete once the boundary conditions are specified. We will only consider the specific boundary conditions introduced by the presence of a wall since they are of most interest here. The Navier–Stokes equation includes the viscous stress, so that the no-slip condition must be satisfied (see Section 1.6). The no-slip condition means that the fluid in contact with a solid surface has no relative velocity with respect to the surface. That is, at the wall

$$\mathbf{V} = \mathbf{V}_w \qquad (6.19)$$

The Euler equation does not include the viscous stress so that it cannot satisfy the same boundary conditions as the Navier–Stokes equation. In particular, it cannot satisfy the no-slip condition, and slip is allowed. However, relative to the solid surface, the velocity of the fluid normal to a solid surface must still go to zero so that there is no flow through the surface. That is, the boundary condition for the Euler equation at the wall is

$$\mathbf{n} \cdot \mathbf{V} = \mathbf{n} \cdot \mathbf{V}_w \qquad (6.20)$$

EXAMPLE 6.4   *Euler Equation*

Find the x-component of the acceleration of an inviscid fluid under the action of a pressure gradient $\nabla p = x^2\mathbf{i} + 2z\mathbf{j}$. The x-direction is horizontal.

Solution: The flow of an inviscid fluid is described by the Euler equation (equation 6.11).

$$\rho\frac{D\mathbf{V}}{Dt} = -\nabla p + \rho\mathbf{g}$$

We take the x-component by forming the dot product with the unit vector $\mathbf{i}$.

$$\rho\frac{D\mathbf{i}\cdot\mathbf{V}}{Dt} = -\mathbf{i}\cdot\nabla p + \mathbf{i}\cdot\rho\mathbf{g}$$

---

[2] A more formal derivation than the one given here may be found in, for example, *Introduction to Fluid Mechanics,* 4th edition, by R. W. Fox & A. T. McDonald, published by John Wiley & Sons, 1992.

That is,

$$\frac{Du}{Dt} = -\frac{1}{\rho}\frac{\partial p}{\partial x} + 0$$

For the flowfield given here

$$\frac{Du}{Dt} = -\frac{x^2}{\rho}$$

■

EXAMPLE 6.5   *Navier–Stokes Equation*

Consider the steady flow of a viscous fluid in a long horizontal duct where $\mathbf{V} = ay^2\mathbf{i}$. Find the corresponding pressure gradient.

Solution: The flow of an viscous fluid is described by the Navier–Stokes equation (equation 6.18).

$$\rho\frac{D\mathbf{V}}{Dt} = -\nabla p + \rho\mathbf{g} + \mu\nabla^2\mathbf{V}$$

The acceleration for this particular flow is given by

$$\frac{D\mathbf{V}}{Dt} = \frac{D(ay^2)}{Dt}\mathbf{i} = 0$$

(flows where the acceleration is zero are called fully developed). Since the channel is horizontal, the Navier–Stokes equation reduces to

$$0 = -\nabla p + \mu\nabla^2\mathbf{V}$$

The viscous term becomes

$$\mu\nabla^2\mathbf{V} = \mu\nabla^2(ay^2\mathbf{i}) = 2a\mu\mathbf{i}$$

and therefore

$$\nabla p = 2a\mu\mathbf{i}$$

The pressure gradient only has an x-component, and therefore the pressure is only a function of x. Hence,

$$\frac{\partial p}{\partial x} = \frac{dp}{dx} = 2a\mu$$

■

6.4   *RIGID-BODY MOTION REVISITED

When a fluid is in rigid-body motion, there are no relative motions in the fluid (Section 2.11). If the fluid is moving as a rigid body with an acceleration $\mathbf{a}$, then $D\mathbf{V}/Dt = \mathbf{a}$, and the momentum equation (equation 6.11) becomes

$$\mathbf{a} = -\frac{1}{\rho}\nabla p + g\nabla z$$

or

$$\nabla p = \rho(g\nabla z - \mathbf{a})$$

where $z = -h$, so that z is positive going down. For the z-direction we obtain

$$\frac{\partial p}{\partial z} = \rho(g - a_z)$$

since the z-component of $\nabla z = \mathbf{k} \cdot \nabla z = 1$. For the x-direction

$$\frac{\partial p}{\partial y} = -\rho a_x$$

which are the same equations obtained in Section 2.11 for rigid-body motion in the y-z plane.

When the acceleration is zero, the momentum equation reduces to

$$0 = -\frac{1}{\rho}\nabla p + \mathbf{g}$$

That is,

$$\nabla p = \rho\mathbf{g}$$

so that ∇p is aligned with $\mathbf{g}$ and it points in the same direction. Therefore, the maximum rate of change of pressure is in the direction of the gravitational vector. For the coordinate system used here, where $\mathbf{g}$ is in the negative z-direction, we obtain

$$\frac{dp}{dz} = -\rho g$$

which is identical to the hydrostatic equation. We see that the hydrostatic equation is just a special case of the momentum equation for a fluid with zero acceleration and no relative motions.

6.5 ONE-DIMENSIONAL UNSTEADY FLOW

As we saw in Chapter 3, the one-dimensional forms of the continuity and momentum equations are very useful because they are much simpler to solve than the full equations, and they can help to develop physical insight. A flow is one-dimensional if the velocity varies only along the flow direction. We could interpret this statement to say that, for a one-dimensional constant density flow in the x-direction, $u = u(x)$, and $v = w = 0$. However, the continuity equation (equation 6.9) would then give $\partial u/\partial x = 0$. That is, the velocity u is constant. This is clearly not what we mean by our usual interpretation of one-dimensional flow.

If we consider first the case where $w = 0$, then the continuity equation indicates that

$$\frac{\partial u}{\partial x} + \frac{\partial v}{\partial y} = 0$$

and so for any flow where the area of the streamtube is varying with x, so that $\partial u/\partial x \neq 0$, the velocity v must vary across the streamtube. Furthermore, even if u is constant over the area of a streamtube, v need not be. For example, in a symmetric diverging flow, $v = 0$ on the centerline, but v will be positive above the centerline and negative below (Figure 6.7).

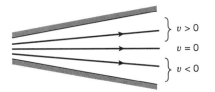

FIGURE 6.7 Flow in a symmetrically diverging duct.

Our definition of one-dimensional flow can be made more precise as follows.

A flow has a one-dimensional velocity field when the streamwise velocity is constant across the flow area.

6.5.1 Continuity Equation

The differential continuity equation for such "quasi"-one-dimensional flows can be found by applying the integral form of the continuity equation.

$$\frac{\partial}{\partial t} \int \rho \, d\forall + \int \mathbf{n} \cdot \rho \mathbf{V} \, dA = 0 \tag{6.21}$$

(equation 5.3) to the elemental volume shown in figure 6.8.

In this case, the area of the duct varies slowly with streamwise distance and all flow quantities such as the velocity, pressure, and density can be taken as constant across the area, and they vary only in the streamwise direction x. We can use this information to simplify the continuity equation. The unsteady term in equation 6.21 represents the rate of change of mass contained in the control volume. Since the control volume is fixed

$$\frac{\partial}{\partial t} \int \rho \, d\forall = \int \frac{\partial \rho}{\partial t} \, d\forall$$

$$= \frac{\partial \rho}{\partial t} \frac{1}{2} \left(A + \left(A + \frac{\partial A}{\partial x} \, dx \right) \right) dx$$

$$= \frac{\partial \rho}{\partial t} A \, dx$$

where we have used the approximation that $d\forall$ equals the average area times the length of the control volume, and higher order terms have been neglected.

For the mass flux term in equation 6.21

$$\int \mathbf{n} \cdot \rho \mathbf{V} \, dA = -\rho u A + \left(\rho + \frac{\partial \rho}{\partial x} \, dx \right) \left(u + \frac{\partial u}{\partial x} \, dx \right) \left(A + \frac{\partial A}{\partial x} \, dx \right)$$

$$= \rho u \frac{\partial A}{\partial x} \, dx + \rho A \frac{\partial u}{\partial x} \, dx + u A \frac{\partial \rho}{\partial x} \, dx$$

$$= \frac{\partial \rho u A}{\partial x} \, dx$$

where u is the streamwise velocity component. Note that v does not contribute to the mass flow across the area of the streamtube, and therefore it does not appear in the result. Hence,

$$A \frac{\partial \rho}{\partial t} + \frac{\partial}{\partial x}(\rho u A) = 0 \tag{6.22}$$

where the density, velocity, and area can vary with x and t. For steady flows (or constant density flows), this reduces to saying that the mass flow must remain constant along the streamtube (see also Section 3.5).

6.5.2 Momentum Equation

We can perform a similar analysis for the one-dimensional momentum equation. The integral form of the momentum equation for inviscid flow with negligible body forces through a fixed control volume (equation 5.16) is given by:

$$\frac{\partial}{\partial t} \int \rho V \, d\forall + \int (\mathbf{n} \cdot \rho \mathbf{V}) \mathbf{V} \, dA = - \int \mathbf{n} \, p \, dA \tag{6.23}$$

For the flow shown in Figure 6.8 the unsteady term becomes

$$\frac{\partial}{\partial t} \int \rho u \, d\forall = \int \frac{\partial \rho u}{\partial t} \, d\forall$$

$$= \frac{\partial \rho u}{\partial t} \frac{1}{2} \left(A + \left(A + \frac{\partial A}{\partial x} dx \right) \right) dx$$

$$= A \frac{\partial \rho u}{\partial t} dx$$

Similarly, the momentum flux term in equation 6.23 becomes

$$\int (\mathbf{n} \cdot \rho \mathbf{V}) \mathbf{V} \, dA = - \int \rho u_1^2 \, dA_1 + \int \rho u_2^2 \, dA_2$$

$$= -\rho u^2 A + \left(\rho + \frac{\partial \rho}{\partial x} dx \right) \left(u + \frac{\partial u}{\partial x} dx \right)^2 \left(A + \frac{\partial A}{\partial x} dx \right)$$

$$= 2\rho u A \frac{\partial u}{\partial x} dx + u^2 A \frac{\partial \rho}{\partial x} dx + \rho u^2 \frac{\partial A}{\partial x} dx$$

$$= \frac{\partial}{\partial x}(\rho u^2 A) \, dx$$

after higher order terms are neglected.

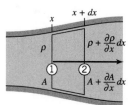

FIGURE 6.8 Elemental control volume for derivation of the differential form of the one-dimensional equations.

The force due to pressure differences becomes

$$-\int \mathbf{n}p\, dA = \int p\, dA_1 - \int p\, dA_2$$

$$= pA - \left(p + \frac{\partial p}{\partial x} dx\right)\left(A + \frac{\partial A}{\partial x} dx\right)$$

$$= -A \frac{\partial p}{\partial x} dx$$

Equation 6.23 then reduces to

$$A \frac{\partial \rho u}{\partial t} + \frac{\partial}{\partial x}(\rho u^2 A) = -A \frac{\partial p}{\partial x}$$

By expanding the terms,

$$Au \frac{\partial \rho}{\partial t} + A\rho \frac{\partial u}{\partial t} + u \frac{\partial}{\partial x}(\rho u A) + \rho u A \frac{\partial u}{\partial x} + A \frac{\partial p}{\partial x} = 0$$

That is,

$$u\left(A \frac{\partial \rho}{\partial t} + \frac{\partial}{\partial x}(\rho u A)\right) + A\rho \frac{\partial u}{\partial t} + \rho u A \frac{\partial u}{\partial x} + A \frac{\partial p}{\partial x} = 0$$

Using the continuity equation (equation 6.22) to eliminate the term in parentheses, we obtain the one-dimensional, unsteady form of the momentum equation for inviscid flows and negligible body forces

$$\boxed{\frac{\partial u}{\partial t} = u \frac{\partial u}{\partial x} + \frac{1}{\rho} \frac{\partial p}{\partial x}} \tag{6.24}$$

This result can also be obtained directly from the x-component of the Euler equation (equation 6.12) by remembering that for a one-dimensional flow the streamwise velocity u is a function of x only.

6.5.3 *Energy Equation

Finally, we derive the differential form of the one-dimensional, unsteady energy equation. The integral form of the energy equation for inviscid flow through a fixed control volume (equation 5.23) is given by:

$$\dot{Q} + \dot{W}_{shaft} = \frac{\partial}{\partial t} \int \rho(\hat{u} + \tfrac{1}{2}V^2 + gz)\, d\forall + \int (\mathbf{n} \cdot \rho \mathbf{V})\left(\hat{u} + \frac{p}{\rho} + \tfrac{1}{2}V^2 + gz\right) dA \tag{6.25}$$

where $V = |\mathbf{V}|$. The procedure for applying the equation in integral form to the quasi-one-dimensional flow shown in Figure 6.8 is very similar to that given for the continuity and momentum equations. The unsteady term becomes

$$\frac{\partial}{\partial t} \int \rho(\hat{u} + \tfrac{1}{2}V^2 + gz)\, d\forall = \frac{\partial}{\partial t} \int \rho e\, d\forall = A \frac{\partial \rho e}{\partial t} dx$$

after higher order terms are neglected. Similarly, the flux term in equation 6.25 becomes

$$\int (\mathbf{n} \cdot \rho \mathbf{V}) \left(\hat{u} + \frac{p}{\rho} + \tfrac{1}{2}V^2 + gz \right) dA = \int (\mathbf{n} \cdot \rho \mathbf{V})(e + pv)\, dA$$

$$= \frac{\partial}{\partial x}[\rho u A(e + pv)]\, dx$$

The term on the left hand side of equation 6.25 will be zero if there is no shaft work done, and no heat transferred (adiabatic flow). Hence, the one-dimensional, unsteady form of the energy equation for inviscid flows, in the absence of heat transfer and shaft work, is given by

$$\boxed{A \frac{\partial \rho e}{\partial t} + \frac{\partial}{\partial x}[\rho u A(e + pv)] = 0} \qquad (6.26)$$

(see also Section 4.7).

EXAMPLE 6.6   *Eulerian Velocity Field*

An Eulerian, Cartesian velocity field is given by $\mathbf{V} = ax^2\mathbf{i} - 2axy\mathbf{j}$.
(a) Is it one-, two-, or three-dimensional?
(b) Is it steady or unsteady?
(c) Is it incompressible?
(d) Find the slope of the streamline passing through the point $[1, -1]$.

Solution   For parts **(a)** and **(b)**, we see that the velocity field is described by two space coordinates (x and y), and it does not depend on time, so that it is two-dimensional and steady.
For part **(c)**, we have

$$\nabla \cdot \mathbf{V} = \frac{\partial u}{\partial x} + \frac{\partial v}{\partial y} = \frac{\partial ax^2}{\partial x} - \frac{\partial 2axy}{\partial y} = 2ax - 2ax = 0$$

so the flowfield is incompressible.
For part **(d)**, we know from the definition of a streamline that its slope in the $[x,y]$-plane is given by the angle α, where

$$\tan \alpha = \frac{v}{u} = \frac{-2axy}{ax^2} - \frac{-2y}{x} = -2$$

Therefore, at the point $[1, -1]$ the streamline makes an angle of $63.4°$ with the x-axis.   ∎

EXAMPLE 6.7   *Unsteady, One-Dimensional Flow*

A gas-filled pneumatic strut in an automobile suspension system can be modeled as a piston in a cylinder filled with a gas of uniform density (see Figure 6.9).[3] As the piston moves, the gas moves also, and we will assume that the velocity distribution in the gas is approximately one-dimensional and linear so that $u = Vx/L$, where V is the speed of the piston, x is the distance measured from the closed end of the cylinder, and L is the position of the piston from the closed end. (Note that this

---

[3] This example was adapted from one given in *Introduction to Fluid Mechanics,* by R. W. Fox & A. T. MacDonald, published by John Wiley & Sons, 1992.

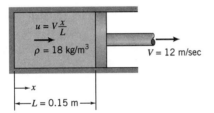

FIGURE 6.9 A gas-filled pneumatic strut.

velocity distribution satisfies the no-slip condition: the velocity at the closed end is zero, and at the surface of the piston it is equal to the speed of the piston.) If $V = $ constant, and if $x = L_0$, $\rho = \rho_0$, at $t = 0$, find the density as a function of time.

Solution: The one-dimensional form of the continuity equation (equation 6.9) is given by

$$\frac{\partial(\rho u)}{\partial x} + \frac{\partial \rho}{\partial t} = 0$$

That is

$$\rho \frac{\partial u}{\partial x} + u \frac{\partial \rho}{\partial x} + \frac{\partial \rho}{\partial t} = 0$$

Since $\rho \neq \rho(x)$,

$$\frac{d\rho}{dt} = -\rho \frac{\partial u}{\partial x}$$

Now,

$$u = V\frac{x}{L}, \quad \frac{\partial u}{\partial x} = \frac{V}{L}, \quad \text{and} \quad L = L_0 + Vt,$$

so that

$$\frac{d\rho}{dt} = -\rho \frac{V}{L_0 + Vt}$$

Separate the variables and integrate:

$$\int_{\rho 0}^{\rho} \frac{d\rho}{\rho} = \int_{0}^{t} \frac{V}{L_0 + Vt} dt$$

Hence:

$$\ln \frac{\rho}{\rho_0} = \ln \frac{L_0}{L_0 + Vt} \quad \text{and} \quad \frac{\rho}{\rho_0} = \frac{L_0}{L_0 + Vt}$$ ∎

PROBLEMS

6.1 State the definition of the total derivative. Define all symbols and notations, and describe the meaning of all terms in words.

6.2 Write down the x-component of the acceleration following a fluid particle given that $\mathbf{V} = u\mathbf{i} + v\mathbf{j} + w\mathbf{k}$, where $\mathbf{V}$ is the velocity and $\mathbf{i}$, $\mathbf{j}$, and $\mathbf{k}$ are the unit vectors in a Cartesian coordinate system.

6.3   State the continuity equation in differential form. Define all symbols and notations, and describe the meaning of all terms in words.

6.4   Write down the vector differential form of the momentum equation for an inviscid, incompressible fluid. Define all symbols and notations, and describe the meaning of all terms in words.

6.5   When is a flow steady? Incompressible?

6.6   Write down the differential form of the continuity equation for:
   (a) Steady flow
   (b) Constant density flow;

   Then write down the integral form of the continuity equation for:
   (a) Steady flow
   (b) Constant density flow

6.7   For a velocity field described by $\mathbf{V} = 2x^2\mathbf{i} - zy\mathbf{k}$, is the flow two- or three-dimensional? Incompressible?

6.8   For an Eulerian flowfield described by $u = 2xyt$, $v = y^3x/3$, $w = 0$, find the slope of the streamline passing through the point [2, 4] at $t = 2$.

6.9   Given the Eulerian description of a two-dimensional fluid flow (in Cartesian coordinates): $u = 2xyt$, $v = y^3x/3$, $w = 0$,
   (a) Find the rate of change of density following a fluid particle as a function of x, y, and t.
   (b) Find the x-component of the acceleration vector as a function of x, y, and t.

6.10   A velocity field is described (in Cartesian coordinates) by $u = 2 - x^3/3$, $v = x^2y - zt$, $w = 0$.
   (a) Write down the y-component of the acceleration of a fluid particle (in the Eulerian system) for this flowfield.
   (b) Is this flowfield incompressible?

6.11   A flowfield is described (in a Cartesian coordinate system) by

$$u = \frac{4x}{t}, \quad v = \frac{y^2}{t}, \quad w = 0, \quad \rho = 3 + \frac{t}{x}$$

Find the rate of change of density following a fluid particle two different ways. Is this flowfield possible?

6.12   A flowfield is described (in Cartesian coordinates) by $\mathbf{V} = (2x^2 + 6z^2x)\mathbf{i} + (y^2 - 4xy)\mathbf{j} - (2z^3 + 2yz)\mathbf{k}$. Is it incompressible?

6.13   Given the fluid flow field (in Cartesian coordinates) $u = 2x$, $v = 16(y + x)$, $w = 0$, $\rho = at^2 + xy$, find the rate of change of density of a particle of fluid with respect to time t two different ways. Is this flowfield possible?

6.14   Consider the following velocity field (in Cartesian coordinates): $u = xt + 2y$, $v = xt^2 - yt$, $w = 0$. Is this flow incompressible?

6.15   Is the flowfield (expressed in Cartesian coordinates) $\mathbf{V} = (2x^2 - xy + z^2)\mathbf{i} + (x^2 - 4xy + y^2)\mathbf{j} + (-2xy - yz + y^2)\mathbf{k}$, compressible or incompressible?

6.16   A Cartesian velocity field is defined by $\mathbf{V} = 2x\mathbf{i} + 5yz^2\mathbf{j} - t^3\mathbf{k}$. Find the divergence of the velocity field. Why is this an important quantity in fluid mechanics?

6.17   Given the following Eulerian description of a fluid flow (in Cartesian coordinates): $u = 2xt$, $v = y^2x/2$, $w = 0$
   (a) How many dimensions does the flow field have?
   (b) Find the rate of change of density (per unit mass) following a fluid particle as a function of x, y, and t.
   (c) Find the x-component of the acceleration as a function of x, y, and t.

6.18   For the velocity field given by $\mathbf{V} = 6x\mathbf{i} - 2yz\mathbf{j} + 3\mathbf{k}$, determine where the flowfield is incompressible.

6.19   In Cartesian coordinates, a particular velocity field is defined by $\mathbf{V} = -2x^2\mathbf{i} + 4xy\mathbf{j} + 3\mathbf{k}$.
   (a) Is this flow field compressible or incompressible?
   (b) Find the acceleration of the fluid at the point $(1, 3, 0)$.
   (c) Find the volume flux passing through area A shown in Figure P6.19.
   (d) What are the dimensions of volume flux?

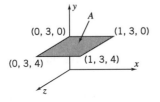

FIGURE P6.19

6.20   For the Cartesian velocity field $\mathbf{V} = 2x^2y\mathbf{i} + 3\mathbf{j} + 4y\mathbf{k}$
   (a) Find $\frac{1}{\rho}\frac{D\rho}{Dt}$.
   (b) Find the rate of change of velocity following a fluid particle.

6.21   A Cartesian velocity field is defined by $\mathbf{V} = 3x^2\mathbf{i} + 4zxt\mathbf{k}$.
   (a) Is the flowfield steady?
   (b) Is the flowfield two- or three-dimensional?
   (c) Find the rate of change of velocity following a fluid particle.
   (d) Is the flowfield incompressible?

6.22   Given the following Eulerian description of a fluid flow (in Cartesian coordinates) $u = 2xyt$, $v = y^3x/3$, $w = 0$
   (a) Is this flow one-, two-, or three-dimensional?
   (b) Is this flow steady?
   (c) Is this flow incompressible?
   (d) Find the x-component of the acceleration vector.

6.23   A velocity field is described (in Cartesian coordinates) by $\mathbf{V} = 2xy\mathbf{i} - 3y^2\mathbf{j}$.
   (a) Is the flowfield incompressible?
   (b) Is the flowfield steady?
   (c) Is the flowfield two- or three-dimensional?
   (d) Find the angle the streamline makes that passes through the point $(3, -2)$.
   (e) Find the acceleration of the flowfield.

6.24   Given the following Eulerian description of a fluid flow (in Cartesian coordinates) $u = 2$, $v = yz^2t$, $w = -z^3t/3$
   (a) Is this flow one-, two-, or three-dimensional?
   (b) Is this flow steady?
   (c) Is this flow incompressible?
   (d) Find the z-component of the acceleration vector.

6.25   Given the following Eulerian description of a fluid flow (in Cartesian coordinates) $u = xyz$, $v = t^2$, $w = 3$
   (a) Is this flow one-, two-, or three-dimensional?
   (b) Is this flow steady?
   (c) Is this flow incompressible?
   (d) Find the x-component of the acceleration vector.

6.26 A velocity field is described by $\mathbf{V} = 2xyz\mathbf{i} - y^2z\mathbf{j}.$
(a) Is the flowfield one-, two-, or three-dimensional?
(b) Is the flowfield steady?
(c) Find the acceleration at the point $[1, -1, 1]$.
(d) Find the slope of the streamline passing through the point $[1, -1, 1]$.

6.27 An Eulerian flowfield is described in Cartesian coordinates by $\mathbf{V} = 4\mathbf{i} + xz\mathbf{j} + 5y^3t\,\mathbf{k}.$
(a) Is it compressible?
(b) Is it steady?
(c) Is the flow one-, two-, or three-dimensional?
(d) Find the y-component of the acceleration.
(e) Find the y-component of the pressure gradient if the fluid is inviscid and gravity can be neglected.

6.28 A fluid flow is described (in Cartesian coordinates) by $u = 2 - x^3/3$, $v = x^2y - zt$, $w = 0$
(a) Is this flow two- or three-dimensional?
(b) Write down the x-component of the acceleration.
(c) Is this flowfield incompressible?

6.29 For a two-dimensional, incompressible flow, the x-component of velocity is given by $u = xy^2$. Find the simplest y-component of the velocity that will satisfy the continuity equation.

6.30 Find the y-component of velocity of an incompressible two-dimensional flow if the x-component is given by $u = 15 - 2xy$. Along the x-axis, $v = 0$.

6.31 The flow of an incompressible fluid in cylindrical coordinates is given by

$$u_\theta = \left(1 + \frac{4}{r^2}\right)\sin\theta - \frac{1}{r}$$

Find u_r if u_r is zero at $r = 2$ for all θ. The flow does not depend on z.

6.32 The velocity in a one-dimensional compressible flow is given by $u = 10x^2$. Find the most general variation of the density with x.

6.33 The x-, y-, and z-components of a velocity field are given by $u = ax + by + cz$, $v = dx + ey + fz$, and $w = gx + hy + jz$. Find the relationship among the coefficients a through j if the flowfield is incompressible.

6.34 For a flow in the xy-plane, the y-component of velocity is given by $v = y^2 - 2x + 2y$. Find a possible x-component for steady, incompressible flow. Is it also valid for unsteady, incompressible flow? Why?

6.35 The x-component of velocity in a steady, incompressible flowfield in the xy-plane is $u = A/x$. Find the simplest y-component of velocity for this flowfield.

6.36 An air bearing is constructed from a circular disk that issues air from many small holes in its lower surface.
(a) Find an expression for the radial velocity under the bearing, assuming that the flow is uniform, steady, and incompressible.
(b) The bearing floats 1.5 mm above the table and the air flows through the bearing with an average velocity of 2 m/s. If the bearing is 1 m in diameter, find the magnitude and location of the maximum radial acceleration experienced by a fluid particle in the gap.

6.37 Show that the velocity distribution in the linear Couette flow illustrated in Figure 1.20 is an exact solution to the incompressible Navier–Stokes equation (the pressure is constant everywhere). Find all the components of the viscous stress for this flow.

6.38   An incompressible fluid flows through a horizontal duct so that

$$V = \frac{V_0}{(1 - \frac{x}{2\ell})}$$

where V_0 is the entrance velocity at $x = 0$, and ℓ is the length of the duct. Find the pressure change along the duct when $U_0 = 5$ m/s and $\ell = 3.0$ m. Assume that the streamwise velocity is uniform over all cross-sections.

6.39   A 1 ft diameter pipe carries an unsteady flow of water. The pipe is connected to a nozzle 0.5 ft in diameter, which exits to the atmosphere. Find the velocity and the acceleration of the parallel stream leaving the nozzle at the instant when the velocity and acceleration in the pipe are 10 ft/s and 2 ft/s², respectively. Assume one-dimensional flow.

6.40   The unsteady flow in a two-dimensional duct 1 m wide passes through a contraction, as shown in Figure P6.40. At a certain time, the volume flow is 1.5 m³/s and it is increasing at the rate of 1.0 m³/s². For quasi-one-dimensional incompressible flow, find the acceleration of the fluid:

(a) At the exit of the contraction

(b) Halfway along the contraction

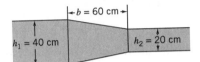

FIGURE P6.40

CHAPTER 7

INCOMPRESSIBLE, IRROTATIONAL FLOWS

When the Mach number is small, or when the fluid is a liquid, we often make the approximation that the flow is incompressible. The flow behavior is then described completely by the incompressible continuity equation

$$\nabla \cdot \mathbf{V} = 0$$

and the Navier–Stokes equation,

$$\frac{D\mathbf{V}}{Dt} = -\frac{1}{\rho}\nabla p + \mathbf{g} + \nu\nabla^2\mathbf{V}$$

together with appropriate boundary conditions such as $\mathbf{V} = \mathbf{V}_w$ (see Section 6.3.4). Although these equations are complex and difficult to solve, it is possible to find solutions under some conditions. Specifically, it is possible to solve the equations when the flow is irrotational, and in this chapter we study some particular solutions for incompressible, irrotational flows.

What does *irrotational* mean? If a flow is irrotational, it does not mean that the flow is not rotating: an irrotational flow can be in rectilinear or rotating motion relative to some reference frame. What is important is the rotation of fluid particles, and

> A flow is irrotational if the average angular velocity of every fluid particle is zero.

We will show that this occurs when the curl of the velocity is zero, that is, when $\nabla \times \mathbf{V} = 0$.

When the flow is irrotational, we can express the vector velocity field in terms of a scalar function ϕ, called the *velocity potential*. As we will see, this greatly simplifies the equations of motion. Lines of constant velocity potential are somewhat like contour lines on a map, and the flow moves from higher to lower potential, and the magnitude of the velocity increases as the equipotential lines get closer together. If the flow is also incompressible, the function ϕ can be found by solving a linear partial differential equation (Laplace's equation, $\nabla^2\phi = 0$), which makes it possible to find analytical solutions for many useful flow cases.

Alternatively, if the flow is incompressible and two-dimensional, the velocity can be expressed in terms of another scalar function ψ, called the *stream function*. Lines of constant ψ correspond to streamlines. If the flow is also irrotational, the function ψ can be found by solving another Laplace's equation, $\nabla^2\psi = 0$.

To know whether a flow is irrotational or not, it is necessary to understand the role of the viscous stresses. In particular, whenever viscous stresses are important the flow will be rotational (that is, not irrotational). Boundary layer flows, fully developed pipe and duct flows, wakes and free shear layers are all rotational. Outside these viscous regions, however, the flow is often irrotational. For example, the flow over a streamlined shape such as an airfoil is rotational inside the boundary layer and wake, but irrotational in the rest of the flowfield.

For irrotational flows, velocity potentials and stream functions can often be used to find solutions to the governing equations. It is clear, however, that we can only find solutions for inviscid flows. Since viscosity is responsible for the energy loss in the system, we cannot find the drag force acting on the airfoil. However, the lift force is due primarily to the pressure difference between the upper and lower surfaces of the airfoil, and these pressure differences are not affected significantly by the presence of the boundary layers and wake (except in the sense that they help to set boundary conditions), so that when viscosity is neglected and the irrotational assumption is made, the pressure distribution can often be found quite accurately. Similarly, the three-dimensional flow through a duct can be calculated with good accuracy using the irrotational assumption, as long as the boundary layers are thin (the flow must be far from being fully developed) and the flow does not separate.

7.1 VORTICITY AND ROTATION

The rotation of a fluid particle is defined by the rotation vector $\boldsymbol{\omega}$, which is the average angular velocity of two originally perpendicular lines "attached" to the fluid particle. Consider a fluid particle in a flow with velocity gradients in all directions. In a short time, dt, the originally square particle will move and distort in the x-y plane as shown in Figure 7.1. For the particle shown, the component of the average angular rotation about the z-axis is given by $\frac{1}{2}(d\alpha - d\beta)$. Therefore

$$\omega_z = \frac{1}{2\,dt}(d\alpha - d\beta)$$

For small angles,

$$d\alpha \approx \tan d\alpha = \frac{\left(v_0 + \frac{\partial v}{\partial x}\Big|_0 dx\right)dt - v_0\,dt}{dx - u_0\,dt + \left(u_0 + \frac{\partial u}{\partial x}\Big|_0 dx\right)dt} = \frac{\partial v}{\partial x}\Big|_0 dt$$

The term $(\partial u/\partial x)_0\,dxdt$ is second order and it is therefore neglected. Similarly, we can show that

$$d\beta \approx \tan d\beta = \frac{\partial u}{\partial y}\Big|_0 dt$$

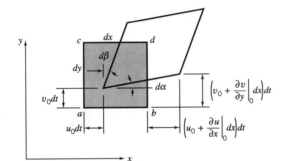

FIGURE 7.1 Rotation of a fluid particle.

Dropping the subscript because the location of the origin is arbitrary gives

$$\omega_z = \frac{1}{2}\left(\frac{\partial v}{\partial x} - \frac{\partial u}{\partial y}\right)$$

Similarly, for the rotation about the y- and x-directions

$$\omega_y = \frac{1}{2}\left(\frac{\partial u}{\partial z} - \frac{\partial w}{\partial x}\right)$$

and

$$\omega_x = \frac{1}{2}\left(\frac{\partial w}{\partial y} - \frac{\partial v}{\partial z}\right)$$

Therefore

$$\boxed{\boldsymbol{\omega} = \tfrac{1}{2}\nabla \times \mathbf{V} = \text{rotation vector}} \tag{7.1}$$

When $\boldsymbol{\omega} = 0$, the average angular velocity of the fluid particle is zero, and the flow is irrotational. The factor $\frac{1}{2}$ is suppressed by defining a vorticity vector such that

$$\boxed{\boldsymbol{\Omega} = 2\boldsymbol{\omega} = \nabla \times \mathbf{V}} \tag{7.2}$$

that is, the curl of the velocity field is the vorticity. When $\boldsymbol{\omega} = 0$, then $\nabla \times \mathbf{V} = 0$, and the flow is irrotational.

In Cartesian coordinates,

$$\nabla \times \mathbf{V} = \left(\frac{\partial w}{\partial y} - \frac{\partial v}{\partial z}\right)\mathbf{i} - \left(\frac{\partial w}{\partial x} - \frac{\partial u}{\partial z}\right)\mathbf{j} + \left(\frac{\partial v}{\partial x} - \frac{\partial u}{\partial y}\right)\mathbf{k} \tag{7.3}$$

This is often written in shorthand using the determinant form

$$\nabla \times \mathbf{V} = \begin{vmatrix} \mathbf{i} & \mathbf{j} & \mathbf{k} \\ \frac{\partial}{\partial x} & \frac{\partial}{\partial y} & \frac{\partial}{\partial z} \\ u & v & w \end{vmatrix}$$

(see Section A.7).

Viscous fluids in contact with a solid surface cannot be irrotational. A boundary layer forms near the surface, and inside this layer the flow is irrotational. For simplicity, consider a two-dimensional boundary layer such as that shown in Figure 1.16. The vorticity vector has only one component, given by

$$\nabla \times \mathbf{V} = \left(\frac{\partial v}{\partial x} - \frac{\partial u}{\partial y}\right)\mathbf{k}$$

Since the boundary layer thickness grows with distance along the surface, the streamlines inside the boundary layer gradually get displaced further from the surface, so that, in addition to the principal velocity gradient $\partial u/\partial y$, v varies with x, so that $\partial v/\partial x \neq 0$, Now, $\partial v/\partial x$ is small compared to $\partial u/\partial y$ so that $\nabla \times \mathbf{V} \neq 0$. That is, the flow in the boundary layer is rotational. Outside this narrow region near the surface, however, the flow can often be considered irrotational. As shown below, it is possible to represent the vector velocity field in that region in terms of a scalar function ϕ, which allows us to find the velocity distribution everywhere.

Once the velocity is known, the pressures can be found using Bernoulli's equation if the flow is steady and incompressible. The Bernoulli constant (Section 4.7.3) has the same value along and across streamlines since the flow is irrotational (see Section 4.2).

7.2   THE VELOCITY POTENTIAL ϕ

The condition of irrotational flow ($\nabla \times \mathbf{V} = 0$) means that, in Cartesian coordinates (see equation 7.3):

$$\frac{\partial u}{\partial y} = \frac{\partial v}{\partial x}, \quad \frac{\partial v}{\partial z} = \frac{\partial w}{\partial y}, \quad \text{and} \quad \frac{\partial w}{\partial x} = \frac{\partial u}{\partial z} \tag{7.4}$$

It is possible to satisfy these equations by defining a potential function ϕ, such that

$$u = \frac{\partial \phi}{\partial x}, \quad v = \frac{\partial \phi}{\partial y}, \quad \text{and} \quad w = \frac{\partial \phi}{\partial z}$$

so that

$$\frac{\partial u}{\partial y} = \frac{\partial^2 \phi}{\partial x \partial y}, \quad \frac{\partial v}{\partial x} = \frac{\partial^2 \phi}{\partial y \partial x}$$

and so on. Therefore, the function ϕ will always satisfy equation 7.4 since the order of differentiation does not matter as long as ϕ is well-behaved. In terms of ϕ the velocity may be written as

$$\mathbf{V} = u\mathbf{i} + v\mathbf{j} + w\mathbf{k} = \frac{\partial \phi}{\partial x}\mathbf{i} + \frac{\partial \phi}{\partial y}\mathbf{j} + \frac{\partial \phi}{\partial z}\mathbf{k}$$

We see that for irrotational flow it is always possible to write the velocity as the gradient of a scalar quantity, that is,

$$\boxed{\mathbf{V} = \nabla \phi} \tag{7.5}$$

In cylindrical coordinates, the condition for irrotational flow requires that

$$\frac{1}{r}\frac{\partial u_z}{\partial \theta} = \frac{\partial u_\theta}{\partial z}, \quad \frac{\partial u_r}{\partial z} = \frac{\partial u_z}{\partial r}, \quad \text{and} \quad \frac{\partial r u_\theta}{\partial r} = \frac{\partial u_r}{\partial \theta}$$

These equations are satisfied when

$$u_r = \frac{\partial \phi}{\partial r}, \quad u_\theta = \frac{1}{r}\frac{\partial \phi}{\partial \theta}, \quad \text{and} \quad u_z = \frac{\partial \phi}{\partial z}$$

and again we have

$$\mathbf{V} = \nabla \phi$$

We can compare ϕ with the gravitational potential defined by

$$\mathbf{g} = -g\nabla \phi_g$$

(Section 6.3.1), where the potential function ϕ_g was identified with the altitude or elevation. Similarly, ϕ is a potential function and it is called the *velocity potential.* Irrotational flows are sometimes called *potential* flows because we can define the velocity in terms of a potential function.

Introducing the function ϕ has two major advantages.

1. The flow can be described completely by the scalar function ϕ, instead of three separate functions, u, v, and w.

2. The function ϕ can be found from a *linear* equation, as will be shown, so that complex flow fields can be constructed from the linear addition of simpler flow fields.

7.3   THE STREAM FUNCTION ψ

For a two-dimensional, incompressible flow (which can be unsteady, and/or rotational) the continuity equation in Cartesian coordinates is given by

$$\frac{\partial u}{\partial x} + \frac{\partial v}{\partial y} = 0$$

We can define a function ψ such that

$$u = \frac{\partial \psi}{\partial y}$$

and

$$v = -\frac{\partial \psi}{\partial x}$$

and therefore

$$\frac{\partial}{\partial x}\left(\frac{\partial \psi}{\partial y}\right) + \frac{\partial}{\partial y}\left(-\frac{\partial \psi}{\partial x}\right) = 0$$

We see that the function ψ was defined so that the continuity equation is always satisfied. In terms of ψ the velocity may be written as

$$\mathbf{V} = u\mathbf{i} + v\mathbf{j} = \frac{\partial \psi}{\partial y}\mathbf{i} - \frac{\partial \psi}{\partial x}\mathbf{j}$$

We see that for two-dimensional incompressible flows, the vector velocity field can be described by a single scalar function ψ, instead of three separate functions u, v, and w. We will show that for irrotational flows ψ can be found from a linear equation. In this way, ψ is similar to ϕ. In addition, ψ has a most important property: lines of constant ψ are streamlines of the flow. The function ψ is therefore called the *stream function*.

To illustrate this aspect, we can write

$$d\psi = \frac{\partial \psi}{\partial x}dx + \frac{\partial \psi}{\partial y}dy$$

$$= -v\,dx + u\,dy$$

When ψ is constant, $d\psi = 0$. Then

$$\frac{dy}{dx} = \frac{u}{v} \tag{7.6}$$

which shows that lines of constant ψ have a direction that is tangential to the instantaneous flow direction. Hence, lines of constant ψ are streamlines.

A stream function can also be defined for a cylindrical coordinate system. In cylindrical coordinates, the continuity equation for a two-dimensional incompressible flow is given by

$$\nabla \cdot \mathbf{V} = \frac{\partial r u_r}{\partial r} + \frac{\partial u_\theta}{\partial \theta} = 0$$

With a stream function defined by

$$u_r = \frac{1}{r}\frac{\partial \psi}{\partial \theta}$$

and

$$u_\theta = -\frac{\partial \psi}{\partial r}$$

we have

$$\frac{\partial r u_r}{\partial r} = \frac{\partial}{\partial r}\frac{\partial \psi}{\partial \theta}$$

$$\frac{\partial u_\theta}{\partial \theta} = -\frac{\partial}{\partial \theta}\frac{\partial \psi}{\partial r}$$

and we see that the continuity equation is again automatically satisfied.

7.4 FLOWS WHERE BOTH ψ AND ϕ EXIST

The velocity potential can be defined for any irrotational flowfield, and the stream function can be defined for any two-dimensional, incompressible flowfield.[1] Therefore when the flow is two-dimensional, irrotational and incompressible, both ϕ and ψ exist (there are no restrictions on steadiness: the flow can be steady or unsteady).

Under these conditions, it can be shown that lines of constant ϕ intersect lines of constant ψ at right angles. We can write

$$d\phi = \frac{\partial \phi}{\partial x}\,dx + \frac{\partial \phi}{\partial y}\,dy$$

$$= u\,dx + v\,dy$$

When ϕ is constant, $d\phi = 0$. Then

$$\frac{dy}{dx} = -\frac{u}{v} \tag{7.7}$$

Hence, from equations 7.6 and 7.7

$$\left.\frac{dy}{dx}\right|_\phi \times \left.\frac{dy}{dx}\right|_\varphi = -1$$

which shows that ϕ and ψ form an orthogonal flownet. This result may have been anticipated from the fact that, since $\mathbf{V} = \nabla\phi$, the flow velocity points in the direction of the maximum rate of change of ϕ, which is at right angles to lines of constant velocity potential (see Section A.4 for a discussion of the gradient operator).

---

[1] The stream function ψ can also be defined for a two-dimensional, steady, compressible flow. See, for example, Milne-Thomson *Theoretical Aerodynamics,* 4th edition, published by Macmillan, 1966.

7.5 SUMMARY OF DEFINITIONS AND RESTRICTIONS

We have defined a potential function ϕ such that

$$\mathbf{V} = \nabla\phi \tag{7.8}$$

where ϕ satisfies the condition of irrotationality, and so

$$u = \frac{\partial\phi}{\partial x} \quad \text{and} \quad v = \frac{\partial\phi}{\partial y} \tag{7.9}$$

or

$$u_r = \frac{\partial\phi}{\partial r} \quad \text{and} \quad u_\theta = \frac{1}{r}\frac{\partial\phi}{\partial\theta} \tag{7.10}$$

Similarly, a stream function ψ was defined such that the two-dimensional continuity equation is automatically satisfied, and so, for an incompressible flow,

$$u = \frac{\partial\psi}{\partial y} \quad \text{and} \quad v = \frac{\partial\psi}{\partial x} \tag{7.11}$$

or

$$u_r = \frac{1}{r}\frac{\partial\psi}{\partial\theta} \quad \text{and} \quad u_\theta = -\frac{\partial\psi}{\partial r} \tag{7.12}$$

If either ϕ or ψ can be found, $\mathbf{V}$ is known, and the pressure can be obtained from Bernoulli's equation.

The velocity potential was defined so that the flow was irrotational, and the stream function was defined so that the two-dimensional incompressible continuity equation was satisfied. This leaves open the question of signs. For example, the stream function could be defined so that u and v in equation 7.12 had opposite signs from those adopted here, and the continuity equation would still be satisfied. There is no general agreement on this question, and therefore there is considerable possibility for confusion. When consulting other textbooks it is necessary to pay close attention to how ϕ and ψ are defined, although the sign convention adopted here seems to be the convention used by most recent textbooks.

It is clear that flows may exist where a velocity potential can be defined, but not a stream function. In what follows, we will assume that the flow is incompressible, irrotational, and two-dimensional, so that both a potential function and a stream function exist.

EXAMPLE 7.1   *Stream Functions and Velocity Potentials*

Show that $\phi = x^3 - 3xy^2$ is a valid velocity potential, and that it describes an incompressible flowfield. Determine the corresponding stream function. Find the stagnation points, and the pressure distribution.

Solution: To be a valid velocity potential, ϕ must satisfy the condition of irrotationality. For this two-dimensional, Cartesian flowfield, we have

$$u = \frac{\partial\phi}{\partial x} \quad \text{and} \quad v = \frac{\partial\phi}{\partial y}$$

Therefore,

$$u = 3x^2 - 3y^2 \quad \text{and} \quad v = -6xy \tag{7.13}$$

The curl of the velocity is given by

$$\nabla \times \mathbf{V} = \left(\frac{\partial v}{\partial x} - \frac{\partial u}{\partial y}\right) \mathbf{k} = (-6y + 6y)\,\mathbf{k} = 0$$

and so ϕ is a valid velocity potential.

The divergence of the velocity field is given by

$$\nabla \cdot \mathbf{V} = \frac{\partial u}{\partial x} + \frac{\partial v}{\partial y} = 6x - 6x = 0$$

and so the flowfield is incompressible.

From the definition of the stream function

$$u = \frac{\partial \psi}{\partial y} \quad \text{and} \quad v = -\frac{\partial \psi}{\partial x}$$

Using the result given in equation 7.13, we obtain by integration

$$\psi = 3x^2 y - y^3 + f(x) + C_1$$

and

$$\psi = 3x^2 y + g(y) + C_2$$

where C_1 and C_2 are constants of integration and f and g are unknown functions of x and y, respectively. By comparing the two results for ψ, we get

$$\psi = 3x^2 y - y^3 + C$$

The constant C is arbitrary (we are only interested in the derivatives of ψ), and we can set it to zero by choosing $\psi = 0$ at the origin. Finally,

$$\psi = 3x^2 y - y^3$$

The stagnation points can be found by finding the points where $u = v = 0$. For the velocity field given by equation 7.13, this happens only at the origin, so there is only one stagnation point.

The pressure distribution is given by Bernoulli's equation, so that

$$p + \tfrac{1}{2}(u^2 + v^2) = p_0$$

where p_0 is a constant for the entire flowfield since it is irrotational. We find that

$$p - p_0 = 9(x^2 + y^2)^2 \qquad \blacksquare$$

7.6 EXAMPLES OF POTENTIAL FLOW

Here we consider a number of simple flows that can be used as a building blocks to construct more complex flows. The results are summarized in Table 7.1.

TABLE 7.1 Summary of Potential Functions and Stream Functions for Simple Flows

	Cartesian coordinates		Cylindrical coordinates	
	ϕ	ψ	ϕ	ψ
Uniform flow	Ux	Uy	$Ur\cos\theta$	$Ur\sin\theta$
Point source	$\dfrac{q}{2\pi}\ln\sqrt{x^2+y^2}$	$\dfrac{q}{2\pi}\tan^{-1}\left(\dfrac{y}{x}\right)$	$\dfrac{q}{2\pi}\ln r$	$\dfrac{q}{2\pi}\theta$
Point sink	$-\dfrac{q}{2\pi}\ln\sqrt{x^2+y^2}$	$-\dfrac{q}{2\pi}\tan^{-1}\left(\dfrac{y}{x}\right)$	$-\dfrac{q}{2\pi}\ln r$	$-\dfrac{q}{2\pi}\theta$
Potential vortex (counterclockwise)	$\dfrac{\Gamma}{2\pi}\tan^{-1}\left(\dfrac{y}{x}\right)$	$-\dfrac{\Gamma}{2\pi}\ln\sqrt{x^2+y^2}$	$\dfrac{\Gamma}{2\pi}\theta$	$-\dfrac{\Gamma}{2\pi}\ln r$
Doublet	$\dfrac{Kx}{x^2+y^2}$	$-\dfrac{Ky}{x^2+y^2}$	$\dfrac{K\cos\theta}{r}$	$\dfrac{K\sin\theta}{r}$

$$u=\frac{\partial\phi}{\partial x}=\frac{\partial\psi}{\partial y},\quad v=\frac{\partial\phi}{\partial y}=-\frac{\partial\psi}{\partial x};\quad u_r=\frac{\partial\phi}{\partial r}=\frac{1}{r}\frac{\partial\psi}{\partial\theta},\quad u_\theta=\frac{1}{r}\frac{\partial\phi}{\partial\theta}=-\frac{\partial\psi}{\partial r}$$

7.6.1 Uniform Flow

For uniform flow to the right (Figure 7.2), $u = U$, and $v = 0$. From equation 7.10

$$u=\frac{\partial\phi}{\partial x}=U$$

so that

$$\phi = Ux + g(y) + \text{constant}$$

and

$$v=\frac{\partial\phi}{\partial y}=0$$

Hence,

$$\phi = Ux + \text{constant}$$

Since we are ultimately interested in the velocity, and since the velocity is the gradient of ϕ (remember $\mathbf{V} = \nabla\phi$), the constants of integration in both expressions can be given any value we like. For convenience, we make them both

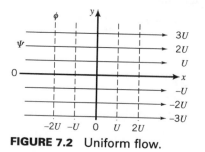

FIGURE 7.2 Uniform flow.

zero. Therefore a uniform flowfield can be expressed in terms of a single scalar function ϕ, where

$$\phi = Ux$$

(We can check this by finding $u = \partial\phi/\partial x$ and $v = \partial\phi/\partial y$, and by making sure that we recover $u = U$, and $v = 0$.)
  Also, from equation 7.12,

$$u = \frac{\partial\psi}{\partial y} = U$$

so that

$$\psi = Uy + f'(x) + \text{constant}$$

and

$$v = -\frac{\partial\psi}{\partial x} = 0$$

Hence,

$$\psi = Uy + \text{constant}$$

Therefore, a uniform flowfield can also be expressed in terms of a scalar function ψ, where

$$\psi = Uy$$

To summarize, for a uniform flow

$$\phi = Ux, \quad \psi = Uy \qquad (7.14)$$

In cylindrical coordinates,

$$\phi = Ur\cos\phi, \quad \psi = Ur\sin\theta \qquad (7.15)$$

7.6.2 Point Source

A point source in a two-dimensional flow is a point where fluid enters the flowfield and flows out equally in all directions in the plane of the source (Figure 7.3). It is an example of a critical or singular point, where streamlines can meet. The mass flow supplied by the source is constant, and since we have restricted ourselves to incompressible flows the volume flow rate is also constant. The flow is two-dimen-

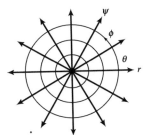

FIGURE 7.3 Source flow. Sink flow has the directions of all the arrows reversed.

sional, and q is the volume flux per unit length (into the page) of source.[2] Cylindrical coordinates are the natural coordinates for this flow, so that $q = 2\pi r u_r$, where u_r is the velocity in the radial direction, and r is the radial coordinate. For a point *source*, q is positive and

$$u_r = \frac{q}{2\pi r}, \quad \text{and} \quad v_\theta = 0$$

Hence,

$$u_r = \frac{q}{2\pi r} = \frac{\partial \phi}{\partial r} = \frac{1}{r} \frac{\partial \psi}{\partial \theta}$$

In cylindrical coordinates, therefore,

$$\phi = \frac{q}{2\pi} \ln r \quad \text{and} \quad \psi = \frac{q}{2\pi} \theta \tag{7.16}$$

In Cartesian coordinates,

$$\phi = \frac{q}{2\pi} \ln \sqrt{x^2 + y^2} \quad \text{and} \quad \psi = \frac{q}{2\pi} \tan^{-1}\left(\frac{y}{x}\right) \tag{7.17}$$

A point *sink* is a point where fluid exits the flow field and fluid flows in equally from all directions. It is a source with a negative volume flux, so that

$$\phi = -\frac{q}{2\pi} \ln r \quad \text{and} \quad \psi = -\frac{q}{2\pi} \theta \tag{7.18}$$

7.6.3 Potential Vortex

A *free* or *potential* vortex is a flow with circular paths around an axis, where the velocity distribution is such that the flow is irrotational (see Figure 7.4). Since $\nabla \times \mathbf{V} = 0$, all components of the vorticity vector must be zero, so that

$$\frac{1}{r} \frac{\partial u_z}{\partial \theta} - \frac{\partial u_\theta}{\partial z} = \frac{\partial u_r}{\partial z} - \frac{\partial u_z}{\partial r} = \frac{1}{r} \frac{\partial r u_\theta}{\partial r} - \frac{1}{r} \frac{\partial u_r}{\partial \theta} = 0$$

A potential vortex has circular symmetry, so that there is no flow in the z- and r-directions, and all derivatives with respect to z and θ are zero. Hence,

$$\frac{\partial r u_\theta}{\partial r} = \frac{d r u_\theta}{d r} = 0$$

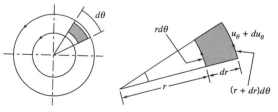

FIGURE 7.4 Potential vortex.

---

[2] The dimensions of q are volume/length per unit time $= L^2 T^{-1}$.

The quantity ru_θ is therefore a constant, and the velocity distribution for a potential vortex is given by

$$u_\theta = \frac{\text{constant}}{r}$$

By convention, we set the constant equal to $\Gamma/2\pi$, where the new constant, Γ, is called the *circulation*, and it is positive for counterclockwise rotation. Therefore

$$u_\theta = \frac{\Gamma}{2\pi r} \tag{7.19}$$

The circulation Γ represents the strength of the vortex. Note that, as r goes to zero, u_θ approaches infinity. This cannot happen in a real fluid. At some point, viscosity will become important in a real fluid, and it will prevent the velocity from becoming infinite. In fact, viscous friction will cause the core of the vortex to start rotating as a solid body, and the flow in this region will no longer be irrotational.

Outside this viscous region, in the irrotational part of the flow, Bernoulli's equation applies between any two points. If the pressure far from the center of the vortex ($r \to \infty$) is p_∞, and we can ignore gravity, Bernoulli's equation gives

$$p + \tfrac{1}{2}\rho V^2 = p_\infty$$

At any position r, $V = u_\theta$, so that

$$p = p_\infty - \frac{\rho \Gamma^2}{8\pi^2 r^2}$$

As the radius decreases, the velocity increases and the pressure decreases. Near the core of a free vortex, very low pressures can be found. It is this low pressure that makes trailing vortices in the wake of airplanes visible by allowing the water vapor in the air to condense and show up as *vapor trails.*

A whirlpool in water behaves very similarly. Here, the free surface is a constant-pressure surface, so that the velocity trades off with the height. The shape of the free surface will be given by

$$h = h_\infty - \frac{\rho \Gamma^2}{8g\pi^2 r^2} \tag{7.20}$$

We see why a whirlpool is marked by a sharp depression in the water surface. However, near its center, viscous friction will become important and the water will begin to rotate as a rigid body. From Section 2.11, we know that rigid-body rotation of a liquid produces a parabolic free surface, so that as r decreases, the surface profile changes from that given by equation 7.20 to a parabolic shape, as shown in Figure 7.5.

In terms of the potential function ϕ, the free vortex has a potential proportional to θ. That is

$$\phi = \frac{\Gamma}{2\pi}\theta$$

To check this result, we find u_r and u_θ

$$u_r = \frac{\partial \phi}{\partial r} = 0$$

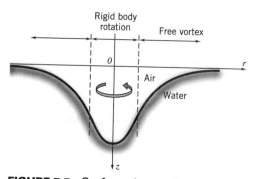

FIGURE 7.5 Surface shape of a whirlpool, showing the transition from free vortex motion at large r to rigid-body rotation at small r.

and

$$u_\theta = \frac{1}{r}\frac{\partial \phi}{\partial \theta} = \frac{\Gamma}{2\pi r}$$

as above.

Equipotential lines are therefore radial lines with $\theta = $ constant, and the flow describes concentric circles. For a positive vortex (counterclockwise flow), the circulation is positive.

In terms of the stream function, we have

$$\psi = -\frac{\Gamma}{2\pi}\ln r$$

and so

$$u_r = \frac{1}{r}\frac{\partial \psi}{\partial \theta} = 0$$

and

$$u_\theta = -\frac{\partial \psi}{\partial r} = \frac{\Gamma}{2\pi r}$$

as above.

To summarize, for a potential vortex in cylindrical coordinates

$$\phi = \frac{\Gamma}{2\pi}\theta, \quad \text{and} \quad \psi = -\frac{\Gamma}{2\pi}\ln r \tag{7.21}$$

where Γ is positive for counterclockwise rotation and negative for clockwise rotation. In Cartesian coordinates,

$$\phi = \frac{\Gamma}{2\pi}\tan^{-1}\left(\frac{y}{x}\right), \quad \text{and} \quad \psi = -\frac{\Gamma}{2\pi}\ln\sqrt{x^2 + y^2} \tag{7.22}$$

7.7 LAPLACE'S EQUATION

We will now show that, under certain conditions, the velocity potential and stream function can be found by solving a linear, partial differential equation called Laplace's equation.

We begin with the velocity potential. Using the continuity equation for incompressible flow ($\nabla \cdot \mathbf{V} = 0$), and the definition of the velocity potential ($\mathbf{V} = \nabla \phi$), we can write

$$\nabla \cdot \mathbf{V} = \nabla \cdot (\nabla \phi) = \nabla^2 \phi = 0$$

where ∇^2 is the Laplacian operator (see Section A.6). That is,

$$\boxed{\nabla^2 \phi = 0} \tag{7.23}$$

This is called Laplace's equation. The velocity potential satisfies Laplace's equation for irrotational, incompressible flow. In Cartesian coordinates,

$$\frac{\partial^2 \phi}{\partial x^2} + \frac{\partial^2 \phi}{\partial y^2} + \frac{\partial^2 \phi}{\partial z^2} = 0 \tag{7.24}$$

and in cylindrical coordinates,

$$\frac{\partial^2 \phi}{\partial r^2} + \frac{1}{r}\frac{\partial \phi}{\partial r} + \frac{1}{r^2}\frac{\partial^2 \phi}{\partial \theta^2} + \frac{\partial^2 \phi}{\partial z^2} = 0 \tag{7.25}$$

Consider the stream function in two-dimensional, irrotational flow. Here,

$$\nabla \times \mathbf{V} = \left(\frac{\partial v}{\partial x} - \frac{\partial u}{\partial y}\right)\mathbf{k} = -\left(\frac{\partial^2 \psi}{\partial x^2} + \frac{\partial^2 \psi}{\partial y^2}\right)\mathbf{k} = 0$$

so that, in Cartesian coordinates,

$$\frac{\partial^2 \psi}{\partial x^2} + \frac{\partial^2 \psi}{\partial y^2} = 0$$

That is,

$$\boxed{\nabla^2 \psi = 0} \tag{7.26}$$

The stream function satisfies Laplace's equation for irrotational, two-dimensional, incompressible flow.

Laplace's equation can be used to find the velocity potential or the stream function (with the appropriate boundary conditions). Once ϕ or ψ is known, $\mathbf{V}$ can be found, and the pressure is found from Bernoulli's equation.

Laplace's equation is useful in many fields and a great deal has been written on its solutions (see, for instance, Lamb's *Hydrodynamics*, reprinted by Dover, 1945). Here, we are concerned with only one aspect of this equation—its linearity. An equation is linear if, when two separate solutions are known, the sum of these two solutions is also a solution of the equation.

If we have, for example, two known solutions to Laplace's equation, ψ_1 and ψ_2, consider $\psi = \psi_1 + \psi_2$. In Cartesian coordinates,

$$\frac{\partial^2 \psi}{\partial x^2} + \frac{\partial^2 \psi}{\partial y^2} = \frac{\partial^2 \psi_1}{\partial x^2} + \frac{\partial^2 \psi_1}{\partial y^2} + \frac{\partial^2 \psi_2}{\partial x^2} + \frac{\partial^2 \psi_2}{\partial y^2}$$

$$= 0 + 0 = 0$$

Therefore $\nabla^2 \psi = 0$, demonstrating that the equation is linear.

The linearity of Laplace's equation means that we can construct new solutions by combining known solutions. For instance, we know the stream functions for a uniform flow ($\psi_1 = Uy$) and for a point source ($\psi_2 = \frac{Q}{2\pi}\theta$), and we can show that each stream function satisfies $\nabla^2\psi = 0$. So their sum also satisfies Laplace's equation, and therefore the combined flowfield is also a valid flow solution. This is the example considered in the next section.

7.8   SOURCE IN A UNIFORM FLOW

To find the flow pattern that results when a source is placed in a uniform flow, we add the stream functions for a source and a uniform flow to obtain the combined stream function ψ.

$$\psi = \psi_1 + \psi_2 = Uy + \frac{q}{2\pi}\theta$$

$$= Uy + \frac{q}{2\pi}\tan^{-1}\frac{y}{x}$$

The streamline pattern for the combination is shown in Figure 7.6. When a *sink* is placed in a uniform flow, we obtain the streamlines shown in Figure 7.7. (Question: what do the equipotential lines look like for this flow?)

The lines of constant ψ represent streamlines. By definition, there can be no flow across streamlines, and therefore a line of constant ψ can be used to represent a solid wall in an inviscid flow. In an inviscid flow, the no-slip boundary condition does not apply, and the boundary condition at a solid surface reduces to the impermeability condition given by equation 6.20, that is,

$$\mathbf{n}\cdot\mathbf{V} = \mathbf{n}\cdot\mathbf{V}_w$$

Since there can be no flow across a streamline, each streamline is equivalent to a solid surface. Figures 7.6 and 7.7, therefore, represent the flow over an infinite number of different half-bodies, one for each streamline. The streamline passing through the point O describes the shape of a particularly interesting body. Furthermore, the straight streamline that acts as a line of symmetry could represent another solid surface, so that the top half of the flowfield can be used to represent the flow over a hill of a particular shape. By changing the relative strengths of the freestream flow and the source or sink, many different shapes can be generated. In this way, we can find the velocity and pressure field around a variety of solid bodies in inviscid flow. We cannot take account of the boundary layers, but it is still a very useful technique.

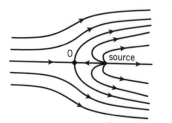

FIGURE 7.6   Source in a uniform flow.

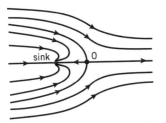

FIGURE 7.7 Sink in uniform flow.

EXAMPLE 7.2 *Solutions Obtained by Superposition*

We have demonstrated that it is possible to generate interesting two-dimensional flows by superposition of some basic building blocks such as uniform flow, sources, sinks, and vortices. Unfortunately, the procedure is rather tedious, especially when a large number of elements are used. To reduce the effort involved in generating such flows, it is possible to use the "Ideal Flow Machine" available on the Web at *http://www.engapplets.vt.edu/.*
    By using this resource, generate the streamline patterns for:
(*a*) a source and a sink of equal strength $q = Ua$, first separated by a distance *a*, and then by a distance $2a$ in the direction of a uniform flow of strength *U*.
(*b*) Repeat this example with the axis joining the source and sink placed at right angles to the uniform flow.
(*c*) For the case in part (*a*), add a vortex of strength $\Gamma = Ua$, located halfway between the source and the sink, and then repeat using a vortex of strength $\Gamma = 2Ua$. ∎

7.9 POTENTIAL FLOW OVER A CYLINDER

One very useful application of potential flow methods is to find the flow over a cylinder. This flow can be generated by placing a sink and a source close together in a uniform flow.
    When a source and a sink of equal strength are placed in a uniform flow, a closed streamline appears (O and o in Figure 7.8). Inside the closed streamline, all the flow originating from the source is absorbed by the sink. The closed streamline acts like a solid body.
    From Figure 7.8, it might be inferred that as the sink and the source come closer together, the shape of the closed streamline will look more and more like a circle. We could guess that the closed streamline becomes a circle when the source

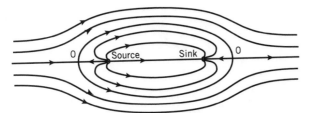

FIGURE 7.8 Sink and source of equal strength in uniform flow.

and the sink occupy the same location. This may sound difficult to accomplish, since we might expect the source and the sink to cancel each other, leaving us with undisturbed, uniform flow. This is not necessarily the case, as we shall see from the following analysis.

When we add the stream functions for a source and a sink, separated by a distance $2a$, we have for the combined source-sink pair

$$\psi = -\frac{q}{2\pi}(\theta_1 - \theta_2) \tag{7.27}$$

where the source is labeled with subscript 2, and the sink is labeled with subscript 1 (Figure 7.9). When the distance a is small, the angle $\theta_1 - \theta_2$ is also small, so that $r(\theta_1 - \theta_2) \approx 2a \sin \theta$, and

$$\theta_1 - \theta_2 \approx \frac{2a \sin \theta}{r}$$

so that

$$\psi = \frac{qa \sin \theta}{\pi r} \tag{7.28}$$

As a becomes very small compared to r, we let q increase so that the product qa remains finite and constant. That is,

$$\psi = -\frac{K \sin \theta}{r} \tag{7.29}$$

Under these conditions, the source-sink pair is called a *doublet,* and $K = qa/\pi$ is called the *strength* of the doublet. The velocity potential of a doublet is given by

$$\phi = \frac{K \cos \theta}{r} \tag{7.30}$$

When a uniform flow is added to the doublet, we obtain the combined stream function

$$\psi = Ux - \frac{K \sin \theta}{r} = Ur\left(1 - \frac{R^2}{r^2}\right)\sin \theta \tag{7.31}$$

(since $x = r \sin \theta$). The streamline pattern is shown in Figure 7.10. We can show that the closed streamline is a circle of radius R $(= \sqrt{K/U})$, and therefore the

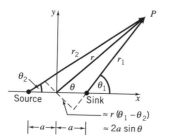

FIGURE 7.9   Notation for a doublet.

FIGURE 7.10 Doublet in a uniform flow.

combination of a doublet and a uniform flow models the inviscid flow over a cylinder.

Note that the stagnation streamlines and the closed streamline describing the cylinder meet at the front and rear stagnation points, A and B respectively. Also, we know that the true flow pattern around a cylinder at reasonable Reynolds numbers does not look like that shown in Figure 7.10, especially in the wake region. For example, the inviscid flow is symmetrical front to back, whereas the viscous flow is not. We will consider these differences further after we have found the velocity and pressure distributions for the inviscid flow.

From equation 7.31, we have

$$u_r = \frac{1}{r}\frac{\partial \psi}{\partial \theta} = U\left(1 - \frac{R^2}{r^2}\right)\cos \theta$$

and

$$u_\theta = -\frac{\partial \psi}{\partial r} = -U\left(1 + \frac{R^2}{r^2}\right)\sin \theta$$

On the surface of the cylinder, $r = R$, and therefore $u_r = 0$ (as expected, since the surface is a streamline), and $u_\theta = u_{\theta s}$, where

$$u_{\theta s} = -2U \sin \theta$$

Note that there are important differences between this inviscid flow solution, and a "real" viscous flow: the inviscid flow does not satisfy the no-slip condition, and boundary layers do not form on the surface of the cylinder.

7.9.1 Pressure Distribution

Since the flow is steady, inviscid, irrotational, and incompressible, we can find the pressure distribution using Bernoulli's equation. If the pressure far from the cylinder is p_∞, then, along the stagnation streamline and around the cylinder,

$$p_\infty + \tfrac{1}{2}\rho U^2 = p_s + \tfrac{1}{2}\rho u_{\theta s}^2$$

where p_s is the surface pressure. Hence, the theoretical pressure distribution over the cylinder surface is given by

$$p_s = p_\infty + \tfrac{1}{2}\rho U^2(1 - 4\sin^2 \theta)$$

Figure 7.11 shows a comparison of the measured and theoretical pressure distributions. The comparison is good over the first 60° or so, but the inviscid solution

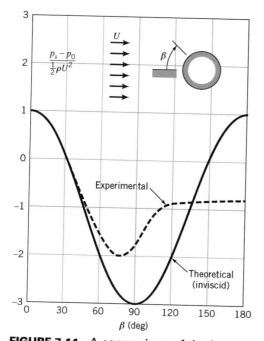

FIGURE 7.11 A comparison of the inviscid pressure distribution on a circular cylinder with a typical experimental distribution. From Munson, Young, & Okiishi, *Fundamentals of Fluid Mechanics,* John Wiley & Sons, 1998.

predicts a much lower pressure at the shoulder ($\beta = 90°$) than what is found by experiment. In the wake ($\beta > 90°$), the analysis predicts a full pressure recovery to its stagnation point value at the rear stagnation point ($\beta = 180°$), whereas the experiment indicates that the pressure at the rear stagnation point recovers to only a fraction of its original value.

We noted earlier that potential flow methods cannot be used to find the drag on a body. For the flow over a cylinder, we see that the inviscid, potential flow solution gives a symmetric pressure distribution, and in the streamwise direction the force due to pressure acting on the front face is exactly balanced by the force due to pressure acting on the back face. The drag on the cylinder is zero. This is obviously not true for the flow of a real fluid where the boundary layers and wake play a major role.

7.9.2 Viscous Effects

Over the front face, the presence of the boundary layer affects the pressure distribution on the cylinder through two principal mechanisms: (a) viscous losses, and (b) a slight displacement of the streamlines caused by the retardation of the flow within the boundary layer (see Section 10.3). Near the shoulder where $\beta = 90°$, the pressure gradient changes from being negative (decreasing pressure, called a *favorable pressure gradient*) to positive (increasing pressure, an *adverse pressure gradient*). The force in the flow direction due to pressure differences changes sign from being an accelerating force to being a retarding force. In response, the flow slows down. However, the fluid in the boundary layer has already given up some energy (in the form of pressure) and momentum (in the form of velocity) because of viscous energy dissipation, and it does not have enough momentum to overcome the retarding force. As the pressure rises, some fluid near the wall actually reverses direction, and the flow separates. Large eddying motions are formed in the wake, and large pressure losses occur. The wake also exerts an upstream influence on the boundary layer, and for the case shown in Figure 7.11, separation actually takes place upstream of the shoulder, somewhere near $\beta = 75°$.

The drag on the cylinder, therefore, is made up of two components: a minor part due to the viscous friction acting on the surface, and a major part due to pressure differences. Bodies where the pressure losses dominate the total drag force are called *bluff* bodies, and the cylinder is a good example of a bluff body. Bodies where the pressure losses are small and the viscous stresses dominate the total drag force are called *streamlined* bodies, and a good example of a streamlined body flow is the flow over an airfoil at small angles of attack.

So, despite the fact that the viscosities of common fluids are very low, a substantial drag force is found on most bodies. This was puzzling to scientists in the nineteenth century who believed that, since the viscosity was so small, the inviscid assumption should hold to a high degree of accuracy. The discrepancy was called "d'Alembert's Paradox," after the famous French scientist who studied this problem (for a historical note on Jean Le Rond d'Alembert, see Section 15.8). The paradox was resolved by Prandtl in 1904, when he described for the first time the nature of the boundary layers that are formed near the surface due to the action of viscosity (for a historical note on Ludwig Prandtl, see Section 15.17). Prandtl concluded that in a thin layer near the wall, strong velocity gradients were present, and that despite the small magnitude of the viscosity, the viscous stress, being the product of the viscosity and the velocity gradient, could become very significant— viscosity could not be ignored. Mathematically, we could say that inviscid flow cannot satisfy the boundary conditions of a real flow: specifically, inviscid flows allow slip at the surface, whereas viscous flows do not.

7.10 LIFT

A very interesting flow is generated when we add a vortex of strength Γ to the uniform flow over a circular cylinder. As seen in Figure 7.12, the vortex moves the stagnation points away from the line of horizontal symmetry. The streamlines over the upper part of the cylinder come closer together, whereas those near the lower part move further apart. At the same time, the pressure decreases on the upper part, and increases on the lower part. A lift is generated. As the strength of the vortex increases, the stagnation points come closer together, and eventually they move off the surface of the cylinder. The lift keeps increasing at the same time.

This hypothetical flowfield helps us to understand two very significant phenomena: the lift generated by a spinning cylinder (the Magnus effect) and the lift generated by an airfoil.

7.10.1 Magnus Effect

The flowfield shown in Figure 7.12 is for inviscid flow. However, in a viscous flow, a spinning cylinder can generate a flowfield that looks remarkably similar. A thin

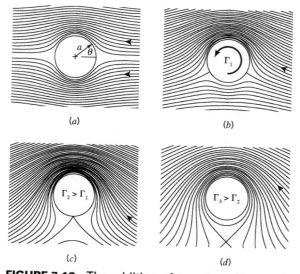

FIGURE 7.12 The addition of a vortex of increasing strength to a uniform flow over a circular cylinder. Adapted from F. M. White, *Fluid Mechanics,* McGraw-Hill, 1986.

layer of air (the boundary layer) is forced to spin with the cylinder because of viscous friction, and it produces a circulation around the cylinder. In regions of the flow where the motion due to spin is opposite to that of the air stream, there is a region of low velocity where the pressure is relatively high. In regions where the direction of motion of the boundary layer is the same as that of the external air stream, the velocities sum and the pressure in this region is relatively low. The cylinder experiences a lift force acting in a direction normal to the freestream. If the spinning cylinder was moving through the air its path would tend to curve.

The appearance of a side force on a spinning cylinder or sphere is called the *Magnus effect,* and it well known to all participants in ball sports, especially baseball, golf, cricket, and tennis players (for a historical note on Gustav Magnus, see Section 15.12). For additional comments on the fluid mechanics of sports balls, see Section 10.7.

The Magnus effect is also seen in spinning cylinders and disks. The effect of rotation on the flowfield around an automobile wheel is shown in Figure 7.13. The rotation is in the sense of a car traveling right to left, and so the motion of the upper surface of the tire is in a direction opposite to the incoming flow direction. As a result, the velocity outside the boundary layer in that region is lower than in the case of a nonspinning wheel, and the pressure is higher. In addition, the fact that the viscous flow in the boundary layer experiences a stronger deceleration moves the upper separation point forward, increasing the size of the wake. As a result, the rate of rotation can strongly affect both the lift and drag forces developed by the wheel.

7.10.2 Airfoils and Wings

If we were to solve the potential flow over an airfoil shape, we would see a streamline pattern similar to that shown in Figure 7.14(*a*). All practical airfoils have rounded

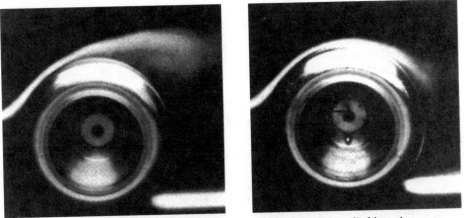

FIGURE 7.13 Visualization of the flow on a counter-rotating (left) and a non-rotating (right) wheel. The Reynolds number is 0.53×10^6. From *Race Car Aerodynamics*, J. Katz, Robert Bentley Publishers, 1995. Copyright AIAA, 1977, with permission.

leading edges and sharp trailing edges. One of the stagnation points is always located near the leading edge, and in a potential flow, the other is typically located somewhere on the upper surface. By analyzing the pressure distribution over the surface, we would find that an airfoil in potential flow generates no lift and no drag, which is completely contrary to our experience.

In fact, the flow shown in Figure 7.14(a) cannot occur in any real fluid. Viscosity always plays a very important role. For example, the potential flow over the sharp trailing edge indicates that the flow in this region needs to change its direction by

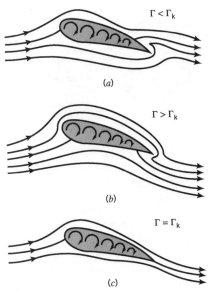

FIGURE 7.14 The addition of a vortex of increasing strength to a uniform flow over an airfoil shape. Adapted from F. M. White, *Fluid Mechanics*, 2nd ed., McGraw-Hill, 1986.

almost 180° *instantaneously* to follow the contour of the body. The slope of the surface changes discontinuously at the trailing edge, which implies that the fluid has to change its velocity infinitely fast, which gives rise to an infinite strain rate. For a viscous fluid, this would be associated with an infinite shear stress.

Clearly, this is unrealistic, and when we look at the real flow over an airfoil (Figure 7.15), we see that the flow near the trailing edge does not behave at all like that shown in Figure 7.14(a). The flow leaves the trailing edge smoothly, which means that the trailing edge is also a stagnation point. Under these conditions, all shear stresses are finite. The question is, how can we modify the potential flow solution to make it resemble the real flow?

If we add a vortex to the flow in Figure 7.14(a) we can move the stagnation points, as we did for the cylinder flow in Figure 7.12. By increasing the strength of the vortex, the flow pattern changes from that shown in Figure 7.14(a), to that shown in Figure 7.14(b). At the same time, the airfoil begins to generate lift. For a particular vortex strength, Γ_K, the rear stagnation point will be located at the trailing edge [Figure 7.14(c)]. The lift force produced by the airfoil under this condition (per unit span) is given by

$$F_L = \rho U_\infty \Gamma_K \tag{7.32}$$

This is called the *Kutta–Joukowski theorem,* after two engineers active in the early part of the twentieth century. The boundary condition imposed on the flow, that is, the requirement that the rear stagnation point coincides with the trailing edge, is called the *Kutta condition.*

The Kutta–Joukowski theorem applies to the flow of an inviscid fluid over a two-dimensional body in steady motion. It holds for bodies of arbitrary shape, and so it also predicts the lift generated by a cylinder with any level of imposed circulation, although the corresponding Kutta condition only applies to bodies with a sharp trailing edge. The Kutta–Joukowski theorem is the basis for the theory of lift on

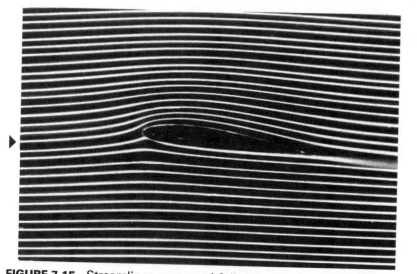

FIGURE 7.15 Streamlines over an airfoil shape made visible using smoke in air. Reynolds number based on chord length $R_c = 2.1 \times 10^5$, angle of attack $\alpha = 5°$. From *Visualized Flow,* Japan Society of Mechanical Engineers, Pergamon Press, 1988.

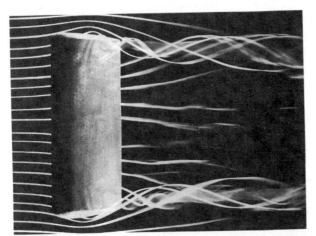

FIGURE 7.16 The formation of trailing vortices from a wing of finite span. In this view, the flow is from left to right, and the top surface has a lower pressure. From M. R. Head, in *Flow Visualization II*, ed. W. Merzkirch, pp. 399–403, published by Hemisphere.

wings and the thrust of fan and propeller blades. Experiments show that it is in very good agreement with measurements of lift on two-dimensional airfoils.

An important parameter is the *angle of attack*, which is the angle the chord line makes with the incoming flow direction.[3] As the angle of attack increases, the circulation required to satisfy the Kutta condition also increases, and therefore the lift increases as well. It can be shown that for small angles of attack α,

$$\frac{F_L}{\rho U_\infty \Gamma_{K0}} = 2\pi\alpha \tag{7.33}$$

where Γ_{K0} is the circulation at zero angle of attack.[4]

The vortex that was added to the potential flow to satisfy the Kutta condition is a theoretical construct. In the case of an airfoil moving through a viscous fluid, we often think of a virtual "bound" vortex contained in the airfoil. It is real in the sense that, on average, the flow over the top surface of the airfoil is faster than that over the bottom surface, and so the sense of the circulation is clockwise if the airfoil is moving from left to right.

So far we have only dealt with two-dimensional airfoils that have, in effect, an infinite wingspan. For an airfoil with finite wingspan, however, a very consistent observation is made. Near the tips of the wing, trailing vortices form which can be made visible using smoke, as in Figure 7.16. Where do they come from? In one interpretation, trailing vortices are formed near the wingtips because the flow tends to spill over the wingtip in response to the pressure difference between the top and bottom surface of the wing. The pressure on the top is lower than on the bottom, so near each wing tip there is a tendency for the flow to move from the bottom to the top of the wing. If we are looking from some point downstream of the wing

---

[3] The airfoil *chord* is the straight line drawn between the leading edge and the trailing edge.
[4] See, for example, *Foundations of Aerodynamics*, 4th edition, by A. M. Kuethe and C.-Y. Chow, published by John Wiley & Sons, 1986.

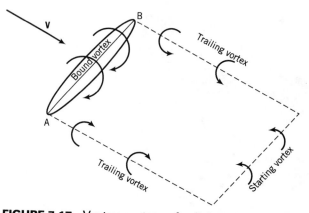

FIGURE 7.17   Vortex system of a finite wing.

back toward the wing, we see the formation of a clockwise vortex near the left wing tip, and a counterclockwise vortex near the right wing tip (see Figure 7.16).

In another interpretation, the trailing vortices are seen as the physical continuation of the virtual vortex "bound" in the wing (see Figure 7.14). Since it is known from theory and experiment that a vortex is shed from the airfoil when it starts its motion, we have a simple picture of the vortex system of a finite wing, comprising a bound vortex, a starting vortex, and two trailing vortices (Figure 7.17). We have a closed loop of circulation Γ_K at all points. This view was developed by Lanchester in England and Prandtl in Germany, and it is often called the Prandtl lifting line theory. In reality, the distribution of vorticity is more complex than that implied by Figure 7.17, but it still represents a powerful concept in understanding three-dimensional airfoil behavior.

7.11   VORTEX INTERACTIONS

The superposition of two vortex flowfields leads to some interesting phenomena. One place where this occurs is in the wake of an airplane (see Figure 7.17). Once the airplane has taken off, the starting vortex has been left far behind. The trailing vortex flow may then be idealized as two infinitely long, line vortices, separated by a distance s, one with a clockwise circulation $-\Gamma$, and one with a counterclockwise circulation Γ. The combined velocity field will be the linear superposition of the two potential vortex flowfields. Each vortex "induces" its own velocity field with a distribution given by equation 7.19

$$u_\theta = \frac{\Gamma}{2\pi r}$$

There will be a velocity induced by the right hand, counterclockwise vortex on the left hand, clockwise vortex, which tends to push the left hand vortex down. The left hand vortex, in turn, is associated with an induced velocity field that tends to push the right hand vortex down. We see that the interaction of a pair of line vortices of opposite sign causes the pair to move down together with a propagation velocity u_p given by

$$u_p = \frac{\Gamma}{\pi s} \tag{7.34}$$

where s is the distance between the vortices. As a result of this induced motion, the trailing vortex system in the wake of a wing moves downward as it is left behind.

This self-induced propagation can be easily seen by moving a flat plate through water in a bath or in a large bucket. Move the plate while holding it vertical and at right angles to the direction of motion for a short distance and then remove it from the liquid. Vortices of opposite sign are produced at the vertical edges, and they will continue to move under their own induced velocity field after the plate is withdrawn.

A similar phenomenon is seen in the motion of a vortex ring. A vortex ring can be generated and visualized using a cardboard cylinder (a handtowel roll works well) sealed off at both ends with paper diaphragms. A 5 *mm* circular hole needs to be made in one diaphragm, and the cylinder should be filled with smoke from a cigarette or a stick of incense. If the end without the hole is lightly tapped, a smoke ring will emanate from the hole in the other end. Each part of the smoke ring induces a velocity that acts on all other parts of the ring, and through this interaction, the ring propagates at a constant velocity through the air. The resulting motion of the smoke-filled vortex ring is quite beautiful to watch.

EXAMPLE 7.3 *Lift*

Consider a wing traveling at a velocity V, with a span equal to 20 c, where c is the chord length. Given a lift coefficient of 2.0 at a zero angle of attack, find the strength of the bound vortex, and the velocity at which the trailing vortices move downward under their own induced velocity field.

Solution: For a two-dimensional wing, the Joukowski lift law (equation 7.32) gives

$$F_L = \rho U_\infty \Gamma_K$$

If we assume that the wing has a sufficiently large wingspan so that the two-dimensional lift estimate is reasonable, we have (since F_L is the force per unit span)

$$C_L = \frac{F_L}{\frac{1}{2}\rho V^2 c} = \frac{\rho V \Gamma_K}{\frac{1}{2}\rho V^2 c} = 2$$

Therefore,

$$\Gamma_K = Vc$$

If the trailing vortices are approximated as a pair of line vortices of infinite extent, the downward propagation velocity is given by equation 7.34:

$$u_p = \frac{\Gamma}{\pi s}$$

Hence,

$$u_p = \frac{Vc}{\pi 10 c} = \frac{V}{\pi}$$

so that

$$\frac{u_p}{V} = \frac{1}{\pi}$$

PROBLEMS

7.1 Define vorticity in terms of the vector velocity field. How is the "rotation" of a fluid particle related to its vorticity? In Cartesian coordinates, write down the general form of the z-component of vorticity. What is the condition on the vector velocity field such that a flow is irrotational?

7.2 For a certain incompressible two-dimensional flow, the stream function, $\psi(x, y)$ is prescribed. Is the continuity equation satisfied?

7.3 If $u = -Ae^{-ky} \cos kx$ and $v = -Ae^{-ky} \sin kx$, find the stream function. Is this flow rotational, or irrotational?

7.4 An inviscid flow is bounded by a wavy wall at $y = H$ and a plane wall at $y = 0$. The stream function is

$$\psi = A(e^{-ky} - e^{ky}) \sin kx$$

where A and k are constants.
(a) Does the flow satisfy the continuity equation?
(b) Is the flow rotational or irrotational?
(c) Find the pressure distribution on the plane wall surface, given that $p = 0$ at $[0, 0]$.

7.5 An inviscid flow is bounded by a wavy wall at $y = H$ and a plane wall at $y = 0$. The stream function is

$$\psi = A(e^{-ky} - e^{ky}) \sin kx + By^2$$

where A, B and k are constants.
(a) Does the flow satisfy the continuity equation?
(b) Is the flow rotational or irrotational?
(c) Find the pressure distribution on the plane wall surface, given that $p = 0$ at $[0, 0]$.

7.6 For the flow defined by the stream function $\psi = V_0 y$:
(a) Plot the streamlines.
(b) Find the x and y components of the velocity at any point.
(c) Find the volume flow rate per unit width flowing between the streamlines $y = 1$ and $y = 2$.

7.7 Find the stream function for a parallel flow of uniform velocity V_0 making an angle α with the x-axis.

7.8 A certain flowfield is described by the stream function $\psi = xy$.
(a) Sketch the flow field.
(b) Find the x and y velocity components at $[0, 0]$; $[1, 1]$; $[\infty, 0]$; $[4, 1]$.
(c) Find the volume flow rate per unit width flowing between the streamlines passing through points $[0, 0]$ and $[1, 1]$; and points $[1, 2]$ and $[5, 3]$.

7.9 Express the stream function $\psi = 3x^2y - y^3$ in cylindrical coordinates (note that $\sin 3\theta = 3 \sin \theta \cos^2 \theta - \sin^3 \theta$). Sketch the streamlines, and find the magnitude of the velocity at any point.

7.10 For the previous four problems, find the velocity potential, and sketch lines of constant ϕ.

7.11 The velocity potential for a steady flowfield is given by $x^2 - y^2$. Find the equation for the streamlines.

7.12 The velocity components of a steady flowfield are $u = 2cxy$ and $v = c(a^2 + x^2 - y^2)$.
(a) Is the flow incompressible?

(b) Is the flow rotational or irrotational?

(c) Find the velocity potential.

(d) Find the stream function.

7.13 The velocity potential for a certain flow is given in cylindrical coordinates by $Cr^2 \cos 2\theta$, where C is a constant. Show that this represents the flow in a right-angle corner. If the velocity at $r = 1\ m$, $\theta = 0$ is $-10\ m/s$, find the velocity at $r = 2\ m$, $\theta = \pi/4$.

7.14 A fluid flows along a flat surface parallel to the x-direction. The velocity u varies linearly with y, the distance from the wall, so that $u = ky$.

(a) Find the stream function for this flow.

(b) Is this flow irrotational?

7.15 Consider the parallel two-dimensional flow shown in Figure P7.15. Is the flow irrotational? Find the stream function, given that $u = 1.5\ m/s$ at $y = 0$, and $u = 4\ m/s$ at $y = 1.2\ m$.

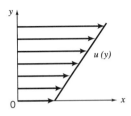

FIGURE P7.15

7.16 Given $u_r = 1/r$ and $u_\theta = 1/r$, find the stream function ψ, and sketch the flowfield.

7.17 Given the irrotational, incompressible velocity field described by the velocity potential $\phi = A\theta\ (A > 0)$.

(a) Sketch lines of constant ϕ.

(b) Find the velocity components u_r and u_θ at any point.

(c) Find ψ, and sketch a few streamlines.

7.18 Given the irrotational, incompressible velocity field described by the stream function $\psi = \frac{2}{3}r^{3/2} \sin \frac{3}{2}\theta$,

(a) Plot the streamline $\psi = 0$.

(b) Find the velocity at the points defined by the cylindrical coordinates $[2, \frac{\pi}{3}]$ and $[3, \frac{\pi}{6}]$.

(c) Find the velocity potential ϕ.

7.19 Consider the two-dimensional flow of an inviscid, incompressible fluid described by the superposition of a parallel flow of velocity V_0, a source of strength q, and a sink of strength $-q$, separated by a distance b in the direction of the parallel flow, the source being upstream of the sink.

(a) Find the resultant stream function and velocity potential.

(b) Sketch the streamline pattern.

(c) Find the location of the upstream stagnation point relative to the source.

7.20 A static pressure probe is constructed with a semi-cylindrical nose, as shown in Figure P7.20. Where should a pressure tap be located so that it reads the same static pressure as that found far from the probe in a uniform flow?

U_∞

p_∞

FIGURE P7.20

7.21 A two-dimensional source is placed in a uniform flow of velocity $2m/s$ in the x-direction. The volume flow rate issuing from the source is $4m^3/s$ per meter.

(a) Find the location of the stagnation point.

(b) Sketch the body shape passing through the stagnation point.

(c) Find the width of the body.

(d) Find the maximum and minimum pressures on the body when the pressure in the uniform flow is atmospheric. The fluid is air and its temperature is $20°C$.

7.22 Repeat the previous problem using the Java-based potential flow solver available at: *http:// www.engapplets.vt.edu,* "The Ideal Flow Machine."

7.23 Using the potential flow solver available at: *http://www.engapplets.vt.edu* ("The Ideal Flow Machine"), place two sinks and one source along the x-direction, each of strength $1m^3/s$ per meter, separated from each other by a distance of $2m$. Plot the streamlines. Locate the stagnation points. Vary the strength of the sinks and source until the stagnation points are a distance of $4m$ apart.

7.24 Using the potential flow solver available at: *http://www.engapplets.vt.edu* ("The Ideal Flow Machine"), place two sources and two sinks alternately along the x-direction, spaced $1m$ apart, each of strength $4m^3/s$ per meter. Add a uniform flow of velocity $2m/s$ in the x-direction. Plot the streamlines. Vary the strength of the sources and sinks (keeping their relative strengths equal) until the streamline defining the largest closed body has a major axis that is twice the minor axis.

7.25 Using the potential flow solver available at: *http://www.engapplets.vt.edu* ("The Ideal Flow Machine"), place a doublet of strength $-8m^3/s$ at the center of the field. Add a uniform flow of velocity $2m/s$ in the x-direction. Place a clockwise vortex at the center of the field. Find the strength of the vortex that will cause the two stagnation points to coincide.

7.26 Using the potential flow solver available at: *http://www.engapplets.vt.edu* ("The Ideal Flow Machine"), place a clockwise and a counterclockwise vortex of strength $10m^3/s$ along the x-direction separated by a distance of $4m$. (a) Find the velocity u_c at the point halfway between them, and check the result against equation 7.19. (b) Add a vertical velocity equal to $0.5u_c$ and find the location of the stagnation points.

CHAPTER 8

DIMENSIONAL ANALYSIS

Dimensional analysis is the process by which the dimensions of equations and physical phenomena are examined to give new insight into their solutions. This analysis can be extremely powerful. Besides being rather elegant, it can greatly simplify problem solving, and for problems where the equations of motion cannot be solved it sets the rules for designing model tests, which can help to reduce the level of experimental effort significantly.

The principal aim of dimensional analysis in fluid mechanics is to identify the important nondimensional parameters that describe a given flow problem. Thus far, we have already encountered a number of nondimensional parameters, each of which has a particular physical interpretation. For example, in Section 1.7 we described the Reynolds number, $Re = UD/\nu$, as the parameter that indicates the onset of turbulent flow. Another nondimensional parameter, the pressure coefficient $C_p = (p - p_\infty)/\frac{1}{2}\rho V^2$ was discussed in Chapter 3, and we see that it is the ratio of static pressure differences to the dynamic pressure. The lift and drag coefficients, C_L and C_D, were defined in Chapter 5, and in Section 1.3.6, we introduced the Mach number M, which we interpreted as the ratio of a wave speed to the flow velocity.

Nondimensional parameters are widely used in fluid mechanics, and there are good reasons for this.

1. Dimensional analysis leads to a reduced variable set. A problem where the "output" variable, such as the lift force, is governed by a set of $(N - 1)$ "input" variables (for example, a length, a velocity, a frequency, the humidity, a roughness height, etc.), can generally be expressed in terms of a total of $(N - 3)$ nondimensional groups (for example, the lift coefficient, the Reynolds number, the Mach number, etc.).

2. When testing a scale model of an object in an airflow, such as a car or an airplane, dimensional analysis provides the guidelines for scaling the results from the model test to the full-scale. In other words, dimensional analysis sets the rules under which full similarity in model tests can be achieved. In this way, we can relate model tests on, for example, a model propellor to the full-scale prototype (see Figure 8.1).

3. Nondimensional parameters are more convenient than dimensional parameters since they are independent of the system of units. In engineering, dimensional equations are sometimes used, and they contribute to confusion, errors, and wasted effort. Dimensional equations depend on using the required units for each of the variables, or the answer will be incorrect. They are common in some areas of engineering, such as in the calculation of heat transfer rates and in describing the performance of turbomachines.

4. Nondimensional equations and data presentations are more elegant than their dimensional counterparts. Engineering solutions need to be practical, but they are always more attractive when they display a sense of style or elegance.

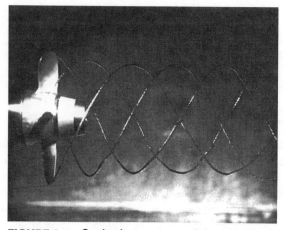

FIGURE 8.1 Cavitation on a model propeller. The bubbles are generated near the tip of each blade, and form a helical pattern in the wake. Photograph courtesy of the Garfield Thomas Water Tunnel, Pennsylvania State University.

The most powerful application of dimensional analysis occurs in situations where the governing equations cannot be solved. This is often the case in fluid mechanics. Very few exact solutions of the equations of motion can be found, and for the vast majority of engineering problems involving fluid flows we need to use an approximate analysis where the full equations are simplified to some extent, or we need to perform experiments to empirically determine the behavior of the system over some range of interest. In both cases, dimensional analysis plays a critical role by reducing the amount of effort involved and by providing physically meaningful interpretations for the answers obtained. Instead of solving the equations directly, we try to identify the important variables (such as force, velocity, density, viscosity, the size of the object, etc.), arrange these variables in nondimensional groups, and write down the functional form of the flow behavior. This procedure establishes the conditions under which similarity occurs, and it always reduces the number of variables that need to be considered.

It is rare for dimensional analysis to actually yield the analytical relationship governing the behavior. Usually, it is just the functional form that can be found, and the actual relationship must be determined by experiment. The experiments will also verify if any parameters neglected in the analysis were indeed negligible. To see how dimensional analysis works, we first need to define what system of dimensions we will use, and what is meant by a "complete physical equation."

8.1 DIMENSIONAL HOMOGENEITY

When we write an algebraic equation in engineering, we are rarely dealing with just numbers. We are usually concerned with quantities such as length, force, or acceleration. These quantities have a dimension (e.g., length or distance) and a unit (e.g., inch or meter). In fluid mechanics, the four fundamental dimensions are usually taken to be mass M, length L, time T, and temperature θ. An alternative system uses force, length, time, and temperature, but it will not be used in this book.

Some common variables and their dimensions are as follows (the square brackets are used as shorthand for "the dimensions of . . . are").

$$[\text{angular velocity}] = \frac{\text{angular measure}}{\text{time}} = T^{-1}$$

$$[\text{density}] = \frac{\text{mass}}{\text{volume}} = ML^{-3}$$

$$[\text{velocity}] = \frac{\text{length}}{\text{time}} = LT^{-1}$$

$$[\text{acceleration}] = \frac{\text{length}}{(\text{time})^2} = LT^{-2}$$

$$[\text{force}] = \text{mass} \times \text{acceleration} = MLT^{-2}$$

$$[\text{pressure}] = \frac{\text{force}}{\text{area}} = ML^{-1}T^{-2}$$

$$[\text{work}] = \text{force} \times \text{distance} = ML^2T^{-2}$$

$$[\text{torque}] = \text{force} \times \text{distance} = ML^2T^{-2}$$

$$[\text{power}] = \text{force} \times \text{velocity} = ML^2T^{-3}$$

$$[\text{dynamic viscosity}] = \frac{\text{stress}}{\text{velocity-gradient}}$$

$$= \frac{(\text{force/area})}{(\text{velocity/length})} = ML^{-1}T^{-1}$$

$$[\text{kinematic viscosity}] = \frac{\text{viscosity}}{\text{density}} = \frac{\text{viscosity}}{(\text{mass/volume})} = L^2T^{-1}$$

$$[\text{surface tension}] = \frac{\text{force}}{\text{length}} = MT^{-2}$$

Some quantities are already dimensionless. These include pure numbers, angular degrees or radians, and strain.

The concept of a dimension is important because we can only add or compare quantities that have similar dimensions: lengths to lengths, and forces to forces. In other words, all parts of an equation must have the same dimension—this is called the principle of dimensional homogeneity, and if an equation satisfies this principle it is called a *complete physical equation*. Take, for example, Bernoulli's equation.

$$\frac{p}{\rho} + \tfrac{1}{2}V^2 + gh = \text{constant}_1 \tag{8.1}$$

We can examine the dimensions of each term in the equation by writing its dimensional equivalent.

$$\frac{M}{L^2}\frac{L}{T^2} \times \frac{L^3}{M} + \frac{L^2}{T^2} + \frac{L}{T^2} \times L = \text{constant}_1$$

(the number $\tfrac{1}{2}$ is just a counting number with no dimensions). That is,

$$\frac{L^2}{T^2} + \frac{L^2}{T^2} + \frac{L^2}{T^2} = \text{constant}_1$$

All the parts on the left hand side have the dimensions of (velocity)^2 and the equation is dimensionally homogeneous. The constant on the right hand side must have the same dimensions as the parts on the left, so that in this case $[\text{constant}_1] = L^2/T^2$.

If we rewrote equation 8.1 as

$$\frac{p}{\rho g} + \tfrac{1}{2}\frac{V^2}{g} + h = \text{constant}_2$$

or

$$p + \tfrac{1}{2}\rho V^2 + \rho g h = \text{constant}_3$$

then in the first case each term has dimensions of length (including constant_2), and in the second case each term has dimensions of pressure (including constant_3).

All physically meaningful equations are dimensionally homogeneous.

To put this another way, in order to measure any physical quantity we must first choose a unit of measurement, the size of which depends solely on our own particular preference. This arbitrariness in selecting a unit size leads to the following postulate: any equation that describes a real physical phenomenon can be formulated so that its validity is independent of the size of the units of the primary quantities. Such equations are therefore called complete physical equations. All equations given in this book are complete. When writing down an equation from memory, it is always a good idea to check the dimensions of all parts of the equation—just to make sure it was remembered correctly. It also helps in an algebraic manipulation or proof where it can be used as a quick check on the answer.

8.2 APPLYING DIMENSIONAL HOMOGENEITY

In this section we will show how the principle of dimensional homogeneity can be used to reduce the number of parameters describing a given problem. To appreciate how this is achieved, we first need to understand how relationships that are expressed in functional form can be manipulated, and what is meant by independent variables. These concepts are illustrated using the following examples.

8.2.1 Example: Hydraulic Jump

Consider a hydraulic jump. This is the name given to standing waves in a water flow. A simple example can be seen by letting a stream of water fall on a flat surface such as a plate. The water spreads out on the plate in a thin layer, and then at some distance from the point of impact, a sudden rise in water level occurs. This is a circular hydraulic jump. A planar jump can be seen at the bottom of a dam spillway, as shown in Figure 8.2. In that case, the sudden change in water height can be described approximately by a simple relationship known as the hydraulic jump relationship

$$\frac{H_2}{H_1} = \frac{1}{2}\left(\sqrt{1 + 8F_1^2} - 1\right)$$

(equation 11.9) where $F_1 = V_1/\sqrt{gH_1}$ is a nondimensional quantity called the Froude number, H is the water depth, and the suffixes 1 and 2 refer to conditions upstream and downstream of the jump. The hydraulic jump relationship can also be written as

$$\frac{H_2}{H_1} = \phi(F_1) \tag{8.2a}$$

where the notation $\phi()$ denotes a functional dependence. This expression states that the ratio of water depths across a hydraulic jump depends only on the Froude number F_1. We can write the same functional dependence in dimensional form, so that

$$H_2 = \phi'(H_1, V_1, g) \tag{8.2b}$$

FIGURE 8.2 A hydraulic jump formed near the bottom of a spillway. From Siemens, with permission.

Four points should now be made. First, the nondimensional form (equation 8.2a) contains only two (nondimensional) parameters, whereas the dimensional form (equation 8.2b) contains four (dimensional) parameters. Second, we have included the dimensional constant g. It is absolutely essential to include all dimensional "constants" (such as g, the speed of sound, the density of water, etc.) in any dimensional analysis because the dimensions of constants are just as important as the dimensions of any of the variables. The only true constants are dimensionless constants.

Third, the relationship given in equation 8.2a can be written in a number of alternative forms. For example, we can write it as

$$\frac{H_2}{H_1} = \phi_1 \left(\frac{1}{F_1} \right)$$

Using the inverse of the Froude number, instead of the Froude number itself, changes the form of the function, but it does not change the fact that the height of the jump depends only on the Froude number.

Fourth, consider the functional relationship

$$\frac{H_2}{H_1} = \phi(F_1, \rho) \tag{8.3}$$

Since the left hand side is dimensionless, the right hand side must also be dimensionless. Any term on the right hand side that involves the Froude number by itself would be fine, but terms that involve some function of density cannot be dimensionless. Therefore Equation 8.3 cannot be correct; since the right hand side must be dimensionless, it cannot depend on the density by itself. This observation leads to two conclusions, based purely on dimensional grounds: either the density is not important and it should be eliminated from the problem, or there are other variables in the problem that we have left out. If we were convinced that the density was important, another parameter is needed that also has the dimensions of density so that the density could be made nondimensional. One parameter that suggests itself is the viscosity μ. This by itself does not have the dimensions of density, but combining the viscosity with some of the existing quantities can make it so. For example, the combination μ/V_1H_1 has the dimensions of mass per unit volume, so that one possibility for a dimensionally correct hydraulic jump relationship that includes density as one of the variables is

$$\frac{H_2}{H_1} = \phi\left(F_1, \frac{\rho V_1 H_1}{\mu}\right)$$

We recognize the new nondimensional group as the Reynolds number based on upstream flow conditions.

8.2.2 Example: Drag on a Sphere

Suppose we need to know the drag force F_D acting on a sphere of diameter D in a fluid flow of velocity V. The equations of motion for this flow can only be solved under very restricted conditions, and the full equations are too complex to allow a general solution. We must resort to dimensional analysis and experiment.

We will assume that the drag force F_D depends only on the fluid density ρ, the stream velocity V, the diameter D, and the fluid viscosity μ. That is,

$$F_D = f(\rho, V, D, \mu) \tag{8.4}$$

Given this functional form, what would it take to find the drag force experimentally over a wide range of variation in velocity, diameter, density, and viscosity? For the sake of illustration, suppose that it takes about 10 points to define an experimental curve. To find the effect of velocity, we need to run the experiment for 10 different values of V, while holding all the other variables constant (Figure 8.3a). To find the effect of diameter at each velocity, we need to find F_D for 10 different values of D for each velocity, while holding the other variables constant, so that the total number of experiments has grown to 100. For each V and each D, we will need 10 values of μ (Figure 8.3b) and 10 values of ρ (Figure 8.3c), making a grand total of 10,000 experiments!

Instead, let us rearrange these variables into nondimensional groups, step by

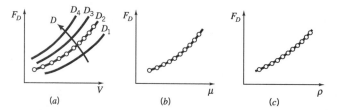

FIGURE 8.3 Experiments to find the drag force on a sphere. (a) Varying V for different D, holding ρ and μ constant. (b) For each point in (a), vary the viscosity, holding ρ constant. (c) For each point in (b), vary the density.

step. We can do this in a number of ways, but the object of each step is to make all the variables in the functional relationship, except one, dimensionless with respect to one of the fundamental dimensions. The dimensional variable that remains can then be deleted, as will be shown. We will use the three fundamental dimensions length L, mass M, and time T.[1]

All the variables except the density are first made dimensionless in mass. There is no need to start with density and mass, but it turns out to be convenient. Divide both sides of equation 8.4 by ρ, in order to change the dependent variable to F_D/ρ, which has dimensions $[MLT^{-2}/ML^{-3}] = [L^4T^{-2}]$. So

$$\frac{F_D}{\rho} = \frac{f(\rho, V, D, \mu)}{\rho} \tag{8.5}$$

or

$$\frac{F_D}{\rho} = f_1(\rho, V, D, \mu) \tag{8.6}$$

The step taken in going from equation 8.5 to 8.6 is perfectly legitimate: equation 8.5 states that F_D/ρ depends on density, velocity, diameter, and viscosity, and equation 8.6 states the same thing, although function f_1 is clearly different from function f.

In equation 8.6, the viscosity μ is the only independent variable (besides ρ) that has nonzero dimensions in mass. Let us define a new independent variable μ/ρ, having dimensions $[ML^{-1}T^{-1}/ML^{-3}] = [L^2T^{-1}]$. So

$$\frac{F_D}{\rho} = f_2\left(\rho, V, D, \frac{\mu}{\rho}\right) \tag{8.7}$$

This change from μ to μ/ρ is permissible, since equations 8.6 and 8.7 still express the same functional dependence because ρ and μ are *independent*. Independence means that in determining the drag we can vary ρ and μ separately (as for equation 8.6), or vary ρ and μ/ρ separately (as for equation 8.7). In other words, we can vary ρ without varying μ/ρ, simply by causing μ to change in a way that is proportional to the change in ρ. The practical difficulties would be the same in either case: to keep either μ or μ/ρ constant while ρ were varied would, in general, require a change of fluid as well as a change of temperature.

Now, the changes of variable made in going from equation 8.6 to equation 8.7 cannot have any effect on the completeness of the equation. The validity of equation 8.7 must therefore be unaffected by a change in the size of the mass unit from slugs to kilograms, for example. The ratio F_D/ρ is unaffected by this change because it is independent of the mass unit. The value of $f_2(\rho, V, D, \frac{\mu}{\rho})$ must likewise be unaltered. The values of V, D, and μ/ρ remain the same, but the value of ρ will be different. Consequently, f_2 cannot depend on ρ, but only on V, D and μ/ρ. This was the same reasoning that led us to conclude that equation 8.3 could not be correct. In that case, the density was eliminated on dimensional grounds, and the same argument applies here. The density is therefore deleted, and

$$\frac{F_D}{\rho} = f_2\left(V, D, \frac{\mu}{\rho}\right) \tag{8.8}$$

---

[1] The following treatment was adapted from *Dimensional Analysis for Engineers* by E. S. Taylor, published by Clarendon Press, Oxford, 1974; and *Engineering Applications of Fluid Mechanics* by Hunsaker & Rightmire, published by McGraw-Hill, 1947.

This procedure is more clearly illustrated by using the *matrix of dimensions*. In this matrix, the columns are the governing parameters, and the rows are the dimensions. The entries are the exponents of the dimensions of each parameter. The matrix of dimensions corresponding to equation 8.6 is

	F_D	ρ	V	D	μ
M	1	1	0	0	1
L	1	−3	1	1	−1
T	−2	0	−1	0	−1

Starting with the mass row, all the variables containing mass are made nondimensional in mass, except the density, by dividing all the variables in that row by the density.

	$\frac{F_D}{\rho}$	ρ	V	D	$\frac{\mu}{\rho}$
M	0	1	0	0	0
L	4	−3	1	1	2
T	−2	0	−1	0	−1

This is the matrix of dimensions corresponding to equation 8.7. The density is alone in having a mass dimension and therefore it can be deleted. Therefore

	$\frac{F_D}{\rho}$	V	D	$\frac{\mu}{\rho}$
M	0	0	0	0
L	4	1	1	2
T	−2	−1	0	−1

The mass row is now empty, and we can delete it. So

	$\frac{F_D}{\rho}$	V	D	$\frac{\mu}{\rho}$
L	4	1	1	2
T	−2	−1	0	−1

which is the matrix of dimensions corresponding to equation 8.8.

A similar procedure can be used to give all the remaining variables, except the velocity, zero dimensions in time. Change F_D/ρ to $F_D/\rho V^2$, which has dimensions $[L^4 T^{-2}/L^2 T^{-2}] = [L^2]$, and change μ/ρ to $\mu/\rho V$, which has dimensions $[L^2 T^{-1}/LT^{-1}] = [L]$. Then

$$\frac{F_D}{\rho V^2} = f_3\left(V, D, \frac{\mu}{\rho V}\right)$$

You might question this result since it appears that the left hand side was divided by V^2, whereas on the right hand side only one term was changed, and it was divided by V. What is really happening, though, is that we are making the quantities on the left and right hand sides independent of time by using V (which contains time as a fundamental dimension) in whatever way necessary. As long as the parameters remain independent, that is, they can be changed in ways that are different from the way the other parameters change, this is a legitimate process and does not alter the basic functional dependence, although it will change the function itself (see also Section 8.4, Step IVc).

According to our previous arguments, we can delete V by virtue of the completeness of the equation (since a change in the unit of time measurement will affect V but no other variable), and we get

$$\frac{F_D}{\rho V^2} = f_3\left(D, \frac{\mu}{\rho V}\right)$$

The matrix of dimensions reduces to

	$F_D/(\rho V^2)$	D	$\mu/(\rho V)$
L	2	1	1

A third application of the same reasoning leads to

$$\frac{F_D}{\rho V^2 D^2} = f_4\left(D, \frac{\mu}{\rho V D}\right) = f_4\left(\frac{\mu}{\rho V D}\right)$$

That is

$$\frac{F_D}{\rho V^2 D^2} = g'\left(\frac{\rho V D}{\mu}\right)$$

Finally:

$$\frac{F_D}{\frac{1}{2}\rho V^2\left(\frac{\pi}{4} D^2\right)} = g\left(\frac{\rho V D}{\mu}\right) \tag{8.9}$$

In the last step we included the nondimensional numbers $\frac{1}{2}$ and $\frac{\pi}{4}$ in the denominator of the left hand side, so that the denominator is of the form dynamic pressure $\times$ cross-sectional area. This allows us to write the functional dependence as

$$C_D = g(Re)$$

that is, the nondimensional drag coefficient C_D is a function of Reynolds number. For a sphere C_D is defined by

$$C_D \equiv \frac{F_D}{\frac{1}{2}\rho V^2\left(\frac{\pi}{4} D^2\right)} \tag{8.10}$$

Equation 8.4, which involves five dimensional variables, has been simplified by purely dimensional reasoning to equation 8.9, which involves only two nondimensional variables. It is clear that experimental data can be correlated more easily by means of equation 8.9 than equation 8.4. Using our earlier methodology, we would need only 10 experiments rather than 10,000 to determine the complete drag behavior.

The experimental results are shown in Figure 8.4. The data were obtained using spheres of different sizes, over a very wide velocity range, and for a number of different fluids. It applies to dust particles settling in the atmosphere, bubbles rising in a glass of beer, droplets formed by a fuel injector, soccer balls, and slow cannon balls (slow because compressibility effects become important at high speed and these were not taken into account). By using a nondimensional representation, all these results can be collapsed onto a single curve.

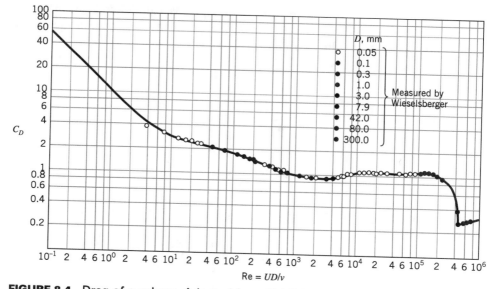

FIGURE 8.4 Drag of a sphere. Adapted from Schlichting, *Boundary Layer Theory,* 7th edition, McGraw-Hill, 1979.

Although this analysis reduced the number of variables from five dimensional ones to two nondimensional ones, the answer is not unique: equation 8.9 is not the only form that can be obtained by starting from equation 8.4. By operating with μ instead of ρ, or by multiplying the left hand side of equation 8.9 by the Reynolds number, we can also get

$$\frac{F}{\mu VD} = g_1\left(\frac{\rho VD}{\mu}\right) \tag{8.11}$$

It is found that, except at very low Reynolds numbers, equation 8.9 is preferable to equation 8.11, because $F/\frac{1}{2}\rho V^2(\frac{\pi}{4}D^2)$ varies less than $F/\mu VD$ in the practical range of the independent variable $\rho VD/\mu$.

We see that more than one result is possible, and the choice of which nondimensional groups to use is entirely arbitrary. We usually try to choose groups which have some meaning, that is, groups that have a physical interpretation such as the Reynolds number, Froude number, and so forth. We will discuss this issue in Section 8.6.

The first formal treatment of this successive application of dimensional homogeneity was given by Buckingham in 1892, in what became known as the "Buckingham Π theorem." Nondimensional parameters are sometimes called Π-products because of this connection with the Buckingham Π theorem.

It is always true that the number of parameters can be reduced by nondimensionalization. Each time we make the equation dimensionless with respect to one of the fundamental dimensions, the number of parameters is reduced by one. So we can expect that, for a total of N variables (including "input" and "output" variables) the number of Π-products = $N -$ (number of fundamental dimensions). The validity of this statement is discussed in the next section.

8.3 THE NUMBER OF DIMENSIONLESS GROUPS

What is the validity of the rule "the number of Π-products $= N -$ (the number of fundamental dimensions)"? It is relatively easy to show that it fails in some cases. For instance, say that we needed to find the functional dependence of the speed of sound a. It could be a function of pressure p and density ρ. We have a total of 3 parameters, and if we choose M, L, and T as fundamental dimensions, our rule would suggest that no dimensionless groups can be found. However, the combination $a/\sqrt{p/\rho}$ is dimensionless, so at least one Π-product exists.

A better rule can be found by examining the matrix of dimensions. For the sphere drag example used in the previous section, we have

	F_D	ρ	V	D	μ
M	1	1	0	0	1
L	1	-3	1	1	-1
T	-2	0	-1	0	-1

The rule that works every time is the one that says:

> The number of dimensionless groups is equal to the total number of parameters minus the rank of the matrix of dimensions.

The rank of a matrix is the order of the largest determinant in the matrix with a nonzero value (see Section A.1). By trying different determinants of order 3 in the matrix for the sphere problem, we can find at least one determinant that is nonzero (for example, try the columns headed by F_D, ρ and V) and so the rank of that matrix is 3. The number of Π-products is therefore $5 - 3 = 2$, as found earlier.

What about the speed of sound example? The matrix of dimensions for that case is

	a	p	ρ
M	0	1	1
L	1	-1	-3
T	-1	-2	0

There is only one determinant of order 3 that can fit, and it is equal to

$$0 \times [(-1 \times 0) - (-3 \times -2)] - 1 \times [(1 \times 0) - (-3 \times -1)]$$
$$+ 1 \times [(1 \times -2) - (-1 \times -1)] = 0$$

Therefore the rank of the matrix cannot be equal to 3. Is it 2? Yes, because it is easy to find a determinant of order 2 which is nonzero. The number of dimensionless groups is therefore $3 - 2 = 1$, and it is given by

$$\Pi_1 = \frac{a}{\sqrt{p/\rho}}$$

When only one nondimensional group exists, as in this case, it cannot be a function of anything else, since that would naturally require another nondimensional parameter. Therefore:

> When only one nondimensional group exists, that group must be a constant.

In the case of the speed of sound problem, the value of the constant depends on the particular fluid under consideration.

Consider another example, based on the hydraulic jump (Section 8.2.1). Let us suppose that

$$H_2 = \psi_1(H_1, V_1, g)$$

Then:

	H_2	H_1	V_1	g
M	0	0	0	0
L	1	1	1	1
T	0	0	-1	-2

The rank of this matrix cannot be 3 because of the zeros in the top row; no parameter contains the mass dimension. We find that the rank is 2 so that the number of dimensionless parameters is $N - 2 = 2$. We already know that this gives

$$\frac{H_2}{H_1} = \psi\left(\frac{V_1}{\sqrt{gH_1}}\right)$$

What if we include the density in the list of parameters? Then

	H_2	H_1	V_1	g'	ρ
M	0	0	0	0	1
L	1	1	1	1	-3
T	0	0	-1	-2	0

The rank of this new matrix is 3, so that the number of dimensionless parameters is $5 - 3 = 2$. But we already have 2; the ratio of depths and the Froude number. The density does not belong since there is no other parameter with a mass dimension that could be used together with the density to make a new nondimensional parameter (we argued this from first principles in Section 8.2.1).

Dimensional analysis, therefore, helps us to find the right list of parameters. It can tell us that a certain parameter does not belong, and it can also tell us if another parameter is missing. For example, if we really believed that the hydraulic jump depends on the density of the fluid, then we need another variable with a nonzero dimension in mass, such as the pressure, to make the density dimensionless in mass. We then end up with 3 nondimensional groups: the ratio of depths, the Froude number, and a pressure coefficient. Alternatively, if we included the viscosity, which also has mass in its dimensions, the Reynolds number would make an appearance.

Dimensional analysis cannot tell us which Π-products are important in any particular flow problem. The results of the analysis should always be tested by experiment, and it is the experiment that will finally show which Π-products are important to describe the flow.

8.4 NON-DIMENSIONALIZING PROBLEMS

In problems where dimensional analysis is likely to be useful:

Step I: Determine the total number of parameters, including input and output quantities (in the sphere example, we had five: F, V, D,

μ, and ρ). This is the most crucial step in the process, since it defines the problem. It is also the most difficult step.

Step II: Select a set of fundamental dimensions, usually mass, length, and time (M, L, and T).

Step III: Write down the matrix of dimensions. In this matrix, the columns are the governing parameters and the rows are the dimensions. The entries are the exponents of the dimensions of each parameter.

Step IV: Find the rank r of the matrix of dimensions (the rank of a matrix is the order of the largest determinant in the matrix with a nonzero value). There will always be $N - r$ dimensionless parameters. In most cases, but not always, the rank of the matrix will be 3 since there are 3 fundamental dimensions (see Section 8.3).

Step V: Establish these $N - r$ dimensionless parameters by:

    (a) Using your intuition. For example, in problems where viscosity is important, the Reynolds number usually makes an appearance. Similarly, if high-speed flow is present, the Mach number should be considered, and if waves or a free surface are present, the Froude number is certain to be useful. In other problems, it is usually possible to take the dependent variable (the output), and choose a number of input parameters that when combined with the dependent variable make a nondimensional group. This process can then be repeated to make dimensionless groups using only input parameters. The matrix of dimensions can be very useful in this regard. Remember, the answer is not unique. This does not mean any answer will do, just that more than one answer is possible. Nevertheless, some answers are "better" than others (those that use standard Π-products, for example).[2]

    (b) Check the result. That is, check that each Π-product really is non-dimensional.

    (c) Check that all the Π-products are independent. That is, if one of the dimensionless parameters is simply a combination of others, or just another raised to some power, it is not independent. A good check for independence is to see that each of the Π-products contains one parameter that none of the other groups does.

Step VI: See if any of the dimensionless parameters are not important, and then neglect them. This requires considerable judgment.

Step VII: Test your results experimentally. This step is very important, since it will verify or refute all your assumptions, especially Steps I, V, and VI.

Finally, remember that dimensional analysis is mathematics, not physics. For example, no amount of manipulation can compensate for neglecting an important physical parameter in Step I.

---

[2] Some texts recommend the *method of indices* to find the Π-products. I have found this approach to be totally nonintuitive and full of opportunities for error, and therefore I strongly advise against its use.

8.5 PIPE FLOW EXAMPLE

Consider flow through a circular, horizontal pipe of length L and diameter D. The average velocity, integrated across the cross-section, is $\overline{V}$ (= volume flow rate/area). Develop a functional dependence for the pressure drop Δp, and express the result nondimensionally.

Step I: Determine the total number of parameters. We expect that Δp will depend on L, D, $\overline{V}$, ρ, μ, but what else? It might depend on the roughness of the surface k, where k is some measure of the average roughness height. It should not depend on gravity since the pipe is horizontal. It might depend on the speed of sound, but if the Mach number is small we can probably neglect it. Let's start with the minimum number of parameters so that

$$\Delta p = \phi_1(L,\, D,\, \overline{V},\, \rho,\, \mu,\, k)$$

Step II: Select a set of fundamental units. Here, we choose M, L and T.

Step III: Write the matrix of dimensions ($N = 7$):

	F	ρ	$\overline{V}$	D	k	L	μ
M	1	1	0	0	0	0	1
L	1	−3	1	1	1	1	−1
T	−2	0	−1	0	0	0	−1

Step IV: Find the number of nondimensional parameters. Since the rank of the matrix is 3, we will have $N - 3 = 7 - 3 = 4$ dimensionless parameters.

Step V: **(a)** Pick dimensionless parameters. Start with the pressure drop. To make it nondimensional, we need a combination of the other parameters that has the dimensions of pressure. One likely candidate is the dynamic pressure $\frac{1}{2}\rho\overline{V}^2$. The first nondimensional group becomes

$$\frac{\Delta p}{\frac{1}{2}\rho\overline{V}^2}$$

We can interpret this group as the ratio of the work done against the pressure rise to the kinetic energy of the motion. The second possible nondimensional group is

$$\frac{L}{D}$$

This ratio is useful because it tells us if the pipe is short or long, compared to its diameter. The third nondimensional group is

$$\frac{k}{D}$$

which is more meaningful than $\frac{k}{L}$ since a large value of $\frac{k}{D}$ can be interpreted immediately as indicating a severe blockage effect due to roughness. What is left? The viscosity has not been used so far, so we can use the Reynolds number as the fourth nondimensional group.

$$\frac{\rho \overline{V} L}{\mu}$$

Finally,

$$\frac{\Delta p}{\frac{1}{2}\rho \overline{V}^2} = \phi_1 \left(\frac{\rho \overline{V} D}{\mu}, \frac{L}{D}, \frac{k}{D} \right) \tag{8.12}$$

(b) Check that the Π-groups are really nondimensional. Here, this step is very straightforward: we recognize the left hand side of equation 8.12 as a pressure coefficient, and the right hand side contains ratios of lengths and a Reynolds number.

(c) Check the independence of the nondimensional groups. None of the dimensionless parameters we found can be formed by a combination of the others, and none of them is just another raised to some power. Also, each Π-product has something which none of the others have: Δp in the first, μ in the second, L in the third, and k in the fourth. So these Π-products are independent.

Step VI: Decide if they are all important. One suggestion might be that, since pipes are typically long compared to their diameter, an asymptotic state may exist where the flow properties do not change with further increase in length. In fact, we know that this happens. After 40 to 100 diameters, the mean flow properties are free of entrance effects and the velocity profile is *fully developed,* which means that it is independent of position along the pipe. In that case, the parameter $\frac{L}{D}$ should no longer be important, and it can be dropped. This would also imply that we should not use the pressure drop along the whole length of the pipe (since a fully developed flow is independent of the distance along the pipe) but instead, we should use the pressure drop per unit length, expressed in terms of the number of diameters of length. A better form might be

$$\frac{\left(\frac{\Delta p}{L}\right) D}{\frac{1}{2}\rho \overline{V}^2} = \phi \left(\frac{\rho \overline{V} D}{\mu}, \frac{k}{D} \right) \tag{8.13}$$

The parameter $\frac{k}{D}$ is called the *relative roughness.* The parameter on the left hand side is called the friction factor f, defined by

$$f \equiv \frac{\left(\frac{\Delta p}{L}\right) D}{\frac{1}{2}\rho \overline{V}^2}$$

The most widely used correlation for the pressure drop in fully developed pipe flow uses the functional dependence expressed in equation 8.13. The experimental data are generally given in the form of a chart (Figure 9.7), where the friction factor f is plotted as a function of Reynolds number based on diameter and average velocity, for different values of the relative roughness $\frac{k}{D}$. The drag correlations shown in Figure 9.7 hold for all Newtonian fluids: water, air, milk, alcohol, natural gas, and so on. A fuller discussion of this figure is given in Section 9.6.

8.6 COMMON NONDIMENSIONAL GROUPS

We have noted that it is possible to have more than one correct answer in dimensional analysis. For example, in the sphere example given in Section 8.2, we started by eliminating the mass dimension by using the density, and we found that

$$\frac{F}{\frac{1}{2}\rho V^2 D^2} = g\left(\frac{\rho V D}{\mu}\right)$$

(8.14)

As was pointed out then, we could have started with the viscosity instead of the density. In that case, we would have found that

$$\frac{F}{\mu V D} = g_1\left(\frac{\rho V D}{\mu}\right)$$

(8.15)

Both answers are correct, but one may be preferable to or "better" than the other for a particular problem. To understand what constitutes a better answer generally requires physical insight into the problem, which may take a considerable amount of experience to develop. In some cases, however, it is reasonably obvious. For example, we could write the sphere drag relationship as

$$\frac{F}{\frac{1}{2}\rho V^2 D^2} = g_3\left(\sqrt{\frac{\mu}{\rho V D}}\right)$$

(8.16)

This does not change its "correctness" (both sides are still nondimensional), but after a little practice with these kinds of problems we would recognize the second Π-product as being a mutation of the Reynolds number, and write equation 8.16 in the form given in equation 8.14. As we pointed out in Section 8.2, equation 8.14 can be "improved" further by changing the left hand side to include the frontal area $A = \frac{\pi}{4}D^2$ instead of D^2, so that

$$\frac{F}{\frac{1}{2}\rho V^2 A} = g_4\left(\frac{\rho V D}{\mu}\right)$$

where we recognize that it is the frontal area that governs the drag. Also, by using A instead of D^2, we can more easily compare the drag coefficients obtained for spheres to those obtained for other shapes.

8.7 NON-DIMENSIONALIZING EQUATIONS

There are a large number of possible Π-products, but only a rather limited set is in common use. The most compelling reason for the popularity of this subset is that these parameters, such as the Reynolds number and the nondimensional force coefficients, often come naturally out of the equations of motion themselves.

For example, consider Bernoulli's equation. If at some point we know the reference static pressure p_0 and velocity V_0, we could use them to nondimensionalize Bernoulli's equation. If we start with equation 8.1,

$$\frac{p}{\rho} + \tfrac{1}{2}V^2 + gh = \text{constant}_1$$

each term has the dimensions of V^2. If we divide through by $\tfrac{1}{2}V_0^2$, we obtain the dimensionless form

$$\frac{p}{\tfrac{1}{2}\rho V_0^2} + \frac{V^2}{V_0^2} + \frac{gh}{\tfrac{1}{2}V_0^2} = \text{constant}_4$$

Each term is now dimensionless. The first term is a pressure coefficient of some sort, the second is a velocity ratio, and the third term contains a Froude number. We can go one step further and subtract the constant quantity $p_0/\tfrac{1}{2}\rho V_0^2$ from both sides of the equation. This gives:

$$\frac{(p - p_0)}{\tfrac{1}{2}\rho V_0^2} + \left(\frac{V}{V_0}\right)^2 + \frac{gh}{\tfrac{1}{2}V_0^2} = \text{constant}_5$$

so that the first term takes the form of the pressure coefficient C_p as defined in Section 4.3.

To take another example, consider the Navier–Stokes equation for constant density flow

$$\rho \frac{D\mathbf{V}}{Dt} = -\nabla p + \rho \mathbf{g} + \mu \nabla^2 \mathbf{V}$$

This is the momentum equation for a viscous fluid in differential form (see Section 6.3, equation 6.18). The symbol ∇^2 represents a second-order differentiation with respect to the space variables (see Section A–A.6).

If at some reference position in the flow, we know the velocity V_0 and we can identify a reference length L, we can form the nondimensional variables

$$x' = \frac{x}{L}$$

$$y' = \frac{y}{L}$$

$$z' = \frac{z}{L}$$

$$t' = \frac{V_0 t}{L}$$

$$\mathbf{V}' = \frac{\mathbf{V}}{V_0}$$

$$p' = \frac{(p - p_0)}{\tfrac{1}{2}\rho V_0^2}$$

We also have $\nabla' = L\nabla$, the nondimensional form of the gradient operator, and $\nabla'^2 = L^2 \nabla^2$, the nondimensional form of the Laplacian operator.

Rewriting the Navier–Stokes equation in terms of these nondimensional variables gives

$$\frac{\rho V_0^2}{L} \frac{D\mathbf{V}'}{Dt'} = -\frac{1}{L}\tfrac{1}{2}\nabla' p' + \rho \mathbf{g} + \frac{\mu V_0}{L^2}\nabla'^2 \mathbf{V}'$$

so that

$$\frac{D\mathbf{V}'}{Dt'} = -\nabla'p' + \frac{\mathbf{g}L}{V_0^2} + \frac{\nu}{V_0 L}\nabla'^2\mathbf{V}' \tag{8.17}$$

Three widely used nondimensional parameters have appeared: a nondimensional time, $t' = V_0 t/L$, which is the inverse of the Strouhal number (see Section 10.5), the square of the Froude number $F_0 = V_0/\sqrt{gL}$, and the Reynolds number $Re_0 = V_0 L/\nu$. We can write equation 8.17 as

$$\frac{D\mathbf{V}'}{Dt'} = -\tfrac{1}{2}\nabla'p' + \frac{1}{F_0}\frac{\mathbf{g}}{g} + \frac{1}{Re_0}\nabla'^2\mathbf{V}'$$

The significance of the Reynolds number is especially interesting. As the Reynolds number becomes very large, the magnitude of the viscous term becomes very small compared to the other terms. In particular, the viscous term becomes small compared to the inertia term on the left hand side. The magnitude of the Reynolds number is therefore often interpreted as a measure of the importance of the inertia force compared to the viscous force. Very close to a solid surface, where boundary layers form, the relevant Reynolds number must be based on some measure of the viscous layer thickness, since that is the proper scale for the velocity gradient. Therefore the viscous term always remains important within the boundary layer. Outside the boundary layer, it can often be ignored.

We can also use the Couette flow shown in Figure 1.20 to illustrate this interpretation of the Reynolds number. The viscous force acting on the fluid at any point in the flow, F_v, is given by the viscous stress times the area, so that

$$F_v = \frac{\mu U}{h}A$$

We can estimate the magnitude of the corresponding inertial force, F_i, by finding the difference in the momentum of the fluid in contact with the top and bottom plates. Since the momentum of the flow in contact with the bottom plate is zero, the difference in momentum is $\rho U^2 A$, and so

$$F_i = \rho U^2 A$$

Hence,

$$\frac{F_i}{F_v} = \frac{\rho U^2 A}{\frac{\mu U}{h}A} = \frac{\rho U h}{\mu} = Re$$

We see again that the Reynolds number can be interpreted as the ratio of a typical inertia force to a typical viscous force.

8.8   SCALE MODELING

In many cases, it is useful to test scale models of full-scale vehicles. To design a new airplane, for example, many hours of wind tunnel tests are performed to test scale models of different wings and fuselage shapes. To help design boats and ships, scale models are often tested in towing tanks (Figure 8.5).

FIGURE 8.5 Testing a model ship in a towing tank at the Naval Ship Research and Development Center, Carderock, Maryland.

Scale models are used to save money since it is much cheaper to build a series of models than a series of full-scale airplanes, and model tests can be used to develop scaling laws. Scaling laws are needed to predict full-scale results from model tests. However, it is not always possible to test a scale model so that the flow conditions are exactly similar to those experienced by the full-scale vehicle. In that case, different size models are often used to give information on how the results might be extrapolated to the full scale.

We have seen how dimensional analysis can be used to find and interpret the nondimensional parameters that govern problems in fluid mechanics. It can also be used to determine the conditions for similarity for scale modeling. A scale model will only give correct results if the flow is "similar" to the full-scale prototype. What does the word similarity mean in this context? We have three levels of similarity that must be satisfied before a model can be said to be completely similar: geometric, kinematic, and dynamic similarity.

8.8.1 Geometric Similarity

The model must have the same shape as the prototype, that is, it must be geometrically similar. This requires that corresponding lengths are related by a constant ratio. A model car, for instance, will be geometrically similar to the full-scale vehicle if

$$\frac{(\text{length})_m}{(\text{length})_p} = \frac{(\text{width})_m}{(\text{width})_p} = \frac{(\text{height})_m}{(\text{height})_p} = \text{constant}_a$$

where the subscripts m and p denote model and full-scale prototype, respectively.

8.8.2 Kinematic Similarity

A flowfield is kinematically similar if, at corresponding points in the flow, the velocities (at corresponding times) are in a constant ratio. For example, the flow over a model airfoil (Figure 8.6) will be kinematically similar to the full-scale airfoil if

$$\frac{V_{1m}}{V_{1p}} = \frac{V_{2m}}{V_{2p}} = \frac{V_{3m}}{V_{3p}} = \text{constant}_b$$

and so on.

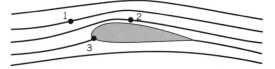

FIGURE 8.6 Flow over an airfoil.

8.8.3 Dynamic Similarity

For dynamic similarity, corresponding forces need to be in a constant ratio. For instance, the forces acting on a fluid element are those due to pressure, viscous stresses, gravity, and inertia. So

$$\frac{(\text{force due to pressure})_m}{(\text{force due to pressure})_p} = \frac{(\text{inertia force})_m}{(\text{inertia force})_p} = \text{constant}_c$$

and so on.

We see that to make a model flow completely similar to the prototype flow, certain nondimensional parameters need to be equal in the two cases. For geometric similarity we require

$$\frac{(\text{length})_m}{(\text{length})_p} = \frac{(\text{width})_m}{(\text{width})_p} = \text{constant}_a$$

and so forth.

We can write this with model quantities on one side and prototype quantities on the other, so that, with l = length, and w = width,

$$\left(\frac{\ell}{w}\right)_m = \left(\frac{\ell}{w}\right)_p \tag{8.18}$$

For dynamic similarity

$$\frac{(\text{inertia force})_m}{(\text{inertia force})_p} = \frac{(\text{force due to friction})_m}{(\text{force due to friction})_p} = \text{constant}_c$$

and so on, so that, for instance

$$\left(\frac{F_i}{F_v}\right)_m = \left(\frac{F_i}{F_v}\right)_p \tag{8.19}$$

We recognize these ratios as nondimensional groups. Equations 8.18 and 8.19 indicate that:

> For two flows to be dynamically similar, all of the nondimensional groups on the input side must have the same value.[3]

This rule is the single most important rule for modeling fluid flows, since we are generally interested in forces and dynamic similarity.

---

[3] We only need the inputs to be the same because the functional relationship can always be expressed as $\Pi_1 = f(\Pi_2, \Pi_3, \Pi_4, \ldots$ etc.), and so if $\Pi_2, \Pi_3, \Pi_4, \ldots$ etc. all have the same values in the model and full scale then Π_1 automatically does too.

We know many of these nondimensional parameters by name. For example

$$\frac{F_i}{F_v} = \frac{\text{inertia force}}{\text{viscous force}} = \text{Reynolds number} = \text{Re}$$

$$\frac{\text{inertia force}}{\text{gravitational force}} = \text{Froude number} = \text{F}$$

$$\frac{\text{pressure force}}{\text{inertia force}} = \text{pressure coefficient} = \text{C}_p$$

$$\frac{\text{lift force}}{\text{inertia force}} = \text{lift coefficient} = \text{C}_L$$

$$\frac{\text{drag force}}{\text{inertia force}} = \text{drag coefficient} = \text{C}_D$$

$$\frac{\text{flow speed}}{\text{sound speed}} = \text{Mach number} = \text{M}$$

$$\frac{\text{wavelength}}{\text{diameter}} = \text{Strouhal number} = \text{St}$$

$$\frac{\text{inertia force}}{\text{surface tension force}} = \text{Weber number} = \text{We}$$

The physical significance of these nondimensional groups is largely derived from the equations of motion, as noted in Section 8.6. It is the concept of similarity, and the basic equations of motion that underlie this concept, which help us to understand the physical significance of the nondimensional parameters in common use.

For any given problem, many different forces could be acting. For example, the flow of a viscous fluid in a narrow open channel, for instance, could be affected by viscous forces, inertia forces, gravitational forces, surface tension, and other such forces. To model this flow accurately, the Reynolds number, Froude number, and Weber number must all be the same as in the full-scale flow. This is not usually possible, but in most cases some approximations can be made. Some forces may be more important than others, and therefore some effects may be neglected. A limited type of similarity can then be achieved which may be accurate enough. If the channel flow considered in this example is not too narrow and the fluid viscosity is not all that important, then a sufficient level of dynamic similarity may be achieved if the ratio of the inertia force to the gravitational force (the Froude number) is the same for model and prototype.

An important point is that each nondimensional group can always be interpreted as the ratio of two dimensional and physically meaningful quantities. For example, the Weber number is often interpreted as the ratio of a typical inertia force to a typical surface tension force. When the Weber number is large, it may mean that surface tension effects are not important. In general, when any nondimensional group is very large or very small, it may not be an important factor in achieving similarity, although experiments will always be necessary to confirm this hypothesis.

EXAMPLE 8.1 *Vortex Shedding*

When the wind blows over a chimney, vortices are shed into the wake (see Figures 8.7 and 10.8). The frequency of vortex shedding f depends on the chimney diameter D, its length L, the wind velocity V, and the kinematic viscosity of air ν.

(a) Express the nondimensional shedding frequency in terms of its dependence on the other nondimensional groups.

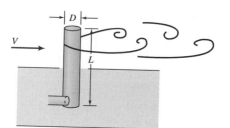

FIGURE 8.7 Vortex shedding from a chimney.

(b) If a $\frac{1}{10}$ scale model were to be tested in a wind tunnel and full dynamic similarity was required:
  (i) What air velocity would be necessary in the wind tunnel compared to the wind velocity experienced by the full scale chimney?
  (ii) What shedding frequency would be observed in the wind tunnel compared to the shedding frequency generated by the full scale chimney?

Solution: We are given that

$$f = \phi(D, L, V, \nu)$$

For part (a), we write down the matrix of dimensions:

	f	D	L	V	ν
L	0	1	1	1	2
T	-1	0	0	-1	-1

The rank of the largest determinant is 2, so we need to find 3 independent dimensionless groups. Two obvious Π-groups are the Reynolds number and the ratio of lengths. We can also make the frequency nondimensional by using the diameter and the velocity. Hence,

$$\frac{fD}{V} = \phi'\left(\frac{VD}{\nu}, \frac{L}{D}\right)$$

The ratio $\frac{fD}{V}$ is called the Strouhal number or the *reduced frequency*, and it always makes an appearance in unsteady problems with a dominant frequency (see Section 10.5).

For part (b), to achieve dynamic similarity the Π-products in the model and full-scale chimneys must be equal. That is,

$$\left(\frac{VD}{\nu}\right)_m = \left(\frac{VD}{\nu}\right)_p, \quad \left(\frac{fD}{V}\right)_m = \left(\frac{fD}{V}\right)_p, \quad \left(\frac{L}{D}\right)_m = \left(\frac{L}{D}\right)_p$$

(Remember: the subscript p indicates the "prototype" or full-scale value). Starting with the Reynolds number, similarity requires that

$$\frac{V_m D_m}{\nu_m} = \frac{V_p D_p}{\nu_p}, \quad \text{or} \quad \frac{V_m}{V_p} = \frac{D_p}{D_m}\frac{\nu_m}{\nu_p}$$

Since $\nu_m = \nu_p$ (the fluid is air in both cases), and $D_p = 10D_m$, we have

$$V_m = 10V_p$$

The Strouhal number similarity gives

$$\frac{f_m D_m}{V_m} = \frac{f_p D_p}{V_p}, \quad \text{or} \quad \frac{f_m}{f_p} = \frac{D_p}{D_m}\frac{V_m}{V_p}$$

Since $V_m = 10V_p$, and $D_p = 10D_m$, we obtain

$$f_m = 100 f_p$$

To have a dynamically similar model, therefore, we will need to run the tunnel at a speed 10 times greater than the natural wind speed, and we expect to see a shedding frequency 100 times greater than that observed for the full-scale chimney. ∎

EXAMPLE 8.2 *Viscometer*

A cone and plate viscometer consists of a cone with a very small angle α, which rotates above a flat surface, as shown in Figure 8.8. The torque required to spin the cone at a constant speed is a direct measure of the viscous resistance, which is how this device can be used to find the fluid viscosity. We see that the torque τ (which has the same dimensions as work) is a function of the radius R, the cone angle α, the fluid viscosity μ, and the angular velocity ω.

(a) Use dimensional analysis to express this information in terms of a functional dependence on nondimensional groups.

(b) If α and R are kept constant, how will the torque change if both the viscosity and the angular velocity are doubled?

Solution: The information given implies that

$$\tau = \phi(R, \alpha, \mu, \omega)$$

The dimensions of torque are force×distance, that is, $MLT^{-2} L = ML^2T^{-2}$, and the dimensions of viscosity are stress over velocity gradient, that is, $MLT^{-2}L^{-2} (LT^{-1})^{-1} = ML^{-1}T^{-1}$. For part (a), we write down the matrix of dimensions

	τ	R	α	μ	ω
M	1	0	0	1	0
L	2	1	0	-1	0
T	-2	0	0	-1	-1

The rank of the largest determinant is 3, so we need to find 2 independent dimensionless groups. One dimensionless number is given by the angle α. The second

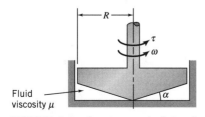

FIGURE 8.8 A cone and plate viscometer.

will be a combination of the other parameters so that the torque becomes nondimensional. Hence,

$$\frac{\tau}{\mu \omega R^3} = \phi'(\alpha)$$

For part (b), if α is constant, the parameter $\tau/(\mu \omega R^3)$ must also remain constant to maintain full similarity. So when μ and ω are both doubled, τ will increase by a factor of 4. ■

EXAMPLE 8.3 *Draining Tank*

A large water tank slowly empties through a small hole under the action of gravity. The flow is steady, and the volume flow rate $\dot{q}$ depends on the exit velocity U, the gravitational acceleration g, the depth of the water h, and the diameter of the nozzle D.

(a) By using dimensional analysis, find the nondimensional groups that govern the behavior of the nondimensional flow rate.

(b) A test is to be made on a $\frac{1}{4}$ scale model. If the test is designed to ensure full dynamic similarity, what is the ratio of model volume flow rate to prototype volume flow rate?

(c) If you now decide that instead of the volume flow rate, you are interested in the mass flow rate $\dot{m}$ (the density must therefore be included in the dimensional analysis), will the number of nondimensional groups change?

(d) If you later discover that the mass flow rate $\dot{m}$ depends on the viscosity μ and the surface tension σ (dimensions of force per unit length), in addition to ρ, U, g, h, and D, find all the relevant nondimensional groups.

Solution: For part (a), we begin with

$$\dot{q} = \phi(U, g, h, D)$$

The dimensions of volume flow rate are $L^3 T^{-1}$, so that the matrix of dimensions becomes

	$\dot{q}$	U	g	h	D
L	3	1	1	1	1
T	−1	−1	−2	0	0

The rank of the largest determinant is 2, so we need to find 3 independent dimensionless groups. One dimensionless number is given by the ratio of lengths D/h. The second will be the nondimensional volume flow rate, for example, $\dot{q}/UD^2$, and the third will be a Froude number, $U/\sqrt{gh}$. Hence,

$$\frac{\dot{q}}{UD^2} = \phi'\left(\frac{D}{h}, \frac{U}{\sqrt{gh}}\right)$$

For part (b) dynamic similarity requires that all the nondimensional parameters take constant values. Therefore

$$\frac{\dot{q}_m}{U_m D_m^2} = \frac{\dot{q}_p}{U_p D_p^2}$$

Since $D_p/D_m = 4$, we have

$$\frac{\dot{q}_m}{\dot{q}_p} = \frac{U_m}{16U_p}$$

We also require

$$\frac{U_m}{\sqrt{gh_m}} = \frac{U_p}{\sqrt{gh_p}}$$

Since $h_p/h_m = 4$,

$$\frac{U_m}{U_p} = \frac{\sqrt{gh_m}}{\sqrt{gh_p}} = \frac{1}{2}$$

and therefore, for dynamic similarity,

$$\frac{\dot{q}_m}{\dot{q}_p} = \frac{1}{32}$$

For part (c), we have:

$$\dot{m} = \phi(U, g, h, D, \rho)$$

The dimensions of mass flow rate are MT^{-1}, so that the matrix of dimensions becomes

	$\dot{m}$	U	g	h	D	ρ
M	1	0	0	0	0	1
L	0	1	1	1	1	-3
T	-1	-1	-2	0	0	0

The rank of the largest determinant is 3, so we will still have 3 independent dimensionless groups: in this case, $\dot{m}/\rho UD^2$, D/h, and $U/\sqrt{gh}$.

For part (d), the functional dependence becomes

$$\dot{m} = \phi(U, g, h, D, \rho, \mu, \sigma)$$

The dimensions of the surface tension are force per unit length, MT^{-2}, so that the matrix of dimensions becomes

	$\dot{m}$	U	g	h	D	ρ	μ	σ
M	0	0	0	0	0	1	1	1
L	3	1	1	1	1	-3	-1	0
T	-1	-1	-2	0	0	0	-1	-2

The rank of the largest determinant is 3, so we will need to find 5 independent dimensionless groups. We already have three: the nondimensional mass flow rate $\dot{m}/\rho UD^2$, the ratio of lengths D/h, and the Froude number, $U/\sqrt{gh}$. The fourth Π-product will be a Reynolds number, $\rho Ud/\mu$, and the fifth will be a nondimensional surface tension such as $\sigma/\rho U^2D$. Hence,

$$\frac{\dot{m}}{\rho UD^2} = \phi'\left(\frac{D}{h}, \frac{U}{\sqrt{gh}}, \frac{\rho UD}{\mu}, \frac{\sigma}{\rho U^2D}\right)$$

■

EXAMPLE 8.4 *Nuclear Explosion*

Imagine that you were given a film of the first atomic bomb explosion in New Mexico. In the movie, there are some images of trucks and other objects that provide a length scale, so that you can plot the radius of the fireball r as a function of time t. Can you estimate E, the energy released by the explosion?

Solution: The most difficult part is choosing the shortest, correct list of parameters. If we had some insight, and some luck, we might suppose that

$$r = f_b(E, t, \rho)$$

where ρ is the ambient density before the explosion takes place. Energy has the dimensions of a force $\times$ length, that is, ML^2T^{-2}, and the matrix of dimensions is

	r	E	t	ρ
M	0	1	0	1
L	1	2	0	-3
T	0	-2	1	0

We see that $N = 4$, the rank of the matrix is 3, and therefore there is only one dimensionless group. So

$$\frac{r^5 E}{t^2 \rho} = \text{constant}$$

or, better,

$$\Pi_b = \frac{r}{t^{2/5}}\left(\frac{E}{\rho}\right)^{1/5} = \text{constant}$$

The constant Π_b must be found by experiment. For example, we can use film of another explosion with a known energy release to find Π_b. This experiment can also be used to check the analysis: if r varies as $t^{2/5}$, as predicted, then our analysis is substantiated and the film of the atomic explosion contains all the information required to find the energy released by the bomb.

This analysis was first performed by the British scientist Sir Geoffrey Ingram Taylor at a time when information on the explosive power of atomic bombs was highly classified. When he tried to publish his findings, which gave a very good estimate of the actual energy released, he found his paper immediately classified at such a level that even he was not allowed to read it. For a historical note on G.I., as he was widely known, see section 15.20. ∎

PROBLEMS

8.1 The velocity of propagation c of surface waves in a shallow channel is assumed to depend on the depth of the liquid h, the density ρ, and the acceleration due to gravity g. By means of dimensional analysis, simplify this problem and express this dependence in nondimensional terms.

8.2 The period of a pendulum T is assumed to depend only on the mass m, the length of the pendulum ℓ, the acceleration due to gravity g, and the angle of swing θ. By means of dimensional analysis, simplify this problem and express this dependence in nondimensional terms.

8.3 The sound pressure level p' generated by a fan is found to depend only on the fan rotational speed ω, the fan diameter D, the air density ρ, and the speed of sound a. Express this dependence in nondimensional terms.

8.4 The power input to a water pump, P, depends on its efficiency η, its discharge (volume flow rate) $\dot{q}$, the pressure increase Δp across the pump, the density of the liquid ρ, and the diameter of the impeller D. Express this dependence in nondimensional terms.

8.5 The thrust T of a marine propeller is assumed to depend only on its diameter D, the fluid density ρ, the viscosity μ, the revolutions per unit time ω, and its forward velocity V. Express this dependence in nondimensional terms.

8.6 In Section 1.4.4, it was argued that the viscosity must depend on the average molecular speed $\bar{v}$, the number density ρ, and the mean free path ℓ. Express this dependence in nondimensional terms.

8.7 A ship 100 m long moves in fresh water at 15°C. Find the kinematic viscosity of a fluid suitable for use with a model 5 m long, if it is required to match the Reynolds number and Froude number. Comment on the feasibility of this requirement.

8.8 The height h to which a column of liquid will rise in a small-bore tube due to surface tension is a function of the density of the liquid ρ, the radius of the tube r, the acceleration due to gravity g, and the surface tension of the liquid σ.

(a) Express this dependence in nondimensional terms.

(b) If the capillary rise for liquid A is 25 mm in a tube of radius 0.5 mm, what will be the rise for a liquid B having the same surface tension but four times the density of liquid A in a dynamically similar system?

8.9 A chimney 100 ft high is being forced to vibrate at a frequency f in a wind of 20 ft/s by vortices that are shed in its wake. This phenomenon depends on the fluid density ρ and viscosity μ and the chimney material modulus of elasticity E, where E = stress/strain. A model is constructed which is geometrically similar to the chimney in every way and is 10 ft high. The mass per unit length, m, of chimney model can be adjusted by attaching dummy masses inside without affecting its elastic behavior. Gravity is not involved in this problem.

(a) Using dimensional analysis, derive all relevant nondimensional groups. Try to use physically meaningful nondimensional groups wherever possible. Make sure these groups are independent.

(b) If the full-scale chimney is made from steel with a modulus of elasticity $E = 30 \times 10^6$ lb_f/in^2, find the necessary modulus of elasticity so as to simulate the correct conditions at the model scale.

(c) At what frequency would you expect the chimney to vibrate if the model vibrates at 5 Hz?

8.10 Two cylinders of equal length are concentric: the outer one is fixed and the inner one can rotate. A viscous incompressible fluid fills the gap between them.

(a) Using dimensional analysis, derive an expression for the torque (= force $\times$ distance) required to maintain constant-speed rotation of the inner cylinder if this torque depends only on the diameters and length of the cylinders, the viscosity and density of the fluid, and the angular speed of the inner cylinder.

(b) To test a full-scale prototype, a half-scale model is built. The fluid used in the prototype is also to be used in the model experiment. If the prototype angular speed is ω_p and the prototype torque is T_p:

(i) At what angular speed must the model be run to obtain full dynamical similarity?

(ii) How is the model torque related to the prototype torque?

8.11 The power P required to drive a propeller is known to depend on the diameter of the propeller D, the density of fluid ρ, the speed of sound c, the angular velocity of the propeller ω, the freestream velocity V, and the viscosity of the fluid μ.

(a) How many dimensionless groups characterize this problem?

(b) If the effects of viscosity are neglected, and if the speed of sound is not an important variable, express the relationship between power and the other variables in nondimensional form.

(c) A one-half scale model of a propeller is built, and it uses P_m horsepower when running at a speed ω_m. If the full-scale propeller in the same fluid runs at $\omega_m/2$, what is its power consumption in terms of P_m if the functional dependence found in part (b) holds? What freestream velocity should be used for the model test?

8.12 The torque T required to rotate a disk of diameter D with angular velocity ω in a fluid is a function of the density ρ and the viscosity μ of the fluid (torque has units of work).

(a) Find a nondimensional relationship between these quantities.

(b) Calculate the angular velocity and the torque required to drive a 750 *mm* diameter disk rotating in air if it requires a torque of 1.2 *Nm* to rotate a similar disk of 230 *mm* diameter in water at a corresponding speed of 1500 *rpm*.

8.13 Consider geometrically similar animals of different linear size, L. Assume that the distance an animal can jump, H, is a function of L, the average density ρ, the average muscle stress σ (that is, muscle force over leg cross-sectional area), and the gravitational constant g.

(a) Find a nondimensional functional expression for H.

(b) In Swift's *Gulliver's Travels,* the Lilliputians were a race of very small people, tiny compared to Gulliver's size. If the Lilliputians had the same ρ and σ as Gulliver, what conclusions can you draw on dimensional grounds? Would they jump higher than Gulliver? The same height as Gulliver? Lower than Gulliver?

8.14 A simple carburetor is sketched in Figure P8.14. Fuel is fed from a reservoir (maintained at a constant level) through a tube so as to discharge into the airstream through an opening where the tube area is a. At this point, the flow area for the air is A and the fuel level is a distance L higher. Let the density and mass flow rate for the air and fuel be ρ_a, $\dot{m}_a$ and ρ_f, $\dot{m}_f$ respectively.

(a) Assuming the flows are inviscid, perform a dimensional analysis to determine the fuel–air ratio, $\dot{m}_f/\dot{m}_a$ as a function of the other dimensionless parameters.

(b) Analyze the problem dynamically and determine a specific relationship between $\dot{m}_f/\dot{m}_a$ and the relevant variables.

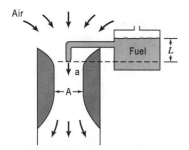

FIGURE P8.14

8.15 A golf ball manufacturer wants to study the effects of dimple size on the performance of a golf ball. A model ball four times the size of a regular ball is installed in a wind tunnel.

(a) What parameters must be controlled to model the golf ball performance?

(b) What should be the speed of the wind tunnel to simulate a golf ball speed of 200 *ft/s*?

(c) What rotational speed must be used if the regular ball rotates at 60 revolutions per second?

8.16 When a river flows at a velocity V past a circular pylon of diameter D, vortices are shed at a frequency f. It is known that f is also a function of the water density ρ and viscosity μ, and the acceleration due to gravity, g.

(a) Use dimensional analysis to express this information in terms of a functional dependence on nondimensional groups.

(b) A test is to be performed on a $\frac{1}{4}$ scale model. If previous tests had shown that viscosity is not important, what velocity must be used to obtain dynamic similarity, and what shedding frequency would you expect to see?

8.17 A propeller of diameter d develops thrust T when operating at N revolutions per minute with a forward velocity V in air of density ρ.

(a) Use dimensional analysis to express this information in terms of a functional dependence on nondimensional groups. Try to choose groups that look familiar.

(b) The single propeller described above is to be replaced by a pair of two propellers of the same shape operating at the same forward velocity and together producing the same thrust in air with the same density. Use the concepts of dynamic similarity to determine the diameter d_2 and the rotational speed N_2 of each of the propellers.

(c) What change in power, if any, is required?

8.18 The lift force F on a high-speed vehicle is a function of its length L, velocity V, diameter D, and angle of attack α (the angle the chord line makes with the flow direction), as well as the density ρ and speed of sound a of air (the speed of sound of air is only a function of temperature).

(a) Express the nondimensional lift force in terms of its dependence on the other nondimensional groups.

(b) If a $\frac{1}{10}$ scale model were to be tested in a wind tunnel at the same pressure and temperature (that is, the same sound speed) as encountered in the flight of the full-scale vehicle, and full dynamic similarity was required:

    (i) What air velocity would be necessary in the wind tunnel compared to the velocity of the full-scale vehicle?

    (ii) What would be the lift force acting on the model compared to the lift force acting on the vehicle in flight?

8.19 The drag of a golf ball F_D depends on its velocity V, its diameter D, its spin rate ω (commonly measured in radians/sec, or revolutions per minute), the air density ρ and viscosity μ, and the speed of sound a.

(a) Express the nondimensional drag force in terms of its dependence on the other nondimensional groups.

(b) If it was decided that the speed of sound was not important, how would this change the dimensional analysis? Under these conditions, if an experiment was carried out in standard air at a velocity of $2V$, what diameter and spin rate would be required to be dynamically similar to an experiment at a velocity of V?

(c) Design an experiment to investigate the influence of the speed of sound on the drag. Start by considering the requirements for dynamic similarity, and then explain how the influence of sound speed on the drag could be isolated by an experiment. Be as specific as you can.

8.20 **(a)** Carry out a dimensional analysis of a hydraulic jump with friction.

(b) Find the minimum number of nondimensional parameters describing this flow, and write the functional relationship governing the ratio of downstream to upstream depth.

(c) If a $\frac{1}{4}$ scale model of a hydraulic jump was to be tested in the laboratory:

    (i) What is the ratio of incoming flow velocities for the model and the full-scale required for dynamic similarity?

    (ii) What is the ratio of kinematic viscosities required for the model and the full-scale?

8.21 A large water tank empties slowly through a small hole under the action of gravity. The flow is steady, and the mass flow rate $\dot{m}$ depends on the exit velocity V, the gravitational acceleration g, the depth of the water h, the diameter of the nozzle D, the viscosity μ, and the surface tension σ (force/unit length).

(a) Express the nondimensional mass flow rate in terms of its dependence on the other nondimensional groups.

(b) If an experiment was carried out using the same fluid (water) in a $\frac{1}{8}$ scale model:
   (i) What is the ratio of the model mass flow rate to prototype mass flow rate that would be needed to obtain dynamical similarity?
   (ii) Do you anticipate any difficulties in obtaining full dynamical similarity?

8.22   Tests on a model propeller in a wind tunnel at sea level (air density $\rho = 1.2 \ kg/m^3$) gave the following results for the thrust at a number of forward velocities.

V (m/s)	0	10	20	30
Thrust (N)	300	278	211	100

The propeller diameter was $0.8 \ m$ and it was spun at 2,000 *rpm*.

(a) Using dimensional analysis find the nondimensional parameters which govern this observed behavior.

(b) Using the experimental data given in the table, find the thrust generated by a geometrically similar propeller of diameter $3 \ m$, spinning at 1,500 *rpm* at a forward velocity of 45 m/s, while operating at an altitude where the density is half that at sea level. You may interpolate from tabulated values.

8.23   In testing the aerodynamic characteristics of golf balls, the scientific officers at the USGA collected the experimental data given below. The diameter is always D, the roughness height is k, the air density is ρ ($= 1.2 \ kg/m^3$), the freestream velocity is V, the lift force L is in Newtons, and the rate of spin ω is measured in revolutions per second.

(a) How many nondimensional parameters describe this problem?

(b) Express all the experimental data in nondimensional form and plot the data in this form.

(c) By comparing the results from Balls 1 and 2, what can you say about the effect of roughness?

Ball 1: D = 42.7 mm, k = 0, V = 100 m/s

ω	20	40	60	80
L	−1.8	+0.9	+7.2	+18

Ball 2: D = 42.7 mm, k = 1 mm, V = 50 m/s

ω	5	15	25	35
L	−0.23	−0.23	+0.68	+2.7

Ball 3: D = 171 mm, k = 4 mm, V = 20 m/s

ω	1	2	3	4
L	−0.87	−0.29	+1.7	+5.8

8.24 The resistance of a sea-going ship is due to wave-making and viscous drag, aı expressed in functional form as

$$F_D = f(V, L, B, \rho, \mu, g)$$

where F_D is the drag force, V is the ship speed, L is its length, B is its width, ρ and μ are the sea water density and viscosity, and g is the gravitational constant.

(a) Find the nondimensional parameters that describe the problem.

(b) If we are to test a model of the ship, what are the requirements for dynamic similarity?

(c) We are going to test a 1/25 scale model of a 100 m long ship. If the maximum velocity of the full-scale ship is 10 m/s, what should the maximum speed of the model be? What should the kinematic viscosity of the model test fluid be compared to the kinematic viscosity of sea water?

8.25 The drag force F_D acting on a ship depends on the forward speed V, the fluid density ρ and viscosity μ, gravity, its length L and width B, and the average roughness height k.

(a) Use dimensional analysis to express this information in terms of a functional dependence on nondimensional groups.

(b) If the ship moved from fresh water to salt water, where the density and viscosity are both increased by 10%, how would its drag force change at a fixed speed?

8.26 The drag D of a ship's hull moving through water depends on its speed V, its width W, length L and depth of immersion H, the water density ρ and viscosity μ, and the gravitational acceleration g.

(a) Use dimensional analysis to express this information in terms of a functional dependence on nondimensional groups. Try to choose groups that look familiar.

(b) The full-scale ship will have a velocity V_1 and it will operate in sea water with a kinematic viscosity ν_1.

(i) If the wave pattern produced by a $\frac{1}{30}$ scale model is to be similar to that observed on the full-scale ship, what must be the model test velocity V_2?

(ii) To obtain full dynamic similarity, what must be the kinematic viscosity ν_2 of the test fluid?

VISCOUS INTERNAL FLOWS

9.1 INTRODUCTION

In this chapter and in Chapter 10, we examine viscous flows. Viscous flows are flows where viscous stresses exert significant forces on the fluid, and viscous energy dissipation is important. Here, we concentrate on *internal* flows, that is, flows through pipes, tubes, and ducts, and in Chapter 10 we examine *external* flows, which include the flow over surfaces of airplanes, ships, and automobiles.

The flow of fluids through pipes and ducts of different shapes is extremely important in all kinds of engineering applications, including the transport of water, oil, natural gas, and sewage, and the design of heating and air conditioning systems in buildings and vehicles, as well as heat exchanger systems of all kinds. In biologic systems, the flow of blood and air through the respiratory and circulatory systems are also dominated by viscous effects.

Three basic phenomena are important to the understanding of viscous flows.

1. The appearance of viscous stresses due to the relative motion of fluid particles.
2. The development of velocity gradients by the no-slip condition whenever the fluid is in contact with a solid.
3. The transition of laminar to turbulent flow as the Reynolds number increases.

For flow through long ducts, there is an additional phenomenon where the velocity profile far from the entry to the duct becomes fully developed, which means it becomes independent of the distance from the entry. Some of these concepts were introduced as early as Chapter 1, but we will now study each phenomenon in detail as a necessary prelude to analyzing laminar and turbulent flow in pipes, ducts, and piping systems.

9.2 VISCOUS STRESSES AND REYNOLDS NUMBER

In Chapter 1, it was noted that the most obvious property of fluids, their capacity to flow and change shape, is a result of their inability to support shearing stresses. Whenever a fluid is deforming its shape, shearing stresses must be present. The magnitude of the shear stress τ is related to the rate of deformation, that is, the rate at which one part of the fluid is moving with respect to another part. In other words, τ depends on the strength of the velocity gradients, so that, for a simple two-dimensional flow,

$$\tau_{yx} = \mu \frac{\partial u}{\partial y}$$

(see Section 1.4.1). Many common fluids obey this kind of "Newtonian" relationship where the stress is proportional to the strain rate, including air and water over very wide ranges of pressures and temperatures.

When the viscous stresses are included in the differential form of the momentum equation we obtain the Navier–Stokes equation. For an incompressible flow, this equation is

$$\frac{D\mathbf{V}}{Dt} = -\frac{1}{\rho}\nabla p + \mathbf{g} + \nu\nabla^2\mathbf{V} \tag{9.1}$$

It was pointed out in Section 6.3.3 that the Navier–Stokes equation is a nonlinear, partial differential equation, and no general solution exists. Nevertheless, some insight may be gained by non-dimensionalizing this equation, as shown in Section 8.6. With a suitable choice of length and velocity scales, the coefficients in the equation indicate the relative importance of each term. The coefficient on the viscous term was given by the reciprocal of a Reynolds number, so that the Reynolds number gives a measure of how important the inertia force is compared to the viscous force.

When the Reynolds number is small (less than one, say), we would expect the viscous terms to be important everywhere in the flow field. Small Reynolds numbers mean very slow flow, or very small characteristic lengths, or very high kinematic viscosity, or all three. For example, a fish of length 1 *mm*, swimming at 1 *mm/s* in water, where $\nu = 10^{-6}\ m^2/s$, has a Reynolds number of one. Dust particles settling in the atmosphere may have Reynolds numbers of order 0.01. The flow of oil in a bearing where the clearances are typically very small, is also dominated by viscous stresses. We call these flows *creeping flows*.

In this chapter, we are interested in flows of more general engineering interest, where the Reynolds numbers are considerably larger, of order 100 to 1000 (at least). Under these conditions, viscous effects only become important in relatively thin regions, such as in boundary layers, fully developed internal flows, jets, and wakes.

9.3 BOUNDARY LAYERS AND FULLY DEVELOPED FLOW

When a viscous fluid flows over a solid surface, the no-slip condition requires that right at the surface there is no relative motion between the fluid and the surface (see Section 1.6). As a result, strong velocity gradients appear in the region near the surface. At Reynolds numbers large compared to one, this region is always very thin and it is called the boundary layer. By definition, the boundary layer is the region where velocity gradients are large enough to produce significant viscous stresses, and significant dissipation of mechanical energy. The region outside the boundary layer is called the *freestream*. It is also sometimes called the inviscid freestream, not because the fluid there is inviscid but because there are no significant velocity gradients in that region and viscous stresses are negligible.

Consider the simple flow of a viscous fluid over a flat surface aligned with the flow direction. At any level in the boundary layer, the viscous stresses tend to decrease the velocity of the flow on the high speed side of the layer, and increase the velocity on the low speed side. At the edge of the boundary layer, therefore, viscous action will also tend to slow the freestream fluid, and as we proceed downstream, more and more of the freestream flow is affected by friction. Therefore the

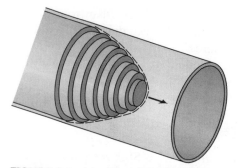

FIGURE 9.1 Idealized fully developed laminar pipe flow. With permission, *Engineering Fluid Mechanics,* A. Mironer, McGraw-Hill, 1979.

thickness of the boundary layer will grow with distance downstream (see Figure 1.21).

For an internal flow, such as that in a pipe or duct, the boundary layers will eventually meet at the center, and at this point they cannot grow any further. The velocity profile reaches an asymptotic state where the flow is said to be fully developed. The effects of viscous friction are then important over the whole cross-section of the tube and all adjacent layers of fluids slide over each other, as shown in Figure 9.1.

For an external flow, such as the flow over the hull of a ship, the velocity profile will not reach an asymptotic state and the boundary layer continues to grow. However, at high Reynolds numbers the boundary layer thickness (that is, the region where appreciable flow deceleration has occurred) is always very small compared to the length of the surface over which the boundary layer develops.

In laminar pipe flow, fully developed flow is attained within about 0.03 Re_D diameters of the entrance, where Re_D is the Reynolds number based on the diameter and the average velocity. For turbulent pipe flow, it takes about 25 to 40 pipe diameters,[1] so that in many pipe flow applications, the flow is fully developed almost along its entire length. It is interesting to note that when the flow is fully developed, the fluid velocity at any distance from the wall is constant with streamwise distance. Since there is no flow acceleration, the viscous force on the fluid must be balanced exactly by some other force (see equation 9.1). This force could come from gravity, but in a horizontal pipe or duct, the viscous force can only be balanced by pressure differences. In any case, work must be done on the system to keep the motion going.

Viscous, frictional stresses also cause energy dissipation in the fluid, which appears as heat. This heat generation is not usually important at subsonic speeds since the resulting temperature changes are typically very small. However, one example of a subsonic flow where this heat generation is important is the flow around a closed circuit wind tunnel, where the air is pumped around and around the same circuit. There is a continual energy input through the work done by the fan on the air, and although some of this energy will be lost by heat transfer to the walls of the tunnel, at high speeds this heat transfer may not be sufficient to keep the air temperature from rising to unacceptable levels. Cooling coils may be necessary to control the temperature.

---

[1] Schlichting, *Boundary Layer Theory,* 7th edition, published by McGraw-Hill, 1979.

At supersonic speeds (where the Mach number $M > 1$), very strong velocity gradients occur within the boundary layer and sufficient heat is generated by viscous dissipation to change the density of the fluid significantly. At hypersonic speeds, where $M \gg 1$, frictional heating can be sufficient to cause molecules to ionize and dissociate.

9.4 TRANSITION AND TURBULENCE

We have described the flow near a solid surface in terms of the slipping of adjacent fluid layers, where viscosity between the layers produces shearing stresses. This type of flow is called laminar flow, and it is seen whenever the Reynolds number is small (here "small" means less than about 1000 to 10,000 based on the freestream velocity and the characteristic length of the flow). At higher Reynolds numbers, the flow changes its nature, and instead of smooth, well-ordered, laminar flow, irregular eddying motions appear indicating the presence of turbulent flow. Because of the high degree of activity associated with turbulent eddies and their fluctuating velocities, the viscous energy dissipation inside a turbulent flow can be very much greater than in a laminar flow.

We can observe this transition from laminar to turbulent flow by examining the flow from a faucet. If the faucet is turned on very slowly, a smooth, orderly flow is observed. If there are no disturbances present, the flow will remain smooth and orderly. This is laminar flow [Figure 9.2(*a*)]. When the faucet is opened a little more, perturbations will start to appear in the surface of the jet. Occasionally, the whole of the jet takes on an irregular appearance, and then it may revert to its laminar state [Figure 9.2(*b*)]. If the faucet is now opened fully, the jet will become

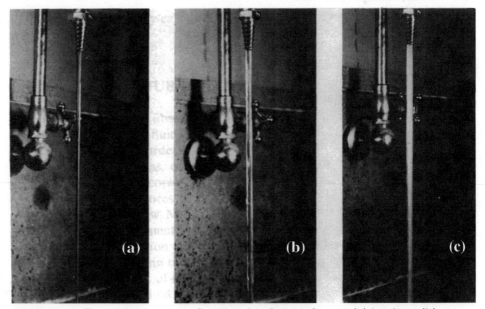

FIGURE 9.2 Transition in water flow issuing from a faucet: (a) laminar, (b) transitional and (c) turbulent. With permission, *Engineering Fluid Mechanics*, A. Mironer, McGraw-Hill, 1979.

fully turbulent [Figure 9.2(c)] and its appearance will change in a seemingly random way. Although its average motion is in one direction, within the flow there are irregularities everywhere. Have you ever watched cigarette smoke rise smoothly for a few inches, then burst into a seemingly chaotic flow? This is another example of transition and turbulence. A smoke particle would, on average, move in the principal flow direction but it would jitter as well, sometimes moving to one side, or up or down. Turbulent flow, while proceeding in a particular direction, like laminar flow, has the added complexity of random velocity fluctuations.

> *Turbulence, or the presence of eddies in a moving fluid, gives rise to the two-fold effect of a pronounced mixing of the fluid and a subsequent dissipation of energy. Like solid friction, fluid turbulence is sometimes a blessing to mankind, and sometimes a curse. Without the mixing which eddies produce, both the water in boiler tubes and the air surrounding the earth would be very poor distributors of heat, steam engines would prove too costly to run, and the atmosphere would be incapable of supporting life. On the other hand, dust storms would not then occur, and rivers would not transport their tremendous loads of silt from the foothills to the sea. Without means of producing eddies in the process of propulsion, moreover, a swimmer—or an ocean linear—could make little headway, just as a car would remain at rest if friction provided no tractive force. Yet, paradoxically, were turbulence not produced by the motion of a body through a fluid, the process of streamlining would be quite unnecessary.*[2]

In the next three sections, we consider laminar, transitional, and turbulent pipe and duct flows, in that order.

9.5 POISEUILLE FLOW

It is possible to find exact solutions for fully developed pipe flow, as long as the flow is laminar and not turbulent (this is called Poiseuille flow). It is also possible to find a solution for fully developed, laminar flow in a rectangular duct (this is called plane Poiseuille flow). Here, we will derive the solutions for these two cases from first principles, starting with duct flow. The results are very similar, and they differ only in the constants that appear in the equations.

9.5.1 Fully Developed Duct Flow

Consider laminar, two-dimensional, steady, constant density duct flow, far from the entrance, so that the flow is fully developed. The duct is rectangular in cross-section, of width W and height D, and it is horizontal, so we do not have to include the forces due to gravity. For a duct, laminar flow requires that the Reynolds number $\overline{V}D/\nu$ is less than about 2000, where $\overline{V}$ is the average velocity. We can find the velocity profile and the friction factor, as follows.

First, we use an integral analysis applied to a large control volume of length dx (Figure 9.3). The pressure varies in the streamwise direction, but it does not vary across the flow since the streamlines are parallel. The pressure over the inflow

---

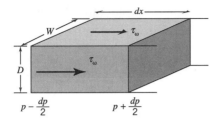

FIGURE 9.3 Integral control volume for fully developed flow in a rectangular duct.

area is $p - \frac{1}{2}dp$, and over the outflow area it is $p + \frac{1}{2}dp$. If the flow is fully developed, the influx and outflux of momentum have the same magnitude and so they cancel. Since there is no change in the flow momentum, the flow must be in equilibrium under the applied forces: the force acting on the fluid due to the pressure drop is balanced by the force due to the viscous stress acting on the interior surface of the duct. The pressures act on the cross-sectional area wD. The viscous stress τ_w will not vary in the streamwise direction since the flow is fully developed and the velocity gradient is independent of x. If we assume that τ_w is the same on each wall of the duct, it acts on the area $2(w + D)dx$. The momentum equation gives

$$(p - \tfrac{1}{2}dp)wD - (p + \tfrac{1}{2}dp)wD = 2\tau_w(w + D)dx$$

and so

$$\frac{dp}{dx} = -2\frac{(w + D)}{wD}\tau_w$$

The duct was taken to be two-dimensional, which implies that $w \gg D$. Hence,

$$\frac{dp}{dx} = -\frac{2}{D}\tau_w \tag{9.2}$$

The pressure gradient is negative because the pressure drops due to viscous friction.

Second, we use a differential analysis applied to a small control volume of size $dx\,dy\,dz$ (Figure 9.4), located at a distance y from the centerline. Since the flow is two-dimensional, there is no variation of the velocity in the spanwise direction. Since the flow is fully developed, there is no variation of velocity in the streamwise direction. The only velocity variation that exists is in the cross-stream direction, that is, in the y-direction. Therefore, the only shearing stresses acting on the elemental volume will act on its top and bottom faces which have an area $dx\,dz$. Let τ be the shear stress in the middle of the control volume. Using a Taylor series expansion,

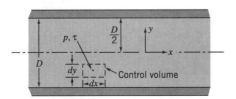

FIGURE 9.4 Differential control volume for fully developed flow in a rectangular duct.

and retaining only the first-order terms, the net viscous force F_v acting in the x-direction is given by

$$F_v = \left(\tau + \frac{\partial \tau}{\partial y}\frac{dy}{2}\right) dxdz - \left(\tau - \frac{\partial \tau}{\partial y}\frac{dy}{2}\right) dxdz$$

$$= \frac{\partial \tau}{\partial y} dxdydz$$

We see that the resultant viscous force is nonzero only if the viscous stress varies through the flowfield (that is, if $\partial \tau/\partial y \neq 0$). The viscous stress itself is given by

$$\tau = \mu \frac{du}{dy}$$

since the velocity only varies in the y-direction. Therefore,

$$F_v = \mu \frac{d^2u}{dy^2} dxdydz$$

We see that the viscous force depends on the second derivative of the velocity field.

Now we evaluate the force due to pressure differences. If the pressure in the middle of the control volume is p, the net force F_p due to pressure acting in the x-direction is given by

$$F_p = \left(p - \frac{\partial p}{\partial x}\frac{dx}{2}\right) dydz - \left(p + \frac{\partial p}{\partial x}\frac{dx}{2}\right) dydz$$

$$= -\frac{\partial p}{\partial x} dxdydz$$

Since the pressure does not vary in the spanwise direction (the flow is two-dimensional), and it cannot vary in the y-direction (the streamlines are straight and forces due to gravity are neglected),

$$\frac{\partial p}{\partial z} = \frac{\partial p}{\partial y} = 0, \quad \text{and} \quad \frac{\partial p}{\partial x} = \frac{dp}{dx}$$

and so

$$F_p = -\frac{dp}{dx} dxdydz$$

Since there is no acceleration, F_v and F_p must sum to zero, so that

$$\boxed{\frac{1}{\rho}\frac{dp}{dx} = \nu \frac{d^2u}{dy^2}} \tag{9.3}$$

This equation may also be found directly from the Navier–Stokes equation (equation 9.1). For negligible body forces and fully-developed flow (where all accelerations are zero), this equation reduces to

$$0 = -\nabla p + \nu \nabla^2 \mathbf{V}$$

The streamlines are parallel, and the flow is two-dimensional, so that $v = w = 0$, and $\mathbf{V} = u\mathbf{i}$. In addition, the pressure gradient can only vary in the streamwise direction, and so $p = p(x)$. It follows that

$$\frac{1}{\rho}\frac{dp}{dx} = \nu \frac{d^2u}{dy^2}$$

in agreement with equation 9.3.

Equation 9.3 holds for all fully developed flows in the absence of body forces. It can be solved by observing that the pressure term only depends on x, whereas the viscous term only depends on y. Therefore the equation can only be satisfied if both terms are equal to a constant, $-K$, say. Hence,

$$\frac{dp}{dx} = -K \tag{9.4}$$

$$\mu \frac{d^2u}{dy^2} = -K \tag{9.5}$$

where K must be positive since the pressure decreases with distance down the duct.

We can integrate equation 9.5 twice to obtain the velocity profile. The boundary conditions are

$$u = 0, \quad \text{at} \quad y = \pm \frac{D}{2} \quad \text{(no-slip)}$$

$$\frac{du}{dy} = 0, \quad \text{at} \quad y = 0 \quad \text{(symmetry)}$$

Hence,

$$u = \frac{KD^2}{8\mu}\left[1 - \left(\frac{2y}{D}\right)^2\right]$$

That is,

$$u = \frac{D^2}{8\mu}\left(-\frac{dp}{dx}\right)\left[1 - \left(\frac{2y}{D}\right)^2\right] \tag{9.6}$$

The average velocity $\overline{V}$ is defined by

$$\overline{V} = \frac{1}{A}\int u\,dA$$

so that $\overline{V}$ is equal to the volume flow rate divided by the cross-sectional area A. For a duct, the average velocity is given by

$$\overline{V} = \frac{1}{WD}\int_{-D/2}^{D/2} uW\,dy = \frac{KD^2}{12\mu} = \frac{D^2}{12\mu}\left(-\frac{dp}{dx}\right) \tag{9.7}$$

From equation 9.6 we obtain,

$$\boxed{\frac{u}{\overline{V}} = \frac{3}{2}\left[1 - \left(\frac{2y}{D}\right)^2\right]} \tag{9.8}$$

We find that the laminar velocity profile is parabolic, as illustrated in Figure 9.4a. The friction factor is given by

$$f = \frac{\frac{\Delta p}{L}D}{\frac{1}{2}\rho\overline{V}^2} = \frac{-\frac{dp}{dx}D}{\frac{1}{2}\rho\overline{V}^2} = -\frac{dp}{dx}\frac{2D}{\rho\overline{V}^2} = \frac{24\mu}{\rho\overline{V}D}$$

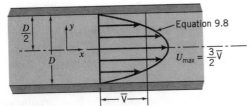

FIGURE 9.4a Parabolic velocity profile in fully-developed laminar duct flow.

where the pressure gradient was eliminated using equation 9.7. That is,

$$f = \frac{24}{Re}$$

(9.9)

which holds for all rectangular ducts where $Re = \rho \overline{V} D / \mu$ is less than about 1400 (the exact Reynold number value depends on the aspect ratio of the duct).

We can find the stress at the wall ($y = -D/2$) by differentiating equation 9.6. That is,

$$\tau_w = \mu \left. \frac{\partial u}{\partial y} \right|_w = -\frac{D}{2} \frac{dp}{dx}$$

which agrees with the result obtained earlier using a large control volume (equation 9.2).

9.5.2 Fully Developed Pipe Flow

Here we consider fully developed laminar flow in a smooth, circular, horizontal pipe. In Chapter 8, this flow was examined using dimensional analysis (Section 8.5). For a smooth pipe of length L and diameter D, we found

$$\frac{\Delta p}{\frac{1}{2} \rho \overline{V}^2} = \phi_1 \left(\frac{\rho \overline{V} D}{\mu}, \frac{L}{D} \right)$$

where the pressure drop over the length L is Δp, and $\overline{V}$ is the average velocity over the cross-section of the pipe. Since pipes are typically long compared to their diameter, an asymptotic, fully developed state exists where the flow properties do not change with further increase in length. The mean velocity profile is then independent of the distance along the pipe. In that case, the parameter L/D is no longer important, and the resistance relationship for fully developed smooth pipe flow becomes

$$f \equiv \frac{\left(\frac{\Delta p}{L} \right) D}{\frac{1}{2} \rho \overline{V}^2} = \phi \left(\frac{\rho \overline{V} D}{\mu} \right) = \phi(Re_D)$$

where f is the friction factor. Therefore, the friction factor depends only on the Reynolds number based on average velocity and pipe diameter, Re_D.

As before, we begin with an integral analysis using a control volume of length dx, as shown in Figure 9.5. The pressure over the inflow area is $p - \frac{1}{2} dp$, and over the outflow area it is $p + \frac{1}{2} dp$. Since the flow is fully developed, the flow is in equilibrium under the applied forces. The pressure drop acts on the cross-sectional area $\pi D^2 / 4$, and the viscous stress acts on the surface area $2\pi D dx$. Hence,

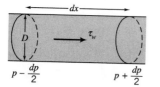

FIGURE 9.5 Integral control volume for fully developed flow in a circular pipe.

$$(p - \tfrac{1}{2}dp)\frac{\pi D^2}{4} - (p + \tfrac{1}{2}dp)\frac{\pi D^2}{4} = \tau_w \pi D \, dx$$

so that

$$\frac{dp}{dx} = -\frac{4}{D}\tau_w$$

Next, we apply a differential analysis using an annular control volume, located at a distance r from the centerline (Figure 9.6). Since the flow is two-dimensional, there is no variation of the velocity in the circumferential direction. Since the flow is fully developed, there is no variation of velocity in the streamwise direction. The only velocity variation that exists is in the radial, that is, in the r-direction. Shearing stresses act on the inner and outer faces of the control volume. If the shear stress at the center of the annular control volume is τ, the net viscous force F_v acting in the x-direction is given by

$$F_v = \left(\tau + \frac{\partial\tau}{\partial r}\frac{dr}{2}\right)2\pi\left(r + \frac{dr}{2}\right)dx - \left(\tau - \frac{\partial\tau}{\partial r}\frac{dr}{2}\right)2\pi\left(r - \frac{dr}{2}\right)dx$$

$$= \tau 2\pi dr dx + \frac{\partial\tau}{\partial r}2\pi r dr dx$$

neglecting second-order terms.

The net force due to pressure differences in the x-direction, F_p, is given by

$$F_p = \left(p - \frac{\partial p}{\partial x}\frac{dx}{2}\right)2\pi r dr - \left(p + \frac{\partial p}{\partial x}\frac{dx}{2}\right)2\pi r dr$$

$$= -\frac{\partial p}{\partial x}2\pi r dr dx$$

where p is the pressure in the center of the control volume. Since there is no acceleration, F_p and F_v must balance, so that

$$\frac{\partial p}{\partial x} = \frac{\tau}{r} + \frac{\partial\tau}{\partial r} = \frac{1}{r}\frac{\partial(r\tau)}{\partial r}$$

The pressure term depends only on x and the viscous term depends only on r. Therefore,

$$\boxed{\frac{dp}{dx} = \frac{1}{r}\frac{d(r\tau)}{dr}} \tag{9.10}$$

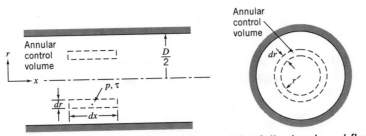

FIGURE 9.6 Differential control volume for fully developed flow in a circular pipe.

This equation can be satisfied only if the left hand side and the right hand side are equal to a constant, $-K'$, say. Hence,

$$\frac{dp}{dx} = -K' \tag{9.11}$$

$$\frac{1}{r}\frac{d(r\tau)}{dr} = -K' \tag{9.12}$$

where K' must be positive since the pressure decreases with distance along the pipe. Integrating equation 9.12 gives

$$r\tau = -\frac{r^2}{2}K' + c_1$$

or

$$\tau = -\frac{r}{2}K' + \frac{c_1}{r}$$

Since the only velocity gradient is in the r-direction, the viscous stress is given by

$$\tau = \mu\frac{du}{dr}$$

and therefore

$$\mu\frac{du}{dr} = -\frac{r}{2}K' + \frac{c_1}{r}$$

and

$$u = -\frac{r^2}{4\mu}K' + \frac{c_1}{\mu}\ln r + c_2$$

Using the boundary conditions

$$u = 0, \quad \text{at} \quad r = \pm D/2$$

$$\frac{du}{dr} = 0, \quad \text{at} \quad r = 0$$

we find $c_1 = 0$, and $c_2 = D^2K'/(16\mu)$, so that

$$u = \frac{K'D^2}{16\mu}\left[1 - \left(\frac{r}{D}\right)^2\right]$$

That is,

$$u = \frac{D^2}{16\mu}\left(-\frac{dp}{dx}\right)\left[1 - \left(\frac{2r}{D}\right)^2\right] \tag{9.13}$$

and we see that the velocity profile for laminar pipe flow is parabolic, as found for the rectangular duct. The average velocity can be found by integration, as before. We find that, for a pipe,

$$\overline{V} = \frac{D^2}{32\mu}\left(-\frac{dp}{dx}\right) \tag{9.14}$$

so that

$$\frac{u}{\overline{V}} = 2\left[1 - \left(\frac{2r}{D}\right)^2\right]$$ (9.15)

The friction factor is then given by

$$f = \frac{\frac{\Delta p}{L} D}{\frac{1}{2}\rho\overline{V}^2} = \frac{-\frac{dp}{dx} D}{\frac{1}{2}\rho\overline{V}^2} = \frac{64\mu}{\rho\overline{V}D}$$

where equation 9.14 was used to eliminate the pressure gradient. Hence,

$$f = \frac{64}{Re}$$ (9.16)

which holds for all circular pipes and tubes with $Re < 2300$.

9.6 TRANSITION IN PIPE FLOW

The experimental friction factors for circular pipes and tubes are shown in Figure 9.7. This figure is called the Moody diagram (for a historical note on Lewis Ferry Moody, see Section 15.18). As expected from the dimensional analysis given in Section 8.5, all the data collapse onto curves that depend on just three nondimen-

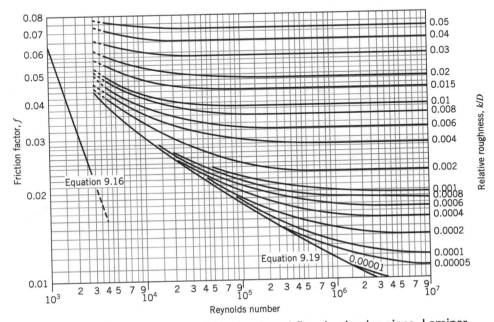

FIGURE 9.7 Moody diagram for fully developed flow in circular pipes. Laminar flow is described by equation 9.16. Prandtl's universal law of friction for turbulent smooth pipes is given by equation 9.19. Adapted from Moody, L. F. "Friction Factors for Pipe Flow," *Trans. of the ASME*, **66**, 671–684, 1944, with permission.

sional groups: the friction factor, the Reynolds number, and the relative roughness k/D. This is true for laminar and turbulent flows.

The Moody diagram displays only one curve for $Re < 2300$, corresponding to equation 9.16: this is the laminar flow regime where the layers of fluid slide over each other, and the pressure drop is due to viscous stresses set up by the velocity gradient. Surface roughness does not affect the drag in the laminar flow regime. However, for Reynolds numbers greater than 2300, transition to turbulence can occur. The precise value of the Reynolds number where transition occurs depends on many factors, including surface roughness, vibrations, noise, and thermal disturbances.

To understand why these factors are important, and to appreciate the role of the Reynolds number in governing the stability of the flow, it is helpful to think in terms of a spring-damper system such as the suspension system of a car. Driving along a bumpy road, the springs act to smooth the movement experienced by the passengers. If there were no shock absorbers, however, there would be no damping of the motion, and the car would continue to oscillate long after the bump has been left behind. So the shock absorbers, through a viscous damping action, dissipate the energy in the oscillations and reduce their amplitude. If the viscous action is strong enough, the oscillations will die out very quickly and the passengers can proceed smoothly. If the shock absorbers are not in good shape, the oscillations may not die out. The oscillations can actually grow if the excitation frequency is in the right range, and the system can experience resonance. The car becomes unstable, and it is then virtually uncontrollable.

The stability of a fluid flow is similarly dependent on the relative strengths of the acceleration and the viscous damping. This is expressed by the Reynolds number, which is the ratio of a typical inertia force to a typical viscous force (see Section 8.6). At low Reynolds numbers, the viscous force is large compared to the inertia force, and the flow behaves in some ways like a car with a good suspension system. Small disturbances in the velocity field, created perhaps by small roughness elements on the surface, or pressure perturbations from external sources such as vibrations in the walls of the pipe, or even the presence of loud noises, will be damped out and not allowed to grow. This is the case for pipe flow at Reynolds numbers smaller than the critical value of 2300. As the Reynolds number increases, however, the viscous damping action becomes comparatively smaller, and at some point, it becomes possible for small perturbations to grow, just as in the case of a car with poor shock absorbers. The flow can become unstable, and it can experience transition to a turbulent state where large variations in the velocity field can be maintained.

The point where the disturbances will grow rather than decay will also depend on the magnitude and frequency of the disturbances. If the disturbances are very small, as in the case where the walls are very smooth, or if the frequency of the disturbance is not close to resonance, transition to turbulence will occur at a Reynolds number higher than the critical value. The point of transition does not correspond to a single Reynolds number, and it is possible to delay transition to relatively large Reynolds numbers by controlling the disturbance environment. At very high Reynolds numbers, however, it is impossible to maintain laminar flow since under these conditions even minute disturbances will be amplified into turbulence.

9.7 TURBULENT PIPE FLOW

In turbulent flow, a significant part of the mechanical energy in the flow goes into forming and maintaining randomly eddying motions, which eventually dissipate their kinetic energy into heat. At a given Reynolds number, therefore, we expect the drag of a turbulent flow to be higher than the drag of a laminar flow, and the results shown in the Moody diagram confirm this expectation in that the friction factor for turbulent flow is larger than for a laminar flow at the same Reynolds number. Also, turbulent flow is affected by surface roughness, so that increasing roughness increases the drag. If the roughness is large enough, and the Reynolds number is large enough, the drag becomes independent of Reynolds number and the friction factor then becomes a function of the relative roughness only. For example, when $k/D = 0.006$ and $Re_D > 3 \times 10^5$, Figure 9.7 shows that f is a constant equal to 0.032.

The unsteady eddying motions in turbulent flow are in constant motion with respect to each other, producing fluctuations in the flow velocity and pressure. If we were to measure the streamwise velocity in turbulent pipe flow, we would see a variation in time as shown in Figure 9.8. We define a time-averaged value $\langle u \rangle$ by

$$\langle u \rangle = \lim_{T \to \infty} \frac{1}{T} \int_t^{t+T} u \, dt \qquad (9.17)$$

and a fluctuating value $u'(= u - \langle u \rangle)$, so that $\langle u \rangle$ is not a function of time, but u' is.

The eddies interact with each other as they move around, and they can exchange momentum and energy. For example, an eddy that is near the centerline of the pipe (and therefore has a relatively high average velocity), may move towards the wall and interact with eddies near the wall (which typically have lower average velocities). As they mix, momentum differences are smoothed out. This process is superficially similar to the action of viscosity, which tends to smooth out momentum gradients by molecular interactions, and therefore turbulent flows are sometimes said to have an equivalent *eddy viscosity*. Because turbulent mixing is such an effective transport process, the eddy viscosity is typically several orders of magnitude larger than the molecular viscosity.

The important point is that turbulent flows are very effective at mixing: the eddying motions can very quickly transport mass, momentum, and energy from one place to another. As a result, velocity, temperature, and concentration differences get smoothed out more effectively than in a laminar flow, and, for example, the time-averaged velocity profile in a turbulent pipe flow is much more uniform

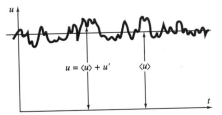

FIGURE 9.8 Velocity at a point in a turbulent flow as a function of time.

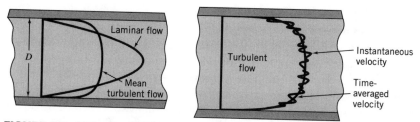

FIGURE 9.9 Velocity distributions in laminar and turbulent pipe flow.

than in a laminar flow (see Figure 9.9). The velocity profile is no longer parabolic, and it is sometimes approximated by a power law, such as

$$\frac{\langle u \rangle}{U_{CL}} = \left(\frac{2y}{D}\right)^{1/n} \tag{9.18}$$

where U_{CL} is the average velocity on the centerline, and the exponent n varies with Reynolds number (for a Reynolds number Re_D of about 100,000, $n = 7$).

As a result of this mixing, the velocity gradient at the wall is higher than that found in a laminar flow at the same Reynolds number, so that the shear stress at the wall is correspondingly larger. This observation agrees with the fact that, since the energy losses in a turbulent flow are higher than in a laminar flow, the pressure drop per unit length will be greater. From the momentum balance considered in Section 9.5.2, we know that there must be a larger frictional stress at the wall, which in turn requires a higher velocity gradient at the wall. Note that we cannot evaluate the wall stress from equation 9.18 since this approximation to the velocity profile actually gives an infinite velocity gradient (and therefore an infinite stress) at the wall (where $y = 0$). Near the wall, equation 9.18 cannot be correct. However, for turbulent flow, there are no exact solutions available for either the velocity profile or the friction factor variation with Reynolds number. Instead, we must always rely on experimental data and scaling arguments based on dimensional analysis. For the friction factor, we often use the semi-empirical relation

$$\frac{1}{\sqrt{f}} = 2.0 \log\left(Re\sqrt{f}\right) - 0.8 \tag{9.19}$$

which is known as Prandtl's universal law of friction for smooth pipes (this is the line corresponding to turbulent flow in a smooth pipe as given in Figure 9.7).

A more accurate form is given by

$$\frac{1}{\sqrt{f}} = 1.873 \log\left(Re\sqrt{f}\right) - 0.2631 - \frac{233}{(Re\sqrt{f})^{0.90}} \tag{9.20}$$

which describes the friction factor to within 1% for Reynolds numbers from 10×10^3 to 35×10^6.[3]

9.8 ENERGY EQUATION FOR PIPE FLOW

So far, we have considered only simple ducts and pipes, without worrying about how these elements fit into a piping or ducting system. Whenever a duct changes

---

[3] M.V. Zagarola & A.J. Smits, *Physical Review Letters*, January 1997.

its size, either through a gradual diffuser or by a sudden expansion, or when a valve is located somewhere along the pipe, or when a pipe flow enters or leaves a tank in a nonideal way, there will be additional losses in the system, manifesting themselves as a drop in pressure, or a reduction in head. Roughness can also be very important. When a pipe system ages, corrosion can roughen the surface of the pipes, and scale can build up, so that the losses due to friction can increase dramatically. As we will show in Section 9.9, for a given available pressure head, increased losses in the system will reduce the flow rate. The Moody diagram can be used to find the friction factor for laminar and turbulent pipe and duct flows but to design piping systems, we need to know the effects of pipe fittings, valves, diffusers, and other components.

Piping and ducting systems are commonly analyzed using the one-dimensional energy equation. Before we can do so, two modifications are necessary. First, viscous internal flows are two-dimensional, and in order to use the one-dimensional energy equation, we introduce the *kinetic energy coefficient, α*. This coefficient allows us to define an average velocity so that a two-dimensional field can be represented as an "equivalent" one-dimensional field.

Second, friction factors and loss coefficients are introduced into the energy equation to represent the kinetic energy losses in the system. Whenever possible, these coefficients are determined from theory, but more commonly they are based on empirical relationships established from experience.

9.8.1 Kinetic Energy Coefficient

Consider the steady flow of an incompressible fluid through a tube. We draw a control volume as shown in Figure 9.10, and apply the steady flow energy equation in integral form. That is,

$$\dot{Q} + \dot{W}_{shaft} = \int (\mathbf{n} \cdot \rho \mathbf{V}) \left(\hat{u} + \frac{p}{\rho} + \tfrac{1}{2}V^2 + gz \right) dA$$

where $\dot{Q}$ is the heat transferred to the fluid, $\dot{W}_{shaft}$ is the rate of doing work on the fluid by a machine (for example, by a pump or turbine), and $\hat{u}$ is the internal energy of the fluid per unit mass (see equation 5.23). If the pressures and internal energy at stations 1 and 2 are uniform,

$$\dot{Q} + \dot{W}_{shaft} = \dot{m}(\hat{u}_2 + \hat{u}_1) + \dot{m} \left(\frac{p_2}{\rho} - \frac{p_1}{\rho} \right) + \dot{m}g(z_2 - z_1)$$

$$+ \int \rho V_2(\tfrac{1}{2}V_2^2) \, dA_2 - \int \rho V_1(\tfrac{1}{2}V_1^2) \, dA_1 \qquad (9.21)$$

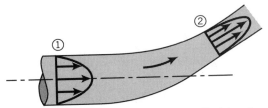

FIGURE 9.10 Control volume applied to steady, two-dimensional flow through a tube of varying cross section.

When the velocity varies across the pipe, the kinetic energy also varies across the pipe. To treat the flow as one-dimensional, we need to introduce an equivalent kinetic energy, based on the average velocity $\overline{V}$ rather than the spatially varying velocities V_1 and V_2. The advantage of using an average velocity $\overline{V}$ is that it is relatively easy to measure (for example, if the fluid was a liquid, we could use a bucket and a stopwatch to find the volume flow rate, and then $\overline{V}$ is found by dividing the volume flow rate by the interior cross-sectional area of the pipe).

How do we achieve this aim of finding an equivalent kinetic energy? The important requirement is that the kinetic energy flux through the pipe be accurately represented. Specifically, we would like to define an *average* kinetic energy flux based on the mean velocity, which is equal to the *actual* kinetic energy flux. For a pipe flow, the actual kinetic energy flux through area A is given by

$$\text{actual KE flux} = \int \tfrac{1}{2}\rho V^3 \, dA$$

where V varies with radial distance. The average kinetic energy flux through the pipe is given by the volume flow rate times the average kinetic energy. The average kinetic energy, in turn, may be written as some multiple of $\tfrac{1}{2}\rho \overline{V}^2$, say $\alpha \tfrac{1}{2}\rho \overline{V}^2$, where the value of α depends only on the shape of the velocity profile. Therefore,

$$\text{average KE flux} = \tfrac{1}{2}\alpha\rho\overline{V}^3 A$$

By equating the actual and the average kinetic energy fluxes we obtain

$$\alpha = \frac{1}{A\overline{V}^3}\int V^3 \, dA \tag{9.22}$$

which defines the kinetic energy coefficient, α. This coefficient can be found if the velocity profile shape is known. In Section 9.5.2, we showed that the velocity profile for laminar pipe flow is parabolic, which gives $\alpha = 2.0$. Turbulent pipe flow has a considerably flatter profile so that the velocity over much of the cross-section is close to the average value, and we find by experiment that $1.08 > \alpha > 1.03$. If the flow were precisely one-dimensional, $V = \overline{V}$, and $\alpha = 1$.

We can now write the energy equation for a two-dimensional flow in an equivalent one-dimensional form, using

$$\text{KE flux} = \int \tfrac{1}{2}\rho V^3 \, dA = \tfrac{1}{2}\alpha\rho\overline{V}^3 A = \rho\overline{V}A(\tfrac{1}{2}\alpha\overline{V}^2)$$

Hence,

$$\dot{Q} + \dot{W}_{shaft} = \dot{m}(\hat{u}_2 - \hat{u}_1) + \dot{m}\left(\frac{p_2}{\rho} - \frac{p_1}{\rho}\right) + \dot{m}g(z_2 - z_1) + \dot{m}(\tfrac{1}{2}\alpha_2\overline{V}_2^2 - \tfrac{1}{2}\alpha_1\overline{V}_1^2)$$

That is,

$$\frac{\dot{Q} + \dot{W}_{shaft}}{\dot{m}} = \hat{u}_2 - \hat{u}_1 + \frac{p_2}{\rho} - \frac{p_1}{\rho} + gz_2 - gz_1 + \tfrac{1}{2}\alpha_2\overline{V}_2^2 - \tfrac{1}{2}\alpha_1\overline{V}_1^2$$

and so

$$\left(\frac{p_1}{\rho} + gz_1 + \tfrac{1}{2}\alpha_1\overline{V}_1^2\right) - \left(\frac{p_2}{\rho} + gz_2 + \tfrac{1}{2}\alpha_2\overline{V}_2^2\right) = (\hat{u}_2 - \hat{u}_1) - \frac{\dot{Q} + \dot{W}_{shaft}}{\dot{m}}$$

The term

$$H = \frac{p}{\rho} + gz + \tfrac{1}{2}\alpha\overline{V}^2$$

represents the mechanical energy per unit mass at any cross-section. If $\alpha = 1$, H is equal to the Bernoulli's constant for incompressible flows (see Section 4.7.3).

The change in internal energy $(\hat{u}_2 - \hat{u}_1)$ represents the (irreversible) conversion of mechanical energy into thermal energy, and $-\dot{Q}/\dot{m}$ is the loss of energy by heat transfer to the surroundings. In the absence of shaft work, the term $(\hat{u}_2 - \hat{u}_1 - \dot{Q}/\dot{m})$ represents the difference in mechanical energy per unit mass between sections 1 and 2, and it is often represented by the term gh_ℓ, where h_ℓ is called the *total head loss* and it has the dimensions of length. The shaft work is the work done *on* the fluid, so that a pump will provide positive shaft work. Finally, we have the energy equation in a form that is very convenient for pipe flow systems.

$$\left(\frac{p_1}{\rho} + gz_1 + \tfrac{1}{2}\alpha_1\overline{V}_1^2\right) - \left(\frac{p_2}{\rho} + gz_2 + \tfrac{1}{2}\alpha_2\overline{V}_2^2\right) = gh_\ell - \frac{\dot{Q} + \dot{W}_{shaft}}{\dot{m}} \tag{9.23}$$

Usually, $\dot{Q}$ is small for a pump or turbine. Also, $\dot{W}_{shaft}$ is the non-thermal power consumed by the pump, and $\dot{W}_{shaft}$ is the non-thermal power developed by the turbine.

9.8.2 Major and Minor Losses

We now consider how the total head loss h_ℓ may be calculated. In a piping network, the total head loss is the sum of *major* losses and *minor* losses. Major losses are due to friction. For a given length L of pipe with a constant diameter D, carrying fluid with an average velocity $\overline{V}$, these losses can be written as

$$\text{major losses} = f\frac{L}{D}\frac{\overline{V}^2}{2g}$$

where f is the friction factor. The Moody diagram gives the value of f as a function of Reynolds number and relative roughness height k/D (see Figure 9.7). The minimum roughness is called *smooth*, or *hydraulically smooth*. Equivalent roughness heights for different pipe materials are listed in Table 9.1.

Minor losses are due to entrances, fittings, area changes, and so forth. Whenever a pipe flow goes around a bend, or changes its cross-sectional area, or it is throttled by a valve, flow separation and secondary flows can occur. When a flow

TABLE 9.1 Roughness Height k of Common Pipe Materials

Type	k (mm)	k (ft)
Glass	Smooth	Smooth
Asphalted cast iron	0.12	0.0004
Galvanized iron	0.15	0.0005
Cast iron	0.26	0.00085
Wood stave	0.18–0.90	0.0006–0.003
Concrete	0.30–3.0	0.001–0.01
Riveted steel	1.0–10	0.003–0.03
Drawn tubing	0.0015	0.000005

SOURCE: *Introduction to Fluid Mechanics*, John & Haberman, Prentice Hall, 1988.

FIGURE 9.11 Flow through round and square bends showing regions of separation. From *Visualized Flow,* Japan Soc. of Mech. Engineers, Pergamon Press, 1988.

separates, the fluid is no longer smoothly flowing in the intended direction. Significant parts of the fluid are eddying and recirculating in a way that absorbs mechanical energy without doing useful work. Some examples are given in Figure 9.11. In addition, when the flow passes through a bend, the path followed often becomes twisted, so that the flow downstream of the bend has two components of velocity: a component in the downstream direction, and a circumferential or "secondary" component. The secondary flow absorbs flow energy, and therefore contributes to the loss of total energy.

Minor losses for inlets, exits, enlargements, and contractions are written as

$$(\text{minor losses})_a = K \frac{\overline{V}^2}{2g}$$

where K is a *loss coefficient,* which depends on the type of fitting. Typical values of K are 0.6 for a sudden two-fold increase in diameter, and 0.5 for a square-edged entry from a reservoir into a pipe (a sudden very large decrease in diameter). Even a minor rounding of the entry to the pipe reduces the loss coefficient significantly. For example, with a radius of curvature at entry equal to just 15% of the pipe diameter (r/D) the loss coefficient reduces to a value less than 0.04 (see Table 9.2).

For pipe bends, tee fittings, and valves, minor losses are sometimes written in terms of an equivalent length of straight pipe, L_e. In that case, for fully developed

TABLE 9.2 Typical Loss Coefficients for Pipe Entrances and Exits (D is the diameter of the pipe, and r is the radius of the rounded entry)

		K
Entrance type	Re-entrant	0.78
	Square-edged	0.5
	Rounded ($r/D = 0.02$)	0.28
	Rounded ($r/D = 0.06$)	0.15
	Rounded ($r/D \geq 0.15$)	0.04
Exit type	Abrupt	1.0

SOURCE: Data from Crane Co., NY, Technical Paper No. 410, 1982.

TABLE 9.3 Typical Loss Coefficients for Pipe Fittings

Valve or fitting	$\frac{L_e}{D}$	K
Gate valve (open)	8	0.20
Globe valve (open)	340	6.4
Angle valve (open)	150	—
Ball valve (open)	3	—
Standard 45° elbow	16	0.35
Standard 90° elbow	30	0.75
Long-radius 90° elbow	—	0.45
Standard tee (flow through run)	20	0.4
Standard tee (branch flow)	60	1.5

SOURCE: Data from Crane Co., NY, Technical Paper No. 410, 1982.

pipe flow with a friction factor f and average velocity V, the losses due to pipe bends, tees, and valves are given by

$$(\text{minor losses})_b = f \frac{L_e}{D} \frac{\overline{V}^2}{2g}$$

Typical values of L_e/D are 3 for a ball valve, 8 for a gate valve, 340 for a globe valve (all valves wide open), and 30 for a standard 90° elbow, where the radius of curvature of the elbow is ten times the diameter of the pipe (see Table 9.3).

Additional information on minor loss coefficients, with an extensive discussion of pipe flow losses, may be found in Munson, Young, & Okiishi *Fundamentals of Fluid Mechanics,* John Wiley & Sons, 1998.

On a practical note, we can write the friction factor f in terms of the volume flow rate $\dot{q} = \frac{\pi}{4} D^2 \overline{V}$:

$$f = \frac{\Delta p D}{\frac{1}{2}\rho \overline{V}^2 L} = \frac{\Delta p \pi^2 D^5}{8\rho \dot{q}^2 L}$$

That is,

$$\Delta p = \frac{8f}{\pi^2} \frac{\rho \dot{q}^2 L}{D^5}$$

This relationship for the pressure drop is very useful for design engineers when the expected Reynolds number is so large that the friction factor is moderately insensitive to changes in the Reynolds number. What happens if your client wants to double the volume flow rate? We see that Δp would quadruple. What happens if your client wants to halve the recommended pipe diameter to save money? The required pressure difference would increase by a factor of 32.

9.9 VALVES AND FAUCETS

For the discharge from a tank, the exit velocity is independent of the exit area but the discharge increases as the exit area increases, provided there are no losses. This result, obtained in Section 4.6, appears to contradict our common experience with valves. Opening a valve does not change the exit area of the pipe and yet it increases the discharge rate. Why? Valves work because they introduce losses into the system.

Imagine a constant pressure supply, such as a large tank of water, connected to a pipe of constant diameter with a valve located somewhere along its length. Assume all the losses are small except in the valve. The pressure difference from the surface of the tank to the exit of the tube is fixed by the depth of water in the tank, because the pressures at the free surface of the tank and at the exit of the tube are equal to atmospheric pressure. This means, in effect, that there is a certain amount of potential energy available to push the fluid through the piping system. If there are no losses anywhere, all of this potential energy can be converted to kinetic energy of fluid motion. The relationship describing the exit velocity in the absence of losses, $V_e = \sqrt{2gH}$ came from the relationship $\frac{1}{2}\rho V_e^2 = \rho gH$ (Section 4.6), which makes this connection between kinetic and potential energies clear. If losses occur, as in a valve or faucet, there will be less potential energy available for conversion to kinetic energy: as a result, the velocity of the outflow will be reduced. More losses mean a lower velocity and a reduced discharge, and so valves control the flow rate by changing the losses in the system.

Where do the losses in a valve come from? Valves come in many shapes and sizes, but a common design such as a gate valve uses a simple sliding gate with a sharp edge that moves in and out of the flow. As the flow passes over the gate, the edge of the gate causes the flow to separate and a recirculating region forms downstream where many losses occur. These losses cause the pressure to drop across the valve, because some of the available mechanical energy has been dissipated. Other valves are designed to have very little energy loss when they are fully opened. In a ball valve, for example, the valve mechanism consists of a ball with a hole drilled through it where the hole has the same diameter as the connecting pipes. By turning the handle, the ball rotates, so that when it is fully opened, the hole in the ball is aligned with the flow direction and there is very little disturbance to the flow. However, when it is partially opened losses occur as the flow passes over the edge of the mis-aligned hole and the flow rate can be controlled.

The throttling effect produced by energy losses can also be observed using a garden hose. Kinking the hose will control the flow rate (at least until the kink is severe enough to stop the flow altogether). A kink produces losses by flow separation, and so it acts like a valve.

Another example is given by the losses due to pipe friction. The friction inside rough pipes can be much greater than in clean, smooth pipes. For a given driving pressure, less potential energy is available in a rough pipe for conversion to kinetic energy than in a smooth pipe. Therefore, the velocity and the discharge will be reduced. For this reason, replacing old copper bathroom pipes that have corroded and developed a high degree of internal roughness with clean plastic pipes can improve the discharge from your shower head dramatically.

These considerations also help to explain why squeezing a hose near its exit will result in a higher exit velocity. According to our earlier analysis, if there are no losses in the system, reducing the exit area of the hose has no effect on the exit velocity. Therefore the explanation for a higher exit velocity is connected with the losses occurring in the system. By squeezing the end, we form a nozzle. The nozzle increases the fluid velocity and drops its pressure, but since the pressure at the exit of the nozzle is atmospheric, the pressure just upstream of the nozzle must be higher than atmospheric. Inside the entire hose (except for the nozzle itself), the driving pressure has decreased (assuming the upstream pressure remains constant), and therefore the velocity in that part of the hose actually decreases. Since the losses are now smaller, more pressure is available to drive the nozzle flow, which can then reach very high exit velocities, depending on the contraction ratio. The discharge, however, is actually smaller than it was when the hose was not being squeezed since the velocity in the main part of the hose has decreased. Squeezing the hose to increase the exit velocity will only work if the losses due to friction in the hose are relatively large. By writing the energy equation for this system, it can be shown that this will only work when $fL/D > 1$, where f is the friction factor.

9.10 HYDRAULIC DIAMETER

To find the friction factor for ducts with noncircular cross-section, there exists a useful concept, called the hydraulic diameter D_H, defined by

$$D_H = \frac{4 \times \text{cross-sectional area}}{\text{perimeter}} \tag{9.24}$$

The hydraulic diameter is used to define equivalent Reynolds numbers $\overline{V}D_H/\nu$, head loss factors $h_L = f(L/D_H)\overline{V}^2/2g$ and relative roughness factors k/D_H, so that the Moody diagram can be used to estimate the relevant friction factor. To show how this works, say we were asked to estimate the friction factor for laminar flow in a smooth rectangular duct by using the known result for a circular pipe (equation 9.16). For a duct, the hydraulic diameter is given by

$$D_H = \frac{4wD}{2(D + w)}$$

When $w \gg D$, $D_H = 2D$. Substituting D_H for the diameter in equation 9.16, we have our friction factor estimate for the duct flow.

$$f = \frac{64}{Re} = \frac{64\nu}{\overline{V}D_H} = \frac{32\nu}{\overline{V}D}$$

This result is 33% larger than the correct result for a two-dimensional duct given in equation 9.9. So the concept of a hydraulic diameter does not work very well for laminar flow in a two-dimensional duct. It tends to work better for flow in ducts with aspect ratios closer to one, such as square ducts, or triangular ones, where the errors for laminar flow are probably less than 10 or 15%. The approach works best for turbulent flows, where the errors are probably less than 10% or 15% for all duct shapes.

EXAMPLE 9.1   *Pipe Flow with Friction*

Consider a large tank filled with water draining through a long pipe of diameter D and length $L = 100D$, located near the bottom of the tank (Figure 9.12). The

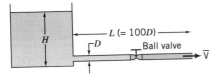

FIGURE 9.12 Tank draining through a pipe, with losses.

entrance to the pipe is square-edged, and there is a ball valve to control the flow rate. The valve is wide open, and the pipe is horizontal. Find an expression for the average exit velocity $\overline{V}$. The Reynolds number of the pipe flow is 2000, so that the flow is laminar.

Solution: We will assume that the flow is quasi-steady. For laminar flow, $\alpha = 2$. The loss coefficient K for a square-edged entrance is 0.5, and at the exit, all the kinetic energy of the flow is lost, so that the loss coefficient at this point is 1.0 (see Table 9.2). The ball valve has an equivalent length $L_e/D = 3$ (Table 9.3). At this Reynolds number, the friction factor $f = 0.032$ (Figure 9.7 and equation 9.16). Along a streamline starting at the free surface of the water in the tank and ending at the entrance to the pipe, there are no losses, and

$$\frac{p_a}{\rho} + 0 + gH = \frac{p_1}{\rho} + \alpha \frac{\overline{V}^2}{2} + 0$$

where p_1 is the pressure at the entrance to the pipe. The one-dimensional energy equation (equation 9.23) applied along the length of the pipe gives

$$\frac{p_1}{\rho} + \alpha \frac{\overline{V}^2}{2} = \frac{p_a}{\rho} + \alpha \frac{\overline{V}^2}{2} + gh_\ell$$

so that

$$gH = \alpha \frac{\overline{V}^2}{2} + K \frac{\overline{V}^2}{2} + f \frac{L_e}{D} \frac{\overline{V}^2}{2} + f \frac{L}{D} \frac{V^2}{2}$$

$$= \tfrac{1}{2}\overline{V}^2(2 + 0.5 + 1.0 + 0.032 \times 3 + 0.032 \times 100)$$

and so

$$\overline{V} = 0.542\sqrt{gH} = \sqrt{0.294gH}$$

We see that the exit velocity is very much less than the ideal value found for frictionless flow $(= \sqrt{2gH})$. ∎

EXAMPLE 9.2 *Pipe Flow with Friction*

Figure 9.13 shows water at $60°F$ flowing in a cast iron pipe. The pressure drop $p_1 - p_2 = 3500\ lb_f/ft^2$, the height difference $z_2 - z_1 = 30\ ft$, the length $L = 150\ ft$, and the diameter $D = 3\ in$. Find the volume discharge rate $\dot{q}$. Ignore minor losses, and assume the flow is turbulent.

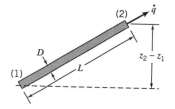

FIGURE 9.13 Water flowing uphill in a smooth pipe.

Solution: To find $\dot{q}$, we need to know the average velocity in the pipe $\overline{V}$. The one-dimensional energy equation gives

$$\frac{p_1}{\rho} + \alpha_1 \frac{\overline{V}^2}{2} + gz_1 = \frac{p_2}{\rho} + \alpha_2 \frac{\overline{V}^2}{2} + gz_2 + gh_\ell$$

Therefore, with $\alpha_2 = \alpha_1 = 1.0$ (the flow is turbulent.)

$$f\frac{L}{D}\frac{\overline{V}^2}{2} = \frac{p_1 - p_2}{\rho} + g(z_1 - z_2)$$

To find $\overline{V}$, we need to know the friction factor. However, the friction factor depends on the Reynolds number, and without knowing $\overline{V}$, we cannot find the Reynolds number. However, if the Reynolds number is large enough, the friction factor for a rough pipe is independent of Reynolds number (see Figure 9.7). From Table 9.1, we see that cast iron pipes typically have $k = 0.00085$ *ft*, so that $k/D = 0.0034$. For large Reynolds numbers, the Moody diagram then gives $f \approx 0.027$. Therefore,

$$\frac{\overline{V}^2}{2} = \frac{1}{f}\frac{D}{L}\left[\frac{p_1 - p_2}{\rho} + g(z_1 - z_2)\right]$$

$$= \frac{1}{0.027}\frac{0.25}{150}\left(\frac{3500\ lb_f}{1.938\ slug/ft^3} - 32.2\ ft/s^2 \times 30\ ft\right)$$

and so

$$\overline{V} = 10.2\ ft/s$$

Hence,

$$\dot{q} = \tfrac{\pi}{4}D^2\overline{V} = 0.50\ ft^3/s$$

Before we accept this answer, we need to check that the Reynolds number is high enough to make the assumption that the pipe was fully rough. We have

$$Re = \frac{\overline{V}D}{\nu} = \frac{10.2 \times 0.25}{1.21 \times 10^{-5}} = 211{,}000$$

The Moody diagram shows that at this Reynolds number the flow is not yet fully rough, and the friction factor is a little higher than we assumed—closer to 0.028 than 0.027. To obtain a more accurate answer for the volume flow rate, we would need to iterate. When we use $f = 0.028$, we find that $\overline{V} = 10.0$ *ft/s*, $\dot{q} = 0.49$ *ft³/s*, and $Re = 207{,}000$. The corresponding friction factor is again 0.028, so this second value of $\dot{q}$ is accurate enough. ∎

EXAMPLE 9.3 *Pipe Flow with Shaft Work*

Consider the previous example, where a 20 *hp* pump is placed halfway along the pipe. We will assume that the pressure drop remains the same at 3500 *lb_f/ft²*, and the friction factor is 0.028. Find the volume flow rate through the pipe.

Solution: The energy equation becomes

$$\frac{p_1}{\rho} + \frac{\overline{V}^2}{2} + gz_1 = \frac{p_2}{\rho} + \frac{\overline{V}^2}{2} + gz_2 + gh_\ell - gH$$

where H is the total head generated by the pump (it is positive since the work is done *on* the fluid by the pump). The head is given by dividing the mechanical shaft power by $\dot{m}g$, so that

$$H = \frac{\dot{W}_{shaft}}{\dot{m}g}$$

(see also Equation 13.8). That is,

$$f\frac{L}{D}\frac{\overline{V}^2}{2} = \frac{p_1 - p_2}{\rho} + g(z_1 - z_2) + gH$$

and $\dot{W}_m = 20\ hp$, so that

$$gH = \frac{20 \times 550}{1.938 \times \frac{\pi}{4}D^2\overline{V}}ft = \frac{115,630}{\overline{V}}ft$$

where $\overline{V}$ is measured in ft/s. The energy equation becomes

$$8.4\overline{V}^2 = 1806 - 966 + \frac{115,630}{\overline{V}}$$

so that

$$\overline{V}^3 - 100\overline{V} - 13,765 = 0$$

This equation has one physically meaningful solution, $\overline{V} = 25.4\ ft/s$. In this case the addition of a 20 hp pump has more than doubled the velocity. Hence, the volume flow rate is given by

$$\dot{q} = \frac{\pi}{4}D^2\overline{V} = 1.25\ ft^3/s$$ ∎

PROBLEMS

9.1 What is the Moody diagram?

9.2 Explain briefly why a faucet acts to control the flow rate.

9.3 What is a reasonable upper limit on the Reynolds number for laminar pipe flow?

9.4 At what Reynolds number would you expect fully developed pipe flow to become turbulent? Write down the definition of the Reynolds number for pipe flow and explain your notation.

9.5 Calculate the Reynolds number for pipe flow with diameter 12 mm, average velocity 50 mm/s, and kinematic viscosity $10^{-6}\ m^2/s$. Will the flow be laminar or turbulent?

9.6 Would you expect the flow of water in an industrial quality pipe of diameter 0.010 m to be laminar or turbulent, when the average velocity was 1 m/s, and the kinematic viscosity was $10^{-6}\ m^2/s$?

9.7 What is the likelihood that a flow with an average velocity of 0.15 ft/s in a 6 $in.$ water pipe is laminar? Find the speed at which the flow will always be laminar.

9.8 If the critical Reynolds number for a river is 2000 based on average velocity and depth, what is the maximum speed for laminar flow in a river 10 ft deep? 2 ft deep? Do you expect any river flow to be laminar?

9.9 Compare the velocity profile, the wall shear stress, and the pressure drop per unit length for laminar and turbulent duct flow (draw some diagrams to illustrate your answer).

9.10 Find the Reynolds number for water at 15°C flowing
   (a) in a tube of 6 *mm* diameter with an average velocity of 10 *cm/s*
   (b) in a pipe of 20 *cm* diameter with an average velocity of 1 *m/s*
   (c) in a tube of 2 *m* diameter with an average velocity of 3 *m/s*
   Indicate whether you expect the flow to be laminar or turbulent.

9.11 A given pipe first carries water and then carries air at 15°C and atmospheric pressure. What is the ratio of the mass discharge rates and the volume discharge rates, if the friction factor were the same for these two flows?

9.12 Two horizontal pipes of the same length and relative roughness carry air and water, respectively, at 60°F. The velocities are such that the Reynolds numbers and pressure drops for each pipe are the same. Find the ratio of the average air velocity to water velocity.

9.13 Water flows steadily through a smooth, circular, horizontal pipe. The discharge rate is 1.5 *ft³/s* and the pipe diameter is 6 *in.* Find the difference in the pressure between two points 400 *ft* apart if the water temperature is 60°F.

9.14 A 2000-gallon swimming pool is to be filled with a 0.75 *in.* diameter garden hose. If the supply pressure is 60 *psig*, find the time required to fill the pool. The hose is 100 *ft* long, with a friction factor equal to 0.02.

9.15 Figure P9.15 shows a pipeline through which water flows at a rate of 0.07 *m³/s*. If the friction factor for the pipe is 0.04 and the loss coefficient for the pipe entrance at point A is 1.0, calculate the pressure at point B.

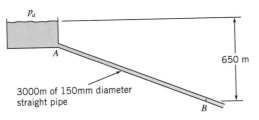

p_a

A

650 m

3000m of 150mm diameter straight pipe

B

FIGURE P9.15

9.16 Water flows from a large reservoir down a straight pipe of 2500 *m* length and 0.2 *m* diameter. At the end of the pipe, at a point 500 *m* below the surface of the reservoir, the water exits to atmosphere with a velocity V_e. If the friction factor for the pipe is 0.03 and the loss coefficient for the pipe entrance is 1.0, calculate V_e.

9.17 Figure P9.17 shows a pipeline connecting two reservoirs through which water flows at a rate of $\dot{q}$ *m³/s*. The pipe is 700 *m* long, it has a diameter of 50 *mm*, and it is straight. If the friction factor for the pipe is 0.001 and the loss coefficients for the pipe entrance and exit are 0.5 and 1.0 respectively, find $\dot{q}$, given that the difference in height between the surfaces of the two reservoirs is 100 *m*.

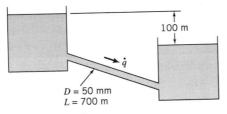

100 m

$\dot{q}$

$D = 50$ mm
$L = 700$ m

FIGURE P9.17

9.18 A pump is capable of delivering a gauge pressure p_1 when pumping water of density ρ, as shown in Figure P9.18. At the exit of the pump, where the diameter of the exit pipe is D and the velocity is V, there is a valve with a loss coefficient of 0.6. At a distance 100 D downstream, the diameter of the pipe smoothly reduces to $D/2$. At a further distance 100 D downstream, the flow exits to atmosphere. The pipes are horizontal. Calculate the pressure p_1 in terms of the density ρ and V, when the friction factor f is taken to be constant and equal to 0.01 everywhere.

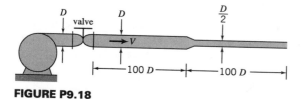

FIGURE P9.18

9.19 Figure P9.19 shows a pipeline through which water flows at a rate of 0.06 m^3/s. The pipe is 3000 m long, it has a diameter of 120 mm, and it is straight. If the friction factor for the pipe is 0.03 and the loss coefficient for the pipe entrance is 1.0, calculate H, the depth of the reservoir, given that the pipe exits to atmosphere.

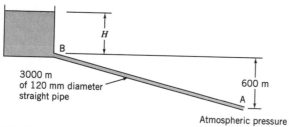

FIGURE P9.19

9.20 A water tank of constant depth H, open to the atmosphere, is connected to the piping system shown in Figure P9.20. After a length of pipe L of diameter D, the diameter decreases smoothly to a value of $D/2$ and then continues on for another length L before exiting to atmosphere. The friction factor f is the same for all piping. C_{D1} and C_{D2} are the loss coefficients for the entry and exit. Calculate the depth of the tank required to produce an exit velocity of V.

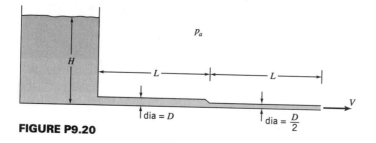

FIGURE P9.20

9.21 A small gap is left between a window and the windowsill, as shown in Figure P9.21. The gap is 0.15 mm by 30 mm, and the width of the window is 1 m. The pressure difference across the window is 60 *Pa.* Estimate the average velocity and the volume flow rate through the gap. The air temperature is 0°*C.* Ignore minor losses.

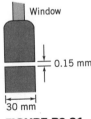

Window

0.15 mm

30 mm

FIGURE P9.21

9.22   For a constant mass flow rate and a constant friction factor, show that the pressure drop in a pipe due to friction varies inversely with the pipe diameter to the fifth power.

9.23   A ½-horsepower fan is to be used to supply air to a class room at 60°*F* through a smooth air-conditioning duct measuring 6 *in.* by 12 *in.* by 50 *ft* long. Find the volume flow rate and the pressure just downstream of the fan. State all of your assumptions.

9.24   Water at 60°*F* is siphoned between two tanks as shown in Figure P9.24. The connecting tube is 2.0 *in.* diameter smooth plastic hose 20 *ft* long. Find the volume flow rate and the pressure at point P, which is 8 *ft* from the entrance to the tube. Ignore minor losses. (To begin, assume a friction factor value, and then iterate.)

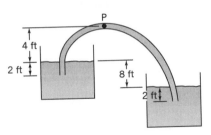

P

4 ft

2 ft

8 ft

2 ft

FIGURE P9.24

9.25   Due to corrosion and scaling, the equivalent roughness height of a pipe *k* increases with its years in service *t,* varying approximately as

$$k = k_0 + \varepsilon t$$

where k_0 is the roughness of the new pipe. For a cast iron pipe, $k_0 \approx 0.26$ *mm,* and $\varepsilon \approx$ 0.00001 *m* per year. Estimate the discharge (volume flow rate) of water through a 20 *cm* diameter cast iron pipe 500 *m* long as a function of time over a 20-year period. Assume the pressure drop is constant and equal to 150 *kPa.*[4]

9.26   Oil of kinematic viscosity $\nu = 4 \times 10^{-4} ft^2/s$ at room temperature flows through an inclined tube of 0.5 *in.* diameter. Find the angle α the tube makes with the horizontal if the pressure inside the tube is constant along its length and the flow rate is 5 ft^2/hr. The flow is laminar.

9.27   Water at 60°*F* flows at a rate of 300 ft^3/s through a rectangular open channel that slopes down at an angle α, as shown in Figure P9.27. The depth of the water is 5 *ft,* and the width

[4] Adapted from John & Haberman, *Introduction to Fluid Mechanics,* published by Prentice Hall, 1988.

of the channel is 8 *ft*. The channel is made of concrete which has a friction factor $f = 0.02$. Find the angle α.

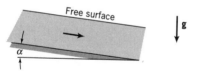

FIGURE P9.27

9.28 You have a choice between a 4 *ft* or a 3 *ft* diameter cast-iron pipe to transport 2000 gallons per minute of water to a power plant. The larger diameter pipe costs more but the losses are smaller. What is the relative decrease in head loss if you choose the larger diameter pipe? The kinematic viscosity is $1 \times 10^{-5} \, ft^2/s$.

9.29 A farmer must pump at least 100 gallons of water per minute from a dam to a field, located 25 *ft* above the dam and 2000 *ft* away. She has a 10-horsepower pump, which is approximately 80% efficient. Find the minimum size of smooth plastic piping she needs to buy, given that the piping is sized by the half-inch. Use $\nu = 10^{-5} \, ft^2/s$.[5] Ignore minor losses.

9.30 Under the action of gravity, water of density ρ passes steadily through a circular funnel into a vertical pipe of diameter d, and then exits from the pipe, falling freely under gravity, as shown in Figure P9.30. Atmospheric pressure acts everywhere outside the funnel and pipe. The entrance to the pipe has a loss coefficient $K_1 = 0.5$, the pipe has a friction factor $f = 0.01$, and the exit from the pipe has a loss coefficient $K_2 = 1.0$. The kinetic energy coefficient $\alpha = 1$. Find the outlet velocity in terms of g and d. You may assume that $D \gg d$.

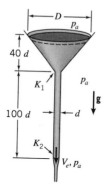

FIGURE P9.30

9.31 Determine the horsepower required to pump water vertically through 300 *ft* of 1.25 *in.* diameter smooth plastic piping at a volume flow rate of 0.1 ft^3/s.

9.32 Water flows with a volume flow rate of 100 *liters/s* from a large reservoir through a pipe 100 *m* long and 10 *cm* diameter, which has a turbomachine near the exit. The outlet is 10 *m* below the level of the reservoir, and it is at atmospheric pressure. If the friction factor of the pipe is 0.025, find the power added to, or provided by the turbomachine. Is it a pump or a turbine? Neglect minor losses.

9.33 Repeat the previous problem for a volume flow rate of 10 *liters/s*.

9.34 Water at 10°C is pumped from a large reservoir through a 500 *m* long, 50 *mm* diameter plastic pipe to a point 100 *m* above the level of the reservoir at a rate of 0.015 m^3/s. If

---

[5] Adapted from Potter & Foss *Fluid Mechanics*, published by Great Lakes Press, Inc., 1982.

the outlet pressure must exceed 10^6 Pa, find the minimum power required. Neglect minor losses.

9.35 The laminar flow of water in a small pipe is given by

$$\frac{u}{U_{max}} = 1 - \left(\frac{r}{R}\right)^2$$

where $R = 0.5$ cm is the pipe radius and $U_{max} = 20$ cm/s.

(a) Find the wall shear stress.

(b) Find the pressure drop in a 10 cm length of the pipe.

(c) Verify that the flow is indeed laminar.

9.36 The velocity distribution in a fully developed laminar pipe flow is given by

$$\frac{u}{U_{CL}} = 1 - \left(\frac{r}{R}\right)^2$$

where U_{CL} is the velocity at the centerline and R is the pipe radius. The fluid density is ρ, and its viscosity is μ.

(a) Find the average velocity $\overline{V}$.

(b) Write down the Reynolds number Re based on average velocity and pipe diameter. At what approximate value of this Reynolds number would you expect the flow to become turbulent? Why is this value only approximate?

(c) Assume that the stress/strain rate relationship for the fluid is Newtonian. Find the wall shear stress τ_w in terms of μ, R, and U_{CL}. Express the local skin friction coefficient C_f in terms of the Reynolds number Re.

9.37 A fully developed two-dimensional duct flow of width W and height D has a parabolic velocity profile as shown in Figure P9.37.

(a) If $w \gg D$ show that the wall shear stress τ_w is related to the pressure gradient dp/dx according to $dp/dx = -2\tau_w/D$.

(b) Express the pressure gradient in terms of the velocity on the centerline, the fluid viscosity μ, and the duct height D.

(c) How does the z-component of vorticity vary with y?

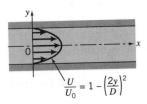

FIGURE P9.37

9.38 Consider fully developed, steady flow of a constant density Newtonian fluid in a circular pipe of diameter D.

(a) Show that the pressure gradient $dp/dx = -4\tau_w/D$, where τ_w is the viscous shear stress at the wall.

(b) If the velocity distribution is triangular, as shown in Figure P9.38, find the average velocity $\overline{V}$ at any cross section, and express the skin friction coefficient

$$C_f = \frac{\tau_w}{\frac{1}{2}\rho\overline{V}^2}$$

in terms of the Reynolds number based on $\overline{V}$.

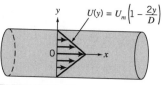

FIGURE P9.38

9.39 Water of density ρ and viscosity μ flows steadily vertically down a circular tube of diameter D, as shown in Figure P9.39. The flow is fully developed, and it has a parabolic velocity profile given by

$$\frac{u}{U_c} = 1 - \left(\frac{2r}{D}\right)^2$$

where the maximum velocity is U_c and the other notation is given in the figure.

(a) Find the shear stress acting at the wall in terms of the density ρ, the gravitational constant g, and the diameter D.

(b) Express the kinematic viscosity in terms of D, U_c, and g.

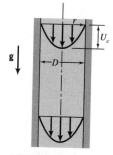

FIGURE P9.39

9.40 A thin layer of water of depth h, flows down a plane inclined at an angle θ to the horizontal, as shown in Figure P9.40. The flow is fully developed (velocity does not change in the x-direction).

(a) Find a theoretical expression for the velocity profile and for the volume flow rate per unit width. Assume the flow is laminar. Use the boundary condition that at the edge of the layer the shear stress is zero.

(b) If the flow is turbulent and the surface fully rough, find the volume flow rate in terms of h, g, and f.

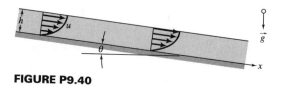

FIGURE P9.40

9.41 Consider the steady, laminar, fully developed flow of water of depth h down a plane inclined at an angle θ to the horizontal. The velocity profile is quadratic, as shown in Figure P9.41, and the velocity at the free surface is U_e. Show that $U_e = gh^2 \sin \theta / 2\nu$, where ν is the kinematic viscosity of water.

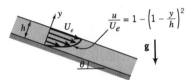

FIGURE P9.41

9.42 Consider the two-dimensional, laminar, fully developed flow of water of depth δ down a plane inclined at an angle θ to the horizontal. The velocity profile may be assumed to be linear, as shown in Figure P9.42, and the velocity at the free surface is V_s.

   (a) Find the shear stress at the wall by:

      (i) using a control volume analysis

      (ii) using the given velocity profile

   (b) Express the skin friction coefficient C_f in terms of:

      (i) the angle θ and a Froude number based on δ and V_s

      (ii) the Reynolds number based on δ and V_s

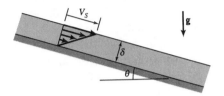

FIGURE P9.42

9.43 In Section 9.9, it is stated that squeezing a hose at its exit will only increase the exit velocity for $fL/D > 1$. By using the energy equation, demonstrate the accuracy of this statement.

CHAPTER 10

VISCOUS EXTERNAL FLOWS

10.1 INTRODUCTION

In Chapter 9, we examined viscous, internal flows, where the flow was fully developed. For fully developed flow, the fluid does not accelerate, and the nonlinear acceleration term in the Navier–Stokes equation is identically zero. For viscous, external flows, fully developed flow does not occur. Boundary layers form, which continue to develop in the streamwise direction, and the acceleration of the fluid cannot be neglected. When a body moves through a fluid, boundary layers produce a viscous resistance to its motion. Near the rear of the body, the boundary layers can separate, and a wake forms which increases the overall resistance. In this chapter, we consider external flows where boundary layers and wakes are important.

10.2 LAMINAR BOUNDARY LAYER

When a fluid flows over a solid surface at reasonable Reynolds numbers, we know that the no-slip condition causes steep velocity gradients to occur in a thin region near the surface. As the flow proceeds downstream, the thickness of this *boundary layer* grows because slower layers of fluid exert friction on faster layers, and more and more fluid is decelerated by viscous stresses. In an internal flow, the layers eventually meet, and further growth stops.

However, in an external flow, such as the flow over the hull of a ship, the fuselage of an airplane, or a flat plate placed in the center of a wind tunnel, the boundary layer continues to grow and there is no fully developed state. The analysis is consequently more difficult because the nonlinear acceleration term in the Navier–Stokes equation is no longer zero. Nevertheless, for a simple flow such as that over a flat plate at zero angle of attack, it is possible to determine approximately the growth rate of the layer and the drag exerted by the flow on the plate by using the integral forms of the continuity and momentum equations.

10.2.1 Control Volume Analysis

We will begin by analyzing a laminar boundary layer, which is the flow that occurs over a plate when the Reynolds number $Re_x = \rho U_e x/\mu$ is less than about 100,000, where x is the distance from the leading edge, and U_e is the velocity of the flow outside the boundary layer. The velocity U_e is taken to be constant so that $\partial p/\partial x = 0$. Also, we will assume that the plate is flat and that the boundary layer is thin, so that the streamlines are nearly parallel and therefore $\partial p/\partial y \approx 0$. The flow is steady and the fluid has a constant density.

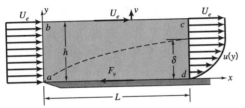

FIGURE 10.1 Control volume for the analysis of a laminar boundary layer.

Consider a rectangular control volume extending a distance L from the leading edge, with a height equal to the extent of the boundary layer at $x = L$ (Figure 10.1). The incoming velocity is uniform, and the outgoing profile is decelerated near the wall because of friction, so that for a distance δ the streamwise velocity $u < U_e$. Actually, u approaches the freestream value asymptotically, so that the *boundary layer thickness* δ is defined to be the distance where the u is virtually indistinguishable from U_e. The most common definition of the boundary layer thickness is

$$u = 0.99U_e \quad \text{at} \quad y = \delta$$

Since the boundary layer thickness grows with downstream distance, $\delta = \delta(x)$.

We begin with the conservation of mass. Since $u(y) < U_e$, there is a greater mass-flow into the control volume over face ab than out of the control volume over face cd, and there must be a mass-flow through face bc. On face bc, the velocity in the x-direction is constant and equal to U_e (remember, $\partial p/\partial x = 0$), but the y-component of the velocity v is unknown. The vector velocities are on face ab: $\mathbf{V} = U_e\mathbf{i}$, face cd: $\mathbf{V} = u\mathbf{i}$, and face bc: $\mathbf{V} = U_e\mathbf{i} + v\mathbf{j}$.

Using the continuity equation, we obtain

$$\int_0^h (\mathbf{n_1} \cdot U_e\mathbf{i})W dy + \int_0^h (\mathbf{n_2} \cdot u\mathbf{i})W dy + \int_0^L [\mathbf{n_3} \cdot (U_e\mathbf{i} + v\mathbf{j})]W dx = 0$$

where W is the width of the plate, and $\mathbf{n_1}$, $\mathbf{n_2}$, and $\mathbf{n_3}$ are the unit normal vectors on faces ab, cd, and bc, respectively. Since $\mathbf{n_1} = -\mathbf{i}$, $\mathbf{n_2} = \mathbf{i}$, $\mathbf{n_3} = \mathbf{j}$,

$$-U_e h + \int_0^h u \, dy + \int_0^L v \, dx = 0 \qquad (10.1)$$

where $u = u(y)$ and $v = v(x)$.

Next, we use the x-component of the momentum equation. It is understood that F_v is the force exerted by the plate on the fluid, so that

$$-F_v = \mathbf{i} \cdot \int_0^h (\mathbf{n_1} \cdot \rho U_e\mathbf{i}) U_e\mathbf{i} W dy + \mathbf{i} \cdot \int_0^h (\mathbf{n_2} \cdot \rho u\mathbf{i}) u\mathbf{i} W dy$$

$$+ \mathbf{i} \cdot \int_0^L [\mathbf{n_3} \cdot \rho(U_e\mathbf{i} + v\mathbf{j})](U_e\mathbf{i} + v\mathbf{j}) W dx$$

That is,

$$-F_v = -\int_0^h \rho U_e^2 W dy + \int_0^h \rho u^2 W dy + \int_0^L \rho v U_e W dx$$

and so

$$-\frac{F_v}{\rho W} = -U_e^2 h + \int_0^h u^2 \, dy + \int_0^L v U_e \, dx$$

The continuity equation can now be used to eliminate the unknown velocity $v(x)$. Multiplying equation 10.1 by U_e and subtracting the result from the momentum equation gives

$$-\frac{F_v}{\rho W} = -U_e^2 h + \int_0^h u^2 \, dy + U_e^2 h - \int_0^h u U_e \, dy$$

$$= \int_0^h (u^2 - u U_e) \, dy$$

Hence,

$$\frac{F_v}{\rho W U_e^2} = \frac{1}{U_e^2} \int_0^h (u U U_e - u U^2) \, dy$$

$$= \int_0^h \left(\frac{u}{U_e} - \frac{u^2}{U_e^2} \right) dy$$

Finally,

$$\frac{F_v}{\rho W U_e^2} = \int_0^h \frac{u}{U_e} \left(1 - \frac{u}{U_e} \right) dy = \int_0^\delta \frac{u}{U_e} \left(1 - \frac{u}{U_e} \right) dy \qquad (10.2)$$

since $u = U_e$ for $y \geq \delta$. To proceed further with the analysis, we need to know (or guess) how u varies with y, and how δ varies with x. This requires some additional input.

10.2.2 Similarity Solution

It was shown by Ludwig Prandtl in 1904[1] that for a boundary layer it is possible to approximate the full Navier–Stokes equation by a simpler equation. The approximations are based on the observation that a boundary layer grows slowly and therefore the streamlines within the layer are nearly parallel. In particular, the pressure across the layer is then nearly constant, as assumed in the control volume analysis given in Section 10.2.1. These approximations yield an equation called the *boundary layer equation,* and Paul Richard Heinrich Blasius (1883–1970), one of Prandtl's students, showed that this equation has a "similarity" solution.

Blasius supposed that the velocity distribution in a laminar boundary layer on a flat plate was a function only of the freestream velocity U_e, the density ρ, the viscosity μ, the distance from the wall y, and the distance along the plate x. Dimensional analysis gives

$$\frac{u}{U_e} = f\left(\frac{U_e x}{\nu}, \frac{y}{x} \right)$$

By an ingenious coordinate transformation of the boundary layer equation, Blasius showed that, instead of depending on two dimensionless variables, the dimensionless velocity distribution was a function of only one composite dimensionless variable η, so that

$$\frac{u}{U_e} = f'\left(\frac{y}{x} \sqrt{\frac{U_e x}{\nu}} \right) = f'(\eta)$$

where $\eta = y\sqrt{U_e/(\nu x)}$. The variables u/U_e and η are called similarity variables, which means that if the velocity distribution is plotted using these nondimensional

---

[1] For a historical note on Prandtl, see Section 15.17.

TABLE 10.1 Dimensionless Velocity Profile for a Laminar Boundary Layer: Tabulated Values

$y(U_e/vx)^{1/2}$	u/U_e	$y(U_e/vx)^{1/2}$	u/U_e
0.0	0.0	2.8	0.81152
0.2	0.06641	3.0	0.84605
0.4	0.13277	3.2	0.87609
0.6	0.19894	3.4	0.90177
0.8	0.26471	3.6	0.92333
1.0	0.32979	3.8	0.94112
1.2	0.39378	4.0	0.95552
1.4	0.45627	4.2	0.96696
1.6	0.51676	4.4	0.97587
1.8	0.57477	4.6	0.98269
2.0	0.62977	4.8	0.98779
2.2	0.68132	5.0	0.99155
2.4	0.72899	∞	1.00000
2.6	0.77246		

SOURCE: Adapted from F. M. White, *Fluid Mechanics,* McGraw-Hill, 1986.

variables (instead of dimensional variables such as u and y), it is defined by a single universal curve, for any Reynolds number and any position along the plate.

These particular similarity variables transform the boundary layer equation (a *PDE*) into an *ODE*, which can be solved numerically. The solution is called the Blasius velocity profile, and the results are usually given in the form of a table (see Table 10.1). The Blasius velocity profile matches the experimental data, such as those shown in Figure 10.2, very well, thereby justifying the assumptions that were made.[2]

The Blasius solution is not an analytical solution, and tabulated values do not reveal the physics very well. It is convenient to have an analytical form for the velocity profile, and it turns out that a parabola is a reasonable curve-fit (see Figure 10.6). That is, we can use the approximation that

$$\frac{u}{U_e} = 2\left(\frac{y}{\delta}\right) - \left(\frac{y}{\delta}\right)^2, \quad \text{for} \quad y \le \delta \tag{10.3}$$

and

$$\frac{u}{U_e} = 1, \quad \text{for} \quad y > \delta$$

which satisfies the boundary conditions $u = 0$ at $y = 0$, and $u = U_e$ at $y = \delta$. Most importantly, it retains the correct similarity scaling, as we will see. For this particular profile, we obtain from equation 10.2 that

$$\frac{F_v}{\rho W U_e^2 \delta} = \int_0^1 \frac{u}{U_e}\left(1 - \frac{u}{U_e}\right) d\left(\frac{y}{\delta}\right) = \frac{2}{15} \tag{10.4}$$

---

[2] Excellent discussions of this topic, and a guide to more general boundary layer problems, may be found in Schlichting, *Boundary Layer Theory,* 7th edition, published by McGraw-Hill, 1979, and White *Viscous Fluid Flow,* 2nd edition, published by McGraw-Hill, 1991.

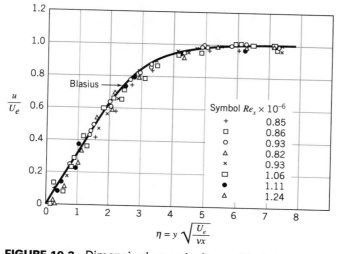

FIGURE 10.2 Dimensionless velocity profile for a laminar boundary layer: comparison with experiments by Liepmann, *NACA Rept. 890,* 1943. Adapted from F. M. White, *Viscous Flow,* McGraw-Hill, 1991.

To eliminate the unknown δ, we use the fact that the total force due to friction F_v is the integral of the shear stress at the wall τ_w over the area of the plate

$$F_v = \int_0^L \tau_w W \, dx$$

Differentiating and using the Newtonian stress-strain rate relationship gives

$$\frac{1}{W}\frac{dF_v}{dx} = \tau_w = \mu \left.\frac{\partial u}{\partial y}\right|_w \tag{10.5}$$

For the parabolic profile approximation used here,

$$\tau_w = \frac{2\mu U_e}{\delta} \tag{10.6}$$

Combining this result with equation 10.4 to eliminate the boundary layer thickness δ, we obtain

$$F_v \frac{dF_v}{dx} = \frac{4}{15}\mu\rho U_e^3 W^2$$

and by integration,

$$F_v = \sqrt{\tfrac{8}{15}\mu\rho U_e^3 W^2 L}$$

Non-dimensionalizing F_v gives

$$C_F = \frac{1.46}{\sqrt{Re_L}}$$

for the parabolic velocity profile. The skin friction coefficient C_F and the Reynolds number Re_L are defined by

$$C_F = \frac{F_v}{\frac{1}{2}\rho U_e^2 LW}, \quad \text{and} \quad Re_L = \frac{\rho U_e L}{\mu} \tag{10.7}$$

C_F is called the *total* skin friction coefficient since it measures the total viscous drag on the plate.

We can find the variation of boundary layer thickness with streamwise distance by using the result for C_F in equation 10.4 to eliminate F_v. For the parabolic profile approximation used here

$$\frac{\delta}{L} = \frac{5.48}{\sqrt{Re_L}}$$

For any position along the plate, therefore,

$$\frac{\delta}{x} = \frac{5.48}{\sqrt{Re_x}} \tag{10.8}$$

so that the boundary layer thickness grows as $\sqrt{x}$.

The total skin friction coefficient C_F expresses the magnitude of the viscous *force* on a plate of width W and length L. We can find the local shear *stress* variation from equations 10.6 and 10.8. For the parabolic velocity profile, we obtain

$$C_f = \frac{0.73}{\sqrt{Re_x}} \tag{10.9}$$

where the local skin friction coefficient C_f is defined as

$$C_f \equiv \frac{\tau_w}{\frac{1}{2}\rho U_e^2} \tag{10.10}$$

C_f is called the *local* skin friction coefficient since it measures the local viscous stress on the plate. The stress at the wall decreases with distance downstream because the velocity gradient at the wall decreases as a result of the boundary layer growth.

We can compare these approximate results with the "exact" results obtained by Blasius, who found that

$$C_F = \frac{1.328}{\sqrt{Re_L}}, \quad \frac{\delta}{x} = \frac{5.0}{\sqrt{Re_x}}, \quad \text{and} \quad C_f = \frac{0.664}{\sqrt{Re_x}} \tag{10.11}$$

We see that the parabolic velocity profile approximation gives the correct dependence on Reynolds number (that is, it gives the correct scaling), but the skin friction coefficients and the boundary layer thickness are about 10% high.

10.3 DISPLACEMENT AND MOMENTUM THICKNESS

As we noted earlier, the velocity u near the edge of the boundary layer approaches the freestream value U_e asymptotically, so that the boundary layer thickness δ is defined to be the distance where the u is "close enough" to U_e. The most common definition of the boundary layer thickness is

$$u = 0.99 U_e, \quad \text{at} \quad y = \delta$$

Two other thicknesses can be defined, the displacement thickness δ^*, and the momentum thickness θ.

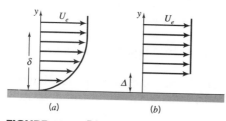

FIGURE 10.3   Displacement thickness. (*a*) Velocity profile. (*b*) Interpretation.

10.3.1 Displacement Thickness

By definition, the displacement thickness is given by

$$\delta^* \equiv \int_0^\infty \left(1 - \frac{u}{U_e}\right) dy \qquad (10.12)$$

What is the purpose of defining the displacement thickness? To answer this question, we rewrite equation 10.12 as

$$\rho U_e \delta^* = \int_0^\infty \rho(U_e - u)\, dy$$

We see that the mass flux passing through the distance δ^* in the absence of a boundary layer is the same as the deficit in mass flux due to the presence of the boundary layer. To make this point further, consider a velocity profile where $u = 0$ for $y \le \Delta$, and $u = U_e$ for $y > \Delta$ (see Figure 10.3). For this profile,

$$\delta^* = \int_0^\infty \left(1 - \frac{u}{U_e}\right) dy = \int_0^\Delta dy = \Delta$$

Therefore, from the point of view of the flow outside the boundary layer, δ^* can be interpreted as the distance that the presence of the boundary layer appears to "displace" the flow outward (hence the name). To the external flow, this streamline displacement also looks like a slight thickening of the body shape.

To illustrate an application of the displacement thickness, consider the entrance region of a two-dimensional duct of width W and height D. The flow is steady and incompressible. Boundary layers grow on the top and bottom surfaces, as shown in Figure 10.4. By continuity, if the flow near the wall is slowed down, the fluid outside the boundary layers (in the "core" region) must speed up. If the incoming velocity has a value U_∞, what is the core velocity U_e at a point where the boundary layer thickness is δ?

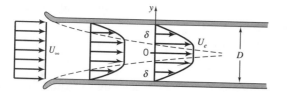

FIGURE 10.4   Flow in the entrance region of a two-dimensional duct.

Consider a control volume extending from the entrance plane to the location of interest. The continuity equation gives

$$-U_\infty WD + \int_{-D/2}^{D/2} uW \, dy = 0$$

So,

$$U_\infty D = \int_{-D/2}^{D/2} U_e \, dy - \int_{-D/2}^{D/2} (U_e - u) \, dy$$

$$= U_e D - 2U_e \int_{(D/2)-\delta}^{D/2} \left(1 - \frac{u}{U_e}\right) dy$$

That is,

$$U_e = U_\infty \left(\frac{D}{D - 2\delta*}\right)$$

where we have made the approximation, that, as far as the contribution to the displacement thickness is concerned, $\delta \approx \infty$. This assumption is commonly made because at $y = \delta$ the velocity is very close to its asymptotic freestream value, and the contribution to the integral for $y > \delta$ is negligible. We see that the increase in the freestream velocity is given by the effective decrease in the cross-sectional area due to the growth of the boundary layers, and this decrease in area is measured by the displacement thickness.

10.3.2 Momentum Thickness

By definition, the momentum thickness is given by

$$\theta \equiv \int_0^\infty \frac{u}{U_e}\left(1 - \frac{u}{U_e}\right) dy \tag{10.13}$$

To interpret the momentum thickness physically, we rewrite equation 10.13 as

$$\rho U_e \theta = \int_0^\infty \rho u(U_e - u) \, dy$$

We see that the momentum flux passing through the distance θ in the absence of a boundary layer is the same as the deficit in momentum flux due to the presence of the boundary layer.

To illustrate this concept, we return to the analysis of a laminar, two-dimensional boundary layer. We start by writing equation 10.2 as

$$\frac{F_v}{\rho W U_e^2} = \int_0^\delta \frac{u}{U_e}\left(1 - \frac{u}{U_e}\right) dy$$

If we make the approximation that, as far as the contribution to the momentum thickness is concerned, $\delta \approx \infty$, then

$$\frac{F_v}{\rho W U_e^2} = \theta$$

so that

$$F_v = \rho U_e^2 W \theta$$

Therefore the momentum flux passing through the cross-sectional area $W\theta$ at $x = L$ in the <u>absence</u> of a boundary layer, is a measure of the drag on a plate of width W and length L due to the <u>presence</u> of the boundary layer.

We can also differentiate this relationship and use equation 10.5 to obtain

$$\frac{d\theta}{dx} = \frac{\tau_w}{\rho U_e^2}$$

That is,

$$\boxed{\frac{d\theta}{dx} = \frac{C_f}{2}} \tag{10.14}$$

where the local skin friction coefficient C_f was defined in equation 10.10. The nondimensional local frictional stress at the wall is therefore equal to the rate of change of the momentum thickness with distance. Equation 10.14 is called the *momentum integral equation* for a boundary layer in a zero pressure gradient. Although we did not show it, this equation applies to laminar and turbulent boundary layers.

10.3.3   Shape Factor

Another term that is often encountered in discussions of boundary layer behavior is the *form,* or *shape,* factor H, where H is defined as the ratio of the displacement thickness to the momentum thickness.

$$H \equiv \frac{\delta^*}{\theta}$$

Since $\delta^* > \theta$, $H > 1$. For a laminar boundary layer the Blasius solution gives $H = 2.59$, and for a turbulent boundary layer experiment shows that $1.4 > H > 1.15$, where H decreases with the Reynolds number. Smaller values of H indicate a "fuller" velocity profile (that is, one that is more filled in near the wall).

10.4   TURBULENT BOUNDARY LAYERS

As in pipe flows (Section 9.7), boundary layer flows become turbulent at high Reynolds number. Transition to turbulence in boundary layers can occur for Re_x greater than about 10^5, where Re_x is the Reynolds number based on the freestream velocity, U_e, and the distance from the leading edge, x.

The major difference between a turbulent fully developed flow and a turbulent boundary layer flow is that a boundary layer is bounded on one side by an external freestream. There exists a well-defined edge between the turbulent fluid in the layer and the nonturbulent fluid in the freestream. An instantaneous cross-section reveals that this edge is highly convoluted (Figure 10.5), and that the instantaneous boundary layer thickness varies with time. We generally prefer to use a time-averaged boundary layer thickness δ defined according to:

$$\langle u \rangle = 0.99 U_e, \quad \text{at} \quad y = \delta$$

where $\langle u \rangle$ is the time-averaged velocity defined by equation 9.17, so that δ has the same broad meaning in laminar and turbulent flows.

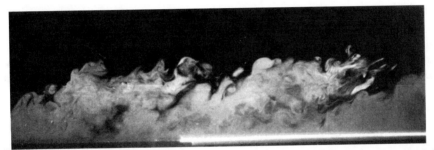

FIGURE 10.5 Streamwise section of a turbulent boundary layer made visible using small oil droplets. Flow is from left to right. The Reynolds number based on momentum thickness is about 4000. R. E. Falco, *Physics of Fluids,* **20,** 1977.

Within the boundary layer, the turbulent fluctuations mix momentum very efficiently, and as a result the velocity profile is fuller than in a laminar flow (see Figure 10.6). No analytical or numerical solution exists for the velocity distribution in a turbulent boundary layer, and so we must rely on experiment. The actual mean flow profile is sometimes approximated by a power law similar to that used to describe turbulent pipe flow. That is,

$$\frac{\langle u \rangle}{U_e} \approx \left(\frac{y}{\delta} \right)^{1/n} \tag{10.15}$$

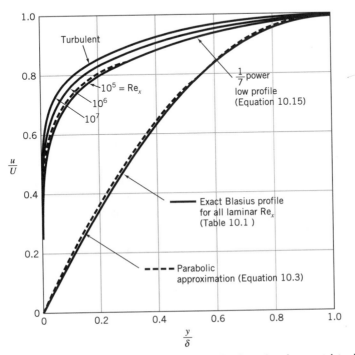

FIGURE 10.6 Comparison of dimensionless laminar and turbulent flat-plate velocity profiles. Adapted from F. M. White, *Fluid Mechanics,* McGraw-Hill, 1986.

where U_e is the freestream velocity, and the exponent n varies with Reynolds number (for a Reynolds number Re_x of about 500,000, $n = 7$).

This power law approximation is reasonably accurate for the outer part of the velocity distribution, as seen in Figure 10.6, but it is a very poor approximation near the wall. Specifically, the velocity gradient at the wall is infinite, implying an infinite wall stress τ_w. Therefore a power law profile cannot be used to find the skin friction in a turbulent flow.

Despite this rather serious limitation, power law profiles can be useful. For example, it can give a good estimate of how the boundary layer thickness varies with Reynolds number. From experiment we find

$$\frac{\delta}{x} = \frac{0.37}{Re_x^{0.2}} \tag{10.16}$$

This result agrees reasonably well with experiment. Also, for a $\frac{1}{n}$ power law profile, we find that

$$\frac{\delta^*}{\delta} = \frac{n+2}{(n+1)(n+2)} \tag{10.17}$$

and

$$\frac{\theta}{\delta} = \frac{n}{(n+1)(n+2)} \tag{10.18}$$

and the shape factor

$$H = \frac{\delta^*}{\theta} = \frac{n+2}{n}$$

For $n = 7$,

$$\frac{\delta^*}{\delta} = \frac{9}{72}, \quad \frac{\theta}{\delta} = \frac{7}{72}, \quad \text{and} \quad H = \frac{9}{7} \tag{10.19}$$

Turbulent boundary layers grow faster than laminar boundary layers. For example, Equation 10.16 can be written as

$$\delta = \frac{0.37x}{Re_x^{0.2}} = \frac{0.37x\nu^{0.2}}{U_e^{0.2}x^{0.2}} = 0.37x^{0.8}\left(\frac{\nu}{U_e}\right)^{0.2}$$

We see that a turbulent boundary layer grows approximately as $x^{0.8}$, which is faster than a laminar boundary layer, where δ grows as $x^{0.5}$ (equation 10.11). This increased growth rate is a consequence of the fact that turbulence diffuses momentum more efficiently than viscosity. For the same reason, the velocity gradient at the wall is higher than that found in a laminar flow at the same Reynolds number, so that the shear stress at the wall is correspondingly larger. Remember that the velocity gradient at the wall cannot be found from equation 10.15; the power law approximation is not valid close to the wall. Instead, the wall stress is given approximately by the empirical relation

$$C_f = \frac{0.0576}{Re_x^{0.2}}, \tag{10.20}$$

where the local skin friction coefficient was defined in equation 10.10. To find the total skin friction coefficient, we need the total viscous drag on the plate F_v, where

$$F_v = \int_0^L \tau_w \, W \, dx$$

where W is the width of the plate, and L its length. Hence,

$$C_F = \frac{F_v}{\frac{1}{2}\rho U_e^2 LW} = \frac{0.072}{Re_L^{0.2}}$$

If the constant is changed to 0.074, this result is in good agreement with experimental data for plates that are turbulent along their entire length.[3] Therefore, by experiment,

$$C_F = \frac{0.074}{Re_L^{0.2}}, \quad \text{for} \quad 5 \times 10^5 < Re_L < 10^7 \tag{10.21}$$

EXAMPLE 10.1 *Boundary Layer Flow*

A flat plate 10 *ft* long is immersed in water at 60°*F*, flowing parallel to the plate at 20 *ft/s*.
(*a*) Find the approximate boundary layer thickness at $x = 5$ *ft* and $x = 10$ *ft*, where x is measured from the leading edge.
(*b*) Find the total drag coefficient C_F.
(*c*) Find the total drag per unit width of the plate if the water covers both sides.

Solution: First, we need to know where transition is likely to occur. If we assume that the boundary layer will be turbulent for $Re_x > 10^5$, the boundary layer will be laminar for

$$x < \frac{\nu \times 10^5}{U_e} = \frac{1.21}{20} \, ft = 0.73 \, in.$$

so that the region of laminar flow is negligible, and the boundary layer is essentially turbulent from the leading edge on.
   For part (*a*), we use the power law approximations for turbulent boundary layers given in equations 10.16 to 10.21. For $x = 5$ *ft*

$$Re_x = \frac{xU_e}{\nu} = \frac{5 \times 20}{1.21 \times 10^{-5}} = 8.26 \times 10^6$$

and

$$\frac{\delta}{x} = \frac{0.37}{Re_x^{0.2}} = 0.0153$$

so that

$$\delta = 0.0765 \, ft = 0.92 \, in.$$

For $x = 10$ *ft*, $Re_x = 16.5 \times 10^6$, and $\delta = 0.133 \, ft = 1.60 \, in.$
   For part (*b*), the total drag coefficient is given by

$$C_F = \frac{0.074}{Re_L^{0.2}}$$

(equation 10.21). With $Re_L = 16.5 \times 10^6$, we find $C_F = 0.00266$.

---

[3] H. Schlichting, *Boundary Layer Theory*, 7th edition, published by McGraw-Hill, 1979.

For part (c), we have a total viscous force acting on the plate equal to $2F_v$, where F_v is the force acting on one surface. From the definition of the total drag coefficient

$$\text{total drag per unit width} = \frac{2F_v}{W} = \rho U_e^2 L C_F$$

$$= 1.938 \, slug/ft^3 \times 400 \, ft^2/s^2 \times 10 \, ft \times 0.00266$$

$$= 20.6 \, lb_f/ft$$ ■

10.5 SEPARATION, REATTACHMENT, AND WAKES

Now we consider external flows that separate and form wakes. In studying internal flows, we noted that sharp corners almost always produce a separated flow. The sudden expansion and contraction flows shown in Figure 3.15 graphically demonstrate the formation of large separated regions whenever a duct suddenly changes its area.

Sharp corners are not the only cause of separation. For example, whenever a flow changes direction suddenly, we can expect to see regions of separated flow. This phenomenon is very common, and it is illustrated in Figure 10.7 for the external flow over the front of a car. Note that the flow is shown relative to an observer traveling with the car, so that it appears to be approaching the car from left to right. The flow is initially uniform and steady, but as it comes into contact with the surface of the car a boundary layer forms, which then separates as the flow makes the turn over the front part of the hood. The flow *reattaches* at some point downstream, so that the boundary layer re-forms, grows some more, and then separates again as the flow meets the windscreen and is forced to turn again.

The flow over a cylinder also produces a large region of separated flow called a wake (Figure 10.8). The flow patterns shown in Figure 10.8 are made visible by illuminating reflective particles suspended in the fluid. A short exposure photograph

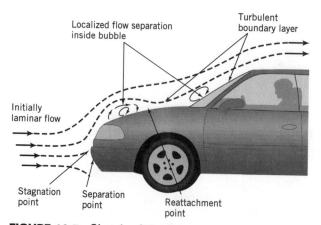

FIGURE 10.7 Sketch of the flow over the front of a car, showing points of separation and reattachment. From *Race Car Aerodynamics*, J. Katz, Robert Bentley Publishers, 1995, with permission.

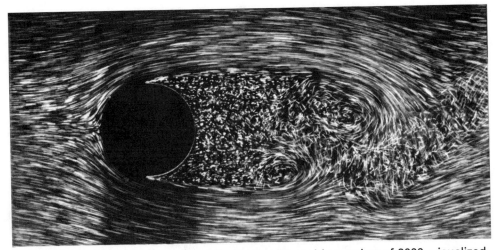

FIGURE 10.8 Flow over a single cylinder at a Reynolds number of 2000, visualized using small bubbles in water. ONERA photograph, Werlé & Gallon, 1972 *Aéronaut. Astronaut.*, **34**, 31–33.

is used to create short streaklines that approximate the instantaneous velocity vector (that is, we see instantaneous streamlines). The flows over the front and rear of the cylinder are quite different. In the front, the flow passes smoothly over the cylinder, but in the rear the wake is highly unsteady and large eddies or vortices are shed downstream. The large eddies are formed at a regular frequency and they produce pressure disturbances in the flow, which we can sometimes hear as sound waves. When we talk of the wind whistling in the trees, it is the sound of eddies being shed by the flow over the smallest branches. An ancient Greek instrument, the Aolian harp, made use of this regular vortex shedding to produce music. In 1911, Theodore von Kármán made an analysis of this flow pattern of alternating vortices, which is now generally known as the Kármán *vortex street* (for a historical note on von Kármán, see Section 15.19). The shedding frequency is determined by the nondimensional Strouhal number, which for a cylinder of diameter D is defined as

$$St = \frac{fD}{V} \tag{10.22}$$

where f is the frequency of vortex shedding from one side of the cylinder (in Hz), and V is the freestream velocity. In general, the Strouhal number is a function of Reynolds number (see Figure 10.9). For Reynolds numbers from about 100 to 10^5, however, the Strouhal number has an almost constant value of about 0.21.

The wake of a body is always characterized by a region of low velocity, which marks a region of momentum loss associated with the extra drag force on the body due to the losses in the wake (see Figure 10.10). In addition, the alternate shedding of vortices produces fluctuating lift and drag forces on the cylinder. If the frequency of the shedding couples to a natural frequency of the cylinder or its supports, large cross-stream oscillations can occur in the cylinder position. This kind of aerodynamic instability was responsible for the destruction of the Tacoma Narrows suspension bridge in the state of Washington in 1940,[4] and it has led to spectacular cooling tower failures in England.

---

[4] A short video is available at *http://www.civeng.carleton.ca:80/Exhibits/Tacoma_Narrows.*

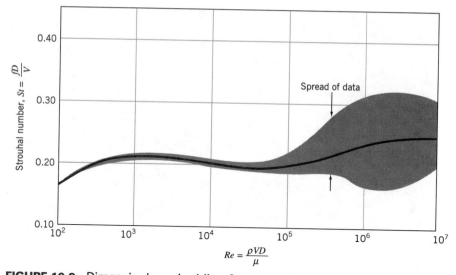

FIGURE 10.9 Dimensionless shedding frequency from a circular cylinder (Strouhal number) as a function of Reynolds number. Adapted from A. Roshko, *Turbulent Wakes from Vortex Streets,* NACA Rept. 1191, 1954.

Wakes almost always contain large eddying motions, which are shed downstream. For some wakes, such as those produced by cylinders, bluff plates, cars and trucks, and other bluff bodies, they often appear at very regular frequencies (see Figures 10.10 and 10.11). In other cases, they are shed more irregularly. For instance, the wake of a boat often contains large eddying motions, but they appear in somewhat random patterns. Similar flow patterns are observed downstream of pylons supporting a bridge, and in the wake of airfoils.

EXAMPLE 10.2 *Vortex Shedding*

In an exposed location, telephone wires will "sing" when the wind blows across them. Find the frequency of the note when the wind velocity is 30 *mph*, and the wire diameter is 0.25 *in*. For air, we assume that $\nu = 15 \times 10^{-6}\,m^2/s$.

Solution: First we need to know the Reynolds number, *Re*, where

$$Re = \frac{VD}{\nu}$$

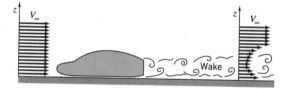

FIGURE 10.10 Wake flow behind a road vehicle (with flow separation and vortex shedding in the base area). Adapted from *Race Car Aerodynamics,* J. Katz, Robert Bentley Publishers, 1995, with permission.

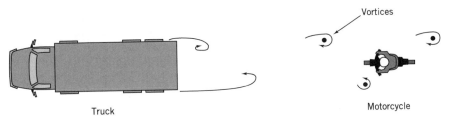

FIGURE 10.11 Periodic vortex formation in the wake of a large truck. Adapted from *Race Car Aerodynamics,* J. Katz, Robert Bentley Publishers, 1995, with permission.

$$= \frac{\left(30\frac{mi}{hr} \times 5280\frac{ft}{mi} \times 12\frac{in.}{ft} \times 0.0254\frac{m}{in.} \times \frac{1}{3600}\frac{hr}{s}\right)\left(0.25\ in. \times 0.0254\frac{m}{in.}\right)}{15 \times 10^{-6}\frac{m^2}{s}}$$

$$= 2774$$

From Figure 10.9, we see that the Strouhal number is approximately equal to 0.21. That is,

$$St = \frac{fD}{V} = 0.21$$

so that

$$f = \frac{0.21V}{D}\ Hz$$

$$= \frac{\left(0.21 \times 30\frac{mi}{hr} \times 5280\frac{ft}{mi} \times 12\frac{in.}{ft} \times 0.0254\frac{m}{in.} \times \frac{1}{3600}\frac{hr}{s}\right)}{0.25\ in. \times 0.0254\frac{m}{in.}}\ Hz$$

$$= 444\ Hz$$

which is very close to the note middle C (= 440 Hz). ∎

10.6 DRAG OF BLUFF AND STREAMLINED BODIES

The drag force experienced by a body as it moves through a fluid is usually divided into two components called *viscous drag,* and *pressure drag.* Viscous drag is associated with the viscous stresses developed within the boundary layers, and it scales with Reynolds number as we have seen above. Pressure drag comes from the eddying motions that are set up in the wake downstream of the body, and it is usually less sensitive to Reynolds number than viscous drag. Formally, both types of drag are due to viscosity (if the body was moving through an inviscid fluid there would be no drag at all), but the distinction is useful because the two types of drag are due to different flow phenomena. Viscous drag is important for attached flows (that is, there is no separation), and it is related to the surface area exposed to the flow. Pressure drag is important for separated flows, and it is related to the cross-sectional area and shape of the body.

We can illustrate the role played by viscous drag (sometimes called frictional drag) and pressure drag (sometimes called form drag) by considering an airfoil at different angles of attack. At small angles of attack, the boundary layers on the top and bottom surfaces experience only mild pressure gradients, and they remain attached along almost the entire chord length. The wake is very small, and the drag is dominated by the viscous stresses inside the boundary layers. At a given Reynolds number, the drag will be higher for turbulent flow than for laminar flow. As the angle of attack increases, however, pressure drag will become more important. With an increased angle of attack, the pressure gradients on the airfoil increase in magnitude. In particular, the pressure increases toward the rear portion of the top surface, so that the pressure gradient in that region is positive. This "adverse" pressure gradient may become sufficiently strong to produce a separated flow (see also Section 7.9). Separation will increase the size of the wake, and increase the magnitude of the pressure losses in the wake due to eddy formation. Therefore the pressure drag increases. At a higher angle of attack, a large fraction of the flow over the top surface of the airfoil may be separated, and the airfoil is said to be *stalled*. At this stage, the pressure drag is much greater than the viscous drag.

When the pressure losses are small and the overall drag is mainly due to viscous drag, we say the body is *streamlined*. When the viscous drag is small and the overall drag is dominated by pressure losses, we describe the body as *bluff*. Whether the flow is viscous-drag dominated or pressure-drag dominated depends entirely on the shape of the body. A fish or an airfoil at small angles of attack behaves like streamlined body, whereas a brick, a cylinder, or an airfoil at a large angle of attack behaves like a bluff body. For a given frontal area and velocity, a streamlined body will always have a lower resistance than a bluff body. For example, the drag of a cylinder of diameter D can be ten times larger than a streamlined shape of the same thickness.

Cylinders and spheres are considered bluff bodies because at Reynolds numbers much greater than one, the drag is dominated by the pressure losses in the wake. The variation of their drag coefficients with Reynolds number is shown in Figure 10.12 and the corresponding flow patterns are shown in Figure 10.13. We

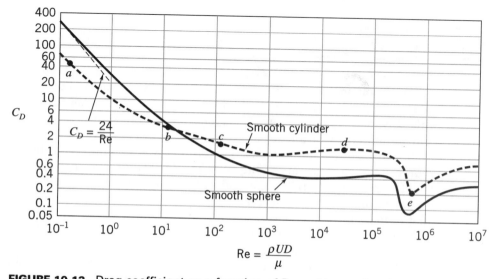

FIGURE 10.12 Drag coefficient as a function of Reynolds number for smooth circular cylinders and smooth spheres. From Munson, Young, & Okiishi, *Fundamentals of Fluid Mechanics*, John Wiley & Sons, 1998.

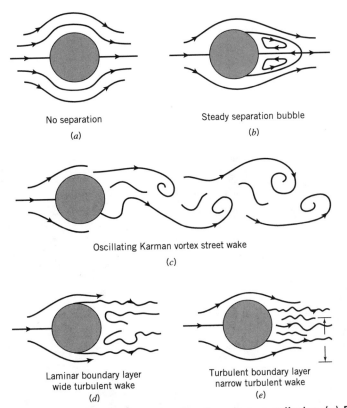

No separation

(a)

Steady separation bubble

(b)

Oscillating Karman vortex street wake

(c)

Laminar boundary layer
wide turbulent wake

(d)

Turbulent boundary layer
narrow turbulent wake

(e)

FIGURE 10.13 Flow patterns for flow over a cylinder. (*a*) Reynolds number = 0.2; (*b*) 12; (*c*) 120; (*d*) 30,000; and (*e*) 500,000. Patterns correspond to the points marked on Figure 10.12. From Munson, Young, & Okiishi, *Fundamentals of Fluid Mechanics,* John Wiley & Sons, 1998.

see that as the Reynolds number increases the variation in the drag coefficient (based on cross-sectional area) decreases, and over a large range in Reynolds number it is nearly constant.

At a Reynolds number between 10^5 and 10^6, the drag coefficient takes a sudden dip. The dip indicates that the pressure losses in the wake have suddenly become smaller, and experiments show that the size of the wake decreases, and that the boundary layer separation on the cylinder or sphere occurs further along the surface than before. What has happened?

The sudden decrease in drag is related to the differences between laminar and turbulent boundary layer. We saw in Section 7.9 that the boundary layer and its interaction with the local pressure gradient plays a major role in affecting the flow over a cylinder. In particular, near the shoulder, the pressure gradient changes from being negative (decreasing pressure, a *favorable* pressure gradient) to positive (increasing pressure, an *adverse* pressure gradient). The force due to pressure differences changes sign from being an accelerating force to being a retarding force. In response, the flow slows down. However, the fluid in the boundary layer has already given up some momentum because of viscous energy dissipation, and it does not have enough momentum to overcome the retarding force. Some fluid near the wall, where the momentum was low to begin with, reverses direction, and the flow separates.

A turbulent boundary layer has more momentum near the wall than a laminar boundary layer (see Figure 10.6) because turbulence is a very effective mixing process. More importantly, turbulent transport of momentum is very effective at replenishing the near-wall momentum. When a turbulent boundary layer enters a region of adverse pressure gradient, it can therefore persist for a longer distance without separating (compared to a laminar flow) because its momentum near the

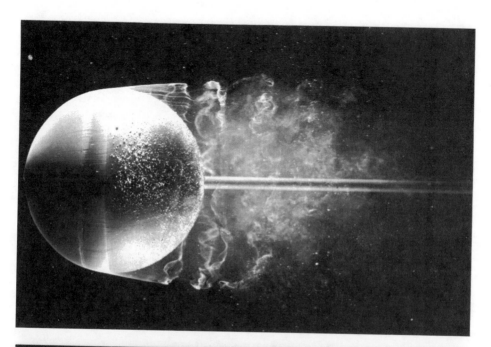

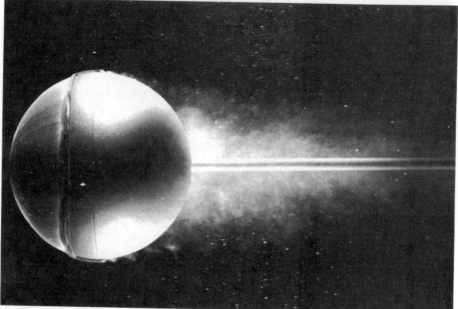

FIGURE 10.14 Drag coefficients of bluff and streamlined bodies. From Munson, Young, & Okiishi, *Fundamentals of Fluid Mechanics*, John Wiley & Sons, 1998.

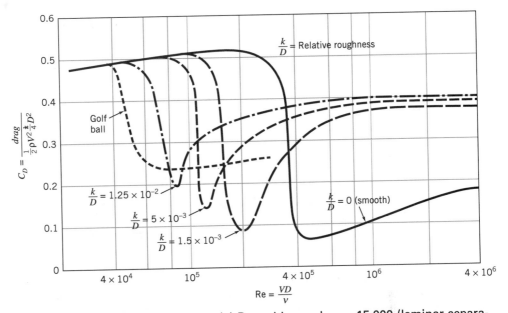

FIGURE 10.15 Flow over a sphere. (*a*) Reynolds number = 15,000 (laminar separation). (*b*) Reynolds number = 30,000, with trip wire (turbulent separation). From Van Dyke, *Album of Fluid Motion*, Parabolic Press, 1982. Original photographs by Werlé, ONERA, 1980.

wall is higher to begin with, and its momentum is continually (and quickly) being replenished by turbulent mixing.

The boundary layer over the front face of a sphere or cylinder is laminar at lower Reynolds numbers, and turbulent at higher Reynolds numbers. The dip in the drag coefficient occurs at the point where the boundary layer changes from laminar to turbulent. When it is laminar ($Re < 10^5$), separation starts almost as soon as the pressure gradient becomes adverse [very near the shoulder, Figure 10.13(*d*)], and a large wake forms. When it is turbulent ($Re > 10^6$), separation is delayed [to a point about 20° past the shoulder, Figure 10.13(*e*)] and the wake is correspondingly smaller. The Reynolds number where the flow switches and the drag suddenly decreases is called the *critical* Reynolds number.

It follows that, if the boundary layer of a sphere could be made turbulent at a Reynolds number lower than the critical value, the drag should also decrease at the same time. This can be demonstrated by using a *trip wire*. A trip wire is simply a wire placed axisymmetrically on the front face of the sphere. As the flow passes over the wire, it introduces a large disturbance into the boundary layer, which causes an early transition to turbulence. Its effect on the size of the wake is quite dramatic, as shown in Figure 10.14 (opposite page). A similar result can be achieved using surface roughness to "trip" the boundary layer, and Figure 10.15 shows the effect of increasing surface roughness on the drag coefficient of a sphere.

Elliptical and airfoil shapes display a similar dip in the drag curve at a critical Reynolds number (see Figure 10.16). Bodies with sharp edges typically do not, at least for Reynolds numbers greater than about 3000, as in the case of a flat plate set at right angles to the flow direction. Here, the points of separation are fixed at the edges, and they do not change position with Reynolds number. The drag coefficients for a number of sharp-edged bodies are given in Table 10.2. Drag coefficients for a wide variety of other interesting shapes are given in Table 10.3.

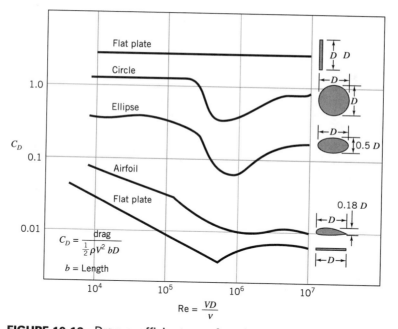

FIGURE 10.16 Drag coefficient as a function of Reynolds number for spheres with different degrees of roughness. k is the equivalent roughness height, and D is the sphere diameter. From Munson, Young, & Okiishi, *Fundamentals of Fluid Mechanics,* John Wiley & Sons, 1998.

TABLE 10.2 Drag Coefficient Data for Sharp-Edged Bodies[a]

Object	Diagrams		C_D ($Re \gtrsim 10^{-1}$)
Square cylinder		$b/h = \infty$	2.05
		$b/h = 1$	1.05
Disk			1.17
Ring			1.20[b]
Hemisphere (open end facing flow)			1.42
Hemisphere (open end facing downstream)			0.38
C-section (open side facing flow)			2.30
C-section (open side facing downstream)			1.20

From Fox & McDonald, *Introduction to Fluid Mechanics,* 4th edition, John Wiley & Sons, 1992.

[a] Original data from Hoerner, Fluid-Dynamic Drag, 2nd edition, Midland Park, NJ, published by the author.

[b] Based on area of ring.

TABLE 10.3 Drag Coefficient Data for Selected Objects

Shape	Reference area	Drag coefficient C_D
Parachute	Frontal area $A = \frac{\pi}{4}D^2$	1.4
Porous parabolic dish	Frontal area $A = \frac{\pi}{4}D^2$	Porosity 0 / 0.2 / 0.5 → 1.42 / 1.20 / 0.82 ← 0.95 / 0.90 / 0.80 Porosity = open area/total area
Average person	Standing Sitting Crouching	$C_DA = 9$ ft^2 $C_DA = 6$ ft^2 $C_DA = 2.5$ ft^2
Fluttering flag	$A = lD$	lD / C_D 1 / 0.07 2 / 0.12 3 / 0.15
Empire State Building	Frontal area	1.4
Six-car passenger train	Frontal area	1.8
Bikes Upright commuter	$A = 5.5$ ft^2	1.1
Racing	$A = 3.9$ ft^2	0.88
Drafting	$A = 3.9$ ft^2	0.50
Streamlined	$A = 5.0$ ft^2	0.12
Tractor-trailor trucks Standard	Frontal area	0.96
With fairing	Frontal area	0.76
With fairing and gap seal	Frontal area	0.70
Tree $U = 10$ m/s $U = 20$ m/s $U = 30$ m/s	Frontal area	0.43 0.26 0.20
Dolphin	Wetted area	0.0036 at Re $= 6 \times 10^6$ (flat plate has $C_{Df} = 0.0031$)
Large birds	Frontal area	0.40

SOURCE: From Munson, Young & Okiishi, *Fundamentals of Fluid Mechanics*, John Wiley & Sons, 1998.

10.7 GOLF BALLS, CRICKET BALLS, AND BASEBALLS

Tripping the boundary layer to reduce the drag of spheres is widely used in sports. For example, golf balls are dimpled. The dimples act as a very effective trip wire, and the consequent delay in separation reduces the drag on the ball and allows the ball to travel further for the same amount of effort. A good golfer can easily make a golf ball carry 250 yards, but the same golfer using a smooth ball will drive it only about 100 yards. Figure 10.15 shows how different degrees of roughness reduce the drag on a sphere, and how effective dimples are.

The same principle is used in cricket. A cricket ball has a single, circumferential seam, which looks remarkably like a trip wire. If a cricket ball is bowled without spin so that its seam is tilted forward on the top of the ball, then the boundary layer over the top surface becomes turbulent, whereas the boundary layer on the bottom surface remains laminar. The wake becomes asymmetric, and a downward force is produced so that the ball dips sharply. A similar effect can be obtained with a baseball or a tennis ball if the seam is held correctly. The seam on a baseball is more convoluted than on a cricket ball, but its tripping effect is similar.

The addition of spin complicates this picture enormously. We saw in Section 4.5 and Figure 7.12 that spin can produce a side force on a ball due to the Magnus effect. The effect depends strongly on the orientation of the seam. Knuckleball pitchers typically pitch a baseball with very little spin, and they do not use the Magnus effect. Instead, they rely largely on the uneven tripping of the boundary layer. Even a little spin will make the direction of the side force change dramatically, however, and it is no wonder that the behavior of a knuckleball is highly unpredictable.[5]

10.8 AUTOMOBILE FLOWFIELDS

As a final topic in this chapter, we examine some aspects of the flow over automobiles. As we have already seen in Figures 7.13, 10.7, and 10.10, boundary layers form, which can separate at points where the flow makes a sudden turn, and where adverse pressure gradients occur. A typical pressure distribution over a car model is shown in Figure 10.17. The region of adverse pressure gradient on the hood is indicated by the negative slope in the upper pressure distribution (where $\partial p/\partial x > 0$). Another region of adverse (or unfavorable) pressure gradient is located near the rear of the car, where the boundary layer is again susceptible to separation. A similar region is found on the rear underside of the car.

The regions of separated flow on the car body, and the presence of a wake, dominate the aerodynamic drag on the vehicle (on most cars, the viscous, boundary layer drag makes only a small contribution to the total drag, and therefore cars are frequently considered to be bluff bodies).

At low speeds, the major source of resistance on a car is its rolling resistance,

---

[5] To learn more about these matters, see the article by Dr. Rabi Mehta of NASA-Ames, entitled "Aerodynamics of sportsballs," *Annual Review of Fluid Mechanics,* **17**:151–189, 1985.

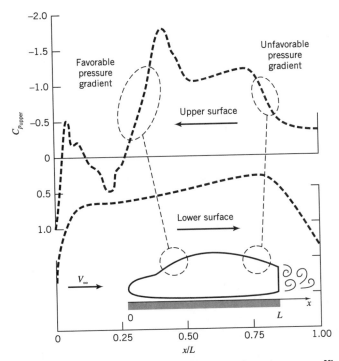

FIGURE 10.17 Distribution of measured pressure coefficients over a two-dimensional automobile shape. From *Race Car Aerodynamics,* J. Katz, Robert Bentley Publishers, 1995, with permission.

due to friction between the moving parts in the drivetrain and the flexing of the tires. As the speed increases, however, the aerodynamic drag increases rapidly, approximately as the square of the speed if the drag coefficient is constant with Reynolds number (see Figure 10.18). For the particular car shown in Figure 10.18, the aerodynamic resistance becomes the major source of drag at about 80 *km/h,*

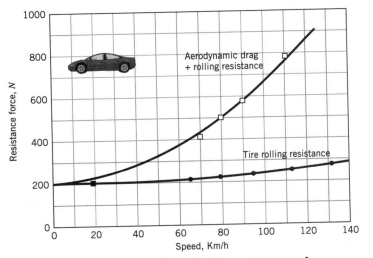

FIGURE 10.18 Aerodynamic and rolling resistances for an average sedan car. From *Race Car Aerodynamics,* J. Katz, Robert Bentley Publishers, 1995, with permission.

TABLE 10.4   Typical Lift and Drag Coefficients

			C_L	C_D
1	Circular plate		0	1.17
2	Circular cylinder $L/D < 1$		0	1.15
3	Circular cylinder $L/D > 2$		0	0.82
4	Low drag body of revolution		0	0.04
5	Low drag vehicle near the ground		0.18	0.15
6	Generic automobile		0.32	0.43
7	Prototype race car		−3.00	0.75

SOURCE: From *Race Car Aerodynamics*, J. Katz, Robert Bentley Publishers, 1995, with permission.

that is, 50 *mph*. At 120 *km/h* (75 *mph*), the aerodynamic resistance accounts for about two-thirds of the total drag.

Typical values of lift and drag coefficients for car-shaped bodies are given in Table 10.4. The most streamlined shape listed in the table has a drag coefficient as low as 0.04, but when this shape is modified to look like a ground vehicle, it increases to about 0.15, which might be taken as a realistic lower limit. Most modern cars have values closer to 0.4. Somewhat surprisingly, race cars have generally higher drag coefficients, primarily because there are other constraints on their shape, such as a requirement for strong negative lift to help cornering performance, and the presence of air inlets for the engine and the water and oil coolers. Also, the downforce generated by wings and other lifting surfaces has a strong effect on the drag coefficient of a race car, with a corresponding effect on its maximum speed. The results shown in Figure 10-19 were obtained by using the maximum engine rpm for each downforce adjustment under full throttle conditions to estimate the drag force.

Another interesting aspect of the flow over cars is the phenomenon of drafting or "tailgating." It is well-known that the air resistance of a car is reduced when it follows another in close proximity, and the leading car acts as a shield for the trailing car. The resulting flowfield is sketched in Figure 10.20. The data shown in Figure 10.21 indicate a significant decrease in drag coefficient for the trailing car when the separation is less than about one car length. Interestingly, the drag coefficient of the leading car is decreased by an even greater margin, suggesting that under race conditions, both cars will travel faster in tandem than they could by themselves.

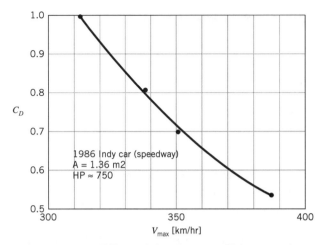

FIGURE 10.19 Effect of the drag coefficient on the maximum speed of an Indy-class speedway car. From *Race Car Aerodynamics,* J. Katz, Robert Bentley Publishers, 1995, with permission.

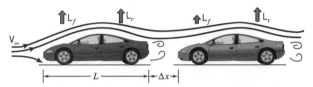

FIGURE 10.20 Sketch of the flow over two cars, separated by a distance Δx. From *Race Car Aerodynamics,* J. Katz, Robert Bentley Publishers, 1995, with permission.

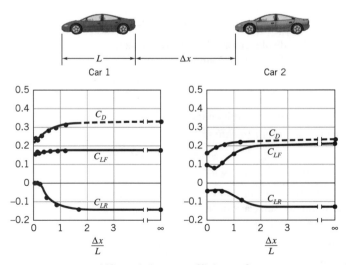

FIGURE 10.21 Lift and drag coefficients for two cars separated by a distance Δx (for notation, See Figure 10.20). From *Race Car Aerodynamics,* J. Katz, Robert Bentley Publishers, 1995, with permission.

EXAMPLE 10.3   *Drag of a Streamlined Body*

A submarine has the shape of an 8:1 ellipsoid. Find the power required to maintain a velocity of 20 *ft/s* fully submerged in the sea. The water temperature is 68°*F*. The submarine has a frontal area of 50 *ft²*, and a drag coefficient $C_D = 0.15$.

Solution: The drag coefficient is given by

$$C_D \equiv \frac{F_D}{\frac{1}{2}\rho V^2 A}$$

(equations 5.17 and 8.10), where A is the cross-sectional area of the body, and F_D is the drag force acting on the body. Hence,

$$F_D = \tfrac{1}{2}\rho V^2 A C_D$$

The density of sea water at 68°*F* is 1025 *kg/m³*, that is, 1.989 *slug/ft³* (Table 1.1). Therefore,

$$F_D = \tfrac{1}{2} \times 1.989 \; slug/ft^3 \times 400 \; ft^2/s^2 \times 50 \; ft^2 \times 0.15$$
$$= 2983 \; lb_f$$

The power required to overcome the drag is the work done per unit time, that is, the drag force times the velocity of motion. Therefore

$$\text{power required} = F_D \times V = 2983 \times 20 \; ft \cdot lb_f/s$$
$$= 59{,}700 \; ft \cdot lb_f/s = 59{,}700 \; ft \cdot lb_f/s \; \frac{1.341 \; hp}{737.5 \; ft \cdot lb_f/s}$$
$$= 108 \; hp \qquad\blacksquare$$

PROBLEMS

10.1   Describe what is meant by the term "boundary layer." Illustrate your answer using a diagram.

10.2   What is the definition of the displacement thickness? Momentum thickness? Shape factor? Give physical interpretations for these parameters.

10.3   Give brief physical explanations and interpretations of the following terms:
(a) the 99% boundary layer thickness
(b) the boundary layer displacement thickness δ^*
(c) the boundary layer momentum thickness θ

10.4   The flow over a leaf is being studied in a wind tunnel. If the wind blows at 8 *mph,* and the leaf is aligned with the flow, is the boundary layer laminar or turbulent?

10.5   Draw the flow around a typical power plant smoke stack. Identify the laminar and turbulent boundary layers, the points of separation, the separated flow, the vortex shedding, and the freestream flow.

10.6   Describe the differences between laminar and turbulent boundary layers in terms of:
(a) the velocity profile
(b) the frictional drag
(c) the behavior in adverse pressure gradients (that is, the tendency to separate)

10.7 Consider the differences between a laminar and a turbulent boundary layer in a zero pressure gradient. If the boundary layer thickness was the same:

(a) Sketch the velocity profile for each flow.

(b) Which flow has the higher wall shear stress? Why?

(c) How do these observations explain why a dimpled golf ball can travel further than a smooth golf ball?

10.8 Consider a laminar and a turbulent boundary layer at the same Reynolds number.

(a) How do the velocity profiles compare?

(b) Which boundary layer grows faster?

(c) Which boundary layer has the larger ratio of displacement thickness to momentum thickness (δ^*/θ)?

(d) When subjected to the same adverse pressure gradient, which boundary layer will separate sooner?

10.9 For the boundary layer velocity profile given by:

$$\frac{u}{U_e} = \frac{y}{\delta}, \quad (y \le \delta)$$

find the wall shear stress, the skin friction coefficient, the displacement thickness, and the momentum thickness.

10.10 Show that for a constant density, constant pressure laminar boundary layer growing on flat plate of width W in a stream of velocity U_e, the total drag force F acting on the plate is given by $F/\rho U_e^2 \delta W = (4 - \pi)/2\pi$ (Figure P10.10) when the velocity profile is given by

$$\frac{u}{U_e} = \sin\left(\frac{\pi y}{2\delta}\right), \quad \text{for} \quad \frac{y}{\delta} \le 1$$

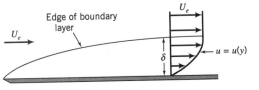

Edge of boundary layer

$u = u(y)$

FIGURE P10.10

10.11 For the boundary layer profile given in the previous problem, show that the displacement thickness $\delta^* = \delta(1 - 2/\pi)$.

10.12 For a turbulent boundary layer velocity profile given by $u/U_e = (y/\delta)^{1/9}$, where U_e is the freestream velocity and δ is the boundary layer thickness, calculate the displacement thickness, the momentum thickness, and the shape factor.

10.13 A flat plate 1 m long and 0.5 m wide is parallel to a flow of air at a temperature of 25°C. The velocity of the air far from the plate is 20 m/s.

(a) Is the flow laminar or turbulent?

(b) Find the maximum thickness of the boundary layer.

(c) Find the overall drag coefficient C_F, assuming that transition occurs at the leading edge.

(d) Find the total drag of the plate if the air covers both sides.

10.14 Repeat Problem 10.13 for the case where the fluid is water at 15°C, assuming the Reynolds number is kept constant by changing the freestream velocity.

10.15 Air at 80°F flows over a flat plate 6 ft long and 3 ft wide at a speed of 60 ft/s. Assume the transition Reynolds number is 5×10^5.

(a) At what distance from the leading edge does transition occur?

(b) Plot the local skin friction coefficient $C_f = \tau_w/(\frac{1}{2}\rho V^2)$ as a function of $Re_x = U_e x/\nu$.

(c) Find the total drag of the plate if the air covers both sides.

10.16 Find the ratio of the friction drags of the front and rear halves of a flat plate of total length ℓ if the boundary layer is turbulent from the leading edge and follows a one-seventh-power law velocity distribution.

10.17 For a particular turbulent boundary layer on a flat plate, the velocity profile is given by $u/U_e = (y/\delta)^{1/7}$, where u is the streamwise velocity, U_e is the freestream velocity, and δ is the boundary layer thickness. Given that the boundary layer thickness grows as in equation 10.16, use the continuity equation to:

(a) Find the distribution of the vertical velocity component $v(y)$.

(b) Calculate the angle the flow vector makes with the flat plate at $y/\delta = 0.05, 0.2$, and 0.8, at $Re_x = 10^6$.

10.18 A ship 200 ft long with a wetted area of 5000 ft^2 moves at 25 ft/s. Find the friction drag, assuming that the ship surface may be modeled as a flat plate, and $\rho = 1.94$ $slugs/ft^3$ and $\nu = 1.2 \times 10^{-5} ft^2/s$. What is the minimum power required to move the ship at this speed?

10.19 Air enters a long two-dimensional duct of constant height h, as shown in Figure P10.19. Identical boundary layers develop on the top and bottom surfaces. In the core region, outside the boundary layers, the flow is inviscid and no losses occur. The flow is steady and the density is constant.

(a) Show that

$$\frac{U_2}{U_1} = \frac{h}{h - 2\delta^*}$$

where δ^* is the displacement thickness at station 2.

(b) Given the information in part **(a)** find the pressure drop between stations 1 and 2 along the central streamline.

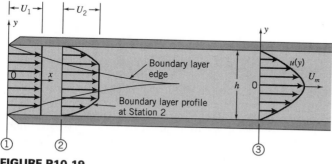

FIGURE P10.19

10.20 In the previous problem, if you were given that velocity u at station 3 varied with y according to

$$\frac{u}{U_m} = 1 - \left(\frac{2y}{h}\right)^2$$

(a) What is the distribution of the viscous stress, and what is its value at the wall?

(b) What is the distribution of the z-component of vorticity, and what is its value at the wall?

10.21 A laminar boundary layer velocity profile may be described approximately by $u/U_e = \sin(\pi y/2\delta)$. Find the shear stress at the wall, τ_w, and express the local skin friction coefficient $C_f = \tau_w/(\frac{1}{2}\rho U_e^2)$ in terms of a Reynolds number based on U_e and δ.

10.22 Consider the entrance section to a circular pipe of diameter D. The incoming velocity is constant over the area and equal to U_1, as shown in Figure P10.22. Downstream, however, a boundary layer grows and this causes the flow in the central region to accelerate. The flow has a constant density ρ. Find U_2/U_1 given that $\delta^* = D/16$ at station 2. (Note that $\delta \ll D$.)

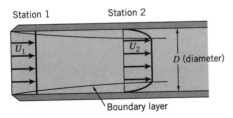

Station 1   Station 2

U_1 U_2 D (diameter)

Boundary layer

FIGURE P10.22

10.23 Assuming that the velocity profile in a laminar boundary layer of thickness δ is given by

$$\frac{u}{U_e} = 2\left(\frac{y}{\delta}\right) - \left(\frac{y}{\delta}\right)^3$$

where u is the velocity at distance y from the surface and U_e is the freestream velocity (as shown in Figure P10.23), demonstrate that

$$\frac{\theta}{\delta} = \frac{31}{420}, \quad \text{and} \quad C_f \equiv \frac{\tau_w}{\frac{1}{2}\rho U_e^2} = \frac{4\nu}{U_e\delta}$$

where θ is the momentum thickness, τ_w is the viscous stress at the wall, C_f is the local skin friction coefficient at a distance x from the leading edge of the plate, ρ is the density, and ν is the kinematic viscosity.

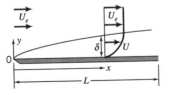

FIGURE P10.23

10.24 For two-dimensional, constant pressure, constant density, laminar flow over a flat plate, the boundary layer velocity profile is crudely described by the following relationship.

$$\frac{u}{U_e} = \frac{y}{\delta}, \quad \text{for } y \leq \delta$$

where u is the velocity parallel to the surface, U_e is the (constant) freestream velocity, y is the distance normal to the surface, and δ is the boundary layer thickness.

(a) Evaluate $d\theta/dx$, where θ is the momentum thickness, and x is the distance along the surface, measured from the leading edge of the plate in terms of the boundary layer growth rate.

(b) Evaluate the skin friction coefficient C_f.

(c) Using the results from parts **(a)** and **(b)**, show that θ grows as $\sqrt{x}$.

10.25 A laminar boundary layer is observed to grow on a flat plate such that the pressure is constant everywhere. If the boundary layer velocity profile is given by

$$\frac{u}{U_e} = \frac{3}{2}\left(\frac{y}{\delta}\right) - \frac{1}{2}\left(\frac{y}{\delta}\right)^3, \quad \text{for } y \leq \delta$$

where u is the velocity at a distance y from the surface, δ is the boundary layer thickness, and U_e is the freestream velocity.

(a) Show that $\theta/\delta = 0.139$, where θ is the momentum thickness.

(b) Find the viscous stress on the plate in terms of the viscosity, the boundary layer thickness, and the freestream velocity.

10.26   For the flow over a body completely immersed in a fluid, what is meant by the terms "pressure drag" and "viscous drag"? Using these terms, explain the difference between a "bluff" body and a "streamlined" body.

10.27   Explain why dimples on a golf ball help to reduce drag.

10.28   Pylons supporting a bridge over a fast-flowing river often have footings (the part of the pylon below and just above the water level) shaped like a wedge, both in the upstream and downstream direction. Why do you think this is done?

10.29   A disk 15 cm in diameter is placed in a wind tunnel at right angles to the incoming flow. The drag of the plate is found to be 3.2 N when the air velocity is 20 m/s and the air temperature is 25°C. Using this information, estimate the drag of a disk 40 cm in diameter in a water flow having a velocity 5 m/s at a temperature of 15°C.

10.30   A rectangular banner 2 ft high by 10 ft long is carried by marchers in a parade that is moving at 2 mph. A 30 mph wind is blowing. If the air density is 2.4×10^{-3} slug/ft³, find the maximum force exerted on the banner by the wind. Estimate how many marchers it might take to hold the banner safely.

10.31   Air flows over a 60 ft high circular smoke stack at a uniform speed of 30 mph. Find the total force acting on the stack if its diameter is 6 ft and the air temperature is 70°F.

10.32   A 0.5 in. diameter cable is strung between poles 120 ft apart. A 60 mph wind is blowing at right angles to the cable. Find the force acting on the cable due to the wind. The air temperature is 40°F.

10.33   A beach ball 20 cm in diameter traveling at a speed of 50 m/s in still air at 30°C is found to have a drag of 8 N.

(a) Find the velocity at which the drag of a 60 cm sphere immersed in water at 15°C can be found from the above data.

(b) What is the drag of the larger sphere at this velocity?

10.34   Find:

(a) The aerodynamic drag force on a car traveling at 75 mph if the drag coefficient is 0.4 and the frontal area is 24 ft².

(b) The maximum possible speed of a minivan, on the level with no wind, if it has a frontal area of 30 ft², a drag coefficient of 0.6, and a 120-horsepower engine that is 80% efficient. How will this speed change if there is a 30 mph headwind?

10.35   A skydiver of mass 75 kg is falling freely, while experiencing a drag force due to air resistance. If her drag coefficient is 1.2, and her frontal area is $= 1 \; m^2$, and ρ is the air density (assume standard atmospheric conditions), find the terminal velocity at 10°C.

10.36   Find the terminal velocity of a parachutist, assuming that the parachute can be modeled as a semicircular cup of diameter 6 m. Use standard atmosphere properties of air corresponding to an altitude of 3000 m. The total mass of the person and parachute is 90 kg.

10.37   Find the terminal velocity of a steel sphere of density 7850 kg/m³ and diameter 0.5 mm diameter, falling freely in SAE 30 motor oil, which has a density of 919 kg/m³ and a viscosity of 0.04 N·s/m².

10.38 A bracing wire used on a biplane to strengthen the wing assembly is found to vibrate at 5000 *Hz*. Estimate the speed of the airplane if the wire has a diameter of 1.2 *mm*. If the natural frequency of the wire is 500 *Hz,* at what speed will the wire vibration resonate?

10.39 An automobile radio aerial consists of three sections having diameters $\frac{1}{8}$ *in.,* $\frac{3}{16}$ *in.,* and $\frac{1}{4}$ *in.* Find the frequency of the vortex shedding when the car is traveling at 35 *mph* and 65 *mph.*

10.40 For problems 10.31 and 10.32 find the vortex shedding frequency. Estimate the wavelength between successive vortices.

OPEN CHANNEL FLOW

11.1 INTRODUCTION

In this chapter, we examine flows where there is a *free surface*. A free surface is the interface between a liquid and a gas, a place where the pressure at the interface is effectively constant. For the case of water flowing in a channel that is open to the atmosphere, the pressure at the free surface is assumed to be constant and equal to atmospheric pressure. If there are no strong disturbances present, such as breaking waves, and the flow is steady, Bernoulli's equation can be used along the surface. Below the free surface, the pressure will vary with depth, even if the flow has parallel streamlines, because hydrostatic pressures still act and the forces due to hydrostatic pressure differences must be taken into account in the momentum equation.

The constant pressure boundary condition imposed by the free surface means that gravity affects the velocity field directly. This has not been true for the flows considered so far. For example, in the flow through a siphon (Section 4.5.3), mass conservation requires that the average velocity in the tube must remain constant since its cross-sectional area does not change. If there are no losses and the flow is steady, Bernoulli's equation then shows that $p + pgz =$ constant, that is, the sum of the fluid pressure and the hydrostatic head remains constant along a streamline. As the height of the water column increases, the pressure decreases. The pressure varies but the velocity does not.

A similar situation occurs when we consider the flow of a liquid about a fully submerged body well away from a free surface. As the liquid flows around the body, fluid particles experience a change in height. As a result, hydrostatic pressure differences are set up in the vertical direction, in addition to those produced by the acceleration of the flow. The hydrostatic head variations exert an extra vertical force on the body, but this force is simply the buoyancy force given by Archimedes' principle. The velocity field is not influenced by this buoyancy force. The velocity is independent of gravity but the pressure is not.

Open-channel flows are different. Since the pressure at the free surface is fixed, regardless of its shape, the sum of the dynamic pressure and the hydrostatic head is constant: as the height of the surface increases, the fluid slows down. Gravity directly influences the velocity field in open channel flows.

One of the most interesting aspects of flows with a free surface is the formation of surface waves. We shall see that the propagation of waves is determined by the Froude number, which is the ratio of the speed of the flow to the speed of propagation of small amplitude waves. As the Froude number changes from subcritical values (where $F < 1$) to supercritical values (where $F > 1$), the wave propagation becomes radically different. This phenomenon has a profound influence on the nature of the flow, and it is very similar to the behavior of sound waves in a gas, where subsonic flows ($M < 1$) behave entirely differently from supersonic flows ($M > 1$). In the next chapter, we examine compressible flows, where the similarity between wave propagation in open channel flows and sound waves in gases will be explored further. Here, we examine open channel flows.

11.2   SMALL AMPLITUDE GRAVITY WAVES

To begin the study of open channel flows, we will consider the behavior of *small amplitude gravity waves* moving in a *shallow, open channel flow*. Small amplitude gravity waves are waves with a height that is small compared to their wavelength. Energy losses are very small and the equations of motion can be linearized (see following discussion). Shallow means that the depth of the water is small compared to the length of the waves, so that the one-dimensional form of the equations of motion can be used. Open channel flow means that the flow is open to the atmosphere (it has a free surface where the pressure is atmospheric).

Consider a long, shallow, open channel containing a stationary liquid. A small disturbance is initiated at one end of the channel, perhaps by using a paddle to raise the water level a small amount and then releasing it. The disturbance will move with velocity c, as shown in Figure 11.1. We call this small-amplitude disturbance a wave, although it is not periodic. It is strictly a *bore*, which is a type of single-sided wave found in shallow channel flows (see Section 11.6 for more details). For the case considered here, the "wave" is assumed to be a plane wave, in that it is straight and does not vary across the channel.

When the control volume is stationary (Figure 11.1), we see that before the arrival of the wave, the control volume contains a motionless body of water of depth y. The wave arrives from the left, and as it enters the control volume the depth of the water behind the wave increases to $y + \delta y$. For a stationary control volume, the flow is unsteady. Also, the fluid behind the wave is no longer at rest but it is moving to the right with velocity δV. To see why this fluid is moving, we note that the mass of water contained in the control volume increases during the passage of the wave through the control volume. To bring that extra mass in, there must be a mass flux through the left hand face of the control volume. That is, the fluid behind the wave must move with a small velocity in the direction of motion of the wave. This velocity δV is called the *drift* velocity, and the shallow channel approximation implies that δV is constant over the depth $y + \delta y$. The drift velocity is typically a small fraction of the wave speed.

The analysis of this flow becomes simpler in a moving frame of reference. Imagine yourself moving with the wave. The wave is now stationary relative to you, but the water in front of the wave is coming towards you at a speed c. Therefore,

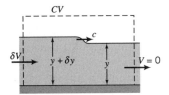

FIGURE 11.1   Control volume for a small amplitude wave in an open channel: unsteady flow.

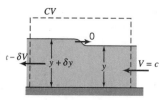

FIGURE 11.2 Control volume for a small amplitude wave in an open channel: steady flow.

if we use a control volume that moves at speed c to the right, the wave will always be at the same location in the control volume and the flow is steady (Figure 11.2). We can then apply the steady one-dimensional continuity equation and Bernoulli's equation. The continuity equation gives

$$(c - \delta V)(y + \delta y) = cy$$

Expanding, and keeping only first order terms ($\delta y \ll y$, and $\delta V \ll c$), we obtain

$$c\delta y = y\delta V \tag{11.1}$$

Applying Bernoulli's equation along the surface (the surface is a streamline where the pressure is constant)

$$\tfrac{1}{2}(c - \delta V)^2 + g(y + \delta y) = \tfrac{1}{2}c^2 + gy$$

Expanding, and keeping only first order terms gives

$$g\, \delta y = c\, \delta V \tag{11.2}$$

Equations 11.1 and 11.2 yield

$$\boxed{c = \sqrt{gy}}$$

Thus the speed of a small amplitude gravity wave moving over the surface of shallow water is $\sqrt{gy}$.

What happens if instead of being motionless with respect to a fixed observer, the water ahead of the wave was moving with velocity V to the right? To make this flow steady, we would have to move with a velocity $V + c$. In this new framework, a similar analysis to that given above would lead to the result that the speed of the wave is $V + \sqrt{gy}$, relative to a stationary observer. Similarly, if the water ahead of the wave was flowing with a velocity V to the left, we would find that the wave moves at the speed $V - \sqrt{gy}$. That is, the wave moves at a speed $\sqrt{gy}$ *relative to the moving liquid.*

Consider a point disturbance on the surface of the water. For example, fill a sink or bath to a depth of about 1/2 *in.* Take a sponge that is soaked with water and, while holding it stationary, squeeze it gently to make it drip onto the water surface. We expect that the ripples made by the drops of water propagate outward in concentric circular patterns at a speed $\sqrt{gy}$ ($\approx$14 *in/s*).[1] You can check this with

---

[1] It can be shown that a circular ripple spreads at the same speed as a planar ripple, that is, at a speed $\sqrt{gy}$, as long as the radial extent of the wave front is small compared to the distance from the center of the disturbance.

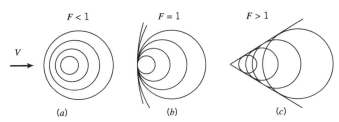

$F < 1$ $F = 1$ $F > 1$

V

(a) (b) (c)

FIGURE 11.3 Wave patterns made by a point disturbance on the free surface of a moving liquid. (*a*) $V < \sqrt{gy}$ (subcritical). (*b*) $V = \sqrt{gy}$ (critical). (*c*) $V > \sqrt{gy}$ (supercritical).

a stopwatch. Now move the sponge slowly as you squeeze it. As each drop hits the water, a circular disturbance travels outward, as before, but the parts of the circles that lie in the same direction as the motion of the sponge are closer together than the parts in the opposite direction. This produces the same effect as holding the dripping sponge stationary over a flowing body of water. If we were stationary, and the water was moving, it would appear that the waves move slower upstream then they do downstream [see Figure 11.3(*a*)]. The wave speed is constant, but the whole of the wave is being carried downstream by the moving fluid. Relative to a stationary observer, it appears that, as the front of the wave moves against the flow, it is also being swept back by the flow. As a result, the front of the wave moves more slowly than the parts of the wave moving downstream where the wave velocity and the fluid velocity are in the same direction. The pattern shown in Figure 11.3(*a*) is observed whenever the fluid velocity is less than the speed of the ripples.

The ratio of fluid speed to wave speed $V/\sqrt{gy}$ is called the Froude number.

$$\text{Froude number} = F \equiv \frac{V}{\sqrt{gy}} \qquad (11.3)$$

The Froude number is named after William Froude who performed many experiments to study the waves produced by ships (see Section 15.13). The pattern shown in Figure 11.3(*a*) is observed when $F < 1$. This phenomenon is similar to the well-known Doppler effect for sound waves. The Doppler effect explains why the sound of a whistle on a train that is moving towards you has a higher pitch (higher frequency, smaller wavelength) than when it is moving away from you, where it has a lower pitch (lower frequency, longer wavelength).

Let us return to our experiment. As we move the sponge faster, there is a point where it will be moving with the same speed as the speed of the ripples. That is, $F = 1$. If the sponge was stationary and the water was moving, the parts of the wave moving upstream at speed $\sqrt{gy}$ will be swept downstream at exactly the same speed, and so all the wavefronts collect along a line [Figure 11.3(*b*)].

When the water moves still faster, so that $F > 1$, we obtain the pattern shown in Figure 11.3(*c*). Here, the wave fronts are being swept downstream by the current faster than they can move upstream, and a wedge-shaped pattern forms with a characteristic angle α (see Section 11.3).

Some interesting observations can be made. For example, consider a fisherman in his boat, moored in a river some distance upstream of the point where a distur-

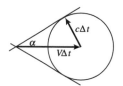

FIGURE 11.4   Wave patterns made at supercritical Froude numbers.

bance is creating waves. When $F < 1$ (called *subcritical* flow), the waves spread out at different speeds relative to the fisherman depending on the direction, but all the waves will reach him eventually. Even if he was not looking at the waves, he would learn of their presence by the rocking of his boat. All this changes when $F = 1$ (called *critical* flow). At this point, the waves collect along a line downstream of the fisherman, and they never reach the boat. He never receives the information that waves are being generated. The line of waves divides the flow into two zones: a downstream zone that is influenced by the presence of waves and an upstream zone that is not. When $F > 1$ (called *supercritical* flow), the zone that is influenced by the waves shrinks to the wedge-shaped region shown in Figure 11.3(*c*).

11.3 FROUDE NUMBER

The Froude number always makes an appearance when we consider flows where the boundaries are not prescribed geometrically. It was defined in equation 11.3 as the ratio of the flow velocity to the speed of a small amplitude disturbance. The Froude number will be small compared to 1 when the velocity is low and/or when the water depth is large. Large rivers are likely to have low Froude numbers. The Froude number will be large compared to 1 when the velocity is large and/or the depth is small. As we indicated, this is called supercritical flow, but it is also sometimes called *shooting flow* because of its appearance.

By squaring the Froude number and multiplying the numerator and denominator by the density, we have

$$F^2 = 2\left(\frac{\frac{1}{2}\rho V^2}{\rho g y}\right) \qquad (11.4)$$

and so the Froude number is related to the ratio of the fluid kinetic energy to its potential energy. Fast flow in a shallow channel has a large Froude number, and its kinetic energy is large compared to its potential energy. Alternatively, equation 11.4 shows that the Froude number is related to the ratio of the dynamic pressure to a typical hydrostatic pressure.

We saw in the previous section that when the Froude number is supercritical ($F > 1$), a wedge-shaped pattern of waves forms, with a characteristic angle α (Figure 11.4). In a given time interval Δt, the center of the disturbance moves with the current a distance $V \Delta t$, while the wave front moves a distance $\sqrt{gy}\Delta t$. Hence,

$$\sin \alpha = \frac{\sqrt{gy}\,\Delta t}{V\,\Delta t} = \frac{\sqrt{gy}}{V} = \frac{1}{F} \qquad (11.5)$$

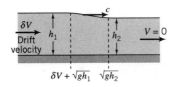

FIGURE 11.5 Notation for a steepening wave (stationary observer).

The Froude number can be estimated, therefore, by measuring the angle the wave fronts make with the flow direction, as long as the waves have small amplitudes. The angle α is sometimes called the *Froude angle*.

11.4 BREAKING WAVES

If we have a long water channel, we can create a large planar disturbance traveling down the channel by using a wave maker (a sort of oscillating airfoil) or a large paddle hinged at the bottom of the channel. As the wave moves down the channel, it can be seen to grow larger and steeper, and possibly "break" like a wave at the beach.

Why does it steepen and break? To a stationary observer, the front of the wave initially moves at a speed close to $\sqrt{gh_2}$, that is, the local speed of a small disturbance (Figure 11.5). The back of the wave moves at a speed very nearly equal to $\sqrt{gh_1} + \delta V$, that is, the local speed of a small amplitude disturbance plus the drift speed δV. Since $\sqrt{gh_1} + \delta V > \sqrt{gh_2}$, the back of the wave catches up with the front and the wave steepens. Eventually, it will break and it is then sometimes called a *positive surge*.

A similar phenomenon occurs at the beach. There, the steepening of the wave is aided by the slope of the bottom: the water depth decreases in the direction of wave movement, accentuating the differences between the speeds of the front and back of the wave (Figures 11.6 and 11.7). In addition, there is significant undertow, which is the result of mass conservation: as the wave brings water up the beach, there must be a current near the bottom to bring it back. This undertow accentuates the effects due to the slope of the beach and hastens the formation of the breaking wave. The most common explanation for the formation of breaking waves on a beach ascribes the phenomenon to "bottom friction." Although there are always effects due to friction, this explanation misses some of the more important mechanisms.

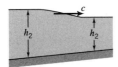

FIGURE 11.6 Waves forming on a sloping beach.

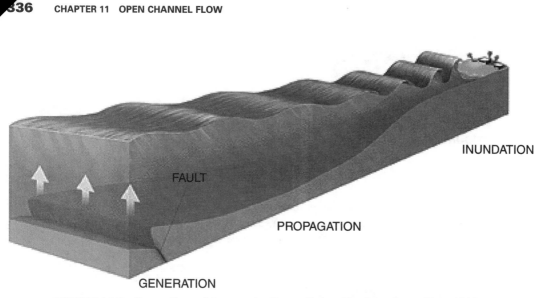

INUNDATION

FAULT

PROPAGATION

GENERATION

FIGURE 11.7   Formation of tsunamis. From *Scientific American,* May 1999.

11.5   TSUNAMIS[2]

ESASHI, Japan, July 14. They can travel as fast as jet planes. They can carry off entire houses. They can inundate coastal communities with violent flooding. Some English speakers call them tidal waves, but they have nothing to do with tides; much of the world recognizes them by their Japanese name, tsunami.

Huge tsunamis inundated northern Japan Monday night, minutes after a powerful earthquake struck the Sea of Japan west of the northern island of Hokkaido. A tsunami contributed heavily to damage along the coast and to the virtual demolition of the Aonae district on Okushiri, a small island known for fishing and resorts.

People were swept away by huge waves and drowned. Cars were flushed into the sea. Ships were thrown onto land where they crashed into buildings. And hundreds of houses were destroyed in a torrent of water. One of the most striking television images of the quake was that of what looked to be an entire house floating out to sea, its roof protruding above the water.

Many things contributed to the damage in the quake, which measured 7.8 on the Richter Scale. There was the shaking, the landslides that ruined roads and buried a hotel, and fires, probably caused by the explosion of ruptured gas lines. But perhaps the most spectacular phenomenon was the tsunami.

Waves outran warning

While Japan, perhaps the world's most earthquake-prone country, has learned how to build structures to withstand earthquakes, it apparently has not yet been able to fully cope with tsunamis.

"Even wooden houses in Japan are built strong enough to withstand the shaking of an earthquake," said Nobuo Shuto, a professor of tsunami engineering at Tohoku University in Sendai.

---

[2] From the article "A Wall of Water Travelling at the Speed of a Jet Plane," by Andrew Pollack, *New York Times,* July 14, 1994.

Japan has a warning system for tsunamis, but on Monday night the waves reached Okushiri at about the same time as the warning, five minutes after the earthquake. "Under this kind of situation, maybe there is little you can do," Professor Shuto said. "The only way to save human lives is to evacuate immediately, even without a warning."

Waves up to 35-feet high

Professor Shuto estimated that the wave that struck Okushiri ranged from 10 to 16 feet high, but noted that he said he had not completed his calculations. A researcher for the Meteorological Agency estimated, based on a survey of the site, that the waves were as high as 35 feet.

Tsunamis are gigantic versions of the ripples produced by a pebble tossed into a still pond. But in tsunamis, the water is displaced not by a pebble, but by an earthquake, volcanic eruption or other violent undersea movement [Figure 11.7]. A huge mass of water can be displaced.

The tsunami's speed depends on the depth of the water above the displaced sea bed. In the case of the tsunami on Monday, where the water was about 6,000 feet deep, the wave travels at 300 miles per hour.

As the wave approaches the land and the ocean becomes shallower, the water in the back of the wave catches up to the water in the front, and the wave height mounts.

Report seeing 10 waves

Depending on the structure of the coastline, a tsunami might strike one time and recede, or reverberate, hitting the shoreline many times. Professor Shuto said some witnesses on Okushiri reported seeing as many as 10 waves.

The professor said that while Japan is most known for tsunamis, they occur elsewhere, and can strike with deadly force thousands of miles from their source. A huge tsunami occurred off the Aleutian Islands of Alaska on April 1, 1946, and traveled to Hawaii. Hawaiians thought warnings of a sea disturbance were an April Fool's joke and ignored them, he said, and 159 people were killed.

Japan began constructing defenses against tsunamis after it was hit by what Professor Shuto said was the strongest tidal wave in its recorded history, with waves up to 40 feet high. The tsunami, which struck in 1960 started off the coast of Chile and took 23 hours to cross the Pacific before slamming into Japan's Pacific Coast.[3]

A graphic animation of the July 17, 1998, Papua New Guinea tsunami is given at *http://walrus.wr.usgs.gov/docs/tsunami/PNGhome.html* (produced by the U.S. Geological Survey). Note the generation of multiple waves by a single earthquake.

11.6 HYDRAULIC JUMPS

To understand breaking waves or tsunamis in more detail, we consider a stationary *hydraulic jump*. Hydraulic jumps may be seen at home by turning on a faucet and

---

[3] With an average Pacific Ocean depth of 15,000 *ft*, the wave speed $\sqrt{gy}$ is about 700 *ft/s*, or 470 *mph*. The distance traveled in 23 hours is therefore about 11,000 miles, which is almost exactly the distance between Chile and Japan. Tidal waves travel as small amplitude disturbances for long distances (the wave height may only be several feet, with a very large wavelength), and they only become destructive near the shore where huge breaking waves can form.

FIGURE 11.8 A moving hydraulic jump (called a *bore*) traveling upstream in the Severn River in England. The bore is caused by the tide moving into the mouth of the river. Photograph by D. H. Peregrine, with permission.

letting the stream of water fall on a flat surface such as a plate. The water spreads out from its point of impact in a thin layer, and then at some point a sudden rise in water level occurs. This is a circular hydraulic jump. Waves breaking on a beach are planar hydraulic jumps, and hydraulic jumps are also often seen in shallow river beds where sudden changes in river depth occur, and just downstream of large rocks (white water enthusiasts know and love hydraulic jumps). In an open channel water tunnel (sometimes called a *flume*), we can form a hydraulic jump by first letting the water flow over a bump (we will see later that this is a mechanism for making the flow supercritical), and then placing an obstacle downstream. The hydraulic jump will form upstream of the obstacle, and it will be stationary with respect to the flume. In the laboratory, we can also produce a moving hydraulic jump by suddenly removing a barrier between two bodies of water of different depth. The motion has two parts: a hydraulic jump moving into the region of lower depth, and a smooth *expansion wave* moving in the opposite direction into the region of higher depth.

Moving hydraulic jumps are sometimes called bores or surges. A tidal bore is caused by a high tide entering a shallow bay or river inlet (see Figure 11.8). As the water proceeds against the current, a bore is formed by the same mechanism responsible for breaking waves. Surges, for example, are formed when a river dam breaks. Bores and surges can be very hazardous, but they can also be useful. For example, they can be used to decelerate a stream of fluid to reduce scouring in a river or channel bed, and Figure 8.2 shows a hydraulic jump formed for this purpose near the bottom of a spillway.

To analyze the features of a hydraulic jump, consider a stationary, planar jump formed in a horizontal, two-dimensional channel flow of width W (Figure 11.9). We will use a control volume that starts well upstream of the jump where the depth of the water is H_1, and that ends well downstream of the jump where

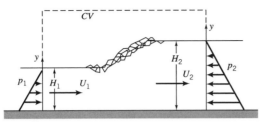

FIGURE 11.9 Control volume for the analysis of a hydraulic jump.

the depth is H_2. Large disturbances such as hydraulic jumps dissipate a lot of energy, so that mechanical energy is not conserved and Bernoulli's equation cannot be used. We can, however, use the continuity and momentum equations.

First, consider the continuity equation. We will assume that the velocities across areas $W H_1$ and $W H_2$ are constant, so that

$$\rho U_1 W H_1 = \rho U_2 W H_2$$

That is,

$$U_1 H_1 = U_2 H_2 \tag{11.6}$$

Next, consider the momentum equation. In the undisturbed part of the channel flow, the streamlines are parallel, so that the pressure varies with depth according to the hydrostatic pressure gradient. In problems involving air or other gases, we have generally ignored these hydrostatic pressures since they were small. With liquids, however, the hydrostatic pressures must be taken into account. Since the depth increases across the jump, the force due to hydrostatic pressure acting on the right hand face of the control volume is larger than that due to hydrostatic pressure acting on the left hand face. This appears as a resultant force due to pressure differences in the momentum equation. If we ignore any viscous forces that might act on the channel bed, there will be no external forces acting on the control volume, and the force due to hydrostatic pressure differences must equal the net outflux of momentum. The x-momentum equation becomes

$$-\mathbf{i} \cdot \int p \mathbf{n} \, dA = \mathbf{i} \cdot \int (\mathbf{n} \cdot \rho \mathbf{V}) \mathbf{V} \, dA$$

By using gauge pressure, we obtain

$$\int_0^{H_1} p_1 W \, dy - \int_0^{H_2} p_2 W \, dy = (-\rho U_1) U_1 W H_1 + (\rho U_2) U_2 W H_2$$

The pressures are hydrostatic, so that

$$\int_0^{H_1} \rho g (H_1 - y) \, dy - \int_0^{H_2} \rho g (H_2 - y) \, dy = -\rho U_1^2 H_1 + \rho U_2^2 H_2$$

That is,

$$\tfrac{1}{2} g H_1^2 - \tfrac{1}{2} g H_2^2 = U_2^2 H_2 - U_1^2 H_1 \tag{11.7}$$

Combining the momentum and continuity equations (equations 11.6 and 11.7) gives

$$\left(\frac{H_2}{H_1}\right)^2 + \left(\frac{H_2}{H_1}\right) - 2F_1^2 = 0 \tag{11.8}$$

where F_1 is the upstream Froude number, that is, $F_1 = U_1/\sqrt{gH_1}$. This equation can be solved for the ratio H_2/H_1

$$\frac{H_2}{H_1} = \frac{-1 \pm \sqrt{1 + 8F_1^2}}{2}$$

We can reject the negative solution because H_2 is always positive. Hence,

$$\boxed{\frac{H_2}{H_1} = \frac{1}{2}\left(\sqrt{1 + 8F_1^2} - 1\right)} \qquad (11.9)$$

This is called the hydraulic jump relationship, and it applies to planar hydraulic jumps in a channel with negligible bottom friction, in a steady frame of reference.

The hydraulic jump relationship states the condition for a jump to take place in a given flow: if the Froude number is F_1, the height of the jump is as given by equation 11.9. Conversely, if the height of the jump is H_2/H_1, the upstream Froude number is as given by equation 11.9.

The hydraulic jump relationship shows that when $F_1 > 1$ then $H_2 > H_1$, and the jump is "up." When $F_1 = 1$ then $H_2 = H_1$, and there is no jump at all. When $F_1 < 1$ then $H_2 < H_1$, and the jump is "down." We will show in Section 11.7, however, that jumps that go down are not possible because of thermodynamic arguments. Therefore

A hydraulic jump occurs only if $F_1 > 1$. As a result, $H_2 > H_1$.

How does a hydraulic jump form? In the upstream flow, $F > 1$, and therefore a small change in flow elevation occurring at some point in the upstream flow cannot travel upstream. If a large bump or a barrier is placed on the channel bottom, the flow cannot adjust smoothly because the "information" about the presence of the obstacle cannot be signalled upstream, and therefore the flow will change abruptly close to the obstacle. If the increased water level caused by the obstacle is such that it satisfies the hydraulic jump relationship, a hydraulic jump will occur just upstream of the obstacle.

What about the Froude number downstream of the hydraulic jump? We first write the continuity equation nondimensionally, as follows. Equation 11.6 gives

$$U_1 H_1 = U_2 H_2$$

Dividing through by $\sqrt{gH_1}$ gives

$$\frac{U_1}{\sqrt{gH_1}} H_1 = \frac{U_2}{\sqrt{gH_1}} H_2$$

$$= \frac{U_2}{\sqrt{gH_1}} \frac{\sqrt{gH_2}}{\sqrt{gH_2}} H_2$$

That is,

$$F_1 H_1 = \frac{U_2}{\sqrt{gH_2}} \frac{\sqrt{gH_2}}{\sqrt{gH_1}} H_2$$

and

$$\boxed{F_2 = \left(\frac{H_1}{H_2}\right)^{3/2} F_1} \qquad (11.10)$$

Equation 11.10 is just another way of writing the continuity equation for flow in a channel of constant width, but it is particularly useful because it is written in terms of nondimensional parameters such as the Froude number. Writing the equations in terms of nondimensional parameters can be very effective in solving problems, particularly in problems concerning open channel flows.

The downstream Froude number can now be found by combining the continuity equation in the form given in equation 11.10 with the hydraulic jump relationship given in equation 11.9.

First,

$$F_2^2 = \left(\frac{H_1}{H_2}\right)^3 F_1^2$$

and so

$$F_2^2 = \left(\frac{H_1}{H_2}\right)^3 \frac{1}{8}\left[\left(\frac{2H_2}{H_1}+1\right)^2 + 1\right]$$

That is,

$$F_2^2 = \frac{1}{2}\frac{H_1}{H_2}\left(\frac{H_1}{H_2}+1\right)$$

For $F_1 > 1$, $H_1/H_2 < 1$ (jumps must go up), and therefore F_2 is always less than 1. That is,

A hydraulic jump occurs only if $F_1 > 1$. As a result, $F_2 < 1$.

11.7 HYDRAULIC DROPS?

The hydraulic jump relationship (equation 11.9) implies that it is possible to have hydraulic jumps that go down ($H_2 < H_1$) when $F_1 < 1$. To show that such hydraulic "drops" are not possible, consider what happens to the Bernoulli constant B through this flow. At the surface of the water upstream and downstream of the jump

$$B_1 = \tfrac{1}{2}U_1^2 + gH_1$$

and

$$B_2 = \tfrac{1}{2}U_2^2 + gH_2 \qquad\qquad (11.11)$$

That is:

$$B_1 = \tfrac{1}{2}U_1^2 + gH_1$$
$$= gH_1(\tfrac{1}{2}F_1^2 + 1)$$

and

$$B_2 = gH_2(\tfrac{1}{2}F_2^2 + 1)$$

We know that there are losses in the hydraulic jump because of energy dissipation due to turbulence and secondary flows, so that $B_2 < B_1$ (see Section 4.7.3). When $F_1 > 1$, then $B_2 < B_1$, as required, so that jumps that go *up* are allowed. However, when $F_1 < 1$, then $B_2 > B_1$, which is not allowed, and so jumps that go *down* are not possible.

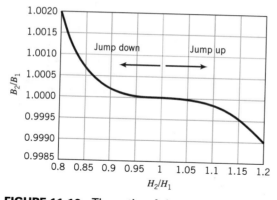

FIGURE 11.10 The ratio of downstream to upstream Bernoulli constants for a hydraulic jump as a function of the ratio of downstream to upstream water depths.

To demonstrate this result, we first form the ratio of the Bernoulli constants.

$$\frac{B_2}{B_1} = \frac{H_2}{H_1} \frac{\left(1 + \frac{F_2^2}{2}\right)}{\left(1 + \frac{F_1^2}{2}\right)}$$

By using the continuity equation (equation 11.10), we obtain

$$\frac{B_2}{B_1} = \frac{H_2}{H_1} \left[\frac{1 + \left(\frac{H_1}{H_2}\right)^3 \frac{F_1^2}{2}}{1 + \frac{F_1^2}{2}} \right] \tag{11.12}$$

We can find an expression for F_1 using the hydraulic jump relationship (equation 11.9)

$$F_1^2 = \frac{1}{2} \frac{H_2}{H_1} \left(\frac{H_2}{H_1} + 1 \right)$$

and use this result to substitute for F_1 in equation 11.12 to obtain

$$\frac{B_2}{B_1} = \frac{H_2}{H_1} \left[\frac{1 + \frac{1}{4}\left(\frac{H_1}{H_2}\right)^2 \left(\frac{H_2}{H_1} + 1\right)}{1 + \frac{1}{4}\left(\frac{H_2}{H_1}\right) \left(\frac{H_2}{H_1} + 1\right)} \right] \tag{11.13}$$

This relationship is plotted in Figure 11.10 for a small range of H_2/H_1. For $H_2 > H_1$, we see that $B_2 < B_1$, indicating that energy is dissipated by the hydraulic jump, as observed. When $H_2 < H_1$, $B_2 > B_1$, implying that energy is created by a hydraulic drop, which is not possible. Only hydraulic jumps are allowed.

11.8 SURGES AND BORES

A surge or a bore is a moving hydraulic jump. It may be caused by a tidal flow entering a shallow estuary (the mouth of a river), as shown in Figure 11.8, or by a sudden increase in water level, as when a dam on a river is breached. They are

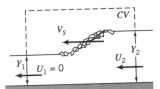

FIGURE 11.11 Surge in a frame of reference where the flow is unsteady.

similar to tsunamis, but tsunamis are coastal features where the breaking waves are usually formed on a sloping beach.

Consider a surge, moving with a constant velocity U_B into a region where the velocity is zero and the bottom is level (Figure 11.11). This is an unsteady flow, but if we adopt a frame of reference moving at the same speed as the wave, it becomes a steady problem (Figure 11.12). The analysis is identical to that followed for the stationary hydraulic jump considered in Section 11.6, with U_B replacing U_1, and $U_B - U_2$ replacing U_2. To see this, imagine riding on the wave as a surfer might. What the surfer sees coming towards her is not a stagnant body of water, but water that appears to be moving towards her at a speed U_B. When she looks back, she sees water moving away from her at a speed $U_B - U_2$. So the relationship for a surge or bore moving into a stagnant pool of water with a level bottom is simply

$$\frac{H_2}{H_1} = \frac{1}{2}\left(\sqrt{1 + 8F_B^2} - 1\right)$$

where $F_B = U_B/\sqrt{gH_1}$, and U_B is the speed of the bore relative to a stationary observer. Since the appropriate Froude number is based on the bore velocity, this Froude number must be supercritical ($F_B > 1$) if a jump is to take place.

11.9 FLOW THROUGH A SMOOTH CONSTRICTION

Consider the steady flow of water through an open channel where the sides of the channel form a symmetric constriction so that the widths of the inlet and outlet are the same, as shown in Figure 11.13. We will assume that the flow is approximately one-dimensional and that there are no losses, so that there are no hydraulic jumps. A flow where there are no losses is sometimes called "smooth," so this is a smooth constriction.

As the flow passes through the constriction, what happens to the water surface? Does the water level go down, or does it go up? To answer this question, we will use the one-dimensional continuity equation and apply Bernoulli's equation along the surface streamline. The continuity equation gives

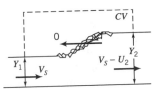

FIGURE 11.12 Surge in a frame of reference where the flow is steady.

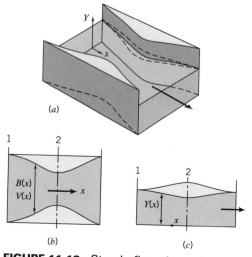

FIGURE 11.13 Steady flow through a smooth constriction. (a) Perspective view. (b) Plan view. (c) Elevation.

$$B_1 Y_1 V_1 = B Y V \tag{11.14}$$

where the subscript 1 denotes the inlet conditions upstream of the constriction, and B, Y, and V are, respectively, the width of the channel, the depth of the stream, and the average velocity in a cross-section of the flow at some location x inside the constriction. Non-dimensionalizing equation 11.14 in the same way as we did earlier (see equation 11.10), we obtain

$$F^2 = \left(\frac{B_1}{B}\right)^2 \left(\frac{Y_1}{Y}\right)^3 F_1^2 \tag{11.15}$$

which is just the continuity equation in nondimensional form.

Bernoulli's equation for steady, incompressible, frictionless flow along the surface streamline, where $p_1 = p_2$, gives

$$\tfrac{1}{2} V_1^2 + g Y_1 = \tfrac{1}{2} V^2 + g Y$$

Non-dimensionalizing this equation by dividing through by $g Y_1$ gives

$$\tfrac{1}{2} F_1^2 + 1 = \tfrac{1}{2} F^2 \frac{Y}{Y_1} + \frac{Y}{Y_1} \tag{11.16}$$

Combining equations 11.15 and 11.16 gives a third-order polynomial for Y/Y_1

$$\left(\frac{Y}{Y_1}\right)^3 - \left(\frac{Y}{Y_1}\right)^2 \left(1 + \frac{F_1^2}{2}\right) + \left(\frac{B_1}{B}\right)^2 \frac{F_1^2}{2} = 0 \tag{11.17}$$

Since this is cubic, there are in general three solutions. One solution is usually nonphysical (for example, it might be negative), so that there are at most two nontrivial real solutions. Now we can answer the question regarding the behavior of the water surface: Y/Y_1 can be greater than 1 or less than 1 (that is, the water level can go up, or it can go down), depending on the values of F_1 and B/B_1.

Instead of solving this equation for a wide range of inputs, we can learn a great deal about the behavior of the water level by examining the slope of the water surface. To find the slope, we differentiate equation 11.17. Before doing so, it is useful to write this equation in terms of the volume flow rate $\dot{q} = B_1 Y_1 V_1 = BYV$, which is a constant. After some algebra, we obtain

$$\frac{dY}{dx} = \frac{\frac{\dot{q}^2}{gB^3 Y^2}}{\left(1 - \frac{\dot{q}^2}{gB^2 Y^3}\right)} \frac{dB}{dx}$$

This can also be written as

$$\frac{dY}{dx} = \frac{\frac{Y}{B} F^2}{1 - F^2} \frac{dB}{dx} \tag{11.18}$$

where $F = V/\sqrt{gY}$ is the local Froude number.

Clearly, this is an interesting result. In particular, the slope of the surface changes sign when the Froude number changes from being less than one to being greater than one. We can also anticipate some problems when $F = 1$. Before we can proceed, we need one more result.

By differentiating the continuity equation for this flow (equation 11.14), we have

$$BY \frac{dV}{dx} + BV \frac{dY}{dx} + YV \frac{dB}{dx} = 0$$

and so

$$\frac{dY}{dx} = -\frac{Y}{B} \frac{dB}{dx} - \frac{Y}{V} \frac{dV}{dx}$$

Equation 11.18 becomes

$$-\frac{Y}{V} \frac{dV}{dx} = \frac{F^2}{1-F^2} \frac{Y}{B} \frac{dB}{dx} + \frac{Y}{B} \frac{dB}{dx} = \frac{1}{1-F^2} \frac{Y}{B} \frac{dB}{dx}$$

Finally,

$$\frac{dV}{dx} = -\frac{\frac{V}{B}}{1-F^2} \frac{dB}{dx} \tag{11.19}$$

which provides information on how the flow accelerates or decelerates as it passes through the smooth constriction.

We will avoid the case where $F = 1$ for now, and consider the two cases where either $F < 1$ everywhere, or $F > 1$ everywhere.

1. When $F < 1$ everywhere, then with
   (a) $\frac{dB}{dx} < 0$, we find $\frac{dY}{dx} < 0$, and $\frac{dV}{dx} > 0$
   (b) $\frac{dB}{dx} > 0$, we find $\frac{dY}{dx} > 0$ and $\frac{dV}{dx} < 0$.

2. When $F > 1$ everywhere, then with
   (a) $\frac{dB}{dx} < 0$, we find $\frac{dY}{dx} > 0$ and $\frac{dV}{dx} < 0$
   (b) $\frac{dB}{dx} > 0$, we find $\frac{dY}{dx} < 0$ and $\frac{dV}{dx} > 0$.

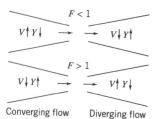

FIGURE 11.14 Summary of flow variation in subcritical and supercritical channels.

These results are summarized in Figure 11.14.

When the Froude number is subcritical everywhere, the water level falls as the flow speeds up in the converging part of the constriction, and it rises as the flow slows down in the diverging part of the constriction. When the Froude number is supercritical everywhere, the water level rises as the flow slows down in the converging part of the constriction, and it falls as the flow speeds up in the diverging part of the constriction (Figure 11.15). It is possible, therefore, to know whether the flow is subcritical or supercritical by observing its behavior as it passes through a constriction.

What happens when the Froude number becomes critical ($F = 1$) somewhere? The initial conclusion from equation 11.18 is that the slope of the water surface becomes infinite. This is not possible in the real world. It is also not the only conclusion that can be drawn. The slope of the water surface depends on the value of dB/dx as well as the Froude number, and it is possible for the Froude number to be critical, as long as $dB/dx = 0$ at the same location. In that case, the numerator and denominator of equation 11.18 are both zero, and although the slope is now indeterminate, it is finite, and physically meaningful solutions can exist. This means that

1. The only place where $F = 1$ must be where $dB/dx = 0$, that is, at the point of minimum area of the constriction. This is called the *throat*. The Froude number cannot be one anywhere else.

2. When $F = 1$ at the throat, the slope of the water surface at the throat cannot be found using equation 11.18. Additional information regarding the downstream flow conditions is required before the solution can be determined, as discussed in the following sections.

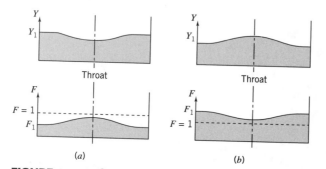

FIGURE 11.15 Steady flow through a smooth constriction. (*a*) Subcritical flow everywhere. (*b*) Supercritical flow everywhere.

11.9.1 Subcritical Flow in Contraction

Consider a stream that is initially subcritical everywhere ($F < 1$). As the water passes through the constriction, its level falls and its velocity increases. Consequently, the local Froude number F increases, and its maximum value will occur at the throat where the area is minimum. Downstream of the throat, the flow recovers so that the water level rises and the Froude number decreases. If the upstream and downstream widths of the constriction are equal, the flow returns to its upstream state (neglecting losses).

When the incoming Froude number F_1 is increased, there comes a point when the Froude number becomes critical at the throat: this will be the first place where F can become critical, and, as we saw previously, it is the *only* place it can occur. Since $dB/dx = 0$ at the throat, the slope of the water surface dY/dx becomes indeterminate (from equation 11.18, $dY/dx = 0/0$). The flow upstream of the throat is subcritical, but we will see that the behavior of the flow downstream of the throat is governed by the downstream conditions.

What happens if we try to increase F_1 beyond the condition where the throat Froude number is 1? Our first suggestion might be that F becomes unity at some point upstream where $dB/dx \neq 0$. But equation 11.18 indicates that when $F = 1$ and $dB/dx \neq 0$, dY/dx becomes infinite. This is not possible. The only place where the flow can become critical is at the throat. If $F = 1$ at the throat, and the upstream flow velocity V_1 is somehow increased, then Y_1 will increase such that F_1 remains as before, and F at the throat remains equal to one.

If we now return to the original equation for the flow through the constriction (equation 11.17), we find that there exist two possible solutions that have $F = 1$ at the throat. That is, two possibilities exist for a continuous surface profile in a steady flow without losses, and they must satisfy one of two particular boundary conditions at the exit from the constriction. In other words, under the conditions where there are no losses, only two flows are possible with $F = 1$ at the throat, and they can only exist if the downstream water level Y_3 is one of two values, Y_3' or Y_3'' in Figure 11.16.

1. When $Y_3 = Y_3'$, the flow returns to subcritical Froude number values, and the water level rises in the diverging part of the constriction.

2. When $Y_3 = Y_3''$, the flow becomes supercritical in the diverging part of the constriction, and the water level falls.

If $F = 1$ at the throat, and the downstream water level Y_3 takes a value not equal to Y_3' or Y_3'', a continuous surface profile without losses cannot be found. For example, if Y_3 was such that $Y_3'' < Y_3 < Y_3'$ (in Figure 11.16, this value of Y_3 is

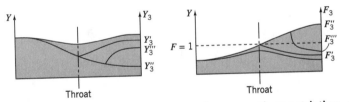

FIGURE 11.16 Steady flow through a smooth constriction, with subcritical flow upstream, and critical flow ($F = 1$) at the throat.

shown as Y_3'''), the flow will be supercritical for some distance downstream of the throat. However, when the Froude number is such that, for the given Y_3 a jump is possible (according to equation 11.9), a hydraulic jump occurs and the flow becomes subcritical. Downstream of the hydraulic jump, the flow will be subcritical, and the water surface begins to rise as it passes through the rest of the constriction until $Y_3 = Y_3''$ at the exit. The actual position of the hydraulic jump depends on the downstream water level.

The lowest value that Y_3''' can take will occur when the hydraulic jump is located at the end of the diverging section. This jump is the strongest jump that can be found in the constriction, since at this point the Froude number has its highest value and the change in water level is the largest possible.

What happens if no hydraulic jump is possible? That is, what happens when the hydraulic jump relationship cannot be satisfied at any location in the diverging part of the constriction? In this case, no solution exists. However, our analysis has been restricted to one-dimensional flow. In practice, oblique hydraulic jumps will appear downstream of the constriction, but the analysis of oblique jumps is beyond the scope of this text.

What happens if $Y_3 > Y_3'$? There is no steady solution: the channel will *drown*. That is, the water will flow back upstream and subcritical flow will be established everywhere.

What happens if $Y_3 < Y_3''$? In this case, there is no one-dimensional solution, and oblique expansion waves will appear downstream of the constriction.

11.9.2 Supercritical Flow in Contraction

When the conditions are such that the Froude number of the flow is supercritical everywhere in the converging part of the constriction, the water level will rise, and the local Froude number will decrease (see Figure 11.17). If it decreases such that the Froude number becomes critical ($F = 1$) at the throat, the flow downstream of the throat has the same two solutions considered in the previous section.

An important point is that, once the Froude number at the throat is equal to one, the upstream and downstream parts of the flow become independent. There is no communication between them, even if they were both subcritical. The location where $F = 1$ acts as a barrier to communication: disturbances in water level such as waves and ripples cannot pass upstream of this location.

If there are no losses and equation 11.17 applies, the downstream part of the flow has the same two possible solutions as before, where $Y_3 = Y_3'$ or $Y_3 = Y_3''$ (see Figure 11.17), no matter what the Froude number is upstream of the throat. If $Y_3'' < Y_3 < Y_3'$, then the flow will be supercritical for some distance downstream of the throat and when the Froude number is such that, for the given Y_3, a jump

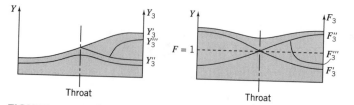

FIGURE 11.17   Steady flow through a smooth constriction, with supercritical flow upstream, and critical flow ($F = 1$) at the throat.

is possible (according to the hydraulic jump relationship, equation 11.9), a hydraulic jump occurs, and the flow becomes subcritical.

11.9.3 Flow Over Bumps

We have seen that when an open channel flow passes through a constriction, the Froude number can become critical at the throat and supercritical in the diverging part of the constriction. A bump on the floor of a channel of constant width can also make the flow become supercritical. For example, if the depth of the water decreases approaching the bump, and then continues to decrease downstream of the bump, we can show that the flow becomes supercritical as it passes over the bump. A throat must have formed near the crest of the bump. This type of flow is often seen in river rapids, where the depth of the water decreases as the water passes over a submerged rock. Usually, a hydraulic jump forms a short distance downstream of the rock, indicating that the flow became supercritical as it passed over the rock. For the flow approaching a bump, if we see that the depth of the water begins to increase, we know that it is initally supercritical. If the depth continues to increase downstream of the bump, we know that the flow became subcritical as it passed over the bump.

SUMMARY

When there are no losses in the constriction

1. When $F < 1$ everywhere, the water level drops in the converging part, and rises in the diverging part (F first increases then decreases).
2. When $F > 1$ everywhere, the water level rises in the converging part, and falls in the diverging part (F first decreases then increases).
3. $F = 1$ only at the throat.
4. The downstream solution is indeterminate when $F = 1$ at the throat. However, when there are no losses, only two possibilities exist: a supercritical solution and a subcritical solution. These solutions are independent of the upstream conditions, and depend only on the downstream conditions. In particular, they correspond to two special values for the downstream depth. If the downstream conditions require that the downstream depth be different from these particular values, hydraulic jumps will appear and losses will occur.

Remember: hydraulic jumps can only occur at places where the hydraulic jump relationship is satisfied. That is, jumps occur where the Froude number is such that the change in height due to the jump is of the right value. Downstream of the jump $F < 1$, and the water surface will continue to rise as the channel expands.

EXAMPLE 11.1 *Flow Under a Sluice Gate*

Water flows steadily under a partially open gate of width W, as shown in Figure 11.18. The flow in the vicinity of the gate is quite complicated but at some distance upstream of the gate, the streamlines are initially straight and the water depth is Y_1. Some distance downstream of the gate, the streamlines are again straight and

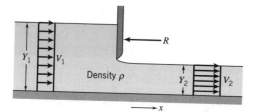

FIGURE 11.18 Partially opened sluice gate.

the water depth is $Y_2 = Y_1/2$. If there are no frictional effects on the flow, that is, the water flows smoothly through the gate and there are no losses:
(a) What is F_1, the Froude number of the upstream flow?
(b) Show that the flow downstream of the gate is supercritical.

Solution: From mass conservation,

$$V_1 Y_1 W = V_2 Y_2 W$$

It is always wise to nondimensionalize the equations. Multiplying and dividing the left hand side of the continuity equation by $\sqrt{gY_1}$, and the right hand side by $\sqrt{gY_2}$, we obtain

$$\frac{V_1}{\sqrt{gY_1}} Y_1 \sqrt{gY_1} = \frac{V_2}{\sqrt{gY_2}} Y_2 \sqrt{gY_2}$$

That is,

$$F_1^2 = F_2^2 \left(\frac{Y_2}{Y_1}\right)^3 \qquad\qquad (11.20)$$

To find F_1, we need to know something about F_2. Since there are no losses, we could use Bernoulli's equation along the surface streamline. This may seem somewhat of a stretch, in that the surface streamline takes two sharp corners and is in contact with a solid surface. By taking the streamline a small distance below the surface, these potential sources of error are minimized, and if this distance is small enough, it will almost coincide with the surface streamline. We will proceed under the assumption that Bernoulli's equation can be used along the surface streamline. Then

$$\tfrac{1}{2}V_1^2 + gY_1 = \tfrac{1}{2}V_2^2 + gY_2$$

Non-dimensionalizing by dividing through by gY_1, we get

$$\tfrac{1}{2}\frac{V_1^2}{gY_1} + 1 = \tfrac{1}{2}\frac{V_2^2}{gY_2}\frac{Y_2}{Y_1} + \frac{Y_2}{Y_1}$$

That is,

$$\tfrac{1}{2}F_1^2 + 1 = \tfrac{1}{2}F_2^2\frac{Y_2}{Y_1} + \frac{Y_2}{Y_1}$$

$$= \frac{Y_2}{Y_1}(\tfrac{1}{2}F_2^2 + 1)$$

Eliminating F_2 using mass conservation (equation 11.20)

$$\tfrac{1}{2}F_1^2 + 1 = \frac{Y_2}{Y_1}\left[\tfrac{1}{2}F_1^2\left(\frac{Y_1}{Y_2}\right)^3 + 1\right]$$

and by collecting terms we obtain

$$F_1^2 = \frac{2 - 2\frac{Y_2}{Y_1}}{\left(\frac{Y_1}{Y_2}\right)^2 - 1}$$

Since $Y_2 = Y_1/2$,

$$F_1 = \frac{1}{\sqrt{3}}$$

In addition, we have from equation 11.20

$$F_2^2 = F_1^2\left(\frac{Y_1}{Y_2}\right)^3 = \sqrt{\frac{8}{3}}$$

so that the upstream Froude number is subcritical and the downstream Froude number is supercritical. ∎

EXAMPLE 11.2 *Force On a Sluice Gate*

For the gate shown in Figure 11.18, find the force required to hold the gate fixed. Express the result nondimensionally.

Solution: The force required to hold the gate fixed is shown in the figure as acting in the negative x-direction. Therefore the force exerted by the fluid on the gate is R, and the force acting on the fluid is $-R$. We also have the force due to differences in hydrostatic pressure, similar to that considered in the analysis of the hydraulic jump (Section 11.6). The x-component momentum equation gives

$$-R + \frac{\rho g Y_1^2 W}{2} - \frac{\rho g Y_2^2 W}{2} = -\rho V_1^2 Y_1 W + \rho V_2^2 Y_2 W$$

where friction has been ignored. By dividing through by $\rho V_1^2 Y_1 W$, and by substituting for V_2 from the continuity equation

$$-\frac{R}{\rho V_1^2 Y_1 W} + \frac{g Y_1^2}{2 V_1^2 Y_1} - \frac{g Y_2^2}{2 V_1^2 Y_1} = -1 + \frac{V_2^2 Y_2}{V_1^2 Y_1} = \frac{V_1^2 Y_1^2}{V_1^2 Y_1 Y_2}$$

Collecting terms, we have

$$\frac{R}{\rho V_1^2 Y_1 W} = \frac{1}{2F_1^2}\left(1 - \frac{Y_2^2}{Y_1^2}\right) + \left(1 - \frac{Y_1}{Y_2}\right)$$

∎

EXAMPLE 11.3 *Flow Over a Bump*

Water flows from left to right in an open channel of constant width, as shown in Figure 11.19. The flow becomes supercritical as it passes over a bump of height H. It remains supercritical for some distance downstream, reaching a maximum Froude number of 1.83 at station 3. It then becomes subcritical by means of a hydraulic jump.

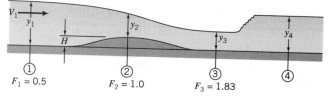

$F_1 = 0.5$ $F_2 = 1.0$ $F_3 = 1.83$

FIGURE 11.19 Flow over a bump.

(a) Find the nondimensional water depth at the throat (Y_2/Y_1).
(b) Find the nondimensional height of the bump (= H/Y_1).
(c) Find the nondimensional water depth before the hydraulic jump (= Y_3/Y_1).
(d) Find the nondimensional water depth after the hydraulic jump (= Y_4/Y_1).

Solution: For part (a) we use mass conservation

$$V_1 Y_1 = V_2 Y_2$$

Non-dimensionalizing, as we have done a number of times, we obtain

$$F_1^2 = F_2^2 \left(\frac{Y_2}{Y_1}\right)^3 \tag{11.21}$$

Now $F_1 = 0.5$, and $F_2 = 1$, so

$$\frac{Y_2}{Y_1} = 0.63$$

For part (b), Bernoulli's equation can be used along the surface streamline in the region where there are no losses. Between stations 1 and 2

$$\tfrac{1}{2} V_1^2 + g Y_1 = \tfrac{1}{2} V_2^2 + g(Y_2 + H)$$

Non-dimensionalizing by dividing through by $g Y_1$,

$$\tfrac{1}{2} F_1^2 + 1 = \tfrac{1}{2} F_2^2 \frac{Y_2}{Y_1} + \frac{Y_2}{Y_1} + \frac{H}{Y_1}$$

Therefore,

$$1.125 = \frac{3}{2}\frac{Y_2}{Y_1} + \frac{H}{Y_1}$$

and

$$\frac{H}{Y_1} = 0.180$$

For part (c) we can use mass conservation again, this time between stations 1 and 3.

$$F_1^2 = F_3^2 \left(\frac{Y_3}{Y_1}\right)^3$$

Since $F_1 = 0.5$, and $F_3 = 1.83$,

$$\frac{Y_3}{Y_1} = 0.421$$

For part (d), we use the hydraulic jump relationship, equation 11.9.

$$\frac{Y_4}{Y_3} = \frac{1}{2}\left(\sqrt{1 + 8F_3^2} - 1\right) = 2.136$$

So

$$\frac{Y_4}{Y_1} = \frac{Y_4}{Y_3}\frac{Y_3}{Y_1} = 2.136 \times 0.421 = 0.899$$ ∎

EXAMPLE 11.4   *Moving Hydraulic Jump*

Water flows in a rectangular channel at a depth of 1 ft and a velocity of 10 ft/s. When a gate is suddenly placed across the end of the channel, a bore travels upstream with velocity V_b as indicated in Figure 11.20. Find V_b when the depth of the water behind the bore is 3 ft.

Solution: To use the hydraulic jump relationship (equation 11.9), we need to move in a frame of reference where the flow is steady. This is accomplished by moving with the bore. Relative to the bore, the incoming water velocity is V_b + 10 ft/s, and the incoming Froude number F_1 is given by

$$F_1 = \frac{V_b + 10}{\sqrt{32.2 \times 1}}$$

where V_b is in ft/s. From the hydraulic jump relationship,

$$\frac{Y_2}{Y_1} = \frac{1}{2}\left(\sqrt{1 + 8F_1^2} - 1\right)$$

we have

$$8F_1^2 = \left(\frac{2Y_2}{Y_1} + 1\right)^2$$

Hence,

$$8\frac{(V_b + 10)}{\sqrt{32.2}} = 48$$

and

$$V_b = 3.90 \; ft/s$$ ∎

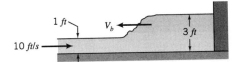

FIGURE 11.20   Bore traveling upstream in a river.

PROBLEMS

11.1   What is the propagation speed of a small planar disturbance in a shallow water basin?

11.2   Consider a small amplitude gravity wave moving at a speed c_m from left to right into a shallow basin of depth y, as shown in Figure P11.2. The water in this basin is moving from right to left at speed U. Find the wave speed c_m in terms of y and U, stating all your approximations clearly. What happens when the Froude number based on y and U equals unity?

FIGURE P11.2.

11.3   Write down the speed of a small amplitude gravity wave in shallow water. Use this result to describe qualitatively the formation of a breaking wave on a beach. How does the slope of the beach affect the formation of this wave?

11.4   To find the drift velocity δv for a small amplitude gravity wave, the linear analysis is not sufficient. By repeating the analysis given in Section 11.2, without linearization, show that

$$\delta v \simeq \frac{2(c^2 - gy)}{3c}$$

11.5   For open channel flow, name one mechanism by which the flow can become supercritical, and one mechanism by which the flow can become subcritical.

11.6   Write down the definition of Froude number. Give two physical interpretations for its significance.

11.7   As an open channel flow enters a smooth constriction, the water level is observed to fall. What can you say about the upstream Froude number?

11.8   Determine the minimum depth in a $3\,m$ wide rectangular channel if the flow is to be subcritical with a flowrate of 30 m^3/s.

11.9   Consider the one-dimensional open channel flow shown in Figure P11.9.
(a) Using the continuity principle and Bernoulli's equation, show that

$$\frac{dy}{dx} = \frac{1}{F^2 - 1}\frac{dh}{dx}$$

where F is the local Froude number.
(b) Discuss the implications of this result for subcritical flow everywhere, supercritical flow everywhere, and for $F = 1$ at station 2.

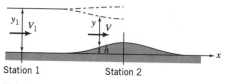

Station 1   Station 2

FIGURE P11.9.

11.10   A rectangular channel has a contraction that smoothly changes width to a minimum of 1.5 ft (the throat). If the flow is critical at the throat, find the volume flow rate when the depth at the throat is 1.5 ft.

11.11 Water in an open channel of constant width flows over a bump of height 1 *ft*, as shown in Figure P11.11. What is the depth of the water Y_2? Assume uniform flow of constant width with no losses.

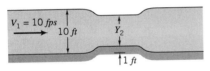

FIGURE P11.11.

11.12 Water flows smoothly over a small bump in a channel of constant width, as shown in Figure P11.12. At any cross-section the velocity may be considered constant over the entire area. If V_1 is the velocity at entry, where the depth is Y_1, and V_2 is the velocity where the bump has its highest point, where the depth is Y_2, find the Froude number of the flow at entry, and the Froude number at the top of the bump, when $Y_1/Y_2 = 1.8$.

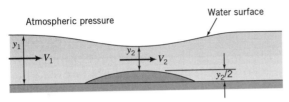

FIGURE P11.12.

11.13 Water in a two-dimensional channel flows smoothly over a submerged obstacle of height H as shown in Figure P11.13. The water depth at the peak of the obstacle is Y_2, where $Y_2 = Y_1/3$.
(a) Find the value of the Froude number of the incoming flow, F_1, given that the Froude number is unity at the point where the depth is Y_2.
(b) Find the nondimensional height of the obstacle, H/Y_1.

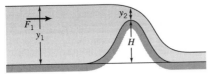

FIGURE P11.13.

11.14 Water in a two-dimensional channel flows over a bump, as shown in Figure P11.14. If $H_2/H_1 = 4$, find the Froude numbers at entry and exit, F_1 and F_2.

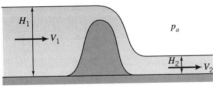

FIGURE P11.14.

11.15 For the flow shown in Figure P11.15:
(a) Find the Froude number at station 1, where the water exits the tank.

(b) Find the water depth at station 2 in terms of h_1 and h_2, given that the Froude number at station 2 is one.

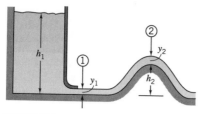

FIGURE P11.15.

11.16   A smooth transition section connects two open channels of the same width, as shown in Figure P11.16. The water depth decreases so that the ratio of the downstream to upstream depths $Y_2/Y_1 = 0.5$. If the upstream Froude number $F_1 = 0.35$, determine the downstream Froude number F_2, and the ratio h/y_1.

FIGURE P11.16.

11.17   An exit of width W allows water to flow from a large tank into an open channel of the same width, as shown in Figure P11.17. The depth of water in the tank is maintained constant at H, which is large compared to h_1, where h_1 is the depth of the water at the exit. As the water flows smoothly over a bump of height b, the depth increases so that the ratio $h_3/h_1 = 4$. There are no losses anywhere. Show that the Froude number F_1 is supercritical. Use Bernoulli's equation and continuity to find the numerical value of F_1 and the ratio b/h_1.

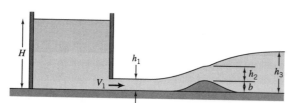

FIGURE P11.17.

11.18   A smooth transition section connects two rectangular channels, as shown in Figure P11.18. The channel width increases from B_1 to B_2 and the water surface elevation is the same in each channel. If the upstream depth of flow is H_1, determine h, the amount the channel bed needs to be raised across the transition section to maintain the same surface elevation.

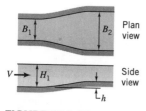

FIGURE P11.18.

11.19 Consider the steady open channel flow of water in the smooth constriction shown in plan view in Figure P11.19. At position 1, the depth is Y_1 and the width is B_1. You may assume that B_1 is much larger than either B_2 or B_3, the widths of the channel at positions 2 and 3, respectively. A hydraulic jump is observed between positions 3 and 4 where the upstream depth $Y_3 = Y_1/3$. You may assume one-dimensional flow, and that $F_1 < 1$.

(a) Find Y_2, the depth at position 2, in terms of Y_1.

(b) Find F_3, the Froude number at position 3.

(c) Find Y_4, the depth just downstream of the hydraulic jump, in terms of Y_1.

(d) Find F_4, the Froude number just downstream of the hydraulic jump.

(e) Find F_5, the Froude number at the exit, given that $Y_5 = 1.1Y_4$.

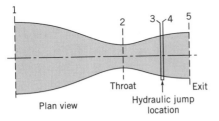

FIGURE P11.19.

11.20 Consider the steady open channel flow of water in the smooth constriction shown in plan view in Figure P11.20. At position 1, the depth is Y_1, the width is B_1, and the Froude number $F_1 = 4$. A similar notation is used at positions 2, 3, and 4. A hydraulic jump is observed just downstream of position 2 where the Froude number $F_2 = 2$. You may assume one-dimensional flow.

(a) Find Y_2, the depth at position 2, in terms of Y_1.

(b) Find Y_3, the depth just downstream of the hydraulic jump, in terms of Y_1.

(c) Find F_3, the Froude number just downstream of the hydraulic jump.

(d) Find F_4, the Froude number at the exit, given that $Y_3/Y_4 = 1.2$.

(e) Find B_4/B_1.

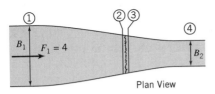

FIGURE P11.20.

11.21 By combining the momentum and continuity equations for a simple hydraulic jump (equations 11.6 and 11.7), obtain equation 11.8 where F_1 is the upstream Froude number, that is, $F_1 = U_1/\sqrt{gH_1}$.

11.22 By using the definition of the Bernoulli constant (equation 11.11), the continuity equation (equation 11.10), and the hydraulic jump relationship (equation 11.9), obtain equation 11.13 for a two-dimensional channel flow.

11.23 The water depth upstream of a stationary hydraulic jump is 1 *m*, while the depth after the jump is 2 *m*. Find the upstream and downstream velocities and Froude numbers.

11.24 The depth of water in a rectangular channel is 1.5 *ft*. The channel is 6 *ft* wide and carries a volume flow rate of 200 *ft³/s*. Find the water depth after a hydraulic jump.

11.25 Draw and label a hydraulic jump (in a channel of constant width).

(a) Indicate regions of supercritical and subcritical flow.

(b) What happens to Bernoulli's constant through the jump?

(c) Write down the relationship between the ratio of upstream and downstream water depths, and the upstream Froude number.

(d) Derive a relationship between the ratio of upstream and downstream water depths, and the downstream Froude number.

(e) A surge is moving at 5 m/s into an estuary of depth 1 m where the water is moving out to sea at 1 m/s. Find the water depth behind the surge.

11.26 A surge is moving at 10 ft/s into a depth 2 ft where the speed is zero. Find the water depth behind the surge.

11.27 A bore is moving with velocity U_b into a stagnant tidal channel of depth Y_1, as shown in Figure P11.27. When $U_b = \sqrt{3gY_1}$, find U_2/U_b.

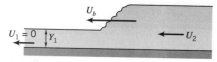

FIGURE P11.27.

11.28 A bore moving at 10 m/s moves upstream into a river, which is moving downstream at 2 m/s. The river has a constant width and a depth of 2 m. Find the speed of the river flow downstream of the bore.

11.29 A tidal bore is moving at 5 m/s into a stagnant basin of depth 1 m. What is the depth of water behind the bore?

11.30 Consider a bore travelling at 6 m/s into tidal basin of depth 1.5 m. Calculate the depth of the water behind the bore.

11.31 A surge is moving at 9 ft/s into a basin of depth 1.4 ft where the speed is zero. Find the water depth behind the surge and the drift velocity behind the surge (that is, the velocity downstream of the surge in a stationary frame of reference).

11.32 Show that the head loss in a hydraulic jump can be expressed as:

$$\text{head loss} = \frac{(Y_2 - Y_1)^3}{4Y_1Y_2}$$

where Y_1 and Y_2 are the water depths upstream and downstream of the jump, respectively. The head loss is the difference between the upstream and downstream values of the Bernoulli constant, and it has the dimensions of length.

11.33 A circular jet of water impinges on a flat plate as shown in Figure P11.33. The water spreads out equally in all directions and at station 1, the flow is essentially uniform over the depth Y_1. The jet exit velocity V is 10 m/s and the exit diameter D is 10 mm. If you are given that $R = 8D$, and $Y_1 = D/4$,

(a) Find the Froude number of the flow at station 1.

(b) If you are given that a circular hydraulic jump forms near station 1, estimate the depth of water after the hydraulic jump using the one-dimensional hydraulic jump relationship. Under what conditions will the accuracy of this estimate improve?

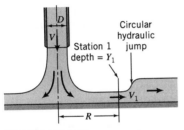

FIGURE P11.33.

11.34 A two-dimensional channel flow flows smoothly over a small bump, as shown in Figure P11.34.
(a) If the Froude number at station 2 equals unity ($F_2 = 1$), find Y_2/Y_1 in terms of F_1 and H/Y_1.
(b) If $Y_3/Y_1 = 0.5$, find F_3 in terms of F_1.
(c) If $F_3 = \sqrt{3}$, find Y_4/Y_1 such that a hydraulic jump is formed in front of the barrier.

FIGURE P11.34.

11.35 Consider the steady flow of water under the sluice gate shown in Figure P11.35. The velocity of the flow at sections 1, 2, and 3 is independent of depth, and the streamlines are parallel.
(a) Show that for $h_2 < h_1$, the flow at section 2 is always supercritical. Assume the flow between sections 1 and 2 occurs without loss.
(b) Between sections 2 and 3 a stationary hydraulic jump is formed. If $h_2/h_1 = 0.5$, find h_3/h_2.

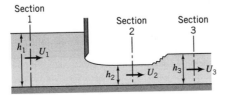

FIGURE P11.35.

11.36 Water flows in an open channel of constant width as shown in Figure P11.36. The upstream Froude number $F_1 = 0.5$. At the point where the water flows over a bump of height H, the Froude number equals one ($H = Y_1/4$).
(a) Find the depth ratio Y_2/Y_1.
(b) Given that a hydraulic jump occurs downstream of the bump such that $Y_4/Y_3 = 8$, find Y_3/Y_1.

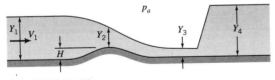

FIGURE P11.36.

11.37   A smooth transition section connects two rectangular channels. In the direction of flow, the channel width increases from B_1 to B_2 and the water surface elevation decreases so that the ratio of the downstream to upstream depths $Y_2/Y_1 = 0.5$

(a) If the upstream Froude number $F_1 = 1.5$, find the downstream Froude number F_2, and the ratio B_2/B_1.

(b) If a stationary hydraulic jump occurs in the channel downstream of the expansion, determine Y_3, the depth downstream of the jump, in terms of Y_1.

11.38   Consider the steady open channel flow of water in the smooth constriction shown in plan view in Figure P11.38. At position 1, the depth is Y_1, the width is B_1, and the Froude number is $F_1 = 1/\sqrt{2}$. A similar notation is used at positions 2, 3, and 4. A hydraulic jump is observed just downstream of position 3 where the depth $Y_3 = Y_1/4$. You may assume one-dimensional flow.

(a) Find F_3, the Froude number at position 3 (this is a number).

(b) Find B_3, the width of the channel at position 3 in terms of B_1.

(c) Find Y_4, the depth of the water at position 4, downstream of the hydraulic jump, in terms of Y_1.

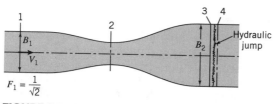

FIGURE P11.38.

11.39   Consider the steady, frictionless open channel flow of water through the constriction, shown in Figure P11.39. At position 1, the depth is Y_1, the velocity is V_1, and the Froude number is $F_1 = V_1/\sqrt{gY_1}$. A similar notation is used at position 2 (the throat), and positions 3, and 4. You may assume one-dimensional flow.

(a) If $Y_1 > Y_2 > Y_3$, is F_1 supercritical or subcritical? Also, is F_3 supercritical or subcritical?

(b) If $Y_1 < Y_2$, and $Y_3 < Y_2$, is F_1 supercritical or subcritical? Also, is F_3 supercritical or subcritical?

(c) Given that $Y_2/Y_1 = 0.75$, and that $Y_3 < Y_2$, find F_1 and B_2/B_1 (these answers are numbers).

(d) What is the smallest value of Y_2/Y_1 that can be achieved for $F_2 = 1$?

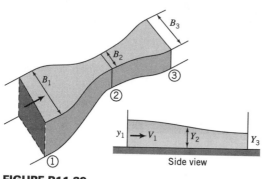

Side view

FIGURE P11.39.

11.40 Water issues from a large reservoir in the form of a jet of width W and depth Y_1, as shown in Figure P11.40. The cross-sectional area of the tank A is much larger than the jet area WY_1, and the depth H is much larger than Y_1.

(a) Calculate the exit Froude number F_1, and show that $F_1 > 1$.

(b) Show that Y_2, the depth at the crest of the bump of height h, is given by the solution to

$$\left(\frac{Y_2}{Y_1}\right)^3 - \left(1 + \frac{F_1^2}{2} - \frac{h}{Y_1}\right)\left(\frac{Y_2}{Y_1}\right)^2 + \frac{F_1^2}{2} = 0$$

Carefully note all your assumptions.

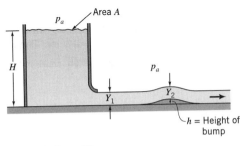

FIGURE P11.40.

11.41 Consider water flowing without friction through a contraction in an open channel. The channel and the surface profile are shown in Figure P11.41.

(a) Is the Froude number at station 1 subcritical or supercritical?

(b) Given that $Y_2/Y_1 = 2$ and $B_2/B_1 = 0.6$, find the Froude number at station 2 (this is a number).

(c) If there was a hydraulic jump downstream of station 2, what would be the height of the jump in terms of Y_1?

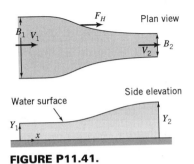

FIGURE P11.41.

11.42 For the flow described in the previous problem, find the force F_H exerted by the water on the flume. Use the continuity equation and the momentum equation to find the nondimensional force coefficient $F_H/(\rho g B_1 Y_1^2)$ in terms of the incoming Froude number and the length ratios Y_2/Y_1 and B_2/B_1.

11.43 An open channel flow of constant width W flows over a small obstruction, as shown in Figure P11.43. Show that

$$\frac{-R}{\rho U_1^2 W h_1} = \frac{h_1}{h_2} - 1 - \frac{1}{2F_1^2}\left(1 - \frac{h_2^2}{h_1^2}\right)$$

where R is the force exerted by the fluid on the obstruction, ρ is the density, and F_1 is the Froude number of the incoming flow. Neglect viscous forces on the channel floor.

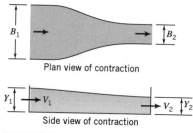

FIGURE P11.43.

11.44 For the open-channel flow in the smooth contraction shown in plan view in Figure P11.44 the depth of the incoming flow is Y_1 and the depth of the exiting flow is Y_2. The flow attains the critical Froude number at the throat. If the upstream Froude number is much less than unity, show that:

(a) $Y_2/Y_1 \approx 2/3$

(b) $\dfrac{F_v}{\frac{1}{2}\rho g Y_1^2 B_1} = 1 - \dfrac{4}{3}\dfrac{B_2}{B_1}$

where F_v is the force acting on the contraction walls, and B_1 and B_2 are the upstream and downstream channel widths, respectively. Ignore all viscous forces.

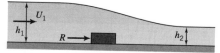

Plan view of contraction

Side view of contraction

FIGURE P11.44.

11.45 Water flows smoothly over a bump of height h in the bottom of a channel of constant width W, as shown in Figure P11.45. If the upstream depth is Y_1, the upstream Froude number F_1 is 0.7, and $Y_1/Y_2 = 2$, find:

(a) The downstream Froude number F_2. Is it supercritical?

(b) The magnitude and direction of the force exerted by the fluid on the bump, in terms of g, W, h, and Y_1.

(c) If the flow was now subcritical everywhere again find the force on the bump.

(d) If a hydraulic jump is located downstream of the bump (and F_1 is 0.7 again), calculate the Froude number of the flow downstream of the jump.

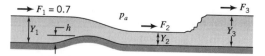

FIGURE P11.45.

11.46 Water flows steadily in an open channel of width W, as shown in Figure P11.46. It passes smoothly over a bump of height h. Initially, the Froude number $F_1 = 0.5$, and the depth of the water is Y_1. The water depth decreases to a depth of Y_2 over the bump, and then continues to decrease downstream of the bump to a depth of Y_3.

(a) What is the Froude number Y_2?

(b) Use Bernoulli's equation and continuity to find the numerical value of Y_2/Y_1, and the numerical value of h/Y_1.

(c) If $Y_3/Y_1 = 0.422$, find the numerical value of the force coefficient $C_D = D/(\rho g W Y_1^2)$ where D is the horizontal force exerted by the fluid on the bump, ρ is the water density, and g is the gravitational acceleration. Ignore friction.

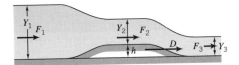

FIGURE P11.46.

11.47 Consider the steady, smooth flow of water in an open channel of constant width W, as shown in Figure P11.47. A deflector plate causes the water to accelerate to a supercritical speed. At position 1, the depth is Y_1 and the Froude number $F_1 = 0.2$. At position 2, the depth is Y_2 and the Froude number $F_2 = 3.38$. Assume one-dimensional flow.

(a) Find the ratio Y_2/Y_1 two different ways (this is a number).

(b) Find the nondimensional ratio $2R/\rho g Y_1^2 W$ where R is the force exerted on the fluid by the deflector plate, in term of Y_2/Y_1, F_1, and F_2.

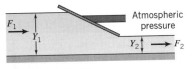

FIGURE P11.47.

11.48 Water in an open-channel flow of constant width flows steadily over a dam as shown in Figure P11.48. The velocity of the water far upstream of the dam is V, the Froude number is F, and the depth is h. Far downstream of the dam the depth is $h/4$.

(a) Find the horizontal force per unit width acting on the dam in terms of V, h, g, and the density ρ. Ignore friction.

(b) There is a hydraulic jump located even further downstream. Find the depth of water downstream of the jump in terms of F and h.

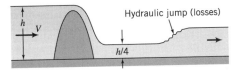

FIGURE P11.48.

11.49 Water in a two-dimensional open channel flows smoothly and steadily over a submerged obstacle of height H and width W as shown in Figure P11.49. The flow over the peak of the obstacle becomes critical at the point where the depth is Y_2, where $Y_2 = Y_1/3$.

(a) Find F_1, the Froude number of the incoming flow.

(b) Find H/Y_1, the nondimensional height of the obstacle.

(c) Find the resultant force acting on the obstacle in terms of ρ, g, W, and Y_1, given that $Y_3/Y_1 = 0.165$.

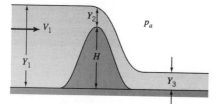

FIGURE P11.49.

CHAPTER 12

COMPRESSIBLE FLOW

12.1 INTRODUCTION

To demonstrate why the compressibility of fluids is important, consider the example of a piston in a long, straight tube filled with gas (Figure 12.1). The piston is initially at rest, and so is the gas. If the piston suddenly starts moving at a constant speed, what happens to the fluid? The fluid in contact with the piston must start to move when the piston does, but what about the fluid further down the tube? If the fluid were incompressible, the gas would behave like a solid body and the entire mass of fluid would move with the piston. As soon as the piston begins to move, all the fluid in the tube must start to move at the same speed, even the fluid far from the piston. In other words, the effect of the piston motion would travel through the gas at an infinite speed.

Real fluids do not behave like this. Real fluids are compressible. As the piston starts to move, the gas near the piston gets compressed first (the gas far away has not started to move yet, so the gas near the piston gets squeezed into a smaller volume), and then the gas a little further away, and so on. The motion of the piston propagates through the tube as a pressure wave at a finite speed.

We can identify the "front" between the compressed gas and the undisturbed gas and measure the speed at which it travels. If the pressure disturbance caused by the piston motion is small compared to the ambient pressure, this compression front travels at the local speed of sound (sound waves are just very weak pressure waves). In fact, the most common evidence of the compressibility of fluids is the propagation of sound waves—if a fluid were truly incompressible, sound waves could not travel through it. In contrast, if the pressure disturbances caused by the piston motion are not small, shock waves will appear. A shock wave is a very thin region where the velocity, pressure, temperature, and density all change significantly. For the case of a shock forming in a tube, the shock is planar, and it travels into the gas at a speed somewhere between the speed of sound in the undisturbed gas ahead of it, and the speed of sound in the compressed gas behind it.

We have all heard shock waves: the thunder that accompanies lightning, the boom that comes with an explosion, and the crack of a whip, are all examples of shock waves. We know from experience that shock waves produce rapid changes in pressure. In fact, they often produce such rapid changes that the changes are said to occur "discontinuously." For example, an explosion generates a very intense pressure and temperature rise, and the pressure disturbance travels out as a shock wave. In this case, the shock wave is spherical and it weakens as it travels outward so that the pressure jump across the shock decreases with distance. When a bullet issues from a barrel, the hot, high pressure gas that is expelled from the muzzle also generates a spherical shock. The bullet itself typically travels at a supersonic speed (that is, at a speed greater than the speed of sound in the undisturbed gas), and its motion will generate shock waves. In fact, whenever a body of finite size travels through a gas at supersonic speed, shocks will appear. For example, the

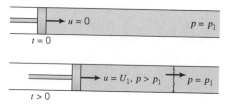

FIGURE 12.1 The motion of a piston in a tube filled with a gas produces a pressure disturbance, which travels down the tube at a finite speed.

crack of a whip is the audible evidence of shock formed by the supersonic motion of its tip, as shown in Figure 12.2.

In this chapter, we focus on high-speed gas flows where compressibility effects are important, and analyze the behavior of shocks and other wave phenomena that occur when bodies travel at supersonic speed.

12.2 PRESSURE PROPAGATION IN A MOVING FLUID

The propagation of sound waves in high-speed flow displays many similarities to the propagation characteristics of small amplitude gravity waves examined in Chapter 11. For example, imagine a body suddenly placed in a subsonic air flow [see Figure 12.3(a)]. The presence of the body creates pressure disturbances, which

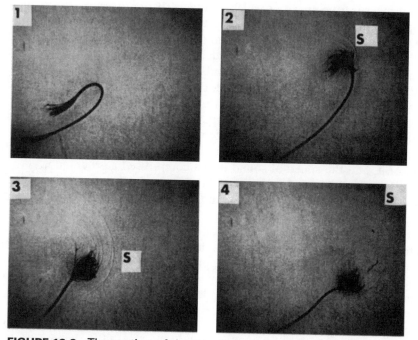

FIGURE 12.2 The motion of the tip of a bull whip is shown to produce shock waves in these four sequential pictures. The shocks are visible as thin lines near the letter S in each picture. The tip speed is 1400 *ft/s,* compared to the sound speed of 1100 *ft/s.* Photographs courtesy of the Naval Research Laboratory.

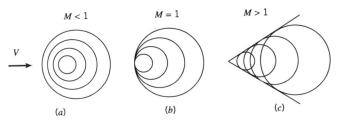

FIGURE 12.3 Wave patterns made by a point disturbance. (*a*) $U < a$ (subsonic). (*b*) $U = a$ (sonic). (*c*) $U > a$ (supersonic).

travel outward from the body at the speed of sound (as long as the pressure waves are small). It is by the transmission of pressure waves that the rest of the flow finds out about the presence of the body, in the same way the fisherman in Section 11.2 found out about the presence of a wave source through the ripples that traveled outward on the free surface of the water.

An important parameter is the Mach number M, defined as the ratio of the bulk velocity of the fluid V to the speed of sound a. That is,

$$M = \frac{V}{a}$$

In a *subsonic* flow ($M < 1$), the whole of the flowfield will be affected by the pressure waves, and this explains why the flow at some distance upstream of the body adjusts to the presence of the body: it "knows" that the body is there because pressure waves transmit information regarding its presence. For a cylinder, this distance is of the order of 10 diameters. In mathematical terms, the flowfield is "elliptic," which means that all parts of the flow are affected by all other parts since information is freely transmitted throughout the flowfield.

If the body was placed in a *sonic* flowfield ($M = 1$), the pressure waves would travel outward at the speed of sound, but they would also be swept downstream by the flow at the same speed [see Figure 12.3(*b*)]. The waves all collect along a line normal to the flow direction, and the flow upstream of the body never feels the presence of the body. The adjustment of the flow to the presence of the body is no longer gradual, but sudden.

When the flow is *supersonic* ($M > 1$), the pressure waves still travel outward at the speed of sound, but they are swept downstream at an even greater speed so that the waves are confined to a wedge-shaped region [see Figure 12.3(*c*)]. Only the flow inside this region feels the presence of the body. The angle made by the envelop of waves is α_M, where

$$\sin \alpha_M = \frac{1}{M} \tag{12.1}$$

(see Figure 12.4). The angle α_M is called the *Mach angle*, and it is the angle a weak pressure wave makes with the flow direction in a supersonic flow. The weak pressure wave is often called a *Mach wave*.

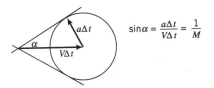

$$\sin \alpha = \frac{a \Delta t}{V \Delta t} = \frac{1}{M}$$

FIGURE 12.4 Wave patterns made at supersonic Mach numbers.

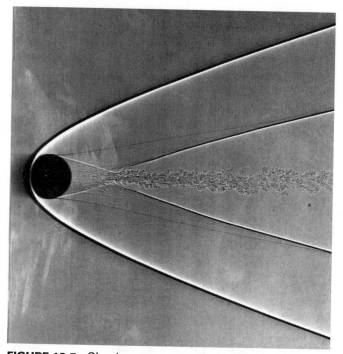

FIGURE 12.5 Shock wave patterns made by a 0.5 *in.* sphere in free flight in atmospheric air at Mach 4.01. A strong shock wave is formed upstream of the sphere. The boundary layer separation just behind the equator is accompanied by a weak shock wave, and a second shock is formed near the point where the separated shear layers meet. Photograph courtesy of A.C. Charters.

If the body has a finite size, the pressure disturbances it generates will no longer be small. At supersonic speeds, the flow cannot adjust gradually to its presence, and a sudden, steep adjustment of pressure (a shock wave) is necessary to let the flow pass over the body. Shock wave patterns made by a sphere moving at Mach 4.01 are shown in Figure 12.5. A similar phenomenon occurs in open channel flows at the critical Froude number ($F = 1$), where an obstruction of finite size forms a sudden change in water level called a hydraulic jump.

For a body in a sonic flow, the shock wave that forms in front of the body is at right angles to the flow direction, and it is called a normal shock. For a supersonic flow, the shock is inclined to the flow direction, and it is called an oblique shock wave.[1] You may have heard shock waves when an airplane flies over you at sonic or supersonic speeds. If you had your eyes closed, the first indication that an airplane was present would have been the passage of the shock wave created by the plane. Until the airplane passed over you, the sound waves generated by the plane could not have reached you because the source of the sound was moving faster than the speed of sound. If you have had this experience, you would also know that the shock wave is a large pressure disturbance and even at considerable distances from the plane, the *sonic boom* can be very unpleasant. In fact, the pressure jump can

---

[1] For very weak shocks, the angle made by an oblique shock β is equal to the Mach angle α_M. This is not true for shocks of finite strength.

be so large that the increased pressure, acting over the area of a window pane, can cause it to break. The wave field generated by a supersonic vehicle is considered further in Section 12.11.

12.3 REGIMES OF FLOW

A flow at very low Mach number can generally be assumed to be incompressible, as long as we are not concerned with the mechanics of sound propagation. As the Mach number increases, there comes a point where compressibility effects become important. At what Mach number does this occur? In Section 1.3.6, it was suggested that density changes due to variations in flow velocity can be neglected, as long as the dynamic pressure is small compared to the static pressure. We see that

$$M^2 = \frac{V^2}{a^2} = 2 \left(\frac{\frac{1}{2}\rho V^2}{\gamma p} \right)$$

so that the Mach number is related to the ratio of the dynamic pressure to the absolute pressure. If we accept a 1% level as a tolerable change in density, then if the temperature remains constant the corresponding change in pressure is 1%.[2] This requires $\frac{1}{2}\rho V^2 < 0.01$, so that at sea level where the density of air is about 1.2 kg/m^3, we are limited to velocities less than 40 m/s (132 ft/s or 90 mph), which correspond to Mach numbers less than about 0.12.

The limit for the assumption of incompressible flow is often given as $M < 0.3$. This corresponds to a maximum change in density of about 5% for an isentropic process. This may seem somewhat generous, but there are good reasons why changes in density are often considered to be less important than changes in pressure. When we consider the forces acting on a wing, for example, the pressure differences acting over the top and bottom surfaces of the wing give rise to a lift force. The pressure differences also generate density differences, which lead to a buoyancy force. This buoyancy force is almost always negligible compared to the lift force, and density differences can be ignored.

At higher Mach numbers, changes in density become more and more important, especially at transonic and supersonic Mach numbers where shocks are formed. Under certain conditions, compressibility effects can even be important at relatively low Mach numbers. Just after take-off, for example, the wings on an airplane are developing very high lift. The flow velocities near the leading edge of the wing can be very high, so that even though the aircraft may be moving at a Mach number of 0.3, the local Mach number can be supersonic and shocks can form.

Based on these considerations, we can identify three regimes of fluid flow.

- Acoustics, where fluid velocities are very small compared with the speed of sound, but the fractional changes in pressure, temperature, and density are important.
- Incompressible flow, where fluid velocities are small compared with the speed of sound, and the fractional changes in density are not significant; however, fractional changes in pressure and temperature may be very important.

---

[2] The process is more likely to be isentropic, but that does not change the argument very much.

- Compressible flow ("gasdynamics"), where fluid velocities are comparable to the speed of sound, and the fractional changes in pressure, temperature, and density are all important.

In this chapter, we will consider flows in the gasdynamics regime. Before we begin this analysis, we need to review some basic thermodynamic principles and relationships.

12.4 THERMODYNAMICS OF COMPRESSIBLE FLOW

In thermodynamics, we often speak of "systems." In fluid mechanics, we talk about "control volumes." They are identical concepts, in that systems and control volumes both describe a specific three-dimensional region bounded by a surface.[3] Mass, momentum, and energy may flow through the control volume surface, and the concept of flux can be used to describe this transport (Section 5.1).

In our earlier considerations of the energy equation, it was always assumed that the system was in "equilibrium." That is, the heat and work interactions were assumed to be slow enough for the thermodynamic state of the system to be described completely by the first law of thermodynamics. What about nonequilibrium states? In compressible flows where high speeds are common and strong velocity and pressure gradients exist, systems may not be in equilibrium. It has been found by experiment, however, that the flow still attains an instantaneous local equilibrium, as long as the temperatures and pressures are not too extreme. This continues to hold even inside shock waves. For many flows, therefore, no new nonequilibrium phenomena need to be considered. For the flows examined here, all systems will be assumed to be in equilibrium at all times.

12.4.1 Ideal Gas Relationships

The behavior of compressible flow depends critically on the properties of the gas we are considering. Many gases, including air at reasonable temperatures and pressures, behave like an "ideal" gas. That is, they obey the ideal gas law (equation 1.5),

$$p = \rho RT \quad \text{or} \quad pv = RT \quad (12.2)$$

where $v = 1/\rho$ is the specific volume. All fluids considered here are assumed to follow the ideal gas law.

It is important to remember that in all thermodynamic and compressible gas relationships, absolute values are used, so that for temperature we use the Kelvin or Rankine scales, and for pressure we always use the absolute value, never the gauge value. Also, in equation 12.2,

$$R = \frac{A}{MW} \quad (12.3)$$

[3] Basic thermodynamic concepts and "state variables" such as internal energy and enthalpy were introduced in Section 4.7.1. An excellent introduction to more general thermodynamic principles is given in *An Introduction to Thermal-Fluid Engineering* by Z. Warhaft, published by CUP, 1997.

where $A = 8,314 \ m^2/s^2K$ (or $8,314 \ J/kgK$) is the universal gas constant, and MW is the molecular weight of the gas. For air, $MW = 28.98$ and $R = 287.03 \ m^2/s^2K = 1716.4 \ ft^2/s^2R$.

12.4.1.1 Specific Heats
The specific heats of a fluid were defined in equation 4.19. For an ideal gas, the internal energy per unit mass $\hat{u}$ is a function of temperature only, and the specific heat of an ideal gas at constant volume, C_v, is then given by

$$C_v \equiv \left(\frac{\partial \hat{u}}{\partial T}\right)_v = \frac{d\hat{u}}{dT}$$

that is,

$$d\hat{u} = C_v \, dT \tag{12.4}$$

For a constant specific heat (a reasonable approximation for moderate changes in temperature),

$$\boxed{\hat{u}_2 - \hat{u}_1 = C_v(T_2 - T_1)} \tag{12.5}$$

For an ideal gas, $\hat{u}$ and p/ρ depend only on temperature. It follows that the enthalpy ($h = \hat{u} + pv$) is a function of temperature only, and the specific heat of an ideal gas at constant pressure, C_p, is then given by

$$C_p = \frac{dh}{dT}$$

that is,

$$dh = C_p \, dT \tag{12.6}$$

For a constant specific heat,

$$\boxed{h_2 - h_1 = C_p(T_2 - T_1)} \tag{12.7}$$

12.4.1.2 Entropy Variations
For compressible flows, changes in *entropy* are important. Entropy is the subject of the second law of thermodynamics. The second law can be stated in a number of different ways, none of which is easy to understand. We will use the second law only indirectly, and it is sufficient for us to regard entropy simply as another variable of state defined by

$$T \, ds = d\hat{u} + p \, dv \tag{12.8}$$

We should also understand that the change in entropy during a process is intimately connected with the concept of *reversibility*. When a process is reversible, it means that no matter what heat and work interactions occur during the process, the original state of the system can be recovered by reversing the directions of all the interactions. For an adiabatic, reversible process, the entropy remains constant, and the process is called *isentropic*. For an adiabatic, irreversible process, the second law indicates that the entropy must increase.

Using the definition of enthalpy, we can write equation 12.8 as

$$T \, ds = dh - v \, dp \tag{12.9}$$

For a *perfect* gas (that is, a gas that obeys the ideal gas law and which has constant specific heats), equation 12.8 becomes

$$ds = C_v \frac{dT}{T} + \frac{R}{v} dv \tag{12.10}$$

and equation 12.9 becomes

$$ds = C_p \frac{dT}{T} - R \frac{dp}{p} \tag{12.11}$$

These two results can be integrated to obtain

$$\boxed{s_2 - s_1 = C_v \ln \frac{T_2}{T_1} - R \ln \frac{\rho_2}{\rho_1}} \tag{12.12}$$

and

$$\boxed{s_2 - s_1 = C_p \ln \frac{T_2}{T_1} - R \ln \frac{p_2}{p_1}} \tag{12.13}$$

12.4.1.3 Specific Heat Relations

We can also derive some useful relationships between C_v and C_p. From the definition of enthalpy and the ideal gas law,

$$h = \hat{u} + RT$$

Differentiating, we obtain

$$dh = d\hat{u} + R\, dT$$

or

$$\frac{dh}{dT} = \frac{d\hat{u}}{dT} + R$$

Hence,

$$\boxed{C_p - C_v = R} \tag{12.14}$$

This relationship holds for all ideal gases, even if the specific heats vary with temperature. The specific heat ratio γ is defined by

$$\gamma = \frac{C_p}{C_v} \tag{12.15}$$

For a perfect gas, γ is a constant. Air behaves as a perfect gas over a relatively wide range of temperatures and pressures, and over this range the specific heat ratio is nearly constant and equal to 1.4 (see Tables C.1 and C.2).

Equations 12.14 and 12.15 lead to

$$C_v = \frac{R}{\gamma - 1} \tag{12.16}$$

and

$$C_p = \frac{\gamma R}{\gamma - 1} \tag{12.17}$$

When a flow is isentropic and obeys the ideal gas law, it follows that

$$\frac{p}{p_r} = \left(\frac{T}{T_r}\right)^{1/(\gamma-1)} \quad \text{and} \quad \frac{p}{p_r} = \left(\frac{T}{T_r}\right)^{\gamma/(\gamma-1)} \tag{12.18}$$

The derivation of these relationships can be found in any thermodynamics textbook.[4] The parameters T_r, ρ_r, and p_r are the values of the temperature, density, and pressure at some reference point. It is common practice to use the "stagnation conditions" as the reference conditions (see Section 12.4.3).

12.4.2 Speed of Sound

Sound waves are pressure disturbances that are small compared to the ambient pressure. For example, sound at 100 *dB,* which is a very noisy sound level,[5] corresponds to a pressure disturbance level of only 1 *Pa* (10^{-5} atmosphere).

The transmission of sound waves is an *isentropic,* compressible phenomenon, and sound waves travel at a speed given by

$$a = \sqrt{\left.\frac{\partial p}{\partial \rho}\right|_s} \tag{12.19}$$

That is, the speed of sound is determined by the rate of change of pressure with respect to density, at a constant entropy, *s.* For an ideal gas, the pressure and density in isentropic flow are related by

$$\frac{p}{\rho^\gamma} = \text{constant}$$

(see equation 12.18). By differentiating this relationship, we find that

$$\frac{dp}{p} - \gamma\frac{d\rho}{\rho} = 0$$

Therefore,

$$\left.\frac{\partial p}{\partial \rho}\right|_s = a^2 = \gamma\frac{p}{\rho}$$

For an ideal gas, therefore, the speed of sound is given by

$$a = \sqrt{\gamma RT} \tag{12.20}$$

For air at 20°C the speed of sound is 343 *m/s* = 1126 *ft/s* = 768 *mph.*[6]

---

[4] See, for instance, *An Introduction to Thermal-Fluid Engineering,* by Z. Warhaft, published by CUP, 1997.

[5] The Occupational Safety and Health Administration (OSHA) requires ear protection for people exposed to noise levels exceeding 90 dB.

[6] It takes sound about 4.7 sec to travel a mile. Since the noise made by thunder travels at approximately the isentropic speed of sound, the time delay between the thunder and the lightning flash can be used as a rough guide to determine how far away a lightning strike has occurred: after the flash, count 5 sec for every mile.

Since we can also write the bulk modulus of a fluid K in terms of the rate of change of pressure with respect to density, that is, from equation 1.4,

$$K = \frac{dp}{d\rho/\rho} = \rho \frac{dp}{d\rho}$$

(12.21)

and so

$$a = \sqrt{K/\rho}$$

For fluids, therefore, the speed of sound and the bulk modulus of the fluid are directly related. For isentropic flow of a gas, it can be shown that $K = \gamma p$ (see Section 12.5), and so

$$a = \sqrt{\gamma p/\rho}$$

(12.22)

For an ideal gas, we obtain $a = \sqrt{\gamma RT}$, as before.

12.4.3 Stagnation Quantities

For steady, adiabatic flow in a streamtube, the one-dimensional energy equation (equation 4.26) reduces to

$$h_1 + \tfrac{1}{2}V_1^2 = h_2 + \tfrac{1}{2}V_2^2$$

as long as the flow is in equilibrium at locations 1 and 2. This relationship can be written as

$$h_0 = h + \tfrac{1}{2}V^2$$

(12.23)

where the constant h_0 is the enthalpy of the fluid at a point where $V = 0$. The quantity h_0 is called the *stagnation enthalpy* or *total enthalpy*. For steady adiabatic flow, the stagnation enthalpy is constant along a streamline. For an ideal gas with constant specific heats (a perfect gas), $h = C_p T$, and

$$C_p T_0 = C_p T + \tfrac{1}{2}V^2$$

(12.24)

where T_0 is the *stagnation temperature* or *total temperature*. Using

$$C_p = \frac{\gamma R}{\gamma - 1}$$

and

$$M^2 = \frac{V^2}{\gamma RT}$$

we obtain

$$\boxed{\frac{T_0}{T} = 1 + \frac{\gamma - 1}{2} M^2}$$

(12.25)

which is a particular form of the one-dimensional energy equation for the steady, adiabatic flow of a perfect gas. Also

For steady, adiabatic flow of a perfect gas, T_0 is constant along a streamline.

Even if the flow does not obey all of these restrictions, it is still possible to compute a stagnation temperature at any point according to equation 12.25, so that this equation may be used as a definition of the stagnation or *total temperature*. For an adiabatic, steady flow of a perfect gas, we know that T_0 is constant along a streamline, but if these conditions are not satisfied no conclusions can be drawn regarding the behavior of T_0.

We can also define stagnation or reservoir conditions for the density and pressure. It is necessary to specify how the flow is brought to rest, and so we define ρ_0 and p_0 as the density and pressure of the gas if it was brought to rest isentropically. The stagnation density and pressure are constant along a streamline only if the flow itself is isentropic.

The isentropic relationships are commonly referred to stagnation conditions, so that

$$\frac{\rho}{\rho_0} = \left(\frac{T}{T_0}\right)^{1/(\gamma-1)}$$

and

$$\frac{p}{p_0} = \left(\frac{T}{T_0}\right)^{\gamma/(\gamma-1)} \tag{12.26}$$

where ρ_0 and p_0 are the density and pressure of the gas if it was brought to rest isentropically. They are called the stagnation or total density and pressure. Using equation 12.25, we find that for isentropic flow

$$\boxed{\frac{\rho_0}{\rho} = \left(1 + \frac{\gamma-1}{2}M^2\right)^{1/(\gamma-1)}} \tag{12.27}$$

and

$$\boxed{\frac{p_0}{p} = \left(1 + \frac{\gamma-1}{2}M^2\right)^{\gamma/(\gamma-1)}} \tag{12.28}$$

These functions are tabulated in Table A-C.10 for $\gamma = 1.4$. Solutions to the isentropic flow relations may also be found using the compressible flow calculator available on the Web at *http: //www.engapplets.vt.edu/*.

EXAMPLE 12.1 *Thermodynamic Properties*

When a fixed mass of air is heated from 20°C to 100°C:
(*a*) What is the change in enthalpy?
(*b*) For a process at constant volume, what is the change in entropy?
(*c*) What is the change in entropy for a process at constant pressure?
(*d*) For an isentropic process, find the changes in density and pressure.
(*e*) Compare the isentropic speed of sound in air to its isothermal value.
Assume that air behaves as a perfect gas.

Solution: For part (*a*), we have $C_p = 1004$ J/kg·K (Table C.1). From equation 12.7

$$h_2 - h_1 = C_p(T_2 - T_1) = 1004(100 - 20) \text{ J/kg} = 80{,}320 \text{ J/kg}$$

For part (*b*), we use equation 12.12. Since the process is at constant volume, $\rho_2 = \rho_1$, and

$$s_2 - s_1 = C_v \ln\left(\frac{T_2}{T_1}\right) = \frac{C_p}{\gamma} \ln\left(\frac{T_2}{T_1}\right)$$

Therefore,

$$s_2 - s_1 = \frac{1004}{1.4} \ln\left(\frac{100 + 273.15}{20 + 273.15}\right) J/kg \cdot K = 173 \, J/kg \cdot K$$

For part (c), we use equation 12.13. Since the process is at constant pressure, $p_2 = p_1$, and

$$s_2 - s_1 = C_p \ln\frac{T_2}{T_1} = 1004 \ln\left(\frac{100 + 273.15}{20 + 273.15}\right) J/kg \cdot K = 242.3 \, J/kg \cdot K$$

For part (d), we use $20°C$ as the reference temperature in the isentropic relationships (equation 12.18). With $\gamma = 1.4$, we obtain

$$\frac{\rho_{100}}{\rho_{20}} = \left(\frac{T_{100}}{T_{20}}\right)^{2.5} = \left(\frac{100 + 273.15}{20 + 273.15}\right)^{2.5} = 1.828$$

and

$$\frac{p_{100}}{p_{20}} = \left(\frac{T_{100}}{T_{20}}\right)^{3.5} = \left(\frac{100 + 273.15}{20 + 273.15}\right)^{3.5} = 2.327$$

For part (e), the isentropic speed of sound is given by equation 12.20

$$a_s = \sqrt{\gamma RT} = \sqrt{1.4 \times 287.03 \times (20 + 273.15)} \, m/s = 343.2 \, m/s$$

The isothermal speed is given by

$$a = \sqrt{\left.\frac{\partial p}{\partial \rho}\right|_T}$$

For an ideal gas at constant temperature

$$\frac{p}{\rho} = RT = \text{constant}$$

By differentiating this relationship, we find that

$$\frac{dp}{p} - \frac{d\rho}{\rho} = 0$$

Therefore,

$$\left.\frac{\partial p}{\partial \rho}\right|_T = a_T^2 = \frac{p}{\rho} = RT$$

That is,

$$a_T = \sqrt{RT} = \sqrt{287.03 \times (20 + 273.15)} \, m/s = 290.07 \, m/s$$

When Newton attempted to compute the speed of sound, he wrongly assumed that the transmission of sound was an isothermal, rather than an isentropic, phenomenon. We see that for air this error leads to an estimate that is about 18% too low. ∎

12.5 COMPRESSIBLE FLOW THROUGH A NOZZLE

We have seen how water passing through a constriction and expansion in an open channel can become supercritical (Chapter 11). Similarly, when a gas passes through a converging and diverging duct, a supersonic flow can be produced. The converging and diverging duct is called a *nozzle,* but the principles of its operation are very similar to what we have already discussed with respect to open channel flow. In a supersonic wind tunnel, for example, a subsonic flow accelerates as it passes through the converging part of the nozzle, and if the downstream pressure is low enough, the flow becomes sonic at the throat and expands to supersonic Mach numbers in the diverging part of the nozzle (see Figure 12.6). The nozzle can then be attached to a test section where experiments in supersonic flow can be performed. When the downstream pressure is not low enough, shock waves can appear in the expansion, in the same way hydraulic jumps can appear in the diverging part of a channel flow. If the downstream pressure is too high, the whole flow will become subsonic and the tunnel is said to *choke* (see Section 12.5.3).

We will now consider the variations in pressure, temperature, density, and velocity experienced by a compressible gas flowing isentropically through a nozzle similar to that shown in Figure 12.6 (often called a Laval nozzle, after the Swedish engineer Carl Gustar Patrick de Laval, 1854–1912, who invented the convergent-divergent nozzle for steam turbine applications in 1888).

12.5.1 Isentropic Flow Analysis

For isentropic flow, there is no heat transfer and all processes are reversible. No shocks are allowed. The flow is steady, and quasi-one-dimensional. We assume a perfect gas. By differentiating equation 12.18, we obtain

$$\frac{d\rho}{\rho} = \frac{1}{\gamma - 1}\frac{dT}{T}$$

and

$$\frac{dp}{p} = \frac{\gamma}{\gamma - 1}\frac{dT}{T} \tag{12.29}$$

The continuity equation gives

$$\frac{d\rho}{\rho} + \frac{dA}{A} + \frac{dV}{V} = 0 \tag{12.30}$$

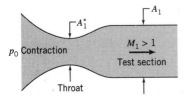

FIGURE 12.6 Isentropic, compressible flow through a converging and di
ing duct.

(see Section 6.5). We also have the one-dimensional Euler equation (equation 6.24)

$$dp + \rho V\, dV = 0 \tag{12.31}$$

which applies to variable density flows. The effects of gravity have been neglected. For isentropic (inviscid, no heat transfer) flow, equation 12.31 can be written as

$$V\, dV = -\frac{dp}{\rho} = -\frac{dp}{d\rho}\frac{d\rho}{\rho} = -a^2 \frac{d\rho}{\rho} \tag{12.32}$$

since the speed of sound (squared) is equal to the rate of change of pressure with density at constant entropy (equation 12.19). Introducing the Mach number (equation 1.7) and using the continuity equation (equation 12.30), we obtain

$$\frac{dV}{dx} = -\frac{\frac{V}{A}}{1 - M^2}\frac{dA}{dx}$$

In addition, by using the isentropic relationships, we find that

$$\frac{d\rho}{dx} = \frac{\frac{\rho}{A}M^2}{1 - M^2}\frac{dA}{dx}$$

$$\frac{dp}{dx} = \frac{\frac{p}{A}\gamma M^2}{1 - M^2}\frac{dA}{dx}$$

and

$$\frac{dT}{dx} = \frac{\frac{p}{A}(\gamma - 1)M^2}{1 - M^2}\frac{dA}{dx}$$

If we avoid the case where $M = 1$, and consider the cases where $M < 1$ everywhere, or $M > 1$ everywhere:

1. When $M < 1$ everywhere, then with
   (a) $\dfrac{dA}{dx} < 0, \dfrac{dp}{dx} < 0, \dfrac{d\rho}{dx} < 0, \dfrac{dT}{dx} < 0$, and $\dfrac{dV}{dx} > 0$
   (b) $\dfrac{dA}{dx} > 0, \dfrac{dp}{dx} > 0, \dfrac{d\rho}{dx} > 0, \dfrac{dT}{dx} > 0$, and $\dfrac{dV}{dx} < 0$

2. When $M > 1$ everywhere, then with
   (a) $\dfrac{dA}{dx} < 0, \dfrac{dp}{dx} > 0, \dfrac{d\rho}{dx} > 0, \dfrac{dT}{dx} > 0$, and $\dfrac{dV}{dx} < 0$
   (b) $\dfrac{dA}{dx} > 0, \dfrac{dp}{dx} < 0, \dfrac{d\rho}{dx} < 0, \dfrac{dT}{dx} < 0$, and $\dfrac{dV}{dx} > 0$

These results are summarized in Figure 12.7. The similarity with respect to the frictionless flow of a liquid in an open channel is clear by comparing Figure 11.14 with Figure 12.7.

We see that when the Mach number is subsonic everywhere, the pressure, density, and temperature drop as the flow speeds up in the converging part of the nozzle, and they rise as the flow slows down in the diverging part of the nozzle. When the Mach number is supersonic everywhere, the pressure, density, and temperature rise as the flow slows down in the converging part of the nozzle, and they fall as it speeds up in the diverging part of the nozzle.

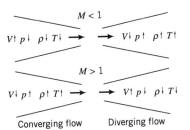

$M < 1$

$V\downarrow$ $p\downarrow$ $\rho\downarrow$ $T\downarrow$ $\rightarrow$ $\rightarrow$ $V\downarrow$ $p\uparrow$ $\rho\uparrow$ $T\uparrow$

$M > 1$

$V\downarrow$ $p\uparrow$ $\rho\uparrow$ $T\uparrow$ $\rightarrow$ $\rightarrow$ $V\uparrow$ $p\downarrow$ $\rho\downarrow$ $T\downarrow$

Converging flow   Diverging flow

FIGURE 12.7 Summary of flow variation in subsonic and supersonic ducts.

In particular, for supersonic flow downstream of the throat (Case 2b above), the velocity increases and the temperature decreases as the area continues to expand. The Mach number is therefore set by the amount the area increases downstream of the throat. That is, the Mach number depends on the area ratio A/A^*, where A^* is the cross-sectional area of the nozzle throat. To obtain a Mach number of 8, for example, an area ratio of about 200 is required (see also Section 12.5.2). Note that the flow in a subsonic diffuser (a diffuser is a duct of increasing area) has a falling velocity and a rising pressure, but the flow in a supersonic diffuser has a rising velocity and a falling pressure.

The downstream pressure level defines a series of flow regimes, much as the downstream water level did in the case of open channel flow through a smooth constriction.

1. Only two solutions exist with $M = 1$ at the throat when there are no losses (Figure 12.8, cases c and j).

2. Normal shocks are found in the nozzle for exit pressures in the range marked by points c and f.

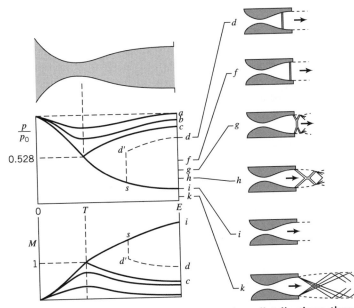

FIGURE 12.8 Pressure and Mach number distributions through a converging and diverging nozzle. From Liepmann and Roshko, *Elements of Gasdynamics*, John Wiley & Sons Inc., 1957.

3. Oblique shocks form outside the nozzle for the range between points f and j.

4. Oblique expansion waves form outside the nozzle for exit pressures below point j.

Shocks and expansion waves are considered further in Sections 12.6 to 12.10.

12.5.2 Area Ratio

We saw that the Mach number downstream of the throat depends on the ratio of the cross-sectional area of the nozzle to the cross-sectional area of the throat. To find this relationship, we write the continuity equation between the throat and a location in the nozzle

$$\rho UA = \rho^* U^* A^*$$

where the asterisk denotes the throat location. We will only consider flows where supersonic flow exists in the test section. At the throat, therefore, $M = 1$, and $U^* = a^*$. That is,

$$\frac{A}{A^*} = \frac{\rho^*}{\rho}\frac{a^*}{U} = \frac{\rho^*}{\rho_0}\frac{\rho_0}{\rho}\frac{a^*}{U}$$

By using the isentropic relations, we can show that

$$\left(\frac{A}{A^*}\right)^2 = \frac{1}{M^2}\left[\frac{2}{\gamma+1}\left(1+\frac{\gamma-1}{2}M^2\right)\right]^{(\gamma+1)/(\gamma-1)}$$

(12.33)

This is the area relation for isentropic, supersonic flow in a nozzle.

12.5.3 Choked Flow

What is the mass flow through the nozzle? For steady, one-dimensional nozzle flow, the mass flow rate $\dot{m}$ is given by

$$\dot{m} = \rho UA = \text{constant}$$

The energy equation gives

$$C_p T_0 = C_p T + \tfrac{1}{2}U^2$$

That is,

$$\tfrac{1}{2}U^2 = C_p(T_0 - T)$$

$$= C_p T_0\left(1 - \frac{T}{T_0}\right)$$

For isentropic flow this becomes

$$\tfrac{1}{2}U^2 = C_p T_0\left[1 - \left(\frac{p}{p_0}\right)^{\gamma-1/\gamma}\right]$$

Hence,

$$\dot{m} = \rho UA = \rho_0\left(\frac{p}{p_0}\right)^{1/\gamma}A\left[2C_p T_0\left(1 - \left(\frac{p}{p_0}\right)^{(\gamma-1)/\gamma}\right)\right]^{1/2}$$

so that

$$\dot{m} = A \left[\frac{2\gamma}{\gamma - 1} p_0 \rho_0 \left(\left(\frac{p}{p_0} \right)^{2/\gamma} - \left(\frac{p}{p_0} \right)^{(\gamma+1)/\gamma} \right) \right]^{1/2}$$ (12.34)

This curve is plotted in Figure 12.9 for a flow that exits to a pressure p_b, called the *back pressure*. We see that the mass flow rate curve has a maximum. To find the pressure ratio at which the maximum mass flow rate occurs, we differentiate equation 12.34 with respect to the pressure ratio p/p_0

$$\left. \frac{\partial \dot{m}}{\partial (p/p_0)} \right|_A = \frac{A^2}{2\dot{m}} \frac{2\gamma}{\gamma - 1} p_0 \rho_0 \left[\frac{2}{\gamma} \left(\frac{p}{p_0} \right)^{2/\gamma - 1} - \frac{\gamma + 1}{\gamma} \left(\frac{p}{p_0} \right)^{(\gamma+1)/\gamma - 1} \right]$$

This derivative is zero when

$$\frac{2}{\gamma} \left(\frac{p}{p_0} \right)^{(2-\gamma)/\gamma} = \frac{\gamma + 1}{\gamma} \left(\frac{p}{p_0} \right)^{1/\gamma}$$

That is, when

$$\left(\frac{p}{p_0} \right)^{(2-\gamma-1)/\gamma} = \frac{\gamma + 1}{\gamma}$$

or

$$\frac{p}{p_0} = \left(\frac{\gamma + 1}{2} \right)^{\gamma/(1-\gamma)} = 0.528 \quad \text{for air}$$ (12.35)

We can now find where this pressure ratio occurs in the nozzle. For isentropic flow, we have

$$\frac{p_0}{p} = \left(1 + \frac{\gamma - 1}{2} M^2 \right)^{\gamma/(\gamma-1)}$$

Substituting the pressure ratio at which the maximum mass flow rate occurs from equation 12.35, we find that this critical pressure ratio occurs when $M = 1$, that is, it occurs at the throat.

Since the maximum mass flow rate occurs when the Mach number at the throat is sonic, it cannot be affected by the pressure distribution downstream of the throat: the pressure changes cannot propagate upstream of the point where the flow is sonic. As a result, lowering the back pressure cannot increase the mass flow

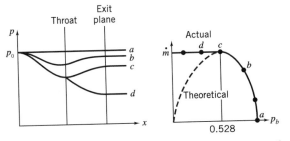

FIGURE 12.9 Variation of mass flow rate as a function of pressure ratio, for steady, isentropic, quasi-one-dimensional flow in a converging and diverging duct. The "theoretical" curve corresponds to equation 12.34. The "actual" curve corresponds to choked flow.

rate. Once the flow at the throat is sonic, the nozzle is said to be "choked." At this point, the mass flow rate has reached its maximum value and it remains fixed at that value even when the downstream back pressure is lowered. Therefore, equation 12.34 does not apply at pressure ratios below the critical value given by equation 12.35. This fact is indicated in Figure 12.9 by the "actual" curve (choked flow) compared to the "theoretical" curve (equation 12.35).

12.6 NORMAL SHOCKS

So far, we have qualitatively discussed the formation of shock waves in compressible flow. To understand this phenomenon more quantitatively, we will now analyze a one-dimensional flow containing a stationary normal shock. In this framework, the flow is steady. The gas is assumed to be "perfect" so that it obeys the ideal gas law and its specific heats are constant.

Consider the control volume shown in Figure 12.10. Stations 1 and 2 are placed well outside the shock so that all gradients are zero. That is, the velocity, pressure, and temperature are all constant at these locations, and the one-dimensional equations of motion apply between stations 1 and 2. With no temperature gradients at the control volume boundaries, the flow is taken to be adiabatic, so that the total temperature is constant (see Section 12.4.3).

> For steady, adiabatic flow of a perfect gas, T_0 is constant across a shock wave.

The flow is not reversible, however, because the entropy is expected to change. The equations describing this flow are

Continuity: $\qquad \rho_1 U_1 = \rho_2 U_2$

Momentum: $\qquad p_1 + \rho_1 U_1^2 = p_2 + \rho_2 U_2^2$

Energy: $\qquad C_p T_1 + \frac{1}{2} U_1^2 = C_p T_2 + \frac{1}{2} U_2^2 = C_p T_0 = \text{constant}$

Second law: $\qquad s_2 - s_1 = C_p \ln (T_2/T_1) - R \ln (p_2/p_1)$

Equation of state: $\qquad p = \rho R T$

We have five equations and five unknowns (ρ, u, p, s, T), which implies that for any given upstream state, a unique downstream state exists.

By using these equations in different ways, a number of results can be obtained, as shown below. It should be emphasized that all these results stem from the one-dimensional equations of motion, and no new phenomena are introduced. We simply manipulate the equations to obtain some useful results called the normal shock relations.

12.6.1 Temperature Ratio

The temperature ratio across the shock can be expressed in terms of the upstream and downstream Mach number and the ratio of specific heats, as follows.

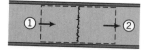

FIGURE 12.10 A normal shock in one-dimensional flow.

$$\frac{T_2}{T_1} = \frac{T_2}{T_{02}} \cdot \frac{T_{02}}{T_{01}} \cdot \frac{T_{01}}{T_1}$$

Since $T_{01} = T_{02}$ across a shock, we use the definition of the stagnation temperature to obtain

$$\frac{T_2}{T_1} = \frac{1 + \frac{\gamma-1}{2} M_1^2}{1 + \frac{\gamma-1}{2} M_2^2} \tag{12.36}$$

Velocity Ratio   From the definition of the Mach number

$$\frac{U_2}{U_1} = \frac{M_2 a_2}{M_1 a_1} = \frac{M_2 \sqrt{\gamma R T_2}}{M_1 \sqrt{\gamma R T_1}}$$

Using equation 12.36,

$$\frac{U_2}{U_1} = \frac{M_2}{M_1} \left[\frac{1 + \frac{\gamma-1}{2} M_1^2}{1 + \frac{\gamma-1}{2} M_2^2} \right]^{1/2} \tag{12.37}$$

12.6.2 Density Ratio

From the continuity equation,

$$\frac{\rho_2}{\rho_1} = \frac{U_1}{U_2} = \frac{M_1}{M_2} \left[\frac{1 + \frac{\gamma-1}{2} M_2^2}{1 + \frac{\gamma-1}{2} M_1^2} \right]^{1/2} \tag{12.38}$$

Pressure Ratio   From the momentum equation and the ideal gas law,

$$p_1 \left(1 + \frac{U_1^2}{R T_1} \right) = p_2 \left(1 + \frac{U_2^2}{R T_2} \right)$$

and since

$$\frac{U^2}{RT} = \frac{M^2 a^2}{RT} = \frac{M^2 \gamma R T}{RT} = M^2 \gamma$$

then

$$\frac{p_2}{p_1} = \frac{1 + \gamma M_1^2}{1 + \gamma M_2^2} \tag{12.39}$$

12.6.3 Mach Number Ratio

From the equation of state

$$\frac{p_2}{p_1} = \frac{\rho_2 T_2}{\rho_1 T_1}$$

Using equations 12.36, 12.38, and 12.39, this relationship can be expressed in terms of M_1 and M_2. Two solutions can be found

$$M_2 = M_1 \tag{12.40}$$

or

$$M_2^2 = \frac{M_1^2 + \frac{2}{\gamma - 1}}{\frac{2\gamma}{\gamma - 1} M_1^2 - 1} \tag{12.41}$$

The first solution implies a shock of zero strength, since we see from equation 12.39 that the pressure does not change across the shock. The second solution indicates that, for $M_1 > 1$, it is possible to have a shock of finite strength in a one-dimensional steady flow. It can also be shown that the downstream Mach number M_2 will always be subsonic. Therefore

A normal shock can occur only if $M_1 > 1$. As a result, $M_2 < 1$.

Substituting for M_2 in equations 12.38 and 12.39, we obtain

$$\boxed{\frac{\rho_2}{\rho_1} = \frac{U_1}{U_2} = \frac{(\gamma + 1)M_1^2}{(\gamma - 1)M_1^2 + 2}} \tag{12.38a}$$

and

$$\boxed{\frac{p_2}{p_1} = 1 + \frac{2\gamma}{\gamma + 1}(M_1^2 - 1)} \tag{12.39a}$$

12.6.4 Stagnation Pressure Ratio

The stagnation pressure ratio across the shock can be found as follows. Using the definitions of the total temperature and pressure (equations 12.25 and 12.28), we have

$$\frac{p_{02}}{p_{01}} = \frac{p_{02}}{p_2} \cdot \frac{p_1}{p_{01}} \cdot \frac{p_2}{p_1} = \frac{p_2}{p_1} \left(\frac{T_{02}}{T_2}\right)^{\gamma/(\gamma-1)} \left(\frac{T_1}{T_{01}}\right)^{\gamma/(\gamma-1)}$$

Since $T_{02} = T_{01}$ across the shock,

$$\frac{p_{02}}{p_1} = \frac{p_2}{p_1} \left(\frac{T_1}{T_2}\right)^{\gamma/(\gamma-1)} \tag{12.42}$$

Alternatively,

$$\frac{p_{02}}{p_{01}} = \frac{p_{02}}{p_2} \cdot \frac{p_2}{p_1} \cdot \frac{p_1}{p_{01}}$$

so that

$$\frac{p_{02}}{p_{01}} = \left(1 + \frac{\gamma - 1}{2} M_2^2\right)^{\gamma/(\gamma-1)} \left(\frac{1 + \gamma M_1^2}{1 + \gamma M_2^2}\right) \left(1 + \frac{\gamma - 1}{2} M_1^2\right)^{-\gamma/(\gamma-1)}$$

Using equation 12.41 to eliminate M_2

$$\frac{p_{02}}{p_{01}} = \left[\frac{\frac{\gamma+1}{2} M_1^2}{1 + \frac{\gamma-1}{2} M_1^2}\right]^{\gamma/(\gamma-1)} \left[\frac{2\gamma}{\gamma+1} M_1^2 - \frac{\gamma-1}{\gamma+1}\right]^{-1/(\gamma-1)} \tag{12.43}$$

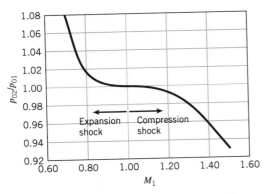

FIGURE 12.11 Stagnation pressure ratio as a function of Mach number, according to equation 12.43.

Equation 12.43 is plotted in Figure 12.11. This figure is very similar to that given for the height ratio across a planar hydraulic jump in Figure 11.10. The interpretation of these results requires a consideration of the entropy changes that occur.

12.6.5 Entropy Changes

Figure 12.11 indicates that for a shock with $M_1 > 1$ (a "compression" shock), $p_{02} < p_{01}$, and for a shock with $M_1 < 1$ (an "expansion" shock), $p_{02} > p_{01}$. We will now use the second law of thermodynamics to show that only compression shocks can occur.[7]

The entropy change across the shock is given by equation 12.13:

$$s_2 - s_1 = C_p \ln \frac{T_2}{T_1} - R \ln \frac{p_2}{p_1}$$

Since $R = C_p (\gamma - 1)/\gamma$,

$$s_2 - s_1 = C_p \ln \left[\frac{T_2/T_1}{(p_2/p_1)^{(\gamma-1)/\gamma}} \right]$$

$$= C_p \ln \left[\frac{\frac{T_2}{T_{02}} \cdot \frac{T_{02}}{T_{01}} \cdot \frac{T_{01}}{T_1}}{\left(\frac{p_2}{p_{02}} \cdot \frac{p_{02}}{p_{01}} \cdot \frac{p_{01}}{p_1} \right)^{(\gamma-1)/\gamma}} \right]$$

By using the definitions of the stagnation temperature (equation 12.25) and stagnation pressure (equation 12.28), this relationship can be simplified to give a very elegant result

$$s_2 - s_1 = C_p \ln \left[\left(\frac{p_{01}}{p_{02}} \right)^{(\gamma-1)/\gamma} \right] = R \ln \frac{p_{01}}{p_{02}} \tag{12.44}$$

Therefore the entropy change is related to the change in stagnation pressure. Since the entropy can only increase across a shock, we see that $p_{02} < p_{01}$, and Figure 12.11 indicates that M_1 must be supersonic. If p_{02} were greater than p_{01}, the entropy would decrease across a shock, which is impossible.

---

[7] This is true for all gases obeying the ideal gas law.

12.6.6 Summary: Normal Shocks

The normal shock relations were derived directly from the equations of motion for one-dimensional flow. Nevertheless, some of the relations are quite complicated, and for convenience they are often presented in tabulated form (see Table C.11). Since we are most commonly interested in air flows, the tables are given only for $\gamma = 1.4$. The compressible flow calculator available on the Web at *http://www.engapplets.vt.edu/*, can also be used.

In many problems the initial conditions are known, but we need to find the downstream conditions. In that case, equation 12.41 represents the starting point of the solution. Once M_2 is found using equation 12.41, all the other downstream variables can be determined using the normal shock relations. Alternatively, we can use equation 12.41 to express the normal shock relations in terms of M_1

$$\frac{\rho_2}{\rho_1} = \frac{U_1}{U_2} = \frac{(\gamma+1)M_1^2}{(\gamma-1)M_1^2+2} \tag{12.45}$$

and

$$\frac{p_2}{p_1} = 1 + \frac{2\gamma}{\gamma+1}(M_1^2-1) \tag{12.46}$$

The temperature rise is most conveniently found using

$$\frac{T_2}{T_1} = \frac{p_2}{p_1}\frac{\rho_1}{\rho_2} \tag{12.47}$$

and the stagnation pressure ratio can be expressed as

$$\frac{p_{02}}{p_{01}} = \frac{p_2}{p_1}\left(\frac{T_1}{T_2}\right)^{\gamma/(\gamma-1)} \tag{12.48}$$

The entropy rise is given by

$$\frac{s_2-s_1}{R} = \ln\left(\frac{p_{01}}{p_{02}}\right) \tag{12.49}$$

EXAMPLE 12.2   *Normal Shock Relations*

A normal shock is observed in an air flow at Mach 3 at $100°K$. The density of the air is $0.8\ kg/m^3$. Find
(*a*) The stagnation temperature.
(*b*) The density, pressure, temperature, and Mach number downstrem of the shock.
(*c*) The entropy rise across the shock.
In the solution, we will use the normal shock relations given here. You should then check the results using the compressible flow calculator available on the web at *http://www.engapplets.vt.edu/*.

Solution: For part (*a*), equation 12.25 (with $\gamma = 1.4$)

$$T_0 = 100 \times (1 + 0.2 \times 3^2) = 100 \times 2.8°K = 280°K$$

For part (*b*), we use equations 12.41, and 12.45 to 12.47. Hence,

$$M_2 = \frac{3^2 + 5}{7 \times 3^2 - 1} = 0.226$$

$$\rho_2 = \frac{0.8 \times 2.4 \times 3^2}{0.4 \times 3^2 + 2} \, kg/m^3 = 3.086 \, kg/m^3$$

and

$$p_2 = \left[1 + \frac{2.8}{2.4}(3^2 - 1)\right] p_1 = 10.33 p_1$$

From the ideal gas law, $p_1 = \rho_1 R T_1 = 0.8 \times 287.03 \times 100 \, Pa = 22,962 \, Pa$, and so

$$p_2 = 237,202 \, Pa$$

Finally,

$$T_2 = \frac{p_2}{p_1}\frac{\rho_1}{\rho_2} = 10.33 \times \frac{0.8}{3.086} \times 100°K = 267.8°K$$

For part (c), we use equations 12.48 and 12.49. Hence,

$$\frac{p_{02}}{p_{01}} = 10.33 \times \left(\frac{100}{267.8}\right)^{3.5} = 0.329$$

and

$$s_2 - s_1 = 287.03 \times \ln\left(\frac{1}{0.329}\right) J/kg \cdot K = 319.4 \, J/kg \cdot K \qquad \blacksquare$$

EXAMPLE 12.3 *Unsteady Shock Motion*

Consider a constant area duct with a normal shock wave moving through it (Figure 12.12). This situation occurs in a *shock tube* where two gases at different pressures are initially separated by a thin diaphragm. When the diaphragm is broken, the high pressure gas propagates into the low pressure gas and this pressure "wave" rapidly forms into a moving shock. This process is very similar to that which generates a moving hydraulic jump (see Section 11.8).

The pressure, temperature, and velocity upstream of the shock are $p_1 = 75$ kPa, $T_1 = 20°C$, and $V_1 = 0 \, m/s$. Downstream of the shock the pressure, temperature and velocity are $p_2 = 180 \, kPa$, $T_2 = 97°C$, and $V_2 = 280 \, m/s$.

Find the shock speed, and the Mach number of the upstream flow relative to an observer moving with the shock.

Solution: To a stationary observer, the flow is not steady: first, nothing is happening, then a shock moves past, followed by a steady flow of gas. If we move with the shock, however, the flow becomes steady. For a control volume moving with the shock, the steady one-dimensional continuity equation gives

$$-\rho_1 V_s A + (V_s - V_2)\rho_2 A = 0$$

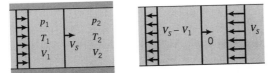

FIGURE 12.12 Moving normal shock in a constant area duct. (*a*) In a stationary frame of reference. (*b*) In a frame of reference moving with the shock.

where V_s is the speed of the shock relative to a stationary observer. Using the ideal gas law (equation 1.5)

$$-\frac{p_1}{p_2}\frac{T_2}{T_1}V_s + V_s - V_2 = 0$$

and so

$$-\frac{80}{180}\frac{380}{293}V_s + V_s - 280 = 0$$

where V_s is in meters per sec. Hence:

$$V_s = 661 \; m/s$$

From Section 12.4.2, $a = \sqrt{\gamma RT}$, so that, relative to the moving shock, the upstream flow has a Mach number of

$$M = \frac{V_s}{\sqrt{\gamma RT_1}} = \frac{661}{\sqrt{1.4 \times 287.03293}} = 1.93$$

∎

12.7 WEAK NORMAL SHOCKS

The pressure ratio across a normal shock is given by equation 12.46. This can be written in terms of the pressure jump Δp, where $\Delta p = p_2 - p_1$, so that

$$\frac{\Delta p}{p_1} = \frac{2\gamma}{\gamma + 1}(M_1^2 - 1) \tag{12.50}$$

The ratio $\Delta p/p_1$ is often called the *shock strength*. When the shock is weak, an interesting relationship may be obtained between the shock strength and the change in entropy. Weak shocks are found when M_1 is close to one, and so the parameter m, defined by $m = M_1^2 - 1$, is small compared to one. This allows a series expansion of the equation 12.49 in the small quantity m (using equations 12.46, 12.47, and 12.48). The final result is[8]

$$\frac{s_2 - s_1}{R} \approx \frac{\gamma + 1}{12\gamma^2}\left(\frac{\Delta p}{p_1}\right)^3 \tag{12.51}$$

For the small pressure change that accompanies a weak normal shock, the entropy rise is therefore very small. A weak shock produces a nearly isentropic change of state. In the limit, a very weak shock becomes an isentropic acoustic wave, that is, a Mach wave (see Section 12.2).

12.8 OBLIQUE SHOCKS

What happens in a supersonic flow that is not one-dimensional? For example, what happens when a supersonic flow is deflected through an angle α? From experiment,

---

[8] For details, see Liepmann and Roshko, *Elements of Gasdynamics,* published by John Wiley & Sons, 1957.

FIGURE 12.13 A cone-cylinder body is shot through still air at Mach 1.84. Oblique shocks are formed as the flow deflects to pass over the body, and expansion fans are formed at the shoulder, and at the start of the wake. A turbulent boundary layer is visible over the main part of the body. Photograph courtesy of A. C. Charters.

we know that an oblique shock forms, as illustrated in Figure 12.13 for a cone-cylinder body.

The flow through a plane oblique shock may be examined using the velocity diagrams shown in Figure 12.14. We will assume that the flows upstream and downstream of the shock are uniform, and that all changes occur discontinuously across the shock.

The incoming velocity V_1 can be resolved into a component normal to the shock, V_{1n}, and a component parallel to the shock, V_{1p}. Similarly, we resolve the downstream velocity V_2 into its components V_{2n} and V_{2p}. The angle between V_1 and the shock wave is β, and the deflection of the flow is α.

Conservation of mass across the shock gives

$$\rho_1 V_{1n} = \rho_2 V_{2n} \tag{12.52}$$

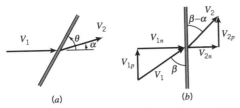

FIGURE 12.14 (*a*) Flow through an oblique shock. (*b*) Resolution of velocity components.

Conservation of momentum normal and parallel to the shock gives

$$p_1 + \rho_1 V_{1n}^2 = p_2 + \rho_2 V_{2n}^2 \tag{12.53}$$

and

$$\rho_1 V_{1n} V_{1p} = \rho_2 V_{2n} V_{2p} \tag{12.54}$$

From equations 12.52 and 12.54 we see that

$$V_{1p} = V_{2p} \tag{12.55}$$

This is a very useful result since it indicates that

> Any velocity change introduced by the oblique shock depends solely on the component normal to it, and the component parallel to the shock is unchanged.

12.8.1 Oblique Shock Relations

The conditions before and after the shock can now be found using the normal shock relations with the normal Mach number $M_1 \sin \beta$ replacing M_1. We obtain the following oblique shock relations

$$\frac{\rho_2}{\rho_1} = \frac{V_{1n}}{V_{2n}} = \frac{(\gamma + 1) M_1^2 \sin^2 \beta}{(\gamma - 1) M_1^2 \sin^2 \beta + 2} \tag{12.56}$$

and

$$\frac{p_2}{p_1} = 1 + \frac{2\gamma}{\gamma + 1} (M_1^2 \sin^2 \beta - 1) \tag{12.57}$$

The change in Mach number may be found by replacing M_2 in the normal shock relation 12.41 with the normal component $M_2 \sin (\beta - \alpha)$. That is

$$M_2^2 \sin^2 (\beta - \alpha) = \frac{M_1^2 \sin^2 \beta + \frac{2}{\gamma - 1}}{\frac{2\gamma}{\gamma - 1} M_1^2 \sin^2 \beta - 1} \tag{12.58}$$

The relationships between the upstream and downstream variables depend only on the normal components of the Mach numbers. The incoming component must be supersonic, so that $M_1 \sin \beta \geq 1$. This condition sets the minimum wave angle for an oblique shock. The maximum angle corresponds to a normal shock, where $\beta = \pi/2$, so, that

$$\sin^{-1} \frac{1}{M} \leq \beta \leq \frac{\pi}{2} \tag{12.59}$$

Also, the outgoing component $M_2 \sin (\beta - \alpha)$ will be subsonic since it is downstream of a normal shock, although M_2 can remain supersonic.

12.8.2 Flow Deflection

From Figure 12.14 we see that

$$\tan \beta = \frac{V_{1n}}{V_{1p}}$$

and

$$\tan (\beta - \alpha) = \frac{V_{2n}}{V_{2p}}$$

Since $V_{1p} = V_{2p}$ (equation 12.55), we can use equation 12.56 to show that

$$\frac{\tan \beta}{\tan (\beta - \alpha)} = \frac{\rho_2}{\rho_1} = \frac{V_{1n}}{V_{2n}} = \frac{(\gamma + 1)M_1^2 \sin^2 \beta}{(\gamma - 1)M_1^2 \sin^2 \beta + 2} \qquad (12.60)$$

or, after a considerable amount of algebra,

$$\tan \alpha = 2 \cot \beta \frac{M_1^2 \sin^2 \beta - 1}{M_1^2(\gamma + \cos 2\beta) + 2} \qquad (12.61)$$

Equation 12.61 has the right limiting behavior, since $\alpha = 0$ for $\beta = \pi/2$ and $\beta = \sin^{-1}(1/M_1)$, as required by the limits given in equation 12.59. Between these two extremes where $\alpha = 0$, the deflection angle α must be positive, and therefore it has a maximum value somewhere, as shown in Figure 12.15. For each value of M_1, there is a corresponding α_{max}, which is the maximum deflection angle for which an oblique shock solution can be found.

When $\alpha > \alpha_{max}$, however, no solution can be found. The analysis given here no longer applies. By experiment, we know that a new phenomenon occurs: the shock *detaches* from the body, so that it moves upstream away from the body, and it curves, so that portions of the shock become normal to the direction of the flow. An example is given by the bow shock shown in Figure 12.5. Downstream of a detached shock, regions of supersonic and subsonic flow occur, and the analysis becomes very complicated. Detached shocks will not be considered any further.

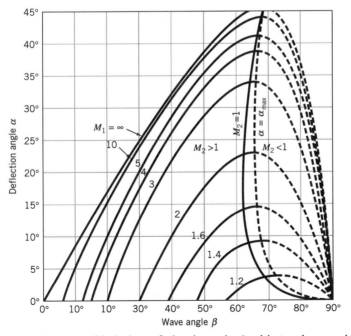

FIGURE 12.15 Variation of shock angle β with turning angle α, as a function of Mach number. Adapted from Liepmann and Roshko, *Elements of Gasdynamics*, John Wiley & Sons, 1957.

In addition, when $\alpha < \alpha_{max}$, there are two solutions for the shock angle β for each value of α and M_1, one to the left of the dashed line in Figure 12.15, called the *weak solution,* and one to the right, called the *strong solution.* In solutions with the stronger shock (on the right), the flow becomes subsonic as it passes through the shock. In solutions with the weaker shock (on the left), the flow remains supersonic, except for a small range of values of α. In practice, the solution adopted by a particular flow depends on the downstream flow conditions. The weak solution is the one that is most commonly observed. Unless otherwise indicated, the weak solution is usually assumed to hold.

12.8.3   Summary: Oblique Shocks

Equation 12.61 is the key to solving many oblique shock problems: when the shock angle can be found for a given deflection angle, the other variables can be found using the normal shock relations (equations 12.45 to 12.49), or the normal shock tables (Table C.11). It is always possible to write a short computer program to solve these relationships, even using a hand-held calculator. The compressible flow calculator provided on the web at *http://www.engapplets.vt.edu/* provides solutions to the relations for isentropic flow, normal shock, and oblique shocks.

EXAMPLE 12.4   *Flow Over a Wedge*

A supersonic airfoil with a symmetrical diamond-shaped cross-section is moving at Mach 3 in air at $100°K$ (see Figure 12.16). Shocks form at the leading edge as the flow is turned over the front part of the airfoil, and also at the trailing edge where the flow is turned back in the freestream direction. At the shoulder, an expansion fan forms, and this will be discussed in Section 12.10. Here, we consider only the flow over the leading edge, which has a half-angle of 10°. The density of the air is $0.8 \ kg/m^3$. Find
(*a*) the stagnation temperature
(*b*) the angle the shock on the leading edge makes with the freestream (assuming the weak solution
(*c*) the density, pressure, temperature, and Mach number downstream of the shock
(*d*) the entropy rise across the shock.

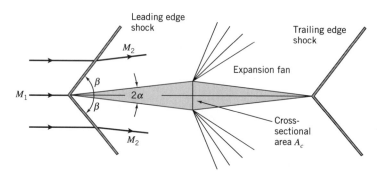

FIGURE 12.16   Diamond-shaped airfoil in a supersonic flow.

Solution: For part (*a*), we use the definition of the total temperature (equation 12.25)

$$\frac{T_0}{T} = 1 + \frac{\gamma - 1}{2} M^2$$

and we find that with $\gamma = 1.4$, $M = 3$, and $T = 100°K$, the stagnation temperature $T_0 = 380°K$.

For part (*b*), we find from Figure 12.15 that the shock angle for the weak solution is $\beta \approx 27°$. A more accurate value can be found by iterating equation 12.61, but 27° is accurate enough for our purposes here.

For part (*c*), we use equations 12.56 to 12.58 and the ideal gas law. With $M_1 = 3$ and $\beta = 27°$, $M_1^2 \sin^2 \beta = 1.855$, so that

1. Density: $\rho_2/\rho_1 = 1.624$, so that $\rho_2 = 1.299\ kg/m^3$.
2. Pressure: $p_2/p_1 = 1.997$. From the ideal gas law, $p_1 = \rho_1 R T_1 = 0.8 \times 287.03 \times 100\ Pa = 22{,}962\ Pa$. Hence, $p_2 = 45{,}866\ Pa$.
3. Temperature: From the ideal gas law, $T_2 = p_2/(\rho_2 R) = 45{,}866/(287.03 \times 1.299)°K = 123°K$.
4. Mach number: From equation 12.58, we see that $M_2^2 \sin^2 (\beta - \alpha) = 0.572$, so that $M_2 = 2.587$.

For part (*d*), we use equations 12.48 and 12.49. Hence, $p_{02}/p_{01} = 0.9676$, and $\Delta s = 9.444\ J/kg \cdot K = 9.444\ m^2/s^2 K$. ∎

12.9 WEAK OBLIQUE SHOCKS AND COMPRESSION WAVES

What happens when the deflection angle for an oblique shock becomes very small? Obviously, the strength of the shock becomes weak, and the changes in temperature, density, and entropy become small. It is particularly interesting to compare the change in entropy to the change in the other variables. We will assume that the downstream Mach number remains supersonic.

If we rearrange equation 12.60, we obtain[9]

$$\frac{1}{M_1^2 \sin^2 \beta} = \frac{\gamma + 1}{2} \frac{\tan(\beta - \alpha)}{\tan \beta} - \frac{\gamma - 1}{2}$$

Further rearrangement gives

$$M_1^2 \sin^2 \beta - 1 = \frac{\gamma + 1}{2} M_1^2 \frac{\sin \beta \sin \alpha}{\cos(\beta - \alpha)} \tag{12.62}$$

When the deflection angle is small, $\sin \alpha \approx \alpha$, and $\cos(\beta - \alpha) \approx \cos \beta$, so that

$$M_1^2 \sin^2 \beta - 1 \approx \left(\frac{\gamma + 1}{2} M_1^2 \tan \beta\right) \alpha \tag{12.63}$$

[9] See Liepmann and Roshko, *Elements of Gasdynamics,* published by John Wiley & Sons, 1957.

Also, from equation 12.61 and Figure 12.15, we see that for the cases where $M_2 > 1$,

$$\tan \beta \approx \tan \alpha_M$$

That is, the angle of the shock approaches the angle of a Mach wave. Since

$$\tan \alpha_M = \frac{1}{\sqrt{M_1^2 - 1}}$$

we have

$$M_1^2 \sin^2 \beta - 1 \approx \frac{\gamma + 1}{2} \frac{M_1^2}{\sqrt{M_1^2 - 1}} \alpha \tag{12.64}$$

Equation 12.64 provides the basis for deriving a number of other relationships that apply to weak oblique shocks. For example, by using equations 12.64 and 12.57, we obtain

$$\frac{p_2 - p_1}{p_1} = \frac{\Delta p}{p_1} = \frac{\gamma M_1^2}{\sqrt{M_1^2 - 1}} \alpha \tag{12.65}$$

We see that the strength of the wave is proportional to the deflection angle. The changes in the other flow quantities, except entropy, are also proportional to α. Entropy, however, is proportional to the third power of the shock strength (equation 12.51). A weak oblique shock, therefore, causes a finite change in pressure, temperature, and flow angle, but these changes occur almost isentropically. In the limit of a very weak wave, all changes are isentropic, and so we can compress a flow isentropically using a succession of weak oblique Mach waves. For example, a supersonic flow can be deflected through a certain angle by a single shock [Figure 12.17(a)], and the stagnation pressure will decrease. However, if we replaced the

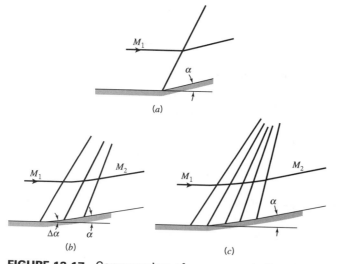

FIGURE 12.17 Compression of a supersonic flow by turning an angle α. (a) Single oblique shock. (b) Series of weak compression waves. (c) Smooth, continuous turning. From Liepmann and Roshko, *Elements of Gasdynamics,* John Wiley & Sons, 1957.

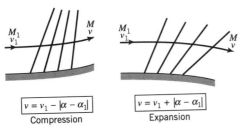

$$\boxed{v = v_1 - |\alpha - \alpha_1|}$$
Compression

$$\boxed{v = v_1 + |\alpha - \alpha_1|}$$
Expansion

FIGURE 12.18 Simple isentropic deflections. (*a*) Compression. (*b*) Expansion. From Liepmann and Roshko, *Elements of Gasdynamics,* John Wiley & Sons, 1957.

single turn by a series of smaller turns [Figure 12.17(*b*)], each generating a weak oblique shock or *compression wave*, there will be almost no change in entropy or stagnation pressure. In the limit of a smooth turn, the flow is exactly isentropic [Figure 12.17(*c*)].

In example 12.4, a Mach 3 flow was deflected by a single shock through an angle of 10°, and we found that the rise in entropy was equal to $0.0334R\ m^2/s^2K$. If instead we accomplished the same turning in 10 steps, each of 1°, we would find that the rise in entropy was equal to $0.00033R\ m^2/s^2K$, that is, 100 times smaller.

We see that compression by Mach waves is an isentropic process. Therefore, if we know the Mach number at any point, all the other flow variables can be found using the isentropic flow relationships derived in Section 12.4.3. The most important relationship that we need to know, therefore, is the relationship between the Mach number and the deflection angle. That is, we need to know the function

$$\alpha = \nu(M) \tag{12.66}$$

This is called the Prandtl–Meyer function. It may be evaluated explicity by finding how an infinitesimal flow deflection is related to the change in flow speed, and integrating the result for a finite deflection. The details are given in textbooks on gasdynamics and compressible flow.[10] Here we give only the final result.

$$\nu(M) = \sqrt{\frac{\gamma + 1}{\gamma - 1}}\ \tan^{-1}\sqrt{\frac{\gamma - 1}{\gamma + 1}(M^2 - 1)} - \tan^{-1}\sqrt{M^2 - 1} \tag{12.67}$$

For convenience, this function is tabulated in Table C.12.

Any given supersonic flow with a Mach number M has a particular value of ν. To find the Mach number M_2 after an isentropic compression by a deflection of α, we begin by finding the value of ν_1 corresponding to the initial Mach number M_1. Then we find ν_2 by *subtracting* the deflection angle from the upstream Prandtl–Meyer function [see Figure 12.18(*a*)]. That is,

$$\boxed{\nu_2 = \nu_1 - (\alpha - \alpha_1)} \qquad \text{isentropic compression} \tag{12.68}$$

which gives M_2, either from equation 12.67 or Table C.12. All other variables can then be found using the isentropic relationships (or the compressible flow calculator available at *http://www.engapplets.vt.edu/*). Note that $\nu_2 < \nu_1$, so that $M_2 < M_1$.

[10] For example, see Liepmann and Roshko, *Elements of Gasdynamics,* published by John Wiley & Sons, 1957.

12.10   EXPANSION WAVES

A most remarkable result can now be obtained. Since the compression by weak oblique shocks (Mach waves) is isentropic, it is *reversible*. In other words, the flow direction can be altered without violating any of the laws of thermodynamics.

> It is therefore possible to describe isentropic expansions using the same Prandtl–Meyer function derived for isentropic compressions.

For an isentropic expansion, the only difference is the sign convention on α. To determine the Mach number M_2 after an isentropic expansion caused by a flow deflection of α [Figure 12.18(b)], we find ν_2 by *adding* the deflection angle to the upstream Prandtl–Meyer function [see Figure 12.18(b)]. That is,

$$\nu_2 = \nu_1 - (\alpha - \alpha_1)$$ isentropic expansion   (12.69)

Note that $\nu_2 > \nu_1$, so that $M_2 > M_1$.

EXAMPLE 12.5   *Isentropic Compression and Expansion*

In example 12.4, we considered a diamond-shaped airfoil traveling in air flow at Mach 3 (See figure 12.16). The deflection angle on the top and bottom of the wedge was 10°.

(a) If instead the compression was achieved by a series of Mach waves, so that the compression was isentropic, find the downstream Mach number and pressure.

(b) If instead of a compression, the flow experienced a 10° isentropic expansion, find the downstream Mach number and pressure.

Solution: For part (a) we find

(i) By interpolation from Table C.12 $\nu_1 = 49.75°$. For an isentropic compression of 10°, $\nu_2 = 49.75° - 10° = 39.75°$. From the table, $M_2 = 2.527$ (compared to 2.587 for the same deflection by a single oblique shock).

(ii) Since the flow is isentropic, $p_{01} = p_{02}$, and equation 12.28 gives

$$\frac{p_2}{p_1} = \left[\frac{1 + \frac{\gamma - 1}{2} M_1^2}{1 + \frac{\gamma - 1}{2} M_2^2} \right]^{\gamma/(\gamma-1)}$$ (12.70)

so that $p_2/p_1 = 2.062$ (compared to 1.997 for the same deflection by a single oblique shock).

For part (b) we find

(i) For an isentropic expansion of 10°, $\nu_2 = 49.75° + 10° = 59.75°$. From Table C.12, $M_2 = 3.578$.

(ii) Equation 12.70 applies to an isentropic expansion as well, so that $p_2/p_1 = 0.786$.

The pressure rises through a compression and falls through an expansion, and the Mach number decreases through a compression and increases through an expansion.   ∎

12.11   WAVE DRAG ON SUPERSONIC VEHICLES

The formation of shocks on vehicles traveling at supersonic speeds gives rise to a drag on the vehicle called *wave drag*. It is most easily demonstrated by considering

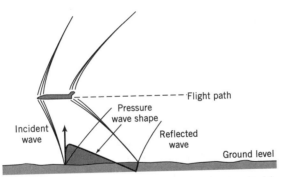

FIGURE 12.19 Sonic boom pattern generated by a supersonic transport. From *Compressible Flow,* M. A. Saad, Prentice-Hall, 1985.

the flow over a diamond-shaped airfoil, such as that shown in Figure 12.16. The oblique shock attached to the leading edge causes the pressure to rise above its ambient value. This pressure acts on the front part of the airfoil, and it has a resultant component in the downstream direction equal to the pressure downstream of the shock times the cross-sectional area of the airfoil, Ac. The flow then expands isentropically through a series of expansion waves centered on the apex of the diamond (this is called an *expansion fan*), so that the pressure drops below the ambient value, thereby augmenting the drag force on the airfoil. The wave drag is equal to the total force due to pressure differences acting on the airfoil in the streamwise direction.

The rise in pressure and its subsequent fall can be computed using the oblique shock relations and the Prandtl–Meyer function, so for the diamond-shaped airfoil the total wave drag can be computed quite easily. For three-dimensional shapes, estimating the wave drag can be very difficult, but it is very important to do it accurately since for supersonic vehicles it is the dominant source of drag.

In addition to causing drag, the wave pattern formed over the vehicle translates into a sonic boom problem on the ground. For the flow over the diamond-shaped airfoil, we see that the wave pattern impinging on the ground would consist of a shock, followed by an expansion, and then another shock attached to the trailing edge (which turns the flow back into its original direction). The pressure signal looks like the letter N (see Figure 12.19), and it is sometimes called the "N-wave." The environmental nuisance caused by sonic booms is one of the principal problems that needs to be solved before supersonic passenger transports become commonplace.

PROBLEMS

12.1 What is the speed of a sound wave in air at $300°K$? What is the speed in helium at the same temperature? Can you explain why a person who has inhaled helium speaks with a high-pitched voice (do not try this yourself)?

12.2 Find the Mach number of an airplane traveling at 2000 *ft/s* at altitudes of 5000 *ft*, 10,000 *ft*, 20,000 *ft*, and 30,000 *ft*. Assume a standard atmosphere (Table C.6).

12.3 Methane (CH_4) at 20°C flows through a pipe at a speed of 400 *m/s*. For methane, $R = 518.3$ $J/(kg \cdot K)$, and $\gamma = 1.32$. Is the flow subsonic, sonic, or supersonic?

12.4 Find the stagnation pressure and temperature of air flowing at 100 *ft/s* if the temperature in the freestream is 60°*F* and the pressure is atmospheric.

12.5 Find the stagnation pressure and temperature of air flowing at 200 *m/s* if the pressure and temperature in the undisturbed flow field are $0.96 \times 10^5 \, Pa$ and 10°*C*, respectively.

12.6 Find the stagnation pressure and temperature of air flowing at 200 *m/s* in a standard atmosphere at sea level, and at heights of 2000 *m* and 10,000 *m*.

12.7 An airplane flies at a speed of 150 *m/s* at an altitude of 500 *m*, where the temperature is 20°*C*. The plane climbs to 12,000 *m* where the temperature is −56.5°*C* and levels off at a speed of 600 *m/s*. Calculate the Mach number for both cases.

12.8 The Lockheed SR-71 reconnaissance airplane is rumored to fly at a Mach number of about 3.5 at an altitude of 90,000 *ft*. Estimate its flight speed under these conditions.

12.9 The National Transonic Facility (NTF) is a wind tunnel designed to operate at Mach and Reynolds numbers comparable to flight conditions. It uses nitrogen at cryogenic temperatures as the working fluid. A schlieren photograph taken in the NTF of a 1/20 scale model of a Concorde airplane (full-scale wingspan of 30 *m*) shows a Mach angle of 28° at a point where the temperature is 100°*K* and the pressure is 9,000 *Pa*. Find the local Mach number and the Reynolds number based on the model wingspan, given that the viscosity of nitrogen under these conditions is $7 \times 10^{-6} \, N \cdot s/m^2$. Compare with the conditions experienced by the full-scale airplane flying at 600 *m/s* at an altitude of 20,000 *m* in a standard atmosphere.

12.10 Show that the air temperature rise in °*K* at the stagnation point of an aircraft is almost exactly

$$\left(\frac{\text{speed in } mph}{100} \right)^2$$

12.11 The working section of a transonic wind tunnel has a cross-sectional area of 0.5 m^2. Upstream, where the cross-section area is 2 m^2, the pressure and temperature are $4 \times 10^5 \, Pa$ and 5°*C*, respectively. Find the pressure, density, and temperature in the working section at the point where the Mach number is 0.8. Assume one-dimensional, isentropic flow.

12.12 Air at 290°*K* and $10^5 \, Pa$ approaches a normal shock. The temperature downstream of the shock is 540°*K*. Find:

(a) The velocity downstream of the shock.

(b) The pressure downstream of the shock, and compare it with that calculated for an isentropic flow with the same deceleration.

12.13 A Pitot tube is placed in a supersonic flow where the freestream temperature is 90°*K* and the Mach number is 2.5. A normal shock forms in front of the probe. The probe indicates a stagnation pressure of $52 \times 10^3 \, Pa$. Find the pressure, density, stagnation pressure, and velocity of the flow upstream of the shock.

12.14 Air with a stagnation temperature of 700°*K* is compressed by a normal shock. If the upstream Mach number is 3.0, find the velocity and temperature downstream of the shock, and the entropy change across the shock.

12.15 Find the maximum increase in density across a shock wave for a gas with $\gamma = 1.4$.

12.16 A blowdown wind tunnel is supplied by a large air reservoir where the pressure is constant at $1.014 \times 10^5 \, Pa$ and the temperature is constant at 15°*C*. The air passes through a working section of area 0.04 m^2 and exits to a large vacuum vessel. Find the pressure, density, velocity, and mass flow rate in the working section if the Mach number there is 4.0.

12.17 A rocket motor is designed to give 10,000 *N* thrust at 10,000 *m* altitude. The combustion chamber pressure and temperature are $2 \times 10^6 \, Pa$ and 2800°*K*, respectively. The gases exit the combustion chamber through a Laval nozzle. Find the exit Mach number and the cross-sectional areas of the exit and the throat of the nozzle. Assume the nozzle flow is isentropic and one-dimensional, and that the ratio of specific heats γ for the combustion gases is 1.32.

12.18 A blowdown wind tunnel exits to atmospheric pressure. In the working section where the cross-sectional area is 0.04 m^2, the flow has a Mach number of 3 and a pressure of 0.3 $\times$ 10^5 Pa.

(a) What is the minimum stagnation pressure required?

(b) What is the minimum stagnation temperature required to avoid air condensation in the working section (the condensation temperature under these conditions is about $70°K$)?

(c) What is the corresponding stagnation density under these conditions?

(d) What is the mass flow rate under these conditions?

12.19 Air flows through a converging and diverging nozzle with an area ratio (exit to throat) of 3.5. The upstream stagnation conditions are atmospheric, and the back pressure is maintained by a vacuum system. Find:

(a) the mass flow rate if the throat area is 500 mm^2.

(b) the range of back pressures for which a normal shock will occur within the nozzle.

12.20 A large reservoir maintains air at 6.8×10^5 Pa and 15°C. The air flows isentropically through a convergent and divergent nozzle to another large reservoir where the back pressure can be varied. The area of the throat is 25 cm^2 and the area of the nozzle exit is 100 cm^2. Find:

(a) The maximum mass flow rate through the nozzle.

(b) The two values of the Mach number at the nozzle exit corresponding to this mass flow rate.

(c) The back pressures required to produce these Mach numbers.

12.21 An air flow with a Mach number of 2.0 passes through an oblique shock wave inclined at an angle of 45°. Find the flow deflection angle α.

12.22 An air flow with a Mach number of 8 is deflected by a wedge through an angle α. What is the maximum value of α for an attached oblique shock?

12.23 A supersonic air flow passes over a symmetrical wedge of semi-angle $\alpha = 10°$. At the leading edge an attached shock of $\beta = 30°$ is observed. Find:

(a) The upstream Mach number.

(b) The downstream Mach number.

(c) The static pressure ratio across the shock.

12.24 Air having an initial Mach number of 2.4, a freestream static pressure of 10^5 Pa, and a static temperature of $270°K$ is deflected by a wedge through an angle of 10°. Find:

(a) the Mach number, pressure, and temperature downstream of the shock.

(b) The change in entropy across the shock.

12.25 The shock in the previous problem "reflects" at the opposite wall, as shown in Figure P12.25. The condition on the second shock is that the flow is turned parallel to the wall, so that the flow deflection through the second shock must be 10°. Find:

(a) The Mach number, pressure, and temperature downstream of the second shock.

(b) The change in entropy $s_3 - s_1$.

(c) The maximum wedge angle for the reflected shock to remain attached.

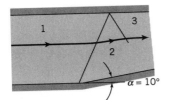

FIGURE P12.25

12.26 An air flow with an initial Mach number of 1.5 and pressure p_1 is expanded isentropically by passing through a deflection angle of 5°. Find the Mach number and pressure ratio after the deflection.

12.27 Air at Mach 3.0 is deflected through 20° by an oblique shock. What isentropic expansion turning angle is required to bring the flow back to

(a) The original Mach number.

(b) The original pressure.

12.28 A flat plate airfoil with a chord length of 1 m and a width of 6 m is required to generate a lift of 400,000 N when flying in air at a Mach number of 2.0, a temperature of $-20°C$ and a pressure of 10^5 Pa. What is the required angle of attack? What is the wave drag at this angle of attack?

12.29 A symmetrical diamond-shaped airfoil is placed at an angle of attack of 2° in a flow at Mach 2 and static pressure of 2×10^3 Pa. The half-angle at the leading and trailing edges is 3°. If its total surface area (top and bottom) is 4 m^2, find the forces due to lift and wave drag acting on the airfoil.

CHAPTER 13

TURBOMACHINES

13.1 INTRODUCTION

In this chapter, we examine the performance and design of turbomachinery. Turbomachines are widely used in engineering applications, and many different types of machines exist. They can be classified according to whether they add mechanical energy to ($+ W_{shaft}$), or extract mechanical energy from, the fluid stream ($- W_{shaft}$). See Figure 13.1. Pumps, fans, blowers, compressors, and propellers are examples of turbomachines that add energy to a fluid, and windmills and water turbines are examples of turbomachines that extract energy from the fluid.

Turbomachines come in many different shapes, sizes and geometries, but their common feature is that they have a *rotor* or *impeller,* that is, a wheel equipped with blades. In the case of a water turbine, the fluid stream acts on the rotor blades to produce a force with a significant component in the circumferential direction, and in the case of a pump, the blades act on the fluid with a significant torque to increase the pressure of the stream.

Turbomachines are further subdivided according to the direction the exit flow takes compared to the entry flow. For example, a propeller turbine is an *axial-flow* machine since the directions of the entry and exit flow are aligned along a common axis [Figure 13.2(*a*)]. The fan shown in Figure 13.2(*b*) is a *radial-flow* or *centrifugal* machine since the directions of the inlet end entry flow are orthogonal. There are also *mixed-flow* devices that fall somewhere in between the axial- and radial-flow machines (Figure 13.3). The design of each machine is adapted to a particular application. Sometimes a high flow rate is required, sometimes a high pressure, and at other times, a high flow rate and a high pressure is needed.

We will now consider some common examples of turbomachines such as pumps, turbines, propellers, and windmills. Before we analyze these particular devices, we will consider the underlying momentum and energy principles that apply to all turbomachinery.

13.2 ANGULAR MOMENTUM EQUATION FOR A TURBINE

Since we are dealing with machines where the blades or rotors spin on an axis, angular momentum is an important variable. In Chapter 5, we applied Newton's second law to the flow of fluid through a fixed control volume to derive a linear momentum equation in integral form. Here, we will derive a similar equation for the angular momentum. The basic angular momentum principle for a system of fixed mass is

$$\mathbf{T} = \frac{d\mathbf{H}}{dt}\bigg|_{system} \tag{13.1}$$

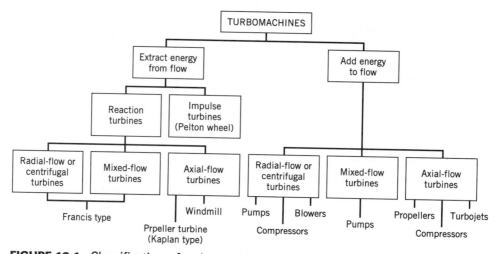

FIGURE 13.1 Classification of turbomachines, with examples.

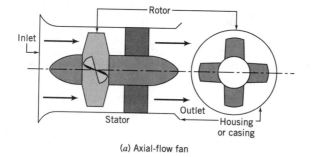

(a) Axial-flow fan

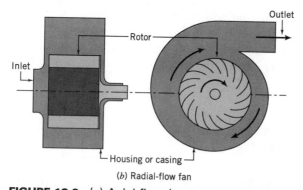

(b) Radial-flow fan

FIGURE 13.2 (a) Axial-flow (or propeller) turbine. (b) Radial-flow (or centrifugal) fan. Adapted from Munson, Young, and Okishii, *Fundamentals of Fluid Mechanics*, John Wiley & Sons, 1998.

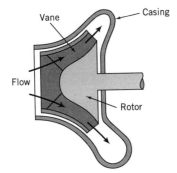

FIGURE 13.3 Mixed-flow pump. From John and Haberman, *Introduction to Fluid Mechanics*, Prentice Hall, 1988.

where **T** is the total torque or moment applied to the system by its surroundings, and **H** is the angular momentum of the system given by the integral of the moment over all the mass in the system. That is

$$\mathbf{H} = \int_{mass} \mathbf{r} \times \mathbf{V} \, dm = \int_{volume} \mathbf{r} \times \mathbf{V} \rho \, d\forall \tag{13.2}$$

(**V** is the velocity of any point, and **r** is the distance of the point from the center of rotation). We can now relate the system formulation to the fixed control volume formulation using the Reynolds transport theorem (equation 5.21), so that

$$\left.\frac{d\mathbf{H}}{dt}\right|_{system} = \frac{\partial}{\partial t} \int \mathbf{r} \times \mathbf{V} \rho \, d\forall + \int (\mathbf{n} \cdot \rho \mathbf{V}) \mathbf{r} \times \mathbf{V} \, dA \tag{13.3}$$

For a fixed control volume, therefore,

$$\mathbf{T} = \frac{\partial}{\partial t} \int \mathbf{r} \times \mathbf{V} \rho \, d\forall + \int (\mathbf{n} \cdot \rho \mathbf{V}) \mathbf{r} \times \mathbf{V} \, dA \tag{13.4}$$

The first term on the right hand side represents, at the instant the system and the control volume occupy the same space, the rate of change of the angular momentum of the fluid contained in the control volume, and the second term represents the net flux of angular momentum leaving the control volume. The sum is equal to the total torque applied to the mass of fluid that is contained in the control volume at time *t*.

In all the applications considered here, the flow is steady and the only external torque is applied by a shaft. In that case

$$\mathbf{T}_{shaft} = \int (\mathbf{n} \cdot \rho \mathbf{V}) \mathbf{r} \times \mathbf{V} \, dA \tag{13.5}$$

A fixed coordinate system is chosen so that the axis of rotation of the machine coincides with the *z*-axis (Figure 13.4). The rotor spins inside an annular control volume at a fixed angular speed, ω (rad/s). One-dimensional flow is assumed, so that the fluid enters the rotor at the radial location r_1 with a uniform absolute velocity V_1, and it leaves the rotor at the radial location r_2 with a uniform absolute velocity V_2. Equation 13.5 reduces to

$$T_{shaft}\mathbf{k} = \int (\mathbf{n} \cdot \rho \mathbf{V}) \mathbf{r} \times \mathbf{V} \, dA_1 + \int (\mathbf{n} \cdot \rho \mathbf{V}) \mathbf{r} \times \mathbf{V} \, dA_2$$
$$= \rho \dot{q}(r_2 V_{t2} - r_1 V_{t1})\mathbf{k}$$

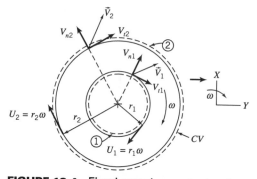

FIGURE 13.4 Fixed annular control volume and absolute velocity components for angular momentum analysis. From Fox and Macdonald, *Introduction to Fluid Mechanics,* John Wiley and Sons, 4th ed., 1992.

where $\dot{q}$ is the volume flow rate and V_{t1} and V_{t2} are the tangential components of the absolute velocities of the fluid crossing the surface of the control volume. In scalar form

$$T_{shaft} = \rho\dot{q}(r_2 V_{t2} - r_1 V_{t1})$$

(13.6)

The tangential velocities are positive when they point in the same direction as the blade speed. This sign convention gives positive torques for machines that do work on the fluid (for example, pumps, fans, blowers, and compressors) and negative torques for machines that extract work from the fluid (for example, hydraulic turbines and windmills).

Equation 13.6 is the basic relationship between torque and angular momentum for all turbomachines. It is sometimes called the *Euler momentum equation.*

The rate of work done on a turbomachine rotor, that is, the mechanical power $\dot{W}_{shaft}$, is given by ωT_{shaft}, so that

$$\dot{W}_{shaft} = \omega T_{shaft} = \rho\dot{q}\omega(r_2 V_{t2} - r_1 V_{t1})$$

(13.7)

We see that the mechanical power increases linearly with the mass flow rate $\dot{m} = \rho\dot{q}$, the rotational speed ω (*rad/s*), and the change in angular momentum.

By dividing the mechanical power by $\dot{m}g$, we obtain a quantity with the dimensions of length, called the *head*

$$H = \frac{\dot{W}_{shaft}}{\dot{m}g} = \frac{1}{g}(r_2 V_{t2} - r_1 V_{t1})$$

(13.8)

The head H represents the mechanical shaft work per unit weight of fluid (positive for a pump, negative for a turbine). V_{t1} and V_{t2} are the absolute tangential velocities at entry and exit, respectively.

13.3 VELOCITY DIAGRAMS

We see that, to find the power produced by or applied to a turbomachine, it is necessary to know the velocity components at the entry and exit sections. *Velocity*

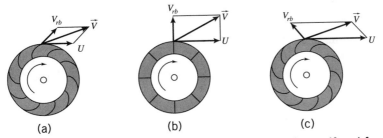

FIGURE 13.5 Velocity diagrams for three types of centrifugal fans. (a) Forward-curved blades; (b) flat blades; (c) backward-curved blades. V is the absolute velocity of air leaving the blade (shown equal for all three blade types). V_{rb} is the velocity of air leaving the blade relative to the blade. U is the velocity of the blade tip.

diagrams are used for this purpose, and some examples are given in Figure 13.5 for different types of centrifugal fan. A velocity diagram is simply a vector diagram showing the relationship between the absolute and relative velocities. The symbol **V** is used for absolute velocities, V_{rb} denotes a velocity relative to the blade, and U is the velocity of the blade tip.

It is always assumed that the flow relative to the rotor enters and leaves tangent to the blade profile. The blade angles, β, are measured relative to the circumferential direction, as shown in Figure 13.6(a). The absolute speed of the fluid is the vector sum of the impeller velocity and the flow velocity relative to the blade, and this sum may be found graphically, as shown in Figure 13.6(b) and 13.6(c). Here, $U_1 = \omega R_1$, and $U_2 = \omega R_2$. The absolute fluid velocity makes an angle α_1 relative to the normal direction at the inlet, and an angle α_2 relative to the normal direction at the outlet. At each section, the normal component of the absolute velocity, V_n, is equal to the normal component of the velocity relative to the blade, V_{rb_n}.

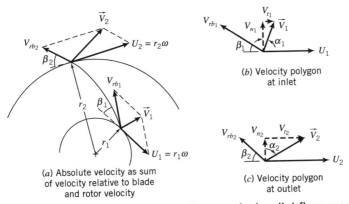

(a) Absolute velocity as sum of velocity relative to blade and rotor velocity

(b) Velocity polygon at inlet

(c) Velocity polygon at outlet

FIGURE 13.6 Velocity diagram for a typical radial-flow machine. From Fox and Macdonald, *Introduction to Fluid Mechanics,* John Wiley & Sons, 4th ed., 1992.

13.4 HYDRAULIC TURBINES

A hydraulic turbine converts gravitational potential energy of water into shaft work. As water passes through the wheel of the turbine, which is fitted with vanes or buckets, its momentum changes. The forces resulting from this momentum change turn the wheel against some external load, causing it to do work. The difference between the initial and final water levels is called the *available head, H,* and it measures the change in potential energy.[1] Different types of turbine are used, depending on the size of the available head. When the head is high (greater or equal to about 300 *m,* or 1000 *ft*), *impulse* turbines are used almost exclusively. In this design, one or more jets of water at atmospheric pressure are directed against buckets on the rim of the wheel (Figure 13.7). Most of the kinetic energy of the water is converted into work, and the discharge has just enough velocity left to clear the buckets and fall into the low-level reservoir or *tail water.* The water loses kinetic energy as it passes through an impulse wheel but it enters and leaves at the same pressure.

For lower heads, *reaction* turbines are used. Here, the flow through the wheel, or "runner," is completely enclosed by a housing, and the pressure and velocity at the exit are different from their values at the entry. There are two basic types of reaction turbines: *radial-flow* turbines and *axial-flow* turbines, which refer to the direction of water movement through the runner.

We will now examine the performance of impulse, radial-flow, and axial-flow turbines in more detail.

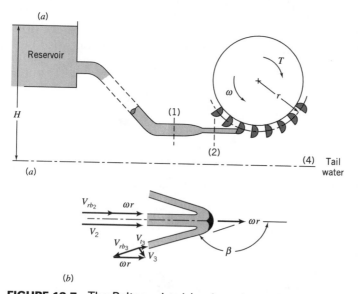

FIGURE 13.7 The Pelton wheel (an impulse turbine). (*a*) Side elevation. (*b*) Section through bucket. From Hunsaker and Rightmire, *Engineering Applications of Fluid Mechanics,* McGraw-Hill, 1947.

---

[1] The available head will always be greater than the head produced by the turbine because the turbine efficiency will always be less than one (see Section 13.6).

13.4.1 Impulse Turbine

The Pelton wheel shown in Figure 13.7 was developed in California about 1880. It is named after Lester A. Pelton (1829–1908), who developed the first efficient design. The concept dates back to the simple water wheels used by the Sumerians. In the case of the Pelton wheel, high velocity water jets strike the buckets, which are shaped like a spoon with a central ridge. The ridge divides the impinging jet in half, and each half is deflected back through an angle of about 165°. The torque exerted on the runner is given by the change in momentum of the water.

The performance of the impulse turbine can be analyzed using the tools developed in Section 13.2. From equation 13.6, the torque is given by

$$T_{shaft} = \rho \dot{q} r (V_2 - V_{t3})$$

(13.9)

where V_2 and V_{t3} are the absolute tangential velocities, which are assumed to have the same moment arm, r. When there are no losses upstream of the nozzle, $V_2 = \sqrt{2gH}$, but in practice V_2 may be significantly less than this value.

We can also use Bernoulli's equation in this flow, but we need to follow a streamline, and we need to move with the buckets so that we are in a steady frame of reference. Since the pressure is atmospheric everywhere, and we neglect changes in height

$$\tfrac{1}{2} V_{rb_2}^2 = \tfrac{1}{2} V_{rb_3}^2$$

so that

$$V_{rb_2} = V_{rb_3}$$

where V_{rb_2} and V_{rb_3} are the magnitudes of the velocities relative to the bucket (see Figure 13.7). That is,

$$V_2 = V_{rb_2} + \omega r$$

and

$$V_{t3} = \omega r - V_{rb_3} \cos(\pi - \beta)$$

Therefore,

$$T_{shaft} = \rho \dot{q} r (V_{rb_2} - V_{rb_3} \cos \beta)$$
$$= \rho \dot{q} r (V_2 - \omega r)(1 - \cos \beta)$$

(13.10)

Due to losses in the system, the actual torque available in practice has a maximum value of about $0.91 T_{shaft}$.

The mechanical power generated by the turbine is given by

$$\dot{W}_{shaft} = \rho \dot{q} r \omega (V_2 - \omega r)(1 - \cos \beta)$$

(13.11)

In nondimensional terms,

$$C_P = \frac{\dot{W}_{shaft}}{\tfrac{1}{2} \rho \dot{q} V_2^2} = 2\xi (1 - \xi)(1 - \cos \beta)$$

(13.12)

where C_P is a power coefficient and the parameter ξ represents the ratio of the speed of the bucket to the absolute speed of the impinging water jet. That is,

$$\xi = \frac{\omega r}{V_2}$$

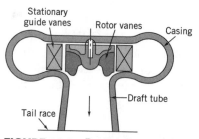

FIGURE 13.8   Radial-flow hydraulic turbine of the Francis type. From Fox and Macdonald, *Introduction to Fluid Mechanics,* John Wiley & Sons, 4th ed., 1992.

For a given value of β, the power coefficient is maximum when the speed of the bucket is half the absolute speed of the water jet, so that $\xi = 0.5$ (this can be shown by differentiating equation 13.12 with respect to ξ). In an actual turbine, the optimum value of ξ is a little less, closer to 0.45, because of losses in the system. We also see that the maximum power coefficient is achieved when β is as close as possible to 180°. In practice, the angle must be less than 180° since the flow must clear the following bucket. As we indicated earlier, a practical maximum is probably close to 165°. With $\xi = 0.45$, and $\beta = 165°$, $C_P = 0.973$.

13.4.2   Radial-Flow Turbine

An example of a radial-flow hydraulic turbine is shown in Figure 13.8. Here, the flow is completely enclosed in a casing, so that the pressure in the machine can be different from atmospheric. The water enters the stationary guide vanes directly from the inlet duct. The guides impart angular momentum to the water before it enters the rotating runner, where its angular momentum is reduced as work is done by the flow. This type of turbine is also known as the Francis turbine after James B. Francis, who developed it in 1849. The purely radial type shown in Figure 13.8 is suited to relatively small volume flow rates and high heads, whereas the mixed-flow type shown in Figure 13.9 works efficiently with larger volume flow rates and

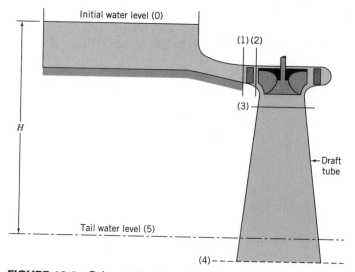

FIGURE 13.9   Schematic of a mixed-flow hydraulic turbine. From Hunsaker and Rightmire, *Engineering Applications of Fluid Mechanics,* McGraw-Hill, 1947.

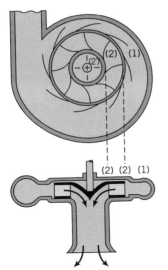

FIGURE 13.10 Notation for a radial-flow turbine. From Hunsaker and Rightmire, *Engineering Applications of Fluid Mechanics*, McGraw-Hill, 1947.

lower heads. Taken together, it is possible to run these machines at heads from as low as 5 *m* to as high as 250 *m*. Efficiencies can be as high as 94%.

To analyze the performance of a radial-flow turbine, we need to find the change in angular momentum. Consider the velocity diagrams for the runner shown in Figure 13.10. For steady, one-dimensional flow, we obtain from equation 13.6:

$$T_{shaft} = \rho \dot{q}(r_2 V_{t2} - r_{2'} V_{t2'}) \qquad (13.13)$$

where V_{t2} and $V_{t2'}$ are the absolute tangential velocities, which are assumed to have the moment arms r_2 and $r_{2'}$, respectively.

13.4.3 Axial-Flow Turbine

An example of an axial-flow turbine is shown in Figure 13.11. Axial-flow (or propeller) turbines are usually fitted with blades that can be adjusted to suit the operating conditions, a feature developed by a Czech engineer named Viktor Kaplan. It is more compact than the Francis type, runs faster, and it maintains a high efficiency over a broad range of conditions because of the flexibility afforded by the adjustable blades. It is more costly than the Francis type because of its greater complexity, but an efficiency of greater than 92% is possible with units up to 60,000 *hp* capacity.

As shown in Figure 13.11, the guide vanes are arranged in the same way as in a Francis turbine, and they serve the same purpose, which is to turn the flow. Before entering the runner, however, the stream turns through a right angle and is assumed to have no radial velocity component as it passes through the runner. It is also assumed that the axial component is uniform across the outlet, and that the flow is steady. In this case, we obtain from equation 13.6 between stations 2' and 3

$$T_{shaft} = \rho \dot{q}(r_{2'} V_{t2'} - r_3 V_{t3}) \qquad (13.14)$$

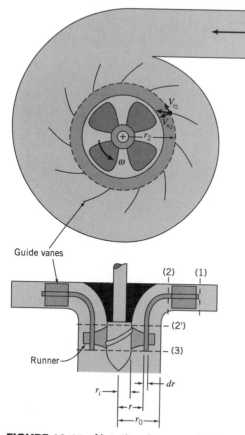

FIGURE 13.11 Notation for an axial-flow turbine of the Kaplan type. From Hunsaker and Rightmire, *Engineering Applications of Fluid Mechanics,* McGraw-Hill, 1947.

where $V_{t2'}$ and V_{t3} are the absolute tangential velocities, which are assumed to have the mean moment arms $r_{2'}$ and r_3, respectively. A more detailed analysis would allow for the deviations from one-dimensional flow, and require the torque to be found by integration over the inlet and outlet areas.

EXAMPLE 13.1 *Axial-Flow Fan*[2]

An axial-flow fan operates at 1200 *rpm*. The blade tip diameter D_t is 1.1 *m* and the hub diameter D_h is 0.8 *m*. The blade inlet and outlet angles are 30° and 60°, respectively. Inlet guide vanes give the absolute flow entering the first stage an angle of 30°. The fluid is air at atmospheric pressure and 15°C, and it may be assumed to be incompressible. The axial velocity of the flow does not change across the rotor. Assume that the relative flow enters and leaves the rotor at the geometric blade angles and use properties at the mean blade radius for calculations.
(*a*) Sketch the rotor blade shapes.

---

[2] This example was adapted from Fox and Macdonald, *Introduction to Fluid Mechanics,* published by John Wiley & Sons, 4th ed., 1992.

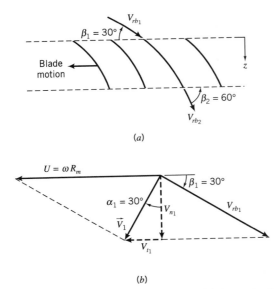

(a)

(b)

FIGURE 13.12 Blade shapes and inlet velocity diagram.

(b) Draw the inlet velocity diagram.
(c) Find the volume flow rate.
(d) Draw the outlet velocity diagram.
(e) Calculate the minimum torque and power needed to drive the fan.

Solution: The blade shapes are as shown in Figure 13.12(a), and the corresponding inlet velocity diagram is shown in Figure 13.12(b). The continuity equation gives

$$\rho V_{n1} A_1 = \rho V_{n2} A_2$$

or

$$\dot{q} = V_{n1} A_1 = V_{n2} A_2$$

Since $A_1 = A_2$, then $V_{n1} = V_{n2}$, and the outlet velocity diagram is as shown in Figure 13.13.

To complete the velocity diagrams, and to find the volume flow rate, we need to find $U, V_{n1}, V_1, V_{t1}, V_{rb_1}, V_2, V_{t2}$, and α_2. We start with U. At the mean blade radius,

$$U = \omega r_m$$

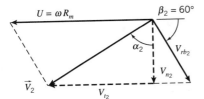

FIGURE 13.13 Outlet velocity diagram.

The mean blade radius $r_m = (D_t + D_h)/4$. Also,

$$\omega = 1200 \, \frac{rev}{min} \times 2\pi \, \frac{rad}{rev} \times \frac{min}{60 \, sec} = 126 \, rad/s$$

Therefore,

$$U = \frac{(1.1 + 0.8) \, m}{4} \times 126 \, \frac{rad}{sec} = 59.7 \, m/s$$

From the inlet velocity diagram,

$$U = V_{n1}(\tan \alpha_1 + \cot \beta_1)$$

where $\alpha_1 = \beta_1 = 30°$, so that

$$V_{n1} = V_{n2} = 25.9 \, m/s$$

Hence,

$$V_1 = \frac{V_{n1}}{\cos \alpha_1} = 29.9 \, m/s$$

$$V_{t1} = V_1 \sin \alpha_1 = 15.0 \, m/s$$

and

$$V_{rb_1} = \frac{V_{n1}}{\sin \beta_1} = 51.8 \, m/s$$

The volume flow rate can now be found

$$\dot{q} = V_{n1} A_1 = V_{n1} \tfrac{\pi}{4}(D_t^2 - D_h^2) = 11.6 \, m^3/s$$

From the outlet velocity diagram,

$$V_{t2} = U - V_{n2}\tan(90° - \beta_2) = (59.7 - 25.9 \tan 30°) \, m/s = 44.7 \, m/s$$

Also

$$\tan \alpha_2 = \frac{V_{t2}}{V_{n2}}$$

so that

$$\alpha_2 = 59.9°$$

and

$$V_2 = \frac{V_{n2}}{\cos \alpha_2} = 51.6 \, m/s$$

To find the torque, we use equation 13.6, with $r_1 = r_2 = r_m$ so that

$$T_{shaft} = \rho \dot{q} r_m (V_{t2} - V_{t1})$$

Hence, using SI units throughout,

$$T_{shaft} = 1.2 \times 11.6 \times \frac{0.95}{2} \times (44.7 - 15.0) \, N \cdot m = 196 \, N \cdot m$$

and the power required is given by equation 13.7

$$\dot{W}_{shaft} = \omega T_{shaft} = 24.7 \, kW$$

■

13.5 PUMPS

Rotary pumps, like turbines, come in three types: radial-flow or centrifugal pumps, mixed-flow or screw pumps, and axial-flow or propeller pumps (Figure 13.14). For centrifugal pumps, the inlet is typically on one side of the impeller, in line with the axis of rotation, but the flow may enter from both sides to balance the thrust. The inlet flow can be assumed to have no swirl, so that its tangential velocity is zero. Centrifugal pumps are used for high head ($\geq 6\ m$), low-volume flow rate applications. Screw pumps have a number of screw-shaped blades, and are used for heads between 3 to 5 m. Propeller pumps are used for low head ($< 6\ m$), high-volume flow rate applications. They also have the advantage of being able to handle solids in suspensions, and they can be used to pump foodstuffs, slurries, and sewage. All rotary pumps generally need *priming,* which means that if they are filled with air they cannot suck up a liquid from below their inlet.

These three types of pump are widely used in industry. However, the centrifugal pump is undoubtedly the most common of the three, and it is the only type considered here.[3]

13.5.1 Centrifugal Pumps

As shown in Figure 13.14, a centrifugal pump has an inpeller rotating within a casing. The fluid enters along the axis of rotation, moves through the impeller radially outward gaining velocity and pressure, and then discharges into the volute or diffuser, where the velocity decreases and the pressure increases. The flow leaves the impeller with a velocity which has an outward and backward component relative to the impeller. The absolute tangential component, V_{t2}, is given by the tip speed of the impeller, $U_2 = \omega r_2$, minus the tangential component of the relative velocity, $V_{rb_2} \cos\beta_2$. A typical velocity diagram is shown in Figure 13.15.

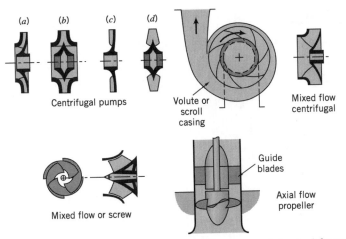

FIGURE 13.14 Typical pump impeller types. Adapted from Duncan, Thom, and Young, *Mechanics of Fluids,* Edward Arnold, 2nd ed., 1970.

---

[3] For a discussion on mixed-flow and axial-flow pump performance, see F. M. White, *Fluid Mechanics,* published by McGraw-Hill, 1986.

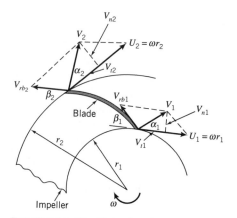

FIGURE 13.15 Velocity diagram at the exit from a centrifugal pump impeller.

If the flow enters the impeller with purely axial absolute velocity, the entering fluid has no angular momentum and $V_{t1} = 0$. The torque is then given by equation 13.6

$$T_{shaft} = \rho \dot{q} r_2 V_{t2} \tag{13.15}$$

The mechanical power input to the pump is given by

$$\dot{W}_{shaft} = \omega T_{shaft} = \rho \dot{q} \omega r_2 V_{t2} \tag{13.16}$$

The performance of a particular centrifugal pump is shown in Figure 13.16 as a function of discharge. Performance charts are usually plotted for a constant shaft rotation speed, as shown in Figure 13.17. The independent variable is the discharge or volume flow rate, $\dot{q}$, and the dependent variables are the output head, H, the input power, $\dot{W}_{shaft}$, and the efficiency, η (see below). The variables are almost always expressed dimensionally.

13.5.2 Cavitation

Cavitation can occur in turbomachines when the local pressure falls below the vapor pressure of the liquid. Small bubbles of gas will form, which can have a devastating

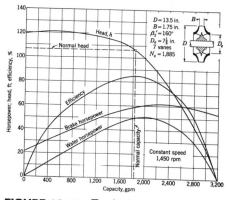

FIGURE 13.16 Typical characteristic performance curves for a centrifugal pump. From Daugherty, Franzini and Finnemore, *Fluid Mechanics with Engineering Applications*, 8th ed., published by McGraw-Hill, 1985.

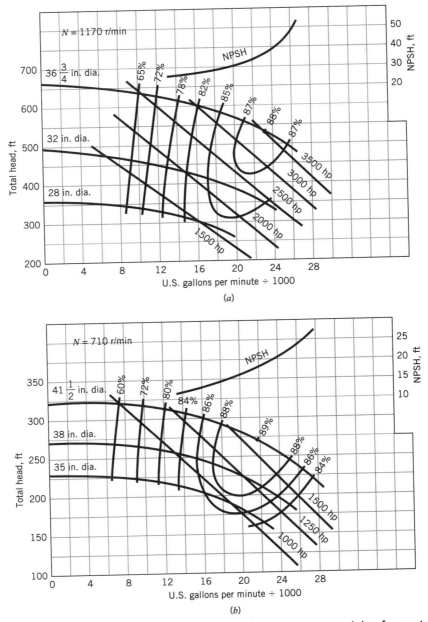

FIGURE 13.17 Measured performance curves for two models of a centrifugal water pump. (*a*) Basic casing with three impeller sizes. (*b*) Twenty percent larger casing with three larger impellers at slower speed. From the Ingersoll-Rand Corporation, Cameron Pump division.

effect on the impeller. As they collapse on the impeller surface, intense pressures are developed, which can cause pitting damage to the impeller with the possibility of eventual failure of the surface material.

There are two sources of cavitation in pumps: the velocities at the tip of the impeller can produce very low pressure regions where bubbles can form, and the inlet pressures may be so low that bubbles are formed on the inlet side. The first

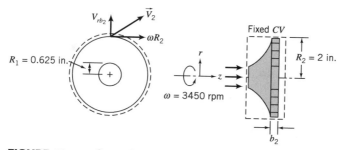

FIGURE 13.18 Centrifugal pump.

source of cavitation can occur in hydraulic turbines, but the second source is restricted to pumps. The pump inlet is the low-pressure side where cavitation will first occur and the bubbles are then entrained into the impeller, where they can collapse and cause erosion.

Near the top of the graphs given in Figure 13.17 for centrifugal water pumps is plotted the *net positive-suction head* (NPSH), which is the head required at the pump inlet to keep the liquid at the inlet from cavitating. The NPSH is defined as

$$\text{NPSH} = \frac{p_i - p_v}{\rho g} + \frac{V_i^2}{2g} \tag{13.17}$$

where p_i and V_i are the pressure and velocity on the inlet side, and p_v is the vapor pressure of the liquid. If the pump inlet is located a distance h_i above the reservoir, and there are no significant losses in the inlet piping, then, by Bernoulli's equation for steady flow,

$$\frac{p_i}{\rho g} + \frac{V_i^2}{2g} + h_i = \frac{p_r}{\rho g}$$

where p_r is the pressure at the surface of the reservoir. That is,

$$\text{NPSH} = \frac{p_r - p_v}{\rho g} - h_i$$

As long as the machine is operated above the NPSH line, there should be no problems with cavitation at the pump inlet.

EXAMPLE 13.2 *Centrifugal Pump*[4]

Water at 150 *gpm* enters a centrifugal pump impeller axially through a 1.25 *in.* diameter inlet (see Figure 13.18). The inlet velocity is axial and uniform. The impeller outlet diameter is 4 *in.* Flow leaves the impeller at 10 *ft/s* relative to the radial blades. The impeller speed is 3450 *rpm*. Determine
(*a*) the impeller exit width, b_2
(*b*) the minimum torque input to the impeller
(*c*) the minimum power required

---

[4] This example was adapted from Fox and Macdonald, *Introduction to Fluid Mechanics,* John Wiley & Sons, 4th ed., 1992.

Solution: The continuity equation gives

$$\rho V_1 A_1 = \rho V_2 A_2$$

or

$$\dot{q} = V_{rb_2} 2\pi R_2 b_2$$

so that

$$b_2 = \frac{\dot{q}}{2\pi R_2 b_2 V_{rb_2}} = \frac{1}{2\pi} \times 150 \frac{gal}{min} \times \frac{1}{2\ in.} \times \frac{sec}{10\ ft} \times \frac{ft^3}{7.48\ gal} \times \frac{min}{60\ sec} \times \frac{12\ in.}{ft}$$

$$= 0.0319\ ft \ \text{or} \ 0.383\ in.$$

The torque is given by equation 13.6

$$T_{shaft} = \rho \dot{q} R_2 V_{t2}$$

since the axial inlet flow has no z-component of angular momentum. That is

$$T_{shaft} = \rho(V_{rb2} 2\pi R_2 b_2) R_2(\omega R_2) = 2\pi \rho \omega R_2^3 b_2 V_{rb2}$$

We have

$$\omega = 3450 \frac{rev}{min} \times 2\pi \frac{rad}{rev} \times \frac{min}{60\ sec} = 361\ rad/s$$

Hence, using BG units throughout,

$$T_{shaft} = 2\pi \times 1.94 \frac{slug}{ft^3} \times 361 \frac{rad}{sec} \times \frac{2^3}{12^3} ft^3 \times 0.0319\ ft \times 10 \frac{ft}{sec} = 6.50\ ft \cdot lb_f$$

and the power required is given by equation 13.7

$$\dot{W}_{shaft} = \omega T_{shaft} = 361 \frac{rad}{sec} \times \frac{6.50}{550} hp = 4.27\ hp$$

The actual input torque and power may be higher than the estimates found here because of energy losses in the system, which are not accounted for in the analysis. ∎

13.6 RELATIVE PERFORMANCE MEASURES

An important parameter is the efficiency of the turbine, which measures how much useful work is done compared to the theoretical maximum. For a hydraulic turbine with a head difference H (the difference between the highest and lowest water levels), there is a certain amount of ideal power available called the hydraulic power $\dot{W}_h$, given by

$$\dot{W}_h = \rho \dot{q} g H \tag{13.18}$$

and the efficiency is defined as $\eta_t = \dot{W}_{shaft}/\dot{W}_h$, so that

$$\eta_t = \frac{\dot{W}_{shaft}}{\dot{W}_h} = \frac{\omega T_{shaft}}{\rho \dot{q} g H} \tag{13.19}$$

For pumps, $\dot{W}_h$ measures the power produced by the pump and $\dot{W}_{shaft}$ the power applied to the pump, so that the efficiency is defined as $\eta_p = \dot{W}_h / \dot{W}_{shaft}$ and

$$\eta_p = \frac{\dot{W}_h}{\dot{W}_{shaft}} = \frac{\rho \dot{q} g H}{\omega T_{shaft}} \tag{13.20}$$

The optimum efficiency of turbines as a function of specific speed is shown in Figure 13.19. Here, the specific speed N'_{sd} is defined by

$$N'_{sd} = \frac{N(rpm)\sqrt{\dot{W}_{shaft}(hp)}}{[H(ft)]^{5/4}} \tag{13.21}$$

and it is calculated in U.S. customary units of *rpm, bhp,* and *ft,* so it is not dimensionless. The figure illustrates that the three different types of hydraulic turbine are suited to different operating ranges, based on the value of the specific speed.

> The specific speed is the principal factor determining the choice of turbine.

These curves may be compared to the optimum efficiencies of different pump types, which are shown as a function of specific speed in Figure 13.20. Here, the specific speed N_s is defined by:

$$N_s = \frac{N(rpm)\sqrt{\dot{q}(gpm)}}{[H(ft)]^{3/4}} \tag{13.22}$$

and it is also calculated in U.S. customary units of *rpm, gpm,* and *ft,* so it is also not dimensionless. We see a very similar behavior based on the value of the specific speed: radial-flow machines are suited to low values, mixed-flow devices are suited to intermediate values, and axial-flow machines are suited to high values.

The efficiency behavior of two models of a particular commercial centrifugal water pump was shown for a range of sizes and discharge values in Figure 13.17. The behavior is quite complex, and the challenge to the designer is to design or specify the pump so that it operates as close as possible to its optimum value for its given duty cycle.

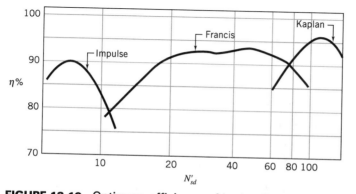

FIGURE 13.19 Optimum efficiency of hydraulic turbines as a function of specific speed. The specific speed N'_{sd} (Equation 13.21) is calculated in U.S. customary units of *rpm, bhp,* and *ft,* so it is not dimensionless. From Munson, Young, and Okishii, *Fundamentals of Fluid Mechanics,* John Wiley & Sons, 1998.

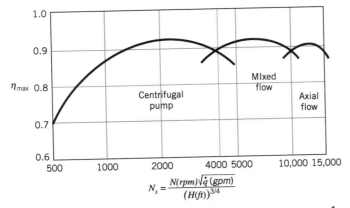

$$N_s = \frac{N(rpm)\sqrt{\dot{q}\,(gpm)}}{(H(ft))^{3/4}}$$

FIGURE 13.20 Optimum efficiency of pump types as a function of specific speed. The specific speed N_s is calculated in U.S. customary units of *rpm, gpm,* and *ft,* so it is not dimensionless. From F. M. White, *Fluid Mechanics,* McGraw-Hill, New York, 1986.

13.7 DIMENSIONAL ANALYSIS

As we have seen, it is possible to derive relatively simple equations to describe the performance of turbomachines. However, the equations were obtained under the assumption of steady, one-dimensional flow, and in most cases, the actual flow through a turbomachine is extremely complex. The actual performance will differ from the ideal performance through the effects of nonuniform velocity distributions, viscous losses, separation losses, and mechanical friction. To estimate the losses in a given machine, the full equations of motion need to be solved. However, an analytical approach is not generally possible. Even computer-generated solutions of the governing equations can provide only a limited set of guidelines for design purposes and therefore, new designs always need to be tested experimentally, usually by building a model. Then, in order to make predictions regarding the full-scale performance, we need to use dimensional analysis. Scale modeling, dimensional analysis, and experiment are often the only practical approaches, although a basic understanding of the underlying physical phenomena is essential.

The performance of a turbomachine is described in terms of the pressure difference across the machine Δp, the flow rate through the machine $\dot{q}$, the size of the machine (indicated by, for example, the diameter of the rotor D), the rotational speed N (usually measured in *rev/s*, or *rps*), and the properties of the fluid, ρ and μ. That is,

$$\Delta p = f''(D, N, \dot{q}, \rho, \mu) \tag{13.23}$$

By dimensional analysis we obtain

$$\frac{\Delta p}{\rho N^2 D^2} = f'\left(\frac{\dot{q}}{ND^3}, \frac{ND^2}{\nu}\right) \tag{13.24}$$

The pressure difference is often expressed in terms of the head difference (gH), so that

$$\frac{gH}{N^2 D^2} = f_1\left(\frac{\dot{q}}{ND^3}, \frac{ND^2}{\nu}\right) \tag{13.25}$$

Alternatively, we could write the left hand side of equation 13.24 in terms of the power of the machine P, where $P \propto \Delta p A V$, where A and V are a representative area and velocity, respectively. In terms of the primary variables D and N, we could write $P \propto \Delta p D^3 N$, so that

$$\frac{P}{\rho N^3 D^5} = f_2 \left(\frac{\dot{q}}{ND^3}, \frac{ND^2}{\nu} \right) \tag{13.26}$$

Similarly, we can write the left hand side of equation 13.24 in terms of the torque of the machine T, where $T \propto \Delta p A D$, so that in terms of the primary variables $T \propto \Delta p D^3$, and

$$\frac{T}{\rho N^2 D^5} = f_3 \left(\frac{\dot{q}}{ND^3}, \frac{ND^2}{\nu} \right) \tag{13.27}$$

As we saw in Section 13.2, the definition of the efficiency η depends on whether the shaft work is done on the fluid or whether the fluid does work on the shaft. For the case of a hydraulic turbine, the mechanical output power is less than the hydraulic power, so that

$$\eta_t = \frac{\rho \dot{q} g H}{TN} = f_t \left(\frac{\dot{q}}{ND^3}, \frac{ND^2}{\nu} \right) \tag{13.28}$$

For a pump, the input power is greater than the hydraulic power produced, so that

$$\eta_p = \frac{TN}{\rho \dot{q} g H} = f_p \left(\frac{\dot{q}}{ND^3}, \frac{ND^2}{\nu} \right) \tag{13.29}$$

The dimensionless parameters developed here have the following names:

$$\frac{gH}{N^2 D^2}, \qquad \text{head coefficient, } C_H$$

$$\frac{P}{\rho N^3 D^5}, \qquad \text{power coefficient, } C_P$$

$$\frac{T}{\rho N^2 D^5}, \qquad \text{torque coefficient}$$

$$\frac{\dot{q}}{ND^3}, \qquad \text{flow coefficient, } C_Q$$

$$\frac{g(NPSH)}{N^2 D^2}, \qquad \text{suction-head coefficient, } C_{HS}$$

$$\frac{ND^2}{\nu}, \qquad \text{Reynolds number}$$

Here, N is usually measured in rev/s, not rad/s. Note that the power coefficient defined here differs from that given in equation 13.12 by the exclusion of some dimensionless constants such as $\frac{1}{2}$ and π.

An additional dimensionless parameter called the *specific speed* N''_s can be formed by dividing the flow coefficient by the head coefficient and taking the square root

$$N''_s = \frac{\sqrt{\frac{\dot{q}}{ND^3}}}{\sqrt{\frac{gH}{N^2 D^2}}} = \frac{N\sqrt{\dot{q}}}{(gH)^{3/4}} \qquad \text{specific speed} \tag{13.30}$$

From equation 13.25 we then have

$$N''_s = \frac{N\sqrt{\dot{q}}}{(gH)^{3/4}} = f_4 \left(\frac{\dot{q}}{ND^3}, \frac{ND^2}{\nu} \right) \tag{13.31}$$

The parameter N''_s is nondimensional, but the quantities N'_{sd} and N_s, defined by equations 13.21 and 13.22, are not.

The Reynolds number for most turbomachines is large, even for scale models, so that turbulent flow will usually occur in the model and the prototype. In that case, it has been found that the effects of viscosity can often be neglected. It is generally assumed, therefore, that

$$\frac{gH}{N^2D^2} = g_1\left(\frac{\dot{q}}{ND^3}\right) \tag{13.32}$$

$$\frac{P}{\rho N^3D^5} = g_2\left(\frac{\dot{q}}{ND^3}\right) \tag{13.33}$$

$$\frac{T}{\rho N^2D^5} = g_3\left(\frac{\dot{q}}{ND^3}\right) \tag{13.34}$$

and

$$\eta_{t,p} = g_{t,p}\left(\frac{\dot{q}}{ND^3}\right) \tag{13.35}$$

These functional relationships provide the basis for model testing. They are also used in presenting experimental data, although, as we noted in Section 13.6, dimensional parameters are commonly used in the industry.

To demonstrate the scaling laws for pump performance, the dimensional data given in Figure 13.17 are replotted non-dimensionally in Figure 13.21. The collapse of the data is reasonably convincing, especially for the power and suction-head coefficients. The nondimensional curves can then be used to scale a wide variety

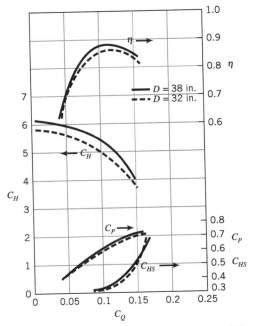

FIGURE 13.21 Nondimensional plot of the pump performance data given in Figure 13.17. These data are not representative of other pump designs. The coefficients are evaluated using N in *revs/s*. From F. M. White, *Fluid Mechanics*, McGraw-Hill, New York, 1986.

of pumps of the same family, as long as the performance does not depart too far from the range covered by the original experimental data.

13.8 PROPELLERS AND WINDMILLS

Another topic of interest is the performance of propellers and windmills. They are considered separately from pumps and turbines, in that they usually operate in a free flow, that is, they are usually unshrouded.

Propellers and windmills are devices that use airfoil sections to change the fluid pressure in order to produce a force, and the change in pressure occurs because the fluid momentum changes. The blade of an airplane propeller, for example, has an airfoil cross-section at each radial position. The speed of rotation is proportional to the radius and therefore, the direction of the resultant velocity also changes with radial distance. The propeller blade is twisted so that the angle of attack stays within the design limits of the airfoil, and that lift is produced efficiently at each section. The blades that drive a simple cooling fan are also a type of propeller, but they are usually made of a twisted sheet of metal or plastic to reduce costs. Ship propellers, on the other hand, need to be designed for maximum thrust while limiting the tip speed to avoid cavitation (Figure 8.1).

A propeller increases the fluid momentum, and a windmill decreases it, but the analysis of their performance is very similar and only the direction of the resultant force is different. We begin with the analysis of a propeller. We will assume that the force due to pressure differences accelerates the flow in the axial direction without swirl, and that flow friction can be neglected. Under these assumptions, we can find the thrust and relative speed V across the propeller. Let F be the force exerted by the propeller on the fluid (Figure 13.22). Relative to the propeller, the velocity of the flow far upstream is V_1. We choose a frame of reference moving with the propeller (we need to move at a velocity V_1), so that the flow is steady with respect to the observer. The fluid is assumed to be incompressible, and the propeller sweeps out an area A.

The propeller draws fluid in over an area larger than A, and expels it over an area smaller than A. If we draw the streamlines that bound the fluid flow through the propeller, we define a *slipstream* boundary. The volume defined by the slipstream boundary and the areas 1 and 4, where the flow velocity is parallel and in the streamwise direction, is often used as the control volume for the analysis of a propeller or windmill (CV1). A second control volume is used to enclose the propeller itself (CV2). This volume is approximated to be rectangular. Under the

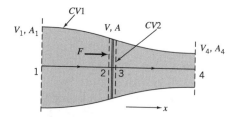

FIGURE 13.22 Control volume for a propeller.

assumption of one-dimensional flow, the continuity and x-momentum equations for $CV2$ are

$$\rho V_2 A = \rho V_3 A$$

and

$$F + p_2 A - p_3 A = -\rho V_2^2 A + \rho V_3^2 A$$

respectively. The continuity equation gives $V_2 = V_3$: there is no jump in velocity across the propeller, only a jump in pressure. Hence,

$$F = (p_3 - p_2)A$$

Bernoulli's equation cannot be applied across $CV2$ because the propeller increases the energy of the flow (the pressure rise does work on the flow). However, for $CV1$, Bernoulli's equation applies between sections 1 and 2, and between sections 3 and 4. So

$$p_1 + \tfrac{1}{2}\rho V_1^2 + \rho g h_1 = p_2 + \tfrac{1}{2}V_2^2 + \rho g h_2$$
$$p_3 + \tfrac{1}{2}\rho V_3^2 + \rho g h_3 = p_4 + \tfrac{1}{2}V_4^2 + \rho g h_4$$

The pressure on the slipstream boundary will be assumed to be equal to atmospheric pressure, so that $p_1 = p_4 = p_a$. Also, from continuity, $V_2 = V_3\ (=V)$, and if we neglect hydrostatic head differences across the streamlines (this is reasonable if they are small compared to $p_3 - p_2$), then

$$\tfrac{1}{2}\rho V_1^2 = p_2 + \tfrac{1}{2}\rho V^2$$

and

$$p_3 + \tfrac{1}{2}\rho V^2 = \tfrac{1}{2}\rho V_4^2$$

Therefore

$$p_3 - p_2 = \tfrac{1}{2}\rho(V_4^2 - V_1^2)$$

and

$$F = \tfrac{1}{2}\rho(V_4^2 - V_1^2)A \tag{13.36}$$

If we now apply the continuity and x-momentum equations between sections 1 and 4 on $CV1$, we obtain, respectively,

$$\rho V_1 A_1 = \rho V_4 A_4 = \dot{m}$$
$$F = (-\dot{m})V_1 + (+\dot{m})V_4 = \dot{m}(V_4 - V_1) \tag{13.37}$$

Eliminating F from equations 13.36 and 13.37, we find

$$V = \tfrac{1}{2}(V_1 + V_4) \tag{13.38}$$

so that the velocity through the propeller disk is just the average of the velocities at stations far upstream and downstream of the propeller. In other words, there is the same increase of velocity ahead of the propeller as there is behind it.

Finally, consider the power required to drive the propeller. The work done by the propeller is equal to the thrust F multiplied by a distance. In a time Δt, the propeller moves a distance $V_1\Delta t$, and so the work done per unit time (= the power P), is given by

$$P = FV_1 = \dot{m}(V_4 - V_1)V_1$$

The rate of change in the kinetic energy of the slipstream between sections 1 and 4, P_{KE}, is given by the mass flow rate times half the difference in the velocities squared, so that

$$P_{KE} = \tfrac{1}{2}\dot{m}(V_4^2 - V_1^2)$$

(see Section 5.1). The ideal efficiency of the propeller is then given by

$$\eta_{prop} = \frac{P}{P_{KE}} = \frac{V_1}{V}$$

and since $V > V_1$, the efficiency of an ideal propeller is always less than 100% (actual ship and airplane propellers approach 80 to 85%).

The magnitude of the thrust produced by a windmill is the same as that calculated for a propeller, but it acts in the opposite direction. The average velocity through the windmill is still given by equation 13.38, but since the windmill reduces the kinetic energy of the flow, the rate of change in the kinetic energy is now given by

$$P_{KE} = \tfrac{1}{2}\dot{m}(V_1^2 - V_4^2)$$

For a windmill, V_1 is the undisturbed wind speed. The efficiency for a windmill is defined by comparing the rate of change of the kinetic energy produced by the windmill P_{KE} to the kinetic energy flux through a streamtube of undisturbed flow, equal in area to the area of the windmill itself. This energy flux, P_{KEF}, is given by

$$P_{KEF} = \tfrac{1}{2}\rho V_1 A V_1^2$$

The ideal efficiency of a windmill is then given by

$$\eta_w = \frac{P_{KE}}{P_{KEF}} = \frac{(V_1 + V_4)(V_1^2 - V_4^2)}{2V_1^3}$$

The maximum efficiency is found by differentiating η_w with respect to V_4/V_1 and setting the result equal to zero. This gives a maximum efficiency of 59.3%.[5] Because of friction and other losses, this efficiency is not found in practice: the highest possible efficiency for a real windmill appears to be about 50% (Figure 13.23), but example 13.3 shows that the traditional Dutch windmill operates at an efficiency of only about 15%.

In this analysis we have made a number of assumptions. In particular, we assumed that the pressure on the slipstream boundary was equal to atmospheric pressure. At the same time, we assumed that the pressures over sections 2 and 3 were different from atmospheric, which leads to a mismatch in pressures at the tips of the propeller. In reality, the pressure cannot be uniform over the propeller disk.

An alternative control volume is sometimes used to avoid these difficulties. Here, a cylindrical control volume is used with a very large diameter (Figure 13.24). The following boundary conditions are then applied: (1) the pressure is atmospheric

---

[5] Glauert, H., Airplane propellers, *Aerodynamic Theory*, Vol. IV, ed. W. F. Durand. Published by Dover Publications, New York, 1963.

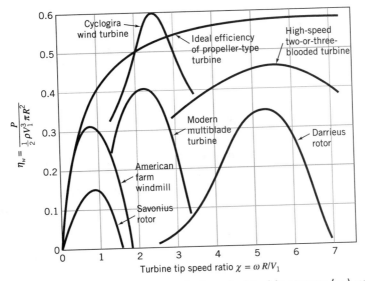

FIGURE 13.23 Efficiency trends for wind turbine types (η_ω) versus tip speed ratio $\chi(= \omega R/V_1)$. From Baumeister, T., Avallone, E. A., and Baumeister, T. III, eds., *Mark's Standard Handbook for Mechanical Engineers,* 8th. ed. McGraw-Hill, New York, 1986.

on the control volume surface, (2) there is negligible flow across the cylindrical part of the control volume surface, and (3) the change of fluid momentum outside the slipstream boundary is negligible. The first assumption seems reasonable, since the effects of pressure die off as velocity squared. However, even if the flow velocities v_c across the area A_c of the cylindrical part of the surface are very small, the product $v_c A_c$ may still be important. Also, the change in momentum of the fluid outside the slipstream boundary may not be negligible. It turns out that the error in the second assumption cancels out the error in the third assumption, and the same answer is obtained as in our original analysis. In fact, it can be shown that these two assumptions are equivalent to that made for the boundary conditions on the slipstream boundary control volume $CV1$.

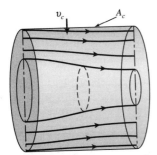

FIGURE 13.24 Alternative control volume for a propeller.

13.9 WIND ENERGY GENERATION

We saw that the theoretical upper limit on the efficiency of a propeller is 100%, so that in the absence of losses in an ideal flow, all the energy supplied by the propeller is converted into flow energy. In contrast, the ideal efficiency of a windmill is limited to about 59%, and in practice it is often much less. This limits the generation of power using wind turbines, and great efforts have been made to improve the efficiency and performance of wind turbines.

The conventional wind turbine is mounted horizontally, and a vane is used to align the propeller with the wind direction. Vertical axis machines, such as the Savonius rotor and the Darrieus turbine, have the advantage that their operation is independent of the wind direction. The Savonius rotor consists of two curved blades forming an S-shaped passage for the airflow, and the Darrieus turbine has two or three airfoils attached to a vertical shaft (Figure 13.25). They are typically shaped so that under load, the stress distribution is constant along the length of the airfoil. The maximum efficiency of a Savonius rotor is only about 15%, whereas the Darrieus turbine has a superior maximum efficiency of about 32%, but the Darrieus turbine is not self-starting.

The efficiency of a wind turbine η_w is a strong function of the tip speed ratio χ, which is the ratio of the maximum speed seen by the blade (which occurs at the tip of a propeller blade) and the speed of the incoming air stream. That is

$$\chi = \frac{\omega R}{V_1} \tag{13.39}$$

where ω is the rotation rate in *rad/s*, R is the radius of the turbine, and V_1 is the wind speed. Figure 13-23 illustrates that each type of wind-powered system has a corresponding tip speed ratio for maximum efficiency.

There are three main difficulties in designing a wind-powered generating system. First, the wind strength and direction continually change. The variation in the wind direction is not a severe problem, since the machine can be designed to be self-aligning, but the variation in the wind strength is. The variation in average wind strength in the United States is illustrated in Figure 13.26, and locations with little or no prevailing wind strength are obviously not suitable for producing wind energy. On the other hand, "runaway" can occur when storms produce unusually

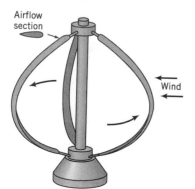

FIGURE. 13.25 Darrieus rotor. From Baumeister, T., Avallone, E. A., and Baumeister, T. III, eds., *Mark's Standard Handbook for Mechanical Engineers*, 8th. ed. McGraw-Hill, New York, 1986.

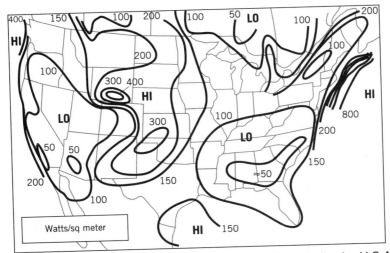

FIGURE 13.26 Annual average available wind power in the U.S.A. in W/m^2. From Baumeister, T., Avallone, E. A., and Baumeister, T. III, eds., *Mark's Standard Handbook for Mechanical Engineers,* 8th. ed. McGraw-Hill, New York, 1986.

strong winds. The rotor can then begin to turn at such high speeds that the material strength limits are exceeded and the blade elements fail. A governor is required to limit the rotation rate, or the blades can be designed to twist, so that at high wing loading corresponding to high wind speeds the blades produce less lift and runaway is prevented.

Perhaps the most severe shortcoming due to variable wind strength is the long-term variations that result in power output. This is not a particular limitation for pumping water from a well, where a large reservoir can be used to average out the fluctuations in the water flow. However, when the windmill is used to generate electricity, a battery storage system is required, or the windmill must be combined with another source of electricity (such as a fossil-fuel driven system) to produce a constant power output. In commercial systems, the wind-driven system is usually connected to a "grid," that is, a common distribution system which receives energy from a variety of sources and regulates and distributes the power supply to the customers. Alternatively, the irregular power output from the windmill can be used to pump water to a reservoir located on top of a nearby hill, and then a water-powered turbine can be used to generate electricity in response to the demand.

The second difficulty in designing a wind-powered generating system is that the flow energy contained in the wind is rather low. For example, if we took a wind speed of 15 *m/s* (about 33 *mph,* a rather stiff breeze) and a rotor diameter of 3 *m* (about 10 *ft*), we find that the maximum power available (with a theoretical maximum efficiency of 59.3%) is given by

$$\text{max theoretical power} = \tfrac{1}{2}\eta_w \rho V_1^3 \pi R^2$$
$$= 8.49\,kW$$

which is enough to run a typical house. However, the average power generated by this system will be much lower because of variability in the wind speed and the much lower efficiencies found in actual systems. There are great benefits to increasing the size of the rotor; doubling the rotor diameter will increase the power by a factor of four. However, the static stresses in the blades due to wind loads will double and their deflection will increase by a factor of eight. With rotation, the static load will be augmented by the inertia force due to the centripetal acceleration. At any location along the blade, this force is proportional to $(\omega R)^2/R = \omega^2 R$, so that it increases linearly with the size and quadratically with the rotational speed. We see that the size and rotation rate of a wind-powered turbine is limited by the possibility of material failure, and to generate significant amounts of electricity, a large number of machines will be required. Such "wind farms" exist, and a large scale example is located east of San Francisco (see Figure 13.27).

The third difficulty is that the rotational speeds generated by common wind turbines are not well matched to generator speeds. Generators in the United States are typically designed to run at 3600 *rpm*, to produce alternating electric current at 60 *Hz*. If we choose a high-speed two-bladed design, operating at peak efficiency, we see from Figure 13.25 that the tip speed ratio χ should equal about 5.5. For the example given earlier, we will need a rotational speed $\omega = V_1 \chi / R = 55 \ rad/s = 525 \ rpm$. So, even if the wind speed was constant at 15 *m/s*, a speed-up gearbox will be required. The variability of the wind speed will mean that the gearbox will need to operate over a large speed range, and that the frequency of the alternating current will vary with wind speed. Consequently, a considerable amount of control and regulation circuitry is necessary for the output to be useful for running standard appliances, or to be acceptable to a grid.

EXAMPLE 13.3 *Windmill Performance*[6]

Calculate the ideal efficiency, actual efficiency, and thrust for a Dutch windmill with $D = 26 \ m$, $\omega = 20 \ rpm$, $V_1 = 36 \ km/hr$, and power output of 41 *kW*. Assume that $V_4/V_1 = 0.5$.

Solution: We have

$$\omega = 20 \ \frac{rev}{min} \times 2\pi \ \frac{rad}{rev} \times \frac{min}{60 \ sec} = 2.09 \ rad/s$$

The tip speed ratio is given by equation 13.39

$$\chi = \frac{\omega R}{V_1} = \frac{2.09 \times 13 \times 3600}{36,000} = 2.72$$

Figure 13.25 indicates that the ideal efficiency for this value of χ is about 0.53. The actual efficiency is given by

$$\eta_{actual} = \frac{P_{actual}}{P_{KEF}} = \frac{P_{actual}}{\frac{1}{2}\rho V_1^3 \frac{\pi}{4} D^2}$$

With $\rho = 1.2 \ kg/m^3$,

$$\eta_{actual} = 0.129$$

which is only 24% of the ideal efficiency at this tip speed ratio.

---

[6] This example was adapted from Fox and Macdonald, *Introduction to Fluid Mechanics*, John Wiley and Sons, 4th ed., 1992.

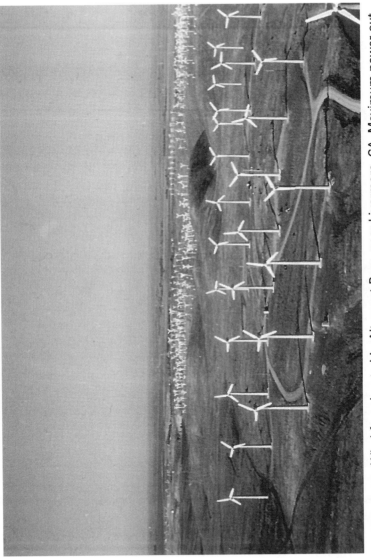

FIGURE 13.27 Wind farm located in Altamont Pass, near Livermore, CA. Maximum power output is 5 *MW* at a wind speed of 15 *mph*. Courtesy U.S. Department of Energy.

The thrust is given by equation 13.36. With $V_4/V_1 = 0.5$, we have

$$F = \tfrac{1}{2}\rho V_1^2 \left(\frac{V_4^2}{V_1^2} - 1 \right) \tfrac{\pi}{4} D^2 = -\tfrac{3\pi}{32} \rho V_1^2 D^2$$

That is,

$$F = -\tfrac{3\pi}{32} \times 1.2 \times 10^2 \times 26^2 N = 23.9 \, kN$$

The thrust is negative in the sense that it is opposite in direction to the thrust developed by a propeller.   ∎

PROBLEMS

13.1   A hydraulic turbine generates 50,000 hp at 75 *rpm* with a head of 100 *ft*. Find the specific speed and use Figure 13.19 to determine what type of turbine is being used.

13.2   A hydraulic turbine operates at a specific speed of 100 (U.S. customary units) and delivers 500 *kW* with a head of 10 *m*. What type of turbine is being used, and what is the rotational speed necessary to give the optimum efficiency?

13.3   When it operates at peak efficiency, a turbine delivers 25,000 hp at a speed of 450 *rpm* with a head of 4500 *ft*. What type of turbine is being used? Estimate the flow rate and the size of the machine, assuming there are no losses in the flow upstream of the turbine.

13.4   A Pelton wheel of 4 *m* diameter operates with a head of 1000 *m*. Estimate the flow rate and power output of the machine, assuming that it operates at peak efficiency and there are no losses in the flow upstream of the turbine.

13.5   A Pelton wheel operates with a head of 400 *m* at 350 *rpm,* and it is powered by a jet 12 *cm* in diameter. Find the specific speed (U.S. customary units) and the wheel diameter, if there are no losses in the flow upstream of the turbine and it operates at peak efficiency.

13.6   A turbine of the Francis type operates with a head of 200 *m* and flow rate of 3 m^3/s. The peak efficiency occurs at a head coefficient C_H of 9, a flow coefficient C_Q of 0.3, and a power coefficient C_P of 2.5 (as defined in Section 13.7). Estimate the size of the machine, the maximum power produced, and the speed.

13.7   A pump delivers 0.7 ft^3/s of water against a head of 50 *ft* at a speed of 1750 *rpm*. Find the specific speed.

13.8   A water pump delivers 0.25 m^3/s against a head of 20 *m* at a speed of 2400 *rpm*. Find the specific speed in U.S. customary units.

13.9   For the previous two problems, determine the type of pump that would give the highest efficiency by using Figure 13.20.

13.10   A radial-flow pump is required to deliver 1000 *gpm* against a total head of 350 *ft*. Find the minimum practical speed using Figure 13.20.

13.11   An axial-flow pump runs at 1,800 *rpm* against a head of 1200 *ft*. Find the flow rate delivered by the most efficient pump using Figure 13.20.

13.12   Axial-flow pumps are to be used to lift water a combined height of 150 *ft* at a rate of 30 ft^3/s. Each pump is designed to operate at 1800 *rpm*. Using Figure 13.20, find the number of pumps required if they are all operating at their maximum efficiency. Neglect losses in the piping.

13.13 Using the curves given in Figure 13.17 for a centrifugal water pump with an impeller of 32 *in.* diameter, operating at 2000 *hp* find:

(a) the volume flow rate, the total head, and the efficiency

(b) the specific speed.

13.14 For the pump described in the previous problem, the inlet is located a distance h_i above the surface of the reservoir, and there are no losses in the inlet piping. The temperature of the water is 50°F. Find the maximum value of h_i for which cavitation will not be a problem.

13.15 Using Figure 13.21, find the effect of a 50% increase in impeller diameter on the rotation rate and the volume flow rate of the centrifugal water pump described in Problem 13.16, if the total head and power input remain the same, and the pump continues to operate at its best efficiency.

13.16 Use Figure 13.21 to find the impeller diameter, power input, and head for a pump of the same family, if it operates at 900 *rpm* with a volume flow rate of 5000 *gpm* at peak efficiency.

13.17 A basement sump pump provides a discharge of 12 *gpm* against a head of 15 *ft*. What is the minimum horsepower required to drive this pump, given that its efficiency is only 60%?

13.18 For a given pump, the gauge pressures at the inlet and outlet are -30×10^3 *Pa* and 200×10^3 *Pa*, respectively. If the flow rate is 0.1 m^3/s, and the power required is 25 *kW*, find the efficiency. The intake and discharge are at the same height.

13.19 The impeller of a radial-flow turbine has an inlet radius of 4 *ft*, turning at 200 *rpm*. The volume flow rate is 1000 ft^3/s, and it discharges in the axial direction. Neglect all losses. If the tangential component of the velocity at the inlet is 5 *ft/s*, find:

(a) the torque exerted on the impeller

(b) the power developed by the machine.

13.20 The diameter of an axial-flow turbine impeller is 2.8 *m*, and the inlet flow makes an angle of 10° with the circumferential direction. The volume flow rate is 24 m^3/s over an area of 8 m^2 with a head of 100 *m*, and it discharges in the axial direction without swirl. Find the speed of the runner. Neglect losses.

13.21 A centrifugal water pump with a 200 *mm* diameter impeller rotates at 1750 *rpm*. Its width is 20 *mm* and the blades are curved backwards so that $\beta_2 = 60°$. If the flow enters the pump in the axial direction, and the volume flow rate is 100 *L/s*, estimate the power input if the pump is 100% efficient.

13.22 A centrifugal pump is designed to have a discharge of 600 *gpm* against a head of 200 *ft*. The impeller outlet diameter is 12 *in.*, and its width is 0.5 *in.* The outlet blade angle is 65°. What is the design speed and the minimum horsepower required to drive this pump?

13.23 A centrifugal pump has inner and outer impeller diameters of 0.5 *m* and 1 *m*, respectively, and a width of 0.15 *m*. The outlet blade angle is 65°. At 350 *rpm*, the volume flow rate is 4 m^3/s. Find:

(a) the exit blade angle so that the water enters the pump in the radial direction

(b) the minimum power required

13.24 A propeller 1 *ft* in diameter rotates at 1200 *rpm* in water and absorbs 20 *hp*. Find the power coefficient, and estimate the power required to increase the speed to 1500 *rpm* if the power coefficient remains constant.

13.25 A high-speed two-bladed wind turbine 35 *m* in diameter operates at its peak efficiency in winds of 30 *km/hr*. Estimate the power generated, the rotor speed, and the wind velocity in the wake.

13.26 A propeller 2 *ft* in diameter moves at 20 *ft/s* in water and produces a thrust of 1000 *lb*$_f$. Find the ratio of the upstream to downstream velocities, and the efficiency.

13.27 An airplane flies at 140 *mph* at sea level at 60°*F*. The propeller diameter is 8 *ft*, and the slipstream has a velocity of 200 *mph* relative to the airplane. Find:
(a) the propeller efficiency
(b) the flow velocity in the plane of the propeller
(c) the power input
(d) the thrust of the propeller
(e) the pressure difference across the propeller disk.

13.28 An airplane flies at 300 *km/hr* at sea level at 20°*C*. The propeller diameter is 1.8 *m*, and the velocity of the air through the propeller is 360 *km/hr*. Find:
(a) the slipstream velocity relative to the airplane
(b) the thrust
(c) the power input
(d) the power output
(e) the efficiency
(f) the pressure difference across the propeller disk.

13.29 On a high-speed submarine the propeller diameters are limited to 15 *ft*, and their maximum rotational speed is set at 200 *rpm* to avoid cavitation. If each propeller has an efficiency of 86% and it is limited to 10,000 *hp*, find the minimum number of propellers required to move at a speed of 35 *mph*, and the torque in each shaft.

13.30 A modern multi-bladed windmill design is adapted to work in a tidal flow. When the water flows at 5 *m/s*, find the maximum power generated for a diameter of 4 *m*. Calculate the tip speed, and determine if cavitation could be a problem.

13.31 An American farm windmill is used to pump water from a 100 *ft* deep well through clean plastic pipe of 2 *in.* diameter. If the rotor diameter is 6 *ft*, estimate the volume flow rate when the winds are blowing at 20 *mph*. Assume that the windmill is working at peak efficiency, and the pump is 90% efficient. Ignore minor losses. What is the expected flow rate when the wind drops to 15 *mph*?

CHAPTER 14

ENVIRONMENTAL FLUID MECHANICS

14.1 ATMOSPHERIC FLOWS

The understanding of wind flows and atmospheric transport processes is crucial to, for example, the ability to predict and quantify the dispersion of pollutants in the atmosphere, estimating the greenhouse effect, and predicting the weather.

The atmosphere is made up of two basic layers: the *troposphere* and the *stratosphere*. In the troposphere, the temperature decreases linearly with height (the slope of the curve is called the *lapse rate*), whereas in the stratosphere, the temperature remains approximately constant with height. The lapse rate, and the height of the troposphere, vary with time and location but the U.S. Weather Service has compiled a set of average conditions over the United States at 40° N latitude called the U.S. Standard Atmosphere.

According to this standard atmosphere, which does not necessarily give an accurate description of the atmosphere at any one time or place, the stratosphere extends from sea level to a height of 11 *km* (36,000 *ft*), with a lapse rate of 6.50°*K/km* (called the *standard lapse rate*). The stratosphere begins at the top of the troposphere and extends to an altitude of 32.2 *km* (106,000 *ft*), and its temperature is constant at 216.7°*K* (−56.5°*C*) up to an altitude of 20.1 *km* (66,000 *ft*).[1] The troposphere contains 80 to 85% of the total mass of the atmosphere and virtually all of its water, and therefore it plays the major role in determining our weather and climate. Above 20.1 km, the temperature gradually increases with height, due to the absorption of the sun's infrared radiation by ozone, which is formed by the intense ultraviolet radiation from the sun. It is also this absorption by ozone that protects living things on earth from the destructive effects of ultraviolet rays. The temperature profile of the U.S. Standard Atmosphere is shown in Figure 14.1.

To put these altitudes into perspective, Mount Everest has a height of about 29,000 *ft* (8840 *m*), long range airplanes travel near the top of the troposphere at about 35,000 *ft* (10,670 *m*), and the Concorde travels well into the stratosphere at about 56,000 *ft* (17,070 *m*). Most clouds appear at heights less than about 10 to 12 *km* so that they are mainly confined to the troposphere, although occasionally the top of particularly large clouds may extend to 18 *km* or higher.

The atmosphere extends well beyond the stratosphere, but the air densities become very small. For example, at the top of the stratosphere the density is only 1% of the sea level value. Vehicles traveling in the upper reaches of the atmosphere (called the *ionosphere*) experience very low density conditions where the continuum approximation may no longer apply.

---

[1] Some authors consider the stratosphere to extend only up to 20.1 km, so that it is confined to the region of constant temperature.

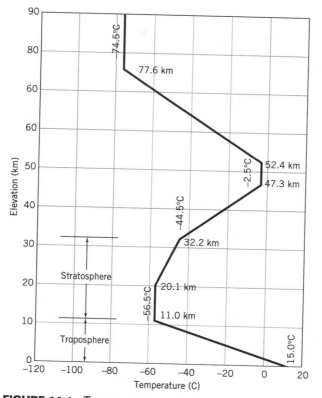

FIGURE 14.1 Temperature variation with altitude in the U.S. Standard Atmosphere. Data from *The U.S. Standard Atmosphere,* U.S. Government Printing Office, Washington, D.C., 1976.

14.2 EQUILIBRIUM OF THE ATMOSPHERE

The lower part of the atmosphere, that is, the troposphere, is constantly being mixed by convection.[2] Water vapor is taken up and then precipitated, and there are great movements of polar air and of tropical air into the temperate regions. In the tropics, due to the intense heating of the ground, air is rising. This produces a large scale circulation of air from the temperate regions toward the tropics (*trade winds*) and a drift near the top of the troposphere from the tropics toward the poles. Cold air from the higher levels will settle toward the ground to maintain continuity. This convective mixing of the atmosphere is due to unequal heating of the earth's surface, caused by the variation of solar intensity with latitude and by the unequal heat-absorption characteristics of land and water areas. As a result, the troposphere is kept in some relatively stable thermal and mechanical equilibrium which determines our climate. Departures from this equilibrium are what we know as weather. Weather varies from day to day or even from hour to hour in an irregular manner.

---

[2] The material in this section was adapted from *Engineering Applications of Fluid Mechanics,* by J. C. Hunsaker and B. G. Rightmire, published by McGraw-Hill, 1947.

Climate is determined by the average weather at a given place at a particular time of year, over many seasons, and climatology predicts from past records the safe date for planting crops, for example. Meteorology predicts tomorrow's weather from a knowledge of the present condition of the atmosphere near a given locality.

The fundamental nature of convection suggests that the lower atmosphere should have an approximately adiabatic temperature gradient. Air next to the ground that is heated more in one place than in another will rise like warm air in a chimney. Colder and denser air will flow in to take its place. The rising air will cool almost adiabatically, for air is a poor conductor. To find this cooling rate, we start with the hydrostatic equation ($dp/dz = -\rho g$) and use the thermodynamic relationships for adiabatic, inviscid flow (that is, isentropic flow) given by equation 12.18. We obtain the *adiabatic lapse rate*:

$$\text{adiabatic lapse rate} = \left(-\frac{dT}{dz}\right)_{ad} = \frac{\gamma - 1}{\gamma R} g \tag{14.1}$$

which is equal to $0.0098°C/m$ for dry air.

It is interesting to observe that the natural temperature lapse rate makes for stability of the lower atmosphere; that is, an air mass displaced vertically for some reason, tends to return to its original level. Consider a mass of air rising by convection because it became slightly unstable at ground level due to local heating. This air will cool at about the adiabatic lapse rate of $0.98°C$ per $100\ m$. At any level, however, it will be subject to the same pressure as the surrounding atmosphere. Hence, since the usual atmospheric lapse rate is less than the adiabatic lapse rate (it is only $0.65°C$ per $100\ m$), the displaced air mass will be denser than its surroundings and will tend to fall back.

Usually, the air is stable, and large-scale convection does not occur. However, when the atmospheric lapse rate exceeds the adiabatic rate, the air is unstable. Thunderstorms arise in an unstable atmosphere in which the lapse rate is temporarily excessive from abnormal heating of the ground and the lower levels of the air. This condition frequently occurs in summer. In winter, when the ground may be covered with snow, the lower levels are cold, the lapse rate is small, and the air is likely to be stable. On a clear summer night, the ground cools quickly by radiation and the lower level of air may become cooler than that lying above it. Here the lapse rate may be zero or even reversed. This is called a *temperature inversion* and makes for great stability.

The formation of cumulus clouds and thunderstorms is a result of instability. Warm, moist air next to the ground becomes unstable and rises, cooling adiabatically until the *dew point* is reached, when condensation of water vapor creates a cloud (see Figure 14.2). If the initial uprush is violent, the rising air may overshoot its equilibrium altitude. If there is an inversion or a stable layer of air at this elevation, further upward motion is stopped. However, if the upper air is neutral or slightly unstable, the latent heat released by condensation may be enough to carry the convection current to greater heights. As we indicated earlier, storm clouds have been found to extend up to $50,000\ ft$, although the air is usually clear above $20,000\ ft$.

Air motion is invariably turbulent, and clouds are usually turbulent also. The Reynolds number is based on an approximate characteristic dimension of the cloud such as its height or width (both are generally of the same order). With a cloud dimension of about $500\ m$, and a characteristic internal motion of $5m/s$, then with $\nu = 10^{-5}\ m^2/s$ (it is approximately the same for water vapor as it is for air), the

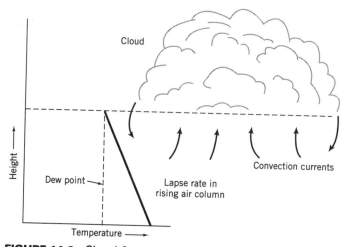

FIGURE 14.2 Cloud formation by condensation of water vapor carried upward in a rising column of air. Hunsaker & Rightmire, *Engineering Applications of Fluid Mechanics,* p. 51, McGraw-Hill, 1947.

Reynolds number $= (500 \times 5)/10^{-5} = 2.5 \times 10^8$. It is no surprise that clouds always have a turbulent appearance.

14.3 CIRCULATORY PATTERNS AND CORIOLIS EFFECTS

The sun, the earth, and the earth's atmosphere form one very large dynamic system. The differential heating of the air gives rise to horizontal pressure gradients, which in turn lead to horizontal motions in the atmosphere.[3] The temperature difference between the atmosphere at the poles and at the equator, and between the atmosphere over the continents and over the oceans, causes the large-scale motions of the atmosphere. Local winds, such as lake breezes, are also caused by temperature differences. The land surface generally heats and cools faster by radiation than a large body of water, and so during the day the air tends to rise over the land, causing the air to move from the water to the land. At night, the land surface cools at a faster rate so that the air over the land gradually becomes cooler and more dense than that over the water, and the movement of air is from the land to the water.

If the earth were not rotating, air would normally tend to flow in the direction of the pressure gradient, that is, from regions of higher pressure to lower pressure. The flow would be perpendicular to the isobars (contours of constant pressure). However, the rotation of the earth prevents this from happening, at least on the scale of large weather patterns. If we are an observer located in a rotating reference frame, such as we are on earth, there are apparent forces that can act, in addition to the effects of pressure gradient, gravity, and viscous forces. The most important is the *Coriolis* force.

---

[3] *Air Pollution: Its Origin and Control,* K. Wark, C. F. Warner, and W. T. Davis, published by Addison-Wesley, 1998.

The "truest" inertial frame is the distant stars. On the earth's surface, we are actually accelerating because of the motion about the sun, the spinning of the earth about its axis of rotation, and by other motions we are not aware of. The most important of these is the spin, being 365 times greater than the angular velocity about the sun. So the center of the earth is a good choice for an *inertial* reference frame ($[X, Y, Z]$ in Figure 14.3).

The positions, velocities, and accelerations that we see (fixed to the earth's surface) are not those seen by an observer fixed to the earth's center. For many applications, the difference between the two reference frames is negligible. The differences are evident only on a large scale, such as in atmospheric and oceanic movements.

We can develop expressions for the equations of motion with reference to the earth's surface, and we will see that we can add a term to Euler's equation that represents an "apparent" acceleration and treat it mathematically and conceptually as a force.

The earth spins on its axis from west to east, so that the rotation looking down from the North Pole is counterclockwise (see Figure 14.3). If the angular velocity vector of the earth's rotation is $\vec{\Omega}$, that is equal to the angular rotation of a particle of fluid at any latitude β, and we know that $\Omega = 2\pi$ *rad/day*. The angular momentum of the particle per unit mass relative to the inertial frame of reference (the earth's center) will be $\mathbf{r} \times \mathbf{V}$, where $\mathbf{r}$ is the radius of rotation and $\mathbf{V}$ its velocity, where $\mathbf{V} = \vec{\Omega} \times \mathbf{r}$. Since $\mathbf{r}$ is measured from the axis of rotation, $r = R \cos \beta$, where R is the radius of the earth ($= 6380$ *km*).

A particle in the northern hemisphere, moving in the northerly direction, will be moving toward the axis of rotation. In the absence of friction, angular momentum will be conserved, and the fluid will acquire a greater angular velocity since its radius is decreasing (β increases). This is the familiar effect seen when a spinning skater pulls in his or her arms to increase his or her rate of rotation. So a north-bound particle will deflect to the right (the east) in the northern hemisphere. A south-bound particle acquires a lower angular velocity to conserve angular momentum, and it will deflect to the west.

We can also consider motions in the east-west direction, at a constant latitude. A particle in the northern hemisphere, moving in the easterly direction, will increase its angular velocity. To conserve its angular momentum, it will experience a tendency to move to a larger radius of rotation; it is deflected to the right (to the south).

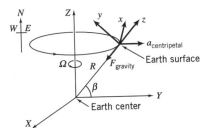

FIGURE 14.3 Coordinate system for the earth spinning around its axis. *XYZ* is the inertial coordinate system with its origin at the center of the earth, and *xyz* is the local coordinate system at the latitude β (at the equator, the latitude is zero). R is the radius of the earth.

Similarly, a west-bound particle will have a lower angular velocity and experience a tendency to deflect to the north to conserve angular momentum.

The overall result of these Coriolis effects is to create clockwise circulatory patterns for the northern hemisphere, and counterclockwise circulatory patterns for the southern hemisphere. This is true for the atmosphere, and for the oceans. Coriolis effects are the dominant mechanisms controlling these circulatory patterns, and therefore they effectively control weather patterns.

To see how Coriolis effects enter into the equations of motion, we note that the acceleration of a particle (fluid or otherwise) near the earth's surface in the earth's inertial reference frame can be written as the sum of a number of accelerations seen by an observer on the earth's surface in the $[x, y, z]$-coordinate system (where the positive z-direction is in the direction of increasing altitude). These include the rectilinear acceleration, the centripetal acceleration, and the Coriolis acceleration, $2\vec{\Omega} \times \mathbf{V}$. The centripetal acceleration is usually very small for motions in the atmosphere and therefore, the acceleration of a fluid particle in the inertial reference frame (based on the earth's center) is given by the sum of its acceleration in the local reference frame plus the Coriolis acceleration. That is,

$$\frac{D\mathbf{V}}{Dt} + 2\vec{\Omega} \times \mathbf{V}$$

If we ignore viscous effects, we obtain

$$\frac{D\mathbf{V}}{Dt} + 2\vec{\Omega} \times \mathbf{V} = -\frac{1}{\rho}\nabla p - \mathbf{g}$$

or

$$\frac{D\mathbf{V}}{Dt} = -2\vec{\Omega} \times \mathbf{V} - \frac{1}{\rho}\nabla p - \mathbf{g} \qquad (14.2)$$

An apparent force, the Coriolis force, appears in the equations of motion due to the earth's rotation. Since the vertical velocities are typically much smaller than the velocities in the horizontal plane, we have

$$2\vec{\Omega} \times \mathbf{V} = -fv\mathbf{i} + fu\mathbf{j}$$

where f is the Coriolis parameter defined by

$$f = 2\Omega \sin \beta \qquad (14.3)$$

On earth, $0 \leq f \leq 1.44 \times 10^{-4} s^{-1}$.

If we non-dimensionalize equation 14.2 (as we did with the Navier–Stokes equation in Section 8.7), we find that the Coriolis term becomes

$$\left(\frac{2 \sin \beta}{Ro}\right)\mathbf{V'}$$

where $\mathbf{V'}$ is the velocity non-dimensionalized by the characteristic velocity scale V_0, and

$$Ro = \frac{V_0}{\Omega L}$$

This is the Rossby number, which measures the importance of the inertia force relative to the Coriolis force. For a typical $40 \ km/hr$ wind, for the Rossby number to be of order one, the Coriolis force must act over a characteristic distance L of about 150 *km*. As it acts over a greater and greater distance, the Coriolis effect becomes more and more important.

For example, in the atmosphere the boundary layer thickness varies between 200 *m* and 1 *km* (see Section 14.4). Outside this region, in the *free layer,* friction is negligible under "normal" conditions. In the free layer, without Coriolis effects, the acceleration term in the momentum equation is balanced by the force due to pressure differences, and the flow tends to be in the direction of the pressure gradient (from high to low pressure). However, in the simplest possible wind system, that of straight uniform flow, there is no acceleration and the force due to pressure differences is balanced by the Coriolis force. So the velocity vector is turned at right angles to the direction of the pressure gradient. Streamlines are therefore almost *parallel* to the isobars. This is called *geostrophic balance.* It follows that in the atmosphere the pressure in a uniform flow is always higher to the right (looking downstream).

Geostrophic balance approximates the conditions found at altitudes of a few hundred meters or more above the surface of the earth. Except in the case of very light winds, the direction and magnitude of the actual winds at this height generally do not differ by more than 10° and 20%, respectively, from their geostrophic values. For this reason, isobaric maps can be used to determine wind speed and direction. The isobars yield direction, and the magnitude of the gradient (the distance between the isobars, which are also streamlines) determines the speed.

When the wind follows a curved path, the centripetal acceleration needs to be taken into account. This effect is only important near the center of high- or low-pressure regions where the isobars have significant curvature. The wind then deviates from its geostrophic value and is called the *gradient wind.*

Within the earth's boundary layer, that is, at heights less than a few hundred meters, viscous effects become important. In particular, the variation of horizontal velocity with height also produces Coriolis effects which vary with height, and the wind direction near the earth's surface will be skewed from its geostrophic or gradient direction.

14.4 PLANETARY BOUNDARY LAYER

The nature of the terrain, the location and density of trees, the location and size of lakes, rivers, hills, and buildings strongly affects the velocity distribution in the atmospheric or *planetary* boundary layer.[4] The depth of the boundary layer can vary from a few hundred meters to several kilometers, and it is greater for unstable conditions than for stable conditions.

The general effect of terrain roughness on the velocity profile is shown in Figure 14.4. In this particular example, the change in the overall boundary layer thickness is from approximately 500 *m* to 280 *m,* for decreasing roughness. Because of the change in wind speed with height, any wind speed value must be quoted with respect to the elevation at which it is measured. The international standard height for surface wind measurements is 10 *m*. The profile is often modeled as a power law similar to that given in Section 10.4 for a flat plate turbulent boundary

---

[4] *Air Pollution: Its Origin and Control,* K. Wark, C. F. Warner & W. T. Davis, published by Addison-Wesley, 1998.

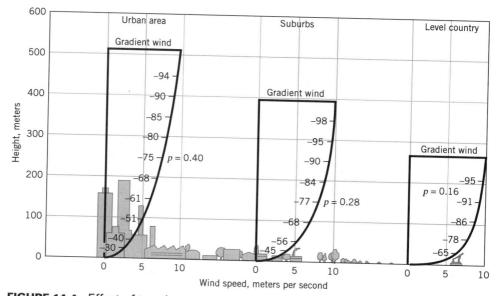

FIGURE 14.4 Effect of terrain roughness on the planetary boundary layer velocity profile. From *Air Pollution: Its Origin and Control,* K. Wark, C. F. Warner, and W. T. Davis, Addison-Wesley, 1998.

layer. When the lapse rate is approximately adiabatic and the terrain is generally level with little surface cover, the exponent p is approximately 0.15, which is close to the $\frac{1}{7}$ value used in Section 10.4, but it can vary widely depending on the roughness and the stability of the atmosphere.

The variation in wind profile is important for a number of reasons, but it is felt most directly in the approach of an airplane to a landing strip. The speed of the airplane relative to the wind speed determines its lift and drag performance, but the speed of the airplane relative to the ground dictates its landing pattern. The wind speed naturally decreases with altitude due to the presence of the planetary boundary layer, and this needs to be taken into account by the pilot. The rapid variation in the wind profile close to the ground adds to the difficulty in judging the approach.

14.5 PREVAILING WIND STRENGTH AND DIRECTION

The large-scale circulatory patterns in the atmosphere are modified by local conditions, which can vary with time. For example, in the mid-latitudes, in the free layer of the atmosphere, the air currents meander, forming the zonal currents which are a type of wave motion. Nevertheless, there are recurring patterns, which help to set the average conditions at any location. At any particular time, in any given season, we have expectations for the average condition including the temperature, precipitation, and the prevailing wind strength and direction. Charts are available for most locations indicating the prevailing wind strength and direction, averaged over many years, as shown by the example given in Figure 14.5. In addition, data are available for maximum wind conditions, rainfall rate, and snow deposits, which

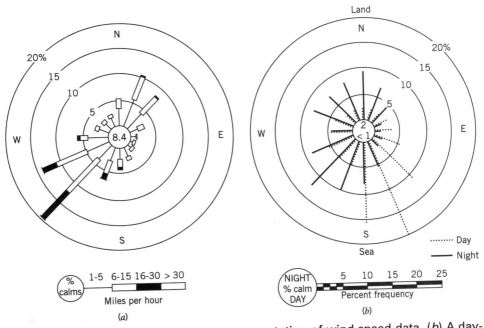

FIGURE 14.5 (*a*) A typical wind rose representation of wind-speed data. (*b*) A day-night wind rose for New York City showing diurnal effect of the sea breeze. From *Air Pollution: Its Origin and Control,* K. Wark, C. F. Warner, and W. T. Davis, Addison-Wesley, 1998.

are important for designing buildings and providing services such as storm drains. Such data are often incorporated in building codes or national standards for drainage systems or other services.

The prevailing wind conditions provide vital information for other purposes as well. Predicting the probable distribution by air currents of a toxic emission from an industrial site requires the use of charts such as that shown in Figure 14.5. Choosing the location of a wind turbine and predicting its likely power output requires similar data, as do long-distance sailing and tracking weather balloons.

14.6 ATMOSPHERIC POLLUTION

Atmospheric pollution is a very broad term, which describes the dispersion of airborne particles and reactive agents and their chemical interaction in the presence of sunlight. Here, we are concerned with the dispersion of airborne particles and other material, of which there is a great variety.

Airborne particles are produced by two very different mechanisms.[5] The larger particles are fragments of still larger ones which, through weathering, mechanical breakage, solution, or some other attrition process, finally become small enough to float in the atmosphere when lifted up by the wind. These particles may sometimes

---

[5] V. J. Schaefer and J. A. Day, *A Field Guide to the Atmosphere,* The Peterson Field Guide Series, published by Houghton Mifflin, 1981.

be reduced in size to 1 μm or smaller, but when found in the air are generally 5 μm or larger, and the smaller ones may be joined to the larger ones by adhesion, electrical forces, or surface tension. Many of these larger particles are of natural origin, although some are of human origin, resulting from processes such as the production of cement, quarrying, and strip mining.

The other major mechanism of small-particle formation begins in the vapor phase and proceeds by condensation, crystalization, or related mechanisms. The cooling of hot saturated or supersaturated vapor, the combination of chemicals, photolytic reactions, and other condensation processes typically produce such particles. Their sizes range from molecular clusters of 0.001 to 0.003 μm to about 1 μm. In the presence of high temperatures and a rich source of condensable material, as in a volcano or the exhaust gases from a coal-burning power plant, some particles can grow to sizes considerably larger than 1 μm.

The largest airborne particles may be in the air for minutes or hours. The intermediate sizes are likely to be suspended for hours or days, and the smallest particles may reside in the atmosphere for weeks, months, and even years. All particles in the atmosphere will tend to fall to the ground under their own weight. In the absence of air resistance, a particle experiences a constant acceleration of g, and its velocity will increase linearly with time. As its velocity increases, however, the aerodynamic drag increases. If we ignore the buoyancy, then at some point the drag force will balance the weight of the particle and the particle reaches a constant velocity called the *terminal velocity*. The terminal velocity V_p of a spherical particle of radius R may be found from an equilibrium force balance, where

$$\rho_p \tfrac{4}{3}\pi R^3 g = \tfrac{1}{2}\rho_a V_p^2 \pi R^2 C_D \tag{14.4}$$

where ρ_p is the density of the particle material, ρ_a is the air density, and C_D is the particle drag coefficient. Small particles settle very slowly and therefore, their Reynolds numbers are typically much less than one. Stokes showed analytically that the drag coefficient for spheres for Reynolds numbers less than one is given by[6]

$$\text{drag force} = 6\pi R \mu V_p$$

or

$$C_D = \frac{24}{Re} \tag{14.5}$$

where the Reynolds number is based on the diameter. Flows at these low Reynolds numbers are dominated by viscous effects, and they are often called *creeping* flows (see Figure 10.12). Equation 14.4 becomes

$$V_p = \frac{2\rho_p g R^2}{9\mu}$$

A particle of soot, for example, with $R = 1$ μm and a density similar to that of pure carbon ($\rho_p = 1600$ kg/m^3) has a terminal velocity in air at sea level of 0.2 mm/s, which is equal to 17 m/day. We see that small particles are extremely slow to settle to the ground once they have become airborne. Their terminal velocity

---

[6] Stokes, G., *Transactions of the Cambridge Philosophical Society*, **8**, 1845; **9**, 1851.

increases with size. Larger particles may have Reynolds considerably larger than one, and Stokes' drag law (Equation 14.5) will not apply. In that case, the terminal velocity may be found using Equation 14.4, in combination with the results given in Figure 10.12.

14.7 DISPERSION OF POLLUTANTS

The dispersion of pollutants in the lower atmosphere is greatly aided by thermal convection and turbulent mixing. Buoyancy effects determine the depth of the convective mixing layer, called the *maximum mixing depth* (MMD). The MMD is the height where a thermally heated fluid particle will stop rising, and it is given approximately by the intersection of the air temperature profile with the adiabatic lapse line. When the atmosphere is very stable, so that the air temperature increases with altitude, the MMD can be very small. These temperature inversions can produce very high pollution levels in urban areas.

Turbulence in the atmosphere is usually defined as the fluctuations in velocity with frequencies greater than 2 cycles per hour ($0.6 \times 10^{-3} Hz$), with the most energetic fluctuations in the 0.01 to 1 Hz range. There are fluctuations due to the "mechanical" turbulence produced by velocity gradients, and there is also turbulence produced by thermal eddies associated with temperature gradients (these eddies are much smaller than the convective motions that transport pollutants up to the MMD). Thermal eddies are prevalent on sunny days when light winds occur and the temperature gradient is unstable. They typically have periods measured in minutes. Mechanical eddies predominate on windy nights with neutral stability, and they have periods on the order of seconds.

The wind, the stability of the atmosphere, and the resulting level of atmospheric turbulence will all affect the shape of the plume produced by a tall smoke stack, as illustrated in Figure 14.6. A wide range of possibilities is shown, and the strong influence of the temperature stability can be seen.

Another significant influence is the terrain surrounding the stack. The proximity of buildings can have a particularly strong effect, as seen in Figure 14.7. If the stack is placed improperly, or it is too low, the backflow in the wake of the building can cause heavy pollutant concentrations in the vicinity of the building.

14.8 DIFFUSION AND MIXING

It is of great interest to know how the properties of a fluid at one point influence the properties at another point. For example, we know that when dye is introduced into a fluid, it slowly diffuses throughout the fluid. How does this process occur, and at what rate? Think of how a scent travels through a room. If a perfume bottle is opened in one corner of the room, there will be a high concentration of perfume molecules near the bottle and a low concentration elsewhere. The molecules will move and mix with the surrounding air by diffusion, even in the complete absence of drafts, so that after a sufficiently long time, they will be distributed uniformly throughout the room. The rate at which this happens depends on the diffusion coefficient (a property of the perfume) and the strength of the concentration gradi-

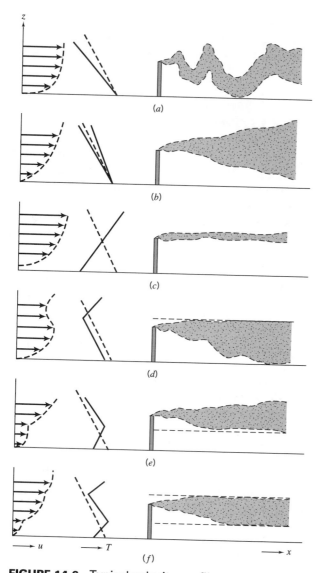

FIGURE 14.6 Typical velocity profile, temperature profile, and plume shape for various atmospheric conditions (— ambient lapse rate; ---- dry adiabatic lapse rate). (*a*) Strong instability. (*b*) Near neutral stability. (*c*) Surface inversion. (*d*) Inversion above stack. (*e*) Inversion below stack. (*f*) Inversion below and above stack. From *Air Pollution: Its Origin and Control*, K. Wark, C. F. Warner, and W. T. Davis, Addison-Wesley, 1998.

ent. So the diffusion will happen quickly to begin with and then slow down as the concentration becomes more uniform and the strength of the gradients decrease (see Figure 14.8). This type of diffusion, where there is no mean flow, is described by Fick's first law of diffusion,

$$J = -D_m \frac{dC}{dx} \tag{14.6}$$

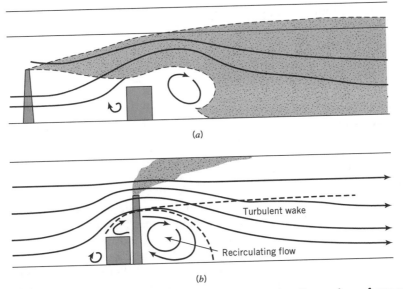

(a)

(b)

FIGURE 14.7 Effect of local flow patterns on the dispersion of gaseous effluent from a stack. From *Air Pollution: Its Origin and Control,* K. Wark, C. F. Warner, and W. T. Davis, Addison-Wesley, 1998.

where J is the mass of perfume crossing a unit area per unit time (the mass flux of perfume), D_m is the mass diffusion coefficient for perfume in air, with dimensions $= L^2 T^{-1}$, and C is the local concentration of perfume (mass per unit volume). The minus sign indicates that the diffusion of perfume is from regions of high to low concentration. Note how similar this equation is to equation 1.9, which described

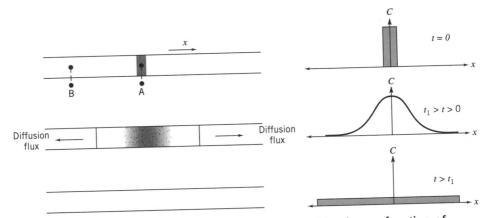

FIGURE 14.8 One-dimensional spreading of perfume in air as a function of time. From Potter and Wiggert, *Mechanics of Fluids,* 2nd ed., Prentice Hall, with permission.

the diffusion of momentum by viscosity due to velocity differences, and also to the Fourier heat conduction equation

$$q = -k\frac{dT}{dx} \tag{14.7}$$

which describes the diffusion of heat due to temperature differences.

Since diffusion is a molecular transport process, the amount of diffusion across an interface is proportional to the surface area. Increasing the surface area will increase the rate of mixing proportionally. This is why we stir a fluid to mix it. When we add sugar to coffee, we could sit and wait for the sugar to diffuse into the coffee. For it to mix thoroughly by diffusion we would have to wait a long time (molecular diffusion processes are typically very slow) and the coffee would undoubtedly be very cold. Instead, we stir the coffee. The resulting turbulence stretches and distorts the fluid particles that contain high concentrations of sugar molecules, which increases their surface area and allows the molecular mixing process to occur in a very short time, before the coffee has cooled.

Mixing also enhances the diffusion of heat and therefore turbulent flows are very efficient for cooling a surface. The turbulent eddies mix the flow very thoroughly, so that hot fluid can be quickly transported to regions where the surrounding fluid is cold, and vice versa. The eddies also stretch and distort the fluid particles, thereby increasing the surface area over which the heat conduction takes place.

In addition, mixing enhances the diffusion of momentum. For example, whenever velocity gradients are present, we also have gradients of momentum. As we saw in Section 1.4.4, viscosity diffuses momentum by molecular exchanges, which cause slower fluid to speed up and faster fluid to speed up. This is typically a slow process. However, in turbulent flows we have strong, large-scale mixing present, and the momentum is redistributed very much faster. Because of this stirring process, the momentum distribution becomes more uniform and the strength of the velocity gradients tends to be reduced. As we indicated in Section 9.7, this process is similar to viscous diffusion, except that it is much more effective. Turbulent "diffusion" is often modeled using an equivalent "eddy" viscosity, which is typically orders of magnitude larger than the molecular viscosity. Since the turbulent diffusion of heat, mass and momentum all depend directly on the fluctuating velocity field, the eddy viscosities for each transport process are all about the same magnitude.

PROBLEMS

14.1 What information is typically available from a wind rose for a given locale?

14.2 Explain the difference between the atmospheric lapse rate and the adiabatic lapse rate.

14.3 How are the environmental (atmospheric) and the adiabatic lapse rate used to classify the degree of stability in the atmosphere?

14.4 Classify the stability of the atmosphere on the basis of the average temperature gradient for the following conditions:

(a) temperature at ground level is $70°F$, temperature at 1500 ft is $80°F$

(b) temperature at ground level is $70°F$, temperature at 2500 ft is $60°F$

(c) temperature at ground level is $60°F$, temperature at 1900 ft is $48°F$

14.5 Classify the stability of the atmosphere on the basis of the average temperature gradient for the following conditions:

    (a) temperature at ground level is $24.6°C$, temperature at 2000 m is $5°C$

    (b) temperature at ground level is $30°C$, temperature at 500 m is $20°C$

    (c) temperature at ground level is $25°C$, temperature at 700 m is $28°C$.

14.6 On a particular day, the wind speed is 2 m/s at a height of 10 m. (a) Estimate the wind speed at heights of 100 m and 300 m if the power law exponent on the velocity profile is 0.15 (level ground, little cover). (b) Repeat the estimates for an exponent of 0.4 (urban area).

14.7 The wind speed over an urban area is 10 m/s at a height of 10 m. Estimate the wind speed at a height of 30 m.

14.8 At a given location the ground-level temperature is $70°F$. At an elevation of 2000 ft, the temperature of the air is $65°F$. (a) What is the maximum mixing depth, in feet, when the normal maximum surface temperature for that time is $90°F$? (b) What if it was $84°F$?

14.9 At a given location the ground-level temperature is $18°C$, while the normal maximum surface temperature for that month is known to be $30°C$. At an elevation of 700 m, the temperature of the air is $15°C$. (a) What is the maximum mixing depth, in meters? (b) What is it if the temperature at 700 m is $20°C$?

14.10 The atmospheric lapse rate on a particular day is constant in the lower part of the atmosphere. At ground level, the pressure is 1020 $mBar$ and the temperature is $15°C$. At a height z_1 the pressure and temperature are 975 $mBar$ and $11.5°C$. Determine the atmospheric temperature gradient, and the height z_1.

14.11 Find the terminal speed of spherical particles falling through atmospheric air at $20°C$. Particle densities of:

    (a) $1.0 \times 10^{-3}\, g/mm^3$

    (b) $2.0 \times 10^{-3}\, g/mm^3$

    should be considered, each with particle diameters of 10, 100, and 1000 μm. Take note of the results given in Figure 10.12.

14.12 Find the terminal speed of spherical particles falling through water at $20°C$. Particle densities of:

    (a) 1000 kg/m^3

    (b) 2000 kg/m^3

    should be considered, each with particle diameters of 10, 100, and 1000 μm. Take note of the results given in Figure 10.2.

14.13 Air at $20°C$ and atmospheric pressure flows horizontally through a collection chamber 3 m high and 5 m long. The air carries particles 70 μm diameter, with a specific gravity of 1.5. What is the maximum air speed that can be used to ensure that all particles have settled out in the chamber?

14.14 Air at $60°F$ and atmospheric pressure flows horizontally at a speed of 1 ft/s through a collection chamber 12 ft high. The air carries particles 0.002 $in.$ diameter, with a specific gravity of 2.0. What is the minimum length of the chamber necessary to ensure that all particles are collected?

14.15 Air at $0°C$ and atmospheric pressure flows horizontally at a speed 0.5 m/s through a collection chamber 4 m high and 20 m long. The air carries particles with a density of 1600 kg/m^3. What is the minimum diameter of particles that will be collected in the chamber?

CHAPTER 15

HISTORICAL NOTES

This chapter presents some short biographical notes on scientists and engineers who have made important contributions to fluid mechanics. The entries are in chronological order, based on date of birth. The material is based on the articles and books given in the list of references.

15.1 ARCHIMEDES OF SYRACUSE

Born: 287 BC in Syracuse, Sicily; Died: 212 BC in Syracuse, Sicily.

"Give me somewhere to stand, and I will move the earth."

Archimedes was fascinated by geometry. In *Measurement of the Circle*, he gave an estimate of π that was accurate to better than one part in a thousand. He also proved, among many other geometric results, that the volume of a sphere is two-thirds of the volume of the cylinder that encloses it. He considered this his most significant accomplishment, requesting that a representation of this result be inscribed on his tomb. Plutarch recorded:

> *Oftimes Archimedes' servants got him against his will to the baths, to wash and anoint him, and yet being there, he would ever be drawing out of the geometrical figures, even in the very embers of the chimney. And while they were anointing of him with oils and sweet savours, with his fingers he drew lines upon his naked body, so far was he taken from himself, and brought into ecstasy or trance, with the delight he had in the study of geometry.*

Archimedes discovered fundamental theorems concerning the center of gravity of plane figures and solids. His most famous theorem in fluid mechanics gives the weight of a body immersed in a liquid, called Archimedes' principle.

He postulated that, by their very nature, fluids cannot have internal "empty spaces," that is they must be continuous. And "if fluid parts are continuous and uniformly distributed, then that of them which is the least compressed is driven along by that which is more compressed." Here we have two important concepts of classical fluid mechanics: pressure applied to any part of a fluid is transmitted to any other part of that fluid, and a fluid flow is caused and maintained by forces due to pressure.

The famous Archimedes' theorems (or Propositions) follow.[1]

1. If a body which is lighter than a fluid is placed in this fluid, a part of the body will remain above the surface (Prop. IV).

[1] From *A History and Philosophy of Fluid Mechanics*, by G.A. Tokaty, published by G.T. Foulis & Co., Henley-on-Thames, 1971. Dover edition, 1994.

FIGURE 15.1 Archimedes of Syracuse.

2. If a body which is lighter than a fluid is placed in the fluid, it will be immersed to such an extent that a volume of fluid which is equal to the volume of the body immersed has the same weight as the whole body (Prop. V).

3. If a body which is lighter than a fluid is totally and forcibly immersed in it, the body will have an upward thrust equal to the difference between its weight and the weight of an equal volume of fluid (Prop. VI).

4. If a body is placed in a fluid which is lighter than itself, because it will be lighter by the amount of the weight of the fluid which has the same volume as the body itself (Prop. VII).

> *One of the most celebrated anecdotes concerning Archimedes' Propositions runs as follows. Hiero, the King of Syracuse, had given some gold to a goldsmith to make into a crown. The crown was delivered, made up, and of the proper weight, but it was suspected that the workman had appropriated some of the gold, replacing it with an equal weight of silver. Archimedes was thereupon consulted. Shortly afterwards, when in the public baths, he noticed that his body was pressed upwards by a force which increased the more completely he was immersed in the water. Recognizing the value of this observation, he rushed out, just as he was, and ran home through the streets, shouting ευρηκα! ευρηκα!—I have found it! I have found it! Indeed he had found it. His experiments showed that water permits an exact computation of solid bodies, because the volume of water displaced weighs precisely the same as the weight lost by the object immersed in the water.*

Archimedes' mechanical skill together with his theoretical knowledge enabled him to construct many ingenious machines. A device known as Archimedes' screw is a type of pump that is still used in many parts of the world.

He was killed during the capture of Syracuse by the Romans in the Second Punic War. According to Plutarch:

> *As fate would have it, Archimedes was intent on working out some problem by a diagram, and having fixed both his mind and eyes upon the subject of his*

speculation, he did not notice the entry of the Romans nor that the city was taken. In this transport of study a soldier unexpectedly came up to him and commanded that he accompany him. When he declined to do this before he had finished his problem, the enraged soldier drew his sword and ran him through.

15.2   LEONARDO DA VINCI

Born: 15 April 1452 in Vinci (near Empolia), Italy; Died: 2 May 1519 in Amboise, France.

"No knowledge can be certain, if it is not based upon mathematics or upon some other knowledge which is itself based upon the mathematical sciences."
"Instrumental; or mechanical science is the noblest and above all others, the most useful."

Leonardo da Vinci had many talents in addition to his painting. He studied biology, physiology, botany, and did extensive work in mechanics, inventing many ingenious devices and machines. He worked for the Duke of Milan and Cesare Borgia as a military architect and engineer. He was also fascinated by geometry. He became interested in the subject when he illustrated Pacioli's *Divina Proportione.* He continued to work with Pacioli and apparently neglected his painting to indulge his interests in geometry.

Leonardo's interests in fluid flow were extremely broad, and were concerned with, among others, the flow of rivers, wavemaking, aerodynamics, helicopters, wind vanes, the flight of birds, parachutes, watermills, blood flow, irrigation, swimming, and ship design. He is also credited with the "velocity-area law," what we know as the one-dimensional continuity equation for steady, incompressible flow. He wrote

FIGURE 15.2   Leonardo da Vinci.

"... where the flow carries a large quantity of water, the speed of the flow is greater and vice versa."

"... where the river becomes shallower, the water flows faster."

Another important observation is his principle of relative motion.

"... a body moving in the motionless air experiences as much air resistance as is experienced by the same body in a motionless state but exposed to an air motion with the same speed."

15.3 EVANGELISTA TORRICELLI

Born: 15 October 1608 in Faenza, Romagna (now Italy); Died: 25 October 1647 in Florence, Tuscany (now Italy).

Torricelli served as secretary and companion to Galileo during the last three months of his life (1641 to 1642) and succeeded him as the court mathematician to Grand Duke Ferdinando II of Tuscany. In 1643, Torricelli proposed an experiment, later performed by his pupil and colleague Vincenzo Viviani (1622–1703), that demonstrated that atmospheric pressure determines the height to which a fluid will rise in a tube inverted over the same liquid. This concept led to the development of the barometer. Torricelli also studied the trajectories of projectiles, and he applied his knowledge to the analysis of a jet issuing from a vessel filled with water to a depth H. In particular, he asked to what height h could the jet rise? He experimented with upward streams and found that $h < H$. He recognized that the difference was due to friction in the orifice, and air resistance to the stream outside the orifice. Neglecting these effects, he used the fact that the distance of free fall is given by $gt^2/2$ and that the velocity of free fall $v = gt$. Equating h to the distance of free fall, Torricelli deduced that $v = \sqrt{2gh}$, a new result, which represented a major milestone in the development of fluid mechanics.

FIGURE 15.3 Evangelista Torricelli

15.4 BLAISE PASCAL

Born: 19 June 1623 in Clermont-Ferrand, France; Died: 19 August 1662 in Paris, France.

"The last thing one knows in constructing a work is what to put first."

"When we see a natural style, we are quite surprised and delighted, for we expected to see an author and we find a man."

"All the misfortunes of men derive from one single thing, which is their inability to be at ease in a room [at home]."

"Man is only a reed, the weakest thing in nature; but he is a thinking reed."

"I have made this letter longer than usual, only because I have not had the time to make it shorter."

According to the MacTutor History of Mathematics Archive, Pascal's father, Étienne Pascal, had unorthodox educational views and decided to teach his son himself. He decided that Pascal was not to study mathematics before the age of 15 and all mathematics texts were removed from their house. Pascal, however, started to work on geometry himself at the age of 12. He discovered that the sum of the angles of a triangle are two right angles and, when his father found out, he relented and allowed Pascal a copy of Euclid. At the age of 14, Pascal started to attend Mersenne's meetings. Mersenne belonged to the religious order of the Minims, and his cell in Paris was a frequent meeting place for Fermat, Pascal, Gassendi, and others. At the age of 16, Pascal presented a single piece of paper to one of Mersenne's meetings. It contained a number of projective geometry theorems, including Pascal's mystic hexagon.

Pascal invented the first digital calculator (1642) to help his father who was a tax collector. The device, called the Pascaline, resembled a mechanical calculator of the 1940s. Further studies in geometry, hydrodynamics, and hydrostatic and atmospheric pressure led him to invent the syringe and hydraulic press, and to discover Pascal's law of pressure. He performed an experiment in which he convinced his brother-in-law to climb the Puy-de-Dôme Mountain in France. He found

FIGURE 15.4 Blaise Pascal

the height of the mercury dropped with altitude, indicating pressure decreases with altitude.

In his *The Equilibrium of Liquids and the Weight of the Mass of Air* (1663), he studied how the weight of liquid varies with depth, the equilibrium of liquids, the equilibrium between a liquid and a solid, bodies wholly immersed in water, immersed compressible bodies, animals in water, and other topics. The unit of measure for pressure in the SI system is named after him.

> . . . *Pascal formulated almost all the laws of aerostatics, e.g., that (1) since every part of the air has weight (mass), it follows that its whole body also has weight (mass); (2) since the mass of the air covers the entire face of the earth, its weight presses upon the earth everywhere; (3) because of the hydrostatic law, the higher parts of the earth, such as summits of mountains, experience less air pressure than the lowlands; (4) bodies in the air are pressed on all sides.*
>
> *He then draws attention to everyday experience: when all the apertures of a bellows are closed, why are they difficult to open; when two polished surfaces are laid one upon another, why are they as difficult to separate as if they had been glued together; when a syringe is dipped in water and the piston drawn back, why does the water follow it as if adhering to the plunger, etc.?*
>
> *"The basic reason for all these phenomena is the weight of the air," wrote Pascal, "although it had hitherto been ascribed to the horror of a vacuum . . . (note: the horror vacui of Aristotle and others)." That is, he knew that a kind of vacuum was indeed possible. But he also knew that the invisible power of the atmosphere could do "wonders of pressure."*
>
> . . . *It had been known, even before Stevinus and Pascal, that the pressure of a fluid is normal to the surface on which it acts. Secondly, Stevinus and especially Pascal showed that pressure applied to the surface of a fluid is almost instantly transmitted to all parts of the fluid. Thirdly, Stevinus assumed, and Pascal proved, that the pressure at any point within the fluid is the same in all directions, and depends only on the depth.[2]*

His most famous work is actually in philosophy. During 1656 to 1658, he wrote *Pensées*, a collection of personal thoughts on human suffering and faith in God (the quotations given at the start of this section are taken from this collection). "Pascal's wager" claims to prove that belief in God is rational with the following argument. "If God does not exist, one will lose nothing by believing in him, while if he does exist, one will lose everything by not believing."

His last work was on the cycloid, the curve traced by a point on the circumference of a rolling circle. He died at the age of 39.

15.5 SIR ISAAC NEWTON

Born: 4 January 1643 in Woolsthorpe, Lincolnshire, England; Died: 31 March 1727 in London, England.

"If I have seen further it is by standing on the shoulders of giants." (Letter to Robert Hooke, 5 February 1675.)

---

[2] From *A History and Philosophy of Fluid Mechanics*, by G.A. Tokaty, published by G.T. Foulis & Co., Henley-on-Thames, 1971. Dover edition, 1994.

FIGURE 15.5 Sir Isaac Newton, at age 24.

"I do not know what I may appear to the world, but to myself I seem to have been only like a boy playing on the sea-shore, and diverting myself in now and then finding a smoother pebble or a prettier shell than ordinary, whilst the great ocean of truth lay all undiscovered before me."

According to the MacTutor History of Mathematics Archive, Newton's life can be divided into three quite distinct periods. The first is his boyhood days from 1643 up to his graduation in 1669. The second period, from 1669 to 1687, was the highly productive period in which he was Lucasian professor at Cambridge. The third period (nearly as long as the other two combined) saw Newton as a highly paid government official in London with little further interest in mathematics.

Isaac Newton was born in the manor house of Woolsthorpe, near Grantham in Lincolnshire. Although he was born on Christmas Day 1642, the date given here (4 January, 1643) is the Gregorian calendar date. (The Gregorian calendar was not adopted in England until 1752.) Newton came from a family of farmers but never knew his father, who died before he was born. His mother remarried, moved to a nearby village, and left him in the care of his grandmother. Upon the death of his stepfather in 1656, Newton's mother removed him from grammar school in Grantham where he had shown little promise in academic work. His school reports described him as "idle" and "inattentive." An uncle decided that he should be prepared for the university, and he entered his uncle's old college, Trinity College, Cambridge, in June 1661, where he studied law.

Instruction at Cambridge was dominated by the philosophy of Aristotle but some freedom of study was allowed in the third year of the course. Newton studied the philosophy of Descartes, Gassendi, and Boyle. The new algebra and analytical geometry of Viète, Descartes, and Wallis, and the mechanics of the Copernican astronomy of Galileo attracted him.

His scientific genius emerged suddenly when the plague closed the University in the summer of 1665 and he had to return to Lincolnshire. There, in a period of less than two years, while Newton was still under 25 years old, he began revolutionary advances in mathematics, optics, physics, and astronomy.

While Newton remained at home, he laid the foundation for differential and integral calculus, several years before its independent discovery by Leibniz. The "method of fluxions," as he termed it, was based on his crucial insight that the integration of a function is merely the inverse procedure to differentiating it. Taking differentiation as the basic operation, Newton produced simple analytical methods that unified many separate techniques previously developed to solve apparently unrelated problems such as finding areas, tangents, the lengths of curves, and the maxima and minima of functions.

Barrow resigned the Lucasian chair in 1669 recommending that Newton (still only 27 years old) be appointed in his place. Newton's first work as Lucasian Professor was on optics. Every scientist since Aristotle had believed that white light was a basic single entity, but the chromatic aberration in a telescope lens convinced Newton otherwise. When he passed a thin beam of sunlight through a glass prism Newton noted the spectrum of colors that was formed. To solve the problem of chromatic aberration in refracting telescopes, he proposed and constructed a reflecting telescope. Newton was an excellent experimentalist as well as theoretician.

Newton was elected a fellow of the Royal Society in 1672 after donating a reflecting telescope. Also in 1672, Newton published his first scientific paper on light and color in the *Philosophical Transactions of the Royal Society*. Newton's paper was well received but Hooke and Huyghens objected to Newton's attempt to prove, by experiment alone, that light consists of the motion of small particles rather than waves. Perhaps because of Newton's already high reputation his corpuscular theory reigned until the wave theory was revived in the nineteenth century.

Newton's relations with Hooke deteriorated and he turned in on himself and away from the Royal Society. He delayed the publication of a full account of his optical researches until after the death of Hooke in 1703, so that Newton's *Opticks* did not appear until 1704.

Probably Newton's greatest achievement was his work in physics and celestial mechanics. By 1666, Newton had early versions of his three laws of motion, and he had also discovered the law giving the centrifugal force on a body moving uniformly in a circular path.

In 1685, Halley persuaded Newton to write a full treatment of his new physics and its application to astronomy. The *Philosophiae Naturalis Principia Mathematica*, or *Principia* as it is always known, appeared in 1687. It is recognized as the greatest scientific book ever written. In it, Newton analyzed the motion of bodies in resisting and nonresisting media under the action of centripetal forces. The results were applied to orbiting bodies, projectiles, pendula, and free-fall near the earth. He further demonstrated that the planets were attracted toward the sun by a force varying as the inverse square of the distance and generalized that all heavenly bodies mutually attract one another.

Further generalization led Newton to the law of universal gravitation: "all matter attracts all other matter with a force proportional to the product of their masses and inversely proportional to the square of the distance between them."

This work explained a wide range of previously unrelated phenomena: the eccentric orbits of comets, the tides and their variations, the precession of the earth's axis, and motion of the moon as perturbed by the gravity of the sun.

After suffering a nervous breakdown in 1693, Newton retired from research to take up a government position in London becoming Warden of the Royal Mint (1696) and Master (1699). In 1703, he was elected president of the Royal Society and was re-elected each year until his death. He was knighted in 1708 by Queen

Anne, the first scientist to be so honored for his work. The unit of measure for force in the SI system is named after him.

His assistant Whiston said "Newton was of the most fearful, cautious and suspicious temper that I ever knew."

Newton's many contributions to fluid mechanics are led first and foremost by his formulation of the basic mathematical understanding of forces and accelerations. More particularly, he developed a drag law based on his particle interpretation of gases and their interaction with a solid body. In brief, Newton assumed that molecules, or particles, move in straight lines until they strike the body surface. When this happens, they lose the component of their momentum normal to the body. For a flat plate of area A, with a unit normal vector $\mathbf{n}$, the drag force R acting on the plate is then given by

$$R = \mathbf{n} \cdot \int (\mathbf{n} \cdot \rho \mathbf{V}) \mathbf{V} \, dA$$

That is,

$$R = \rho A V^2 \sin^2 \alpha$$

where α is its angle of incidence. This is the so-called Newton's Sine-Square Law of air resistance. It is incorrect for most flows since the particles sense the presence of the body by pressure propagation. However, at hypersonic speeds, the region of accommodation becomes very small and Newton's sine-square relationship becomes a reasonable description of the drag on a body.

In addition, Newton recognized the role played by viscosity and friction in fluid flows, and formulated the correct stress–strain-rate relationship for common fluids. Newton recognized that if portions of a mass of fluid are caused to move relative to one another, the motion gradually subsides unless sustained by external forces. Conversely, if a portion of a mass of fluid is kept moving, the motion gradually communicates itself to the rest of the fluid. These effects were ascribed by him to a *defectus lubricitatis*, that is, a lack of slipperiness, or internal friction, or viscosity in modern terminology. The description given in the *Principia* of the effects of fluid friction on a mass of fluid set in rotation by a solid cylinder lead to the conclusion that the shear stress developed by friction is given by the coefficient of viscosity times the velocity gradient. Fluids that obey this stress–strain-rate relationship are known as "Newtonian" fluids, in his honor.

15.6 DANIEL BERNOULLI

Born: 8 February 1700 in Groningen, Netherlands; Died: 17 March 1782 in Basel, Switzerland.

Daniel Bernoulli came from a very talented family, and his father, his uncle, his two brothers, and his own two sons all distinguished themselves in the sciences. Daniel, along with his brother Nikolaus, was invited in 1725 to work at the St. Petersburg Academy of Sciences. There he collaborated with Euler, who came to St. Petersburg in 1727, and shared a house with him. He found the social life in St. Petersburg highly distasteful, and in 1733 Daniel went to Basel where he taught anatomy, botany, physiology, and physics.

His most important work is *Hydrodynamica* (1738), which considered the basic properties of fluid flow, pressure, density, and velocity, and it is commonly assumed that he derived the fundamental relationship between pressure and velocity

FIGURE 15.6 Daniel Bernoulli.

now known as Bernoulli's principle. However, the usual relationship, with the addition of the hydrostatic head,

$$p + \tfrac{1}{2}\rho V^2 + \rho gh = \text{constant}$$

appears nowhere in his writings, although he was probably the first to undertake a fundamental study of the interdependence of pressure and velocity, and the first to conclude that an increase in V leads to a decrease in p, and vice versa. As Tokaty points out, Bernoulli's equation as we know it was actually first derived by Lagrange by integrating Euler's equations of motion.

In his book, *Hydrodynamica,* Bernoulli also gave a theoretical explanation of the pressure of a gas on the walls of a container, and established the basis for the kinetic theory of gases. Between 1725 and 1749, he won ten prizes for work on astronomy, gravity, tides, magnetism, ocean currents, and the behavior of ships at sea.

15.7 LEONHARD EULER

Born: 15 April 1707 in Basel, Switzerland; Died: 18 September 1783 in St Petersburg, Russia.

According to the MacTutor History of Mathematics Archive, Euler's father wanted his son to follow him into the church and sent him to the University of Basel to prepare for the ministry. However, geometry soon became his favorite subject. Euler obtained his father's consent to change to mathematics after Johann Bernoulli (the father of Daniel) had used his persuasion. Johann Bernoulli then became his teacher.

Euler joined the St. Petersburg Academy of Science in 1727, two years after it was founded by Catherine I, the wife of Peter the Great. Euler served as a medical lieutenant in the Russian navy from 1727 to 1730. In St Petersburg, he lived with Daniel Bernoulli. He became professor of physics at the academy in 1730 and professor of mathematics in 1733. He married and left Bernoulli's house in 1733. He had 13 children of which five survived their infancy. He claimed that he made

FIGURE 15.7 Leonhard Euler.

some of his greatest discoveries while holding a baby on his arm with other children playing round his feet.

The publication of many articles and his book *Mechanica* (1736–1737), which presented Newtonian dynamics in the form of mathematical analysis for the first time, started Euler on the way to major mathematical work. In 1741, at the invitation of Frederick the Great, Euler joined the Berlin Academy of Science, where he remained for 25 years. Even while in Berlin he received part of his salary from Russia and never got on well with Frederick. During his time in Berlin, he wrote over 200 articles, three books on mathematical analysis, and a popular scientific publication *Letters to a Princess of Germany* (3 vols., 1768–1772).

In 1766, Euler returned to Russia. He had been arguing with Frederick the Great over academic freedom, and Frederick was greatly angered at his departure. Euler lost the sight of his right eye at the age of 31, and soon after his return to St. Petersburg he became almost entirely blind after a cataract operation. Because of his remarkable memory, he was able to continue with his work on optics, algebra, and lunar motion. Euler produced almost half his works after the age of 58 when he became totally blind.

Euler made decisive and formative contributions to geometry, calculus, and number theory. He was the most prolific writer of mathematics of all time. His complete works contains 886 books and papers. After his death in 1783, the St. Petersburg Academy continued to publish his unpublished work for nearly 50 more years. He introduced the notations $f(x)$ (1734), e for the base of natural logs (1727), i for the square root of -1 (1777), π for pi, Σ for summation (1755), and others. He also introduced beta and gamma functions, integrating factors for differential equations, and so forth. He studied continuum mechanics, lunar theory with Clairaut, the three body problem, elasticity, acoustics, the wave theory of light, hydraulics, music, and other subjects, and he laid the foundation of analytical mechanics, especially in his *Theory of the Motions of Rigid Bodies* (1765).

He is also known for some of the most important ideas in fluid mechanics. For example, he introduced the concept of a "fluid particle," which is imagined as an infinitesimal body, small enough to be treated mathematically as a point, but large enough to have physical properties such as volume, mass, density, inertia, energy, and so on. Using this concept, he derived the differential equation describing a streamline

$$\frac{dx}{u} = \frac{dy}{v} = \frac{dz}{w}$$

and the notion of a streamtube. He formulated the integral and differential forms of the continuity equation, and the latter, in our notation, reads,

$$\mathbf{\nabla} \cdot \mathbf{V} = -\frac{1}{\rho}\frac{D\rho}{Dt}$$

Most importantly perhaps, he derived the equations of motion for the flow of an inviscid, compressible fluid

$$\frac{D\mathbf{V}}{Dt} = -\frac{1}{\rho}\nabla p + \mathbf{g}$$

which is known as the "Euler equation."

15.8 JEAN LE ROND D'ALEMBERT

Born: 17 November 1717 in Paris, France; Died: 29 October 1783 in Paris, France.
D'Alembert was a pioneer in the study of differential equations and introduced their use in physics. He lived all of his life in Paris, where he was a friend of Voltaire. In 1741, he was admitted to the Paris Academy of Science, where he worked for the rest of his life, declining an offer from Frederick II to go to Prussia as President of the Berlin Academy. He also turned down an invitation from Catherine II to go to Russia as a tutor for her son. He helped to resolve the controversy in physics over the conservation of kinetic energy by improving Newton's definition of force in his *Traité de Dynamique* (1742), which articulates d'Alembert's principle of mechanics. In 1744, he applied the results to the equilibrium and motion of fluids.
D'Alembert studied hydrodynamics, the mechanics of rigid bodies, the three-body problem in astronomy, and atmospheric circulation.

FIGURE 15.8 Jean Le Rond d'Alembert.

D'Alembert was the first to apply the relationship between pressure and velocity to the study of the resistance offered by an inviscid fluid to a body moving through it (*Traite de L'equilibre et du Mouvement des Fluides,* Paris 1744, and *Essai d'une Nouvelle Theorie de la Resistance des Fluides,* Paris 1752). The notions of "stagnation points" and "stagnation regions" were introduced by d'Alembert. "But my theory of motion of fluids," he writes, "shows that the velocity components in front of and behind the body are the same, therefore all the other conditions will also be the same. If so, the pressure of the fluid on the forward surface will be equal and opposite to the pressure on the backward surface, therefore the resultant pressure will be absolutely nothing." This is "d' Alembert's Paradox," which concludes that the resistance to motion of a body moving steadily in an inviscid fluid is zero. Viscosity was not included in the problem since the viscosity of common fluids is so low. The fact that bodies moving in "real," that is, viscous, fluids have a nonzero resistance was not satisfactorily explained until Prandtl's boundary layer concept was introduced in 1904.

15.9 JOSEPH-LOUIS LAGRANGE

Born: 25 January 1736 in Turin, Sardinia-Piedmont (now Italy); Died: 10 April 1813 in Paris, France.

"I have always observed that the pretensions of all people are in exact inverse ratio to their merits; this is one of the axioms of morals." (Bell, p. 160.)

According to the MacTutor History of Mathematics Archive, Lagrange's interest in mathematics began at a very early age when he read a copy of a book by Halley. In his career, Lagrange served as professor of geometry at the Royal Artillery School in Turin from 1755 to 1766 and helped to found the Royal Academy of Science there in 1757. In 1764, he was awarded his first prize of many when the Paris Academy awarded him a prize for his essay on the libration of the moon.

FIGURE 15.9 Louis de Lagrange.

When Euler left the Berlin Academy of Science, Lagrange succeeded him as director of mathematics in 1766. In 1787, he left Berlin to become a member of the Paris Academy of Science, where he remained for the rest of his career.

During the 1790s, he worked on the metric system and advocated a decimal base. He also taught at the École Polytechnique, which he helped to found. Napoleon named him to the Legion of Honour and Count of the Empire in 1808.

He excelled in all fields of analysis and number theory, and analytical and celestial mechanics. In 1788, he published *Mécanique Analytique,* which summarized all the work done in the field of mechanics since the time of Newton, and is notable for its use of the theory of differential equations. In it, he transformed mechanics into a branch of mathematical analysis. "My objectives were," wrote Lagrange, "to reduce the theory of mechanics and the art of solving the associated problems to general formulae and to unite the different principles in mechanics."

His contributions were extremely broad, but his most noteworthy achievement in fluid mechanics was the integration of Euler's equations of motion for an irrotational, compressible fluid.

$$\frac{V^2}{2} + \int \frac{dp}{\rho} + \frac{\partial \phi}{\partial t} - gdh = C(t)$$

where h is the altitude, and ϕ is the velocity potential, a concept introduced for the first time by Lagrange. For a steady incompressible flow, the equation simplifies to the usual "Bernoulli equation," although, as Tokaty points out, it is actually Euler and Lagrange who deserve the credit for deriving this relationship.

Lagrange is also known for his system for describing fluid motion. Euler had considered the motion of individual fluid particles along their trajectories, without following each fluid particle. In the Eulerian system, the flow parameters were given as a field variable: that is, the flow parameters were given as a function of space and time. For example, $\mathbf{V} = \mathbf{V}(x, y, z, t)$. A particle at a given position and time would have a certain density, velocity, energy, and so on, and as time went by, different particles would occupy that position. Lagrange felt that the motion of each particle required to be specified in some way. He proposed identifying each particle by its position at a given time, say (x_0, y_0, z_0, t_0), and then following them in time. In the Lagrangian system, $\mathbf{V} = \mathbf{V}(x_0, y_0, z_0, t - t_0)$. However, since the number of particles is infinitely large, an infinite system of equations is necessary to describe the fluid motion, and even Lagrange recognized that Euler's method was more convenient. Nevertheless, there are instances when the Lagrangian system is preferable, as when calculating the diffusion of gases or solid particles.

15.10 CLAUDE LOUIS MARIE HENRI NAVIER

Born: 10 February 1785 in Dijon, France; Died: 21 Aug 1836 in Paris, France.

Navier was educated at the École Polytechnique and became a professor there in 1831. From 1819 until his death, he was also professor at the École des Ponts et Chaussées.

He worked on applied topics such as engineering, elasticity, and fluid mechanics. A specialist in road and bridge building, he was the first to develop a theory of suspension bridges, which before then had been built to empirical principles. He also made contributions to Fourier series and their application. His most important

FIGURE 15.10   Claude Louis Marie Henri Navier.

contribution to fluid mechanics came in a paper read to the Académie des Sciences in Paris on 18th of March, 1822. In this paper, Navier presented the full, three-dimensional equations of motion for an incompressible, viscous fluid. "But although they are based upon Newton's hypothesis $\tau \alpha \, \partial v / \partial y$," he said, "it cannot be assumed that they represent nothing new." In a different form, the same equations were obtained by Stokes 23 years later in 1845, and the equations are therefore called the Navier–Stokes Equations.

15.11   JEAN L. M. POISEUILLE

Born: April 22 1799 in Paris, France; Died: December 26 1869 in Paris, France.
Poiseuille was a physician and studied the flow of blood through small capil-

FIGURE 15.11   Jean L. M. Poiseuille.

laries. As part of his studies, he performed very careful experiments on the flow of water through small diameter tubes (that is, very low Reynolds number or viscous-dominated flow). "I began my studies," wrote Poiseuille in the introduction to *Le Mouvement des Liquides dans des Tubes de Petit Diametres,* "because progress in physiology demands a knowledge of the laws of motion in small diameter (≈ 0.01 *mm*) pipes." His experiments led to the conclusion that

$$\dot{q} = \frac{\pi r^4}{8\mu}\left(-\frac{dp}{dx}\right)$$

which is now known as the Hagen–Poiseuille formula (H. Hagen later confirmed this result). This relationship agrees exactly with the result given in equation 9.14.

15.12 GUSTAV HEINRICH MAGNUS

Born: 1802; Died: 1870.

Gustav Magnus was a German chemist and physicist, a professor of physics at the University of Berlin, and a teacher of Helmholtz. In 1837, he showed that carbon dioxide and oxygen were present in both the arterial and venous systems. He also noticed that the ratio of oxygen to carbon dioxide was higher in the arteries.

In 1853, he succeeded in explaining the deviation of artillery balls from their theoretical trajectories, for which a prize was offered by the Berlin Academy of Sciences in 1794. By experiment, he showed that a spinning rotor will deflect the air flow over it towards the side where the air flow and the rotor have the same direction of motion. The flow becomes asymmetric, and a lift force is generated normal to the rotor in the direction towards the confluent side. The generation of a side force by a spinning body is universally called the "Magnus effect," even though it was observed much earlier, notably by Benjamin Robins (1707–1751) in his 1742 book, *New Principles of Gunnery Containing the Determination of the Force of Gunpowder and Investigation of the Difference in the Resisting Power of the Air to Swift and Slow Motions.* (Robins' observations were not well understood at the time and they were overlooked by later authors.) The Magnus effect is well known to all sports enthusiasts, and it attracted the interest of none other than Lord Rayleigh, who wrote an article "On the irregular flight of a tennis ball," for *Messenger of Mathematics* on July 14, 1877.

15.13 WILLIAM FROUDE

Born: 1810; Died: 1879.

In 1871, Froude constructed an experimental towing tank at Chelston Cross, Torquay, on behalf of the British Admiralty. He then studied frictional and wave-making resistance by towing ship models and planks up to 50 *ft* long with different surface finishes. Using these results, reported in 1872 and 1874, he determined the first coherent set of scaling laws for ship resistance. He developed the "Law of Comparison," which in modern terminology is called Froude number similarity. His correlations for frictional resistance (known as Froude coefficients) are still in use today for interpreting experimental data obtained using ship models.

In 1877, he studied the wave systems generated by ship models and their interference and found the characteristic humps and hollows in the wave drag curve.

FIGURE 15.12 William Froude. Photograph courtesy of Science Museum, London.

His son, R.E. Froude showed in 1881 that the lowest drag was found when a trough of the bow wave occurred at the same time as a crest occurred in the stern wave, and vice versa for the highest drag. William Froude also made important contributions to the understanding of the free rolling behavior of ships (1861, 1862) and the design of propellers (1878).

In 1877, he published the design of a water-brake, or dynamometer, which is still used today. After Froude's death in 1879, his son continued his father's work, constructing a larger and much improved tank at Haslar, near Portsmouth, and continuing to publish systematic series of model experiments, improved correlation procedures, studies on rolling motions, and the design of propellers. Both father and son were Fellows of the Royal Society.

15.14 GEORGE GABRIEL STOKES

Born: 13 August 1819 in Skreen, County Sligo, Ireland; Died: 1 February 1903 in Cambridge, Cambridgeshire, England.

According to the MacTutor History of Mathematics Archive, Stokes published papers on the motion of incompressible fluids in 1842–1843 and on the friction of fluids in motion and the equilibrium and motion of elastic solids in 1845. In 1849, Stokes was appointed Lucasian Professor of Mathematics at Cambridge. In 1851, he was elected to the Royal Society and was secretary of the Society from 1854 to 1884, when he was elected president.

He investigated the wave theory of light, named and explained the phenomenon of fluorescence in 1852, and in 1854 proposed an explanation of the Fraunhofer lines in the solar spectrum. He suggested these were caused by atoms in the outer layers of the sun absorbing certain wavelengths. However, when Kirchhoff later published this explanation, Stokes modestly disclaimed any prior discovery.

Stokes developed mathematical techniques for application to physical prob-

FIGURE 15.13 George Gabriel Stokes.

lems, founded the science of geodesy, and greatly advanced the study of mathematical physics in England. He also formulated the three-dimensional analog of Green's Theorem, known as Stokes's Theorem, and was a pioneer in the use of divergent series.

He is remembered in fluid mechanics particularly for his drag law for spheres in a viscous fluid (Stokes drag), his hypothesis that the normal stresses in a fluid were unchanged by its motion (Stokes hypothesis), the characteristic scale for viscous diffusion (Stokes length), and his co-authorship of the Navier–Stokes Equations for viscous, compressible flows. The unit of measure for kinematic viscosity in the cgs system is named after him.

15.15   ERNST MACH

Born: 1838 in Moravia; Died: 1916.

Ernst Mach, an Austrian physicist and ballistics expert, was one of the founders of supersonic aerodynamics. He made shock waves visible for the first time, using the schlieren system invented by August Joseph Ignaz Töpler (1836–1912), a German physicist who worked in the field of acoustics. He produced the first photographs of projectiles in flight, using an electrified screen to trigger the camera.

Jacob Ackeret suggested that the ratio of the speed of sound to the speed of the fluid flow was an important dimensionless parameter in the study of gasdynamics, and proposed that it be named after Ernst Mach, since Mach was the first to show that compressibility effects in gas flows did not depend on the speed as such, but on its ratio to the sound speed.

Strongly influenced by Fechner, the physicist Mach turned philosopher and unsuccessfully tried to found the science of "psychophysics." The basis of Mach's natural philosophy was that all knowledge is a matter of sensations, so that what we call "laws of nature" are only summaries of experience provided by our own fallible senses.

FIGURE 15.14   Ernst Mach.

15.16   OSBORNE REYNOLDS

Born: 23 August 1842 in Belfast, Ireland; Died: 21 February 1912 in Watchet, Somerset, England.

Reynolds was a mathematics graduate of Cambridge in 1867 and became the first professor of engineering in Manchester in 1868. He held this post until he

FIGURE 15.15   Osborne Reynolds.

retired in 1905. He became a Fellow of the Royal Society in 1877, and 11 years later, won their Royal Medal.

His early work was on magnetism and electricity but he soon concentrated on hydraulics and hydrodynamics. In 1883, in a paper published in the *Philosophical Transactions of the Royal Society,* he identified, for the first time, the differences between laminar and turbulent flow. His flow visualization experiments in glass tubes determined the conditions under which transition took place. This is now described in terms of a "critical Reynolds number." In 1886, he formulated a theory of lubrication and three years later he developed the standard mathematical framework used in the study of turbulence. That is, he suggested splitting a turbulent velocity field into mean and fluctuating components, and wrote down the equations of motion for these two parts. This procedure is still the most common method of analyzing turbulent flows, and it is known as "Reynolds averaging." The "Reynolds number" used in modeling fluid flow is also named after him.

15.17 LUDWIG PRANDTL

Born: 4 February 1875 in Freising, Germany; Died: 15 August 1953 in Göttingen, Germany.

The following information is largely based on the article by Oswatitch and Wieghardt in *Annual Reviews of Fluid Mechanics,* 1987.

Ludwig Prandtl was born on 4 February 1875 in Freising, north of Munich, as son of Alexander Prandtl, a professor of surveying and engineering at the Agricultural Central School at Wehenstephan, near Freising. Starting in 1895, Prandtl studied mechanical engineering at the *Technische Hochschule* (TH) in Munich and

FIGURE 15.16 Ludwig Prandtl. Photography courtesy of California Institute of Technology.

became an assistant of A. Föppl (1854–1924), who was also his doctoral advisor (and later became his father-in-law). His thesis was in mechanics, *On Tilting Phenomena, an Example of Unstable Elastic Equilibrium,* and he graduated to doctor of philosophy on 29 January 1900. He then worked for less than two years at the Maschinenfabrik Augsburg in Nuremberg, where one of his tasks was to improve the suction of sawdust from a wood-cutting machine; he failed because he used a diffuser with too large an opening angle. However, this mishap haunted him for years, even after he had left the firm, until he "invented" boundary layers and separation in 1904.

Before that, he was appointed Professor of Mechanics at the TH in Hanover in 1901 at the age of 26. In 1904, Prandtl was offered a position at Göttingen University. Although it was somewhat of a demotion, he accepted the post because of the strong support and encouragement offered by the mathematician Felix Klein (1849–1925). By 1907, Prandtl became a full professor again, and his chair for mechanics was the only one at a German University for almost a half-century.

Prandtl delivered his famous lecture on flows with very small friction at the Third Mathematical Congress in Heidelberg in 1904. Among the other major research advances made by Prandtl and his group in the early years were the analysis of supersonic Prandtl–Meyer flow around a corner (1907, 1908), the Blasius boundary layer solution for flow over a flat plate (1907), and the Hiemenz solution for a circular cylinder (1911).

Tests with airfoils and wings, together with earlier conversations with Lanchester, who visited in 1908 and 1909, stimulated Prandtl to formulate his wing theory in 1918–1919, leading to the famous Prandtl lifting line and lifting surface theories for calculating lift and induced drag. His student Munk published his thesis on the elliptic lift distribution in 1918, and Betz's thesis on the rectangular wing appeared in 1919. K. Pohlhausen's approximation to laminar boundary layers in pressure gradients was published in 1921, and his brother E. Pohlhausen wrote on the heat transfer on a laminar flat plate boundary layer (the origin of the "Prandtl number," although Prandtl himself always pointed out that it was originally due to Nusselt in 1909). The Prandtl–Glauert rule for computing the compressibility effect on profile lift was first stated by Prandtl in 1922 (without proof) and derived using linearized small perturbation theory by Glauert in 1928.

In 1925, Prandtl became the director of the new Kaiser-Wilhelm Institut für Strömungsforschung, with facilities for research on subsonic and supersonic flow, cavitation, and rotating flow. Prandtl directed this institute with a staff of about 40 until 1945, while continuing to direct his university mechanics institute until 1934.

In 1925, Betz invented his wake-survey method to measure the drag of an airfoil, Ackeret developed his formulae for profile forces in supersonic flow, and Prandtl proposed the concept of the mixing length in turbulent flow. In 1927, Betz gave the first extensive aerodynamics of the windmill, and in 1929, Prandtl and Busemann found the first characteristic method for supersonic flow. Also in 1929, Busemann published his shock-polar diagram, and Tollmien produced his work on the stability of a laminar boundary layer.

Several other well-known researchers worked with Prandtl in Göttingen in the period just before the war, including Nikuradse, Schlichting, Schultz-Grunow, Görtler, Oswatitsch, and Wieghardt. Prandtl himself became more and more interested in dynamic meteorology, including general circulation models, the generation of cyclones, atmospheric lee waves, and Coriolis effects on wind deflection.

Apart from his interests in fluid mechanics, Prandtl also published extensively on solid mechanics. He advanced the soap-film analogy for the elastic stress of a twisted bar in 1903 and 1904, determined the pressures under which the blunt edges of ductile metals yield in 1920, and generalized the concept of the ideally plastic substance the following year. In 1928, he described a simple model for calculating hysteresis in solid bodies, which is now generally known as the Prandtl-Eyring model, and in 1933, he extended this model for the case of time-dependent fracture of a brittle solid. Two co-workers at the mechanics institute, A. Nadai and W. Prager, became very well-known, and later made illustrious careers in the United States.

During his long and productive career, Prandtl supervised over 80 doctoral students. By all accounts he was a dignified, kindhearted person, well-remembered by his students and associates.

15.18 LEWIS FERRY MOODY

Born: 5 January 1880 in Philadelphia, Pennsylvania; Died: April 18 1953 in Plainfield, New Jersey.

The following information was obtained from the Memorial Resolution on the Occasion of the Death of Professor Lewis Ferry Moody, presented to the Princeton Faculty, probably in June, 1953.

Lewis Ferry Moody, received his B.S. degree from the University of Pennsylvania in 1901, and his M.S. degree in Mechanical Engineering in 1902. On account of his high record, he was appointed an instructor in mechanical engineering, a post which he held for two years.

In 1904, he began his remarkable career in research, design, construction, testing, and invention of hydraulic turbines and pumps, as one of the engineering staff of the Hydraulic Division of the I. P. Morris Company of Philadelphia.

In 1908, Lewis Moody became Assistant Professor of Mechanical Engineering at the Rensselaer Polytechnic Institute, assisting in planning of the new curriculum and the design and layout of apparatus for the new mechanical laboratory including that for hydraulics. He was soon advanced to the Professorship of Hydraulic Engineering. Frequently he was consulted on water power problems and from 1911, he devoted part of his time to the I. P. Morris Company as their consultant. They were then building larger and larger Francis turbines for various parts of the United States and Canada, such as Keokuk, Niagara Falls, Laurentide, and the southern states. In 1916, he resigned his professorship and joined the Morris Company full time.

During the next fifteen years, Moody's laboratory designed, installed, and operated tests of models, which improved the performance of turbines and centrifugal pumps. In this laboratory, Moody studied such important subjects as cavitation; the method of transferring results of model tests to the full scale turbines, known as the efficiency step-up formula; the losses in draft tubes; water hammer theory; and the shapes of water passages.

Moody's work resulted in over 90 patents in the period from 1911 to 1946, including those for a diagonal propeller turbine, a spiral draft tube, the Moody spreading draft tube, the Moody spiral pump, and several for high specific speed turbines. For these contributions he was internationally known and appreciated and was awarded the Elliott Cresson Medal of the Franklin Institute in 1945.

In 1930, Lewis Moody came to Princeton University as Professor of Hydraulic Engineering and held this post until his retirement as Professor Emeritus in 1948.

His lectures and conferences reflected his interests in literature, music, and art which continued from his student days at Pennsylvania.

15.19   THEODORE VON KÁRMÁN

Born: 11 May 1881 in Budapest, Hungary; Died: 6 May 1963 in Aachen, Germany.

According to the MacTutor History of Mathematics Archive, Von Kármán was a mathematical prodigy and his father, fearing that his son would become a freak, steered him towards engineering. He graduated in 1902 from Budapest, and from 1903 to 1906 he worked at the Technical University of Budapest.

In 1908, he came to Göttingen where he was greatly influenced by Klein, as was Prandtl, and visited Paris where he watched some pioneering aviation flights, which turned his interest to applying mathematics to aeronautics. He finished his thesis on a buckling problem in 1909 and then he was employed by Prandtl to work on zeppelin problems in their new wind tunnel. A year later, he became *Privatdozent* at the university. Von Kármán's paper on the vortex street was presented at the Göttingen Academy in 1911, and in 1913 he became professor at TH Aachen, where he worked until 1929, except for five years of war service in the Austro-Hungarian air force.

He visited the United States in 1926, and four years later he was offered the post of director of the Aeronautical Laboratory at California Institute of Technology. Despite his love for Aachen, the political events in Germany persuaded him to accept.

In 1933, he founded the U.S. Institute of Aeronautical Sciences (the forerunner of the American Institute for Aeronautics and Astronautics—AIAA). He continued his research on fluid mechanics, turbulence theory, and supersonic flight at Caltech, and traveled and lectured widely. He also studied applications of mathematics to engineering, aircraft structures, and soil erosion. Von Kármán had a profound influence on the development of aeronautics, propulsion and fluid mechanics as a researcher, teacher, and government advisor. His autobiography, *The Wind and Beyond* (written with L. Edson), provides a fascinating account of the early days

FIGURE 15.17   Theodore von Kármán. Photograph courtesy of California Institute of Technology.

of aeronautics in the U.S. The book by W. R. Sears, *Stories from a 20th Century Life,* gives many amusing insights into von Kármán's unique personality.

15.20 GEOFFREY INGRAM TAYLOR

Born: 7 March 1886 in London; Died: 27 June 1975.

The following account is based in large part on the article "Geoffrey Ingram Taylor," by G. K. Batchelor, Journal of Fluid Mechanics, **173,** 1–14, 1986.

Geoffrey Taylor was the grandson of George Boole (1815–1864), but he was attracted to science at an early age, apparently as a consequence of his natural ability and the encouragement of schoolmasters rather than any family influence. A few months before his twelfth birthday, he went to the Christmas lectures for children at the Royal Institution given in that year by Oliver Lodge on "The principles of the electric telegraph," and many years later he wrote, "I wish I could again capture the exquisite thrill those lectures gave me. From that time I knew I wanted to be a scientist." He went to the University of Cambridge in 1905, and studied mathematics for two years and physics for one. At the end of the third year, he obtained first class honors and a major scholarship awarded by Trinity College, which enabled him to stay on for research in the Cavendish Laboratory.

Taylor asked the Cavendish Professor, who at that time was J.J. Thomson, to suggest some possible research projects, and from among them he chose to make a test of the new quantum theory by examining the formation of interference fringes by light waves of very small intensity. Taylor's simple but effective experiment, which involved exposing a photographic plate for as long as three months (during

FIGURE 15.18 Geoffrey Ingram Taylor at age 33. Photograph courtesy of Cambridge University Press.

which time he went sailing!), showed conclusively that light behaves as a wave even for the smallest intensity. Taylor's second published paper—on the structure of a shock wave—provided, for the first time, an estimate for the thickness of a shock wave. This was his debut in fluid mechanics, the field in which he was to publish over 150 papers during the next 60 years, and it helped to gain him a fellowship at Trinity College in 1910, which provided support and freedom to pursue his research for up to six years.

In 1911, he was appointed as Schuster Reader, a temporary post in dynamical meteorology at Cambridge. He studied the vertical transfer of momentum and heat by turbulent velocity fluctuations in the wind at various heights above flat ground with home-made instruments, and began thinking about the nature of turbulence, which was to be his major preoccupation for the next thirty-five years.

In 1912, in the wake of the *Titanic* disaster, Taylor was asked to join the *Scotia* expedition to report on icebergs in the North Atlantic as the meteorologist. He designed instruments to be carried by both tethered balloons and kites flown from the masthead of the ship up to heights as large as 2000 meters. In the famous 1915 paper describing the work, he also showed the possibility of representing the turbulent transfer rates of momentum, heat and water vapor in terms of eddy diffusivities which varied with height, and he introduced the idea of a mixing length (ten years before Prandtl's similar ideas).

His long and varied career spanned many fields. During World War I, he became involved with the design and operation of aircraft. He learned to fly, and made parachute jumps. He made the first measurements of the pressure distribution over a wing in steady flight, and analyzed the stress distribution in propeller shafts under torsion. After a series of definitive experiments carried out in later years, he developed the notion of a dislocation, or detached tip of a shear crack, which was able to account for the observed strength properties of metal crystals. His practical touch was always evident, and with respect to seaplanes he invented a sea anchor (the CQR anchor), which digs itself into the seabed like a ploughshare regardless of the way it falls.

In 1919, Taylor returned to Cambridge to take up a teaching fellowship at Trinity College, and in 1923, he was appointed to a Royal Society research professorship. In the period between the wars, he worked in the Cavendish Laboratory and produced a stream of papers on an extraordinarily wide range of topics in fluid and solid mechanics. Most famous are his seminal series of papers on statistical turbulence, which laid the foundation for the scientific study of turbulence. Also, in 1933, G.I. Taylor and J.W. Maccoll published an analysis of conical supersonic flow, and in 1934 they authored an article, The mechanics of compressible fluids, in Volume III of the six volume series *Aerodynamic Theory*, edited by W.F. Durand. This paper stands as a mini-course in high-speed flow, covering acoustic theory and finite waves, shock wave theory, nozzle flows and the design of high-speed wind tunnels, potential theory, conical flow and the theory of characteristics.

After World War II, in his sixties and seventies, Taylor studied the swimming of very small creatures, flow bounded by porous media, fingering of moving interfaces, waves on thin sheets of moving liquid, and electrohydrodynamics, further enhancing his reputation as a designer of innovative, simple, and effective experiments. Batchelor writes:

Taylor's contributions have a distinctive character, of which three principal features may be identified immediately. First, they show profound insight and abil-

ity to see how things worked physically; secondly, they have the elegance and beauty that is conferred by functional simplicity, simplicity of experimental design and simplicity of mathematical argument, both being sufficient, and no more than sufficient, for the purpose in hand; and thirdly, and most important, they exhibit that uncanny knack common to the greatest scientists of recognizing the essential aspects of a phenomenon or a problem that everyone will see later to be significant and of wide applicability.

BIOGRAPHICAL REFERENCES

1. MacTutor History of Mathematics Archive Web site, at *http://www-history.mcs.st-and.ac.uk/~history/*.
2. Eric's Treasure Trove of Scientific Biography Web site, at *http://www.astro.virginia.edu/~eww6n/bios/*.
3. *Dictionary of Scientific Biography,* ed. C. C. Gillispie, Scribner Book Co., NY, 1970–1980.
4. *Encyclopaedia Britannica,* ed. W. Yust, William Benton, Chicago, London, 1955.
5. Tokaty, G. A. *A History and Philosophy of Fluid Mechanics.* G. T. Foulis & Co., Henley-on-Thames, 1971. Dover, NY, edition, 1994.
6. Moody, L. F. Friction factors for pipe flow. *Transactions of the ASME,* 66, 671–684 1944.
7. Oswatitsch K., and Wieghardt, K. Ludwig Prandtl and his Kaiser-Wilhelm-Institut. *Annual Review of Fluid Mechanics,* 19, 1–25 1987.
8. von Kármán, T. *Aerodynamics: Selected Topics in the Light of Their Historical Development.* Cornell University Press, NY, 1954.
9. Anderson J. D. Jr, *Modern Compressible Flow: With Historical Perspective,* McGraw-Hill series in Mechanical Engineering. McGraw-Hill, NY, 1982.
10. Rott, N. Note on the history of the Reynolds number. *Annual Review of Fluid Mechanics* 22, 1–11 1990.
11. Batchelor, G. K. Geoffrey Ingram Taylor, 7 March 1886–1927 June 1975. *Journal of Fluid Mechanics* 173, 1–14 1986.
12. von Kármán T., and Edson, L. *The Wind and Beyond.* Little, Brown, Boston and Toronto, 1967.

APPENDIX A

ANALYTICAL TOOLS

Here, we consider a number of basic analytical skills necessary for the study of fluid mechanics. This material is by way of summary only, and neither the selection nor its treatment is comprehensive. It is in the nature of a review, and it will be assumed that you have seen this material elsewhere in greater depth.

A.1 RANK OF A MATRIX

The customary notation for a determinant A of order n consists of a square array of n^2 quantities enclosed between vertical bars

$$|A| = |a_{ij}| = \begin{vmatrix} a_{11} & a_{12} & \cdot & a_{1n} \\ a_{21} & a_{22} & \cdot & a_{2n} \\ \cdot & \cdot & \cdot & \cdot \\ a_{n1} & a_{n2} & \cdot & a_{nn} \end{vmatrix}$$

This notation implies a particular summation process that applies to the elements of the determinant. For a determinant of order 2, the summation is

$$\begin{vmatrix} a_{11} & a_{12} \\ a_{21} & a_{22} \end{vmatrix} = a_{11}a_{22} - a_{12}a_{21}$$

For a determinant of order 3, the summation is

$$\begin{vmatrix} a_{11} & a_{12} & a_{13} \\ a_{21} & a_{22} & a_{23} \\ a_{31} & a_{32} & a_{33} \end{vmatrix} = a_{11}(a_{22}a_{33} - a_{23}a_{32}) - a_{12}(a_{21}a_{33} - a_{23}a_{31}) + a_{13}(a_{21}a_{32} - a_{22}a_{31})$$

The order of a determinant gives the rank of a matrix. The rank of a matrix r is given by the largest nonzero determinant that fits in the matrix. For example, a 3 by 6 matrix (3 rows by 6 columns) has a maximum rank of 3, since a determinant is by definition square. However, four different determinants can be found, depending on which column of the matrix corresponds to the first column of the determinant. The first two possibilities for the matrix $\|A\|$ are

$$\begin{Vmatrix} \begin{vmatrix} a_{11} & a_{12} & a_{13} \\ a_{21} & a_{22} & a_{23} \\ a_{31} & a_{32} & a_{33} \end{vmatrix} & \begin{matrix} a_{14} & a_{15} & a_{16} \\ a_{24} & a_{25} & a_{26} \\ a_{34} & a_{35} & a_{36} \end{matrix} \end{Vmatrix}$$

and

$$\begin{Vmatrix} \begin{matrix} a_{11} \\ a_{21} \\ a_{31} \end{matrix} & \begin{vmatrix} a_{12} & a_{13} & a_{14} \\ a_{22} & a_{23} & a_{24} \\ a_{32} & a_{33} & a_{34} \end{vmatrix} & \begin{matrix} a_{15} & a_{16} \\ a_{25} & a_{26} \\ a_{35} & a_{36} \end{matrix} \end{Vmatrix}$$

If any of the four possible determinants of order 3 is nonzero, the rank of this matrix is 3.

A.2 SCALAR PRODUCT

Consider two vectors **A** and **B** in Cartesian coordinates, so that

$$\mathbf{A}(x, y, z, t) = A_x\mathbf{i} + A_y\mathbf{j} + A_z\mathbf{k}$$

and

$$\mathbf{B}(x, y, z, t) = B_x\mathbf{i} + B_y\mathbf{j} + B_z\mathbf{k}$$

The scalar product is given by

$$\mathbf{A}\cdot\mathbf{B} = |\mathbf{A}||\mathbf{B}|\cos\theta = \mathbf{B}\cdot\mathbf{A}$$

where θ is the angle between **A** and **B**. The scalar product, or the *dot product,* of two vectors is a scalar, and it is *commutative,* which means that

$$\mathbf{A}\cdot\mathbf{B} = \mathbf{B}\cdot\mathbf{A}$$

In other words, the scalar product is independent of the sign of θ, as indicated by the cosine; it may be measured from **A** to **B**, or **B** to **A**. We also have

$$\mathbf{A}\cdot\mathbf{B} = A_xB_x + A_yB_y + A_zB_z$$

A very useful scalar product is the one where one of the vectors is a unit vector, say, **n**. Then

$$\mathbf{n}\cdot\mathbf{A} = |\mathbf{n}||\mathbf{A}|\cos\theta = A\cos\theta$$

which gives the component of **A** in the direction of **n**. This particular scalar product can be used whenever we need to find the component of the velocity or momentum in a given direction. For example, the x-component of the velocity **V** is given by $\mathbf{i}\cdot\mathbf{V} = u$, the y-component is given by $\mathbf{j}\cdot\mathbf{V} = v$, and the z-component is given by $\mathbf{k}\cdot\mathbf{V} = w$. In general, the component of **V** in the direction of **n** is given by $\mathbf{n}\cdot\mathbf{V}$.

A.3 VECTOR PRODUCT

For the vectors **A** and **B** we have the vector product

$$\mathbf{A}\times\mathbf{B} = |\mathbf{A}||\mathbf{B}|\sin\theta\,\mathbf{e} = -\mathbf{B}\times\mathbf{A}$$

where θ is the (smaller) angle between **A** and **B**, and the unit vector **e** is at right angles to **A** and **B**, and its direction is given by the right-hand-screw rule. The vector product, or the *cross product,* of two vectors is not commutative, and

$$\mathbf{A}\times\mathbf{B} = -\mathbf{B}\times\mathbf{A}$$

and it depends on the sign of θ, as indicated by the sine function; for $\mathbf{A}\times\mathbf{B}$ it must be measured *from* **A** *to* **B**, and for $\mathbf{B}\times\mathbf{A}$ it must be measured *from* **B** *to* **A**. To find the components of the vector product in Cartesian coordinates, we can use the determinant form

$$\mathbf{A} \times \mathbf{B} = \begin{vmatrix} \mathbf{i} & \mathbf{j} & \mathbf{k} \\ A_x & A_y & A_z \\ B_x & B_y & B_z \end{vmatrix}$$

$$= (A_y B_z - A_z B_y)\mathbf{i} - (A_x B_z - A_z B_x)\mathbf{j} + (A_x B_y - A_y B_x)\mathbf{k}$$

A.4 GRADIENT OPERATOR ∇

The gradient of a scalar quantity ϕ is the rate of change of that quantity in space. The gradient operator, ∇, or *grad*, operates on a scalar so that $\nabla\phi$ is a vector. For example, we can find the gradient of the pressure field, the temperature field, or the density field. The gradient vector $\nabla\phi$ has the property that its projection in any direction is equal to the derivative of ϕ in that direction. That is, for a unit vector $\mathbf{n}$, the scalar product $\mathbf{n} \cdot \nabla\phi$ gives the rate of change of ϕ in the direction of $\mathbf{n}$. Since the maximum projection of a vector is in the direction of the vector itself, it is clear that the vector $\nabla\phi$ points in the direction of the maximum rate of change of ϕ. For example, the gradient of the altitude, ∇h, is a vector that always points in the vertical direction since that is the direction in which the altitude increases most quickly. Also, the gradient of the pressure, ∇p, is a vector that points in the direction normal to the isobars (lines of constant pressure).

Consider a closed volume $\forall$ with a surface area A. Then the gradient of any scalar ϕ is formally defined as

$$\nabla\phi \equiv \lim_{\forall \to 0} \frac{1}{\forall} \int \mathbf{n}\phi \, dA$$

where $\mathbf{n}$ is an outward facing unit normal vector. So the gradient of ϕ is the ratio of the integral of $\mathbf{n}\phi$ over an area A, to the volume $\forall$ enclosed by A, as the volume shrinks to zero about some point.

In Cartesian coordinates,

$$\nabla\phi = \frac{\partial\phi}{\partial x}\mathbf{i} + \frac{\partial\phi}{\partial y}\mathbf{j} + \frac{\partial\phi}{\partial z}\mathbf{k}$$

In cylindrical coordinates,

$$\nabla\phi = \frac{\partial\phi}{\partial r}\mathbf{e_r} + \frac{1}{r}\frac{\partial\phi}{\partial\theta}\mathbf{e_\theta} + \frac{\partial\phi}{\partial z}\mathbf{e_z}$$

Note: the operator ∇ looks like a vector. In fact, it has many properties that are similar to the properties of vectors. However, it differs from a vector in some important ways, as we shall see, and it is best thought of as an *operator* with special properties. We can write $\nabla\phi$ in Cartesian coordinates as,

$$\nabla\phi = \left(\frac{\partial}{\partial x}\mathbf{i} + \frac{\partial}{\partial y}\mathbf{j} + \frac{\partial}{\partial z}\mathbf{k}\right)\phi$$

which emphasizes that $\nabla\phi$ is shorthand for the result obtained when ∇ operates on a scalar ϕ.

A.5 DIVERGENCE OPERATOR $\nabla \cdot$

The divergence of a vector quantity is the outflow of that quantity per unit time per unit volume. The divergence operator $\nabla \cdot$ operates on a vector, and the result is a scalar. That is, the divergence of a vector is a scalar (note the contrast with the gradient operator ∇). The divergence of any vector, such as the velocity $\mathbf{V}$, is formally defined as

$$\nabla \cdot \mathbf{V} \equiv \lim_{\forall \to 0} \frac{1}{\forall} \int \mathbf{n} \cdot \mathbf{V} \, dA$$

So the divergence of the velocity is the ratio of the integral of $\mathbf{n} \cdot \mathbf{V}$ over an area A, to the volume $\forall$ enclosed by A, as the volume shrinks to zero about some point, say P. More physically, $\mathbf{V}$ is the velocity through any small surface element dA, and $\mathbf{n} \cdot \mathbf{V} = |\mathbf{V}| \cos \theta$ is the volume flux per unit area. The area integral is the total volume flux through the surface area A, and the divergence of $\mathbf{V}$ ($= \nabla \cdot \mathbf{V}$) is the volume flux per unit volume at the point P.

In Cartesian coordinates, the divergence of $\mathbf{V}(\mathbf{x}, t) = \mathbf{V}(x, y, z, t)$ is simply

$$\nabla \cdot \mathbf{V} = \left(\frac{\partial}{\partial x} \mathbf{i} + \frac{\partial}{\partial y} \mathbf{j} + \frac{\partial}{\partial z} \mathbf{k} \right) \cdot (u\mathbf{i} + v\mathbf{j} + w\mathbf{k})$$

That is

$$\nabla \cdot \mathbf{V} = \frac{\partial u}{\partial x} + \frac{\partial v}{\partial y} + \frac{\partial w}{\partial z}$$

In cylindrical coordinates,

$$\nabla \cdot \mathbf{V} = \frac{1}{r} \frac{\partial r u_r}{\partial r} + \frac{1}{r} \frac{\partial u_\theta}{\partial \theta} + \frac{\partial u_z}{\partial z}$$

Note: we see that the divergence operator $\nabla \cdot$ behaves in some ways like a scalar product. However, as we have noted, ∇ differs from a vector in some important ways. In particular, $\nabla \cdot$ is not commutative, so that $\nabla \cdot \mathbf{V} \neq \mathbf{V} \cdot \nabla$ (see Section A.9).

Also, a vector field that has zero divergence (for example, $\nabla \cdot \mathbf{A} = 0$) is sometimes called a *solenoidal* vector field. A velocity field that has zero divergence ($\nabla \cdot \mathbf{V} = 0$) is also called an *incompressible* velocity field (see Section 6.2).

A.6 LAPLACIAN OPERATOR ∇^2

The Laplacian operator is composed of second derivatives in space. When it operates on a scalar, the result is a a scalar; when it operates on a vector, the result is a vector. For a scalar ϕ, in Cartesian coordinates,

$$\nabla^2 \phi = \frac{\partial^2 \phi}{\partial x^2} + \frac{\partial^2 \phi}{\partial y^2} + \frac{\partial^2 \phi}{\partial z^2} \tag{A.1}$$

In cylindrical coordinates,

$$\nabla^2 \phi = \frac{\partial^2 \phi}{\partial r^2} + \frac{1}{r} \frac{\partial \phi}{\partial r} + \frac{1}{r^2} \frac{\partial^2 \phi}{\partial \theta^2} + \frac{\partial^2 \phi}{\partial z^2}$$

$$= \frac{1}{r} \frac{\partial}{\partial r} \left(r \frac{\partial \phi}{\partial r} \right) + \frac{1}{r^2} \frac{\partial^2 \phi}{\partial \theta^2} + \frac{\partial^2 \phi}{\partial z^2} \tag{A.2}$$

Note: we see that the Laplacian operator behaves in some ways like the scalar product $\nabla \cdot \nabla$. Remember, however, ∇ is best thought of as an operator with special properties.

Also, when a scalar parameter ϕ is operated on by the Laplacian ∇^2 and the result is zero, it is said to satisfy Laplace's equation $\nabla^2\phi = 0$.

For a vector $\mathbf{V}$, in Cartesian coordinates,

$$\nabla^2\mathbf{V} = \left(\frac{\partial^2 u}{\partial x^2} + \frac{\partial^2 u}{\partial y^2} + \frac{\partial^2 u}{\partial z^2}\right)\mathbf{i}$$

$$+ \left(\frac{\partial^2 v}{\partial x^2} + \frac{\partial^2 v}{\partial y^2} + \frac{\partial^2 v}{\partial z^2}\right)\mathbf{j} + \left(\frac{\partial^2 w}{\partial x^2} + \frac{\partial^2 w}{\partial y^2} + \frac{\partial^2 w}{\partial z^2}\right)\mathbf{k}$$

In cylindrical coordinates,

$$\nabla^2\mathbf{V} = \left[\frac{1}{r}\frac{\partial}{\partial r}\left(r\frac{\partial v_r}{\partial r}\right) - \frac{v_r}{r^2} + \frac{1}{r^2}\frac{\partial^2 v_r}{\partial \theta^2} - \frac{2}{r^2}\frac{\partial v_\theta}{\partial \theta} + \frac{\partial^2 v_r}{\partial z^2}\right]\mathbf{e_r}$$

$$+ \left[\frac{1}{r}\frac{\partial}{\partial r}\left(r\frac{\partial v_\theta}{\partial r}\right) - \frac{v_\theta}{r^2} + \frac{1}{r^2}\frac{\partial^2 v_\theta}{\partial \theta^2} - \frac{2}{r^2}\frac{\partial v_r}{\partial \theta} + \frac{\partial^2 v_\theta}{\partial z^2}\right]\mathbf{e_\theta}$$

$$+ \left[\frac{1}{r}\frac{\partial}{\partial r}\left(r\frac{\partial v_z}{\partial r}\right) + \frac{1}{r^2}\frac{\partial^2 v_z}{\partial \theta^2} + \frac{\partial^2 v_z}{\partial z^2}\right]\mathbf{e_z}$$

A.7   CURL OPERATOR $\nabla\times$

The curl operator $\nabla\times$ operates on a vector and the result is another vector. That is, the curl of a vector is a vector. A physical interpretation relevant to fluid dynamics is that the curl of the velocity field $\nabla \times \mathbf{V}$ is equal to twice the angular velocity of the flow field at a given point (this is not obvious, of course, but it is useful to realize that $\nabla \times \mathbf{V}$ is related to the rotation of the flow; see Section 7.1). The curl of any vector, such as the velocity $\mathbf{V}$, is formally defined as

$$\nabla \times \mathbf{V} \equiv \lim_{\forall \to 0}\frac{1}{\forall}\int \mathbf{n} \times \mathbf{V}\, dA$$

So the curl of the velocity is the ratio of the integral of $\mathbf{n} \times \mathbf{V}$ over an area A, to the volume $\forall$ enclosed by A, as the volume shrinks to zero about some point.

In Cartesian coordinates, the curl of $\mathbf{V}(\mathbf{x}, t) = \mathbf{V}(x, y, z, t)$ is given in the shorthand, determinant form by

$$\nabla \times \mathbf{V} = \begin{vmatrix} \mathbf{i} & \mathbf{j} & \mathbf{k} \\ \frac{\partial}{\partial x} & \frac{\partial}{\partial y} & \frac{\partial}{\partial z} \\ u & v & w \end{vmatrix}$$

That is:

$$\nabla \times \mathbf{V} = \left(\frac{\partial w}{\partial y} - \frac{\partial v}{\partial z}\right)\mathbf{i} - \left(\frac{\partial w}{\partial x} - \frac{\partial u}{\partial z}\right)\mathbf{j} + \left(\frac{\partial v}{\partial x} - \frac{\partial u}{\partial y}\right)\mathbf{k}$$

In cylindrical coordinates,

$$\nabla \times \mathbf{V} = \left(\frac{1}{r}\frac{\partial u_z}{\partial \theta} - \frac{\partial u_\theta}{\partial z}\right)\mathbf{e_r} + \left(\frac{\partial u_r}{\partial z} - \frac{\partial u_z}{\partial r}\right)\mathbf{e_\theta} + \left(\frac{1}{r}\frac{\partial ru_\theta}{\partial r} - \frac{1}{r}\frac{\partial u_r}{\partial \theta}\right)\mathbf{e_z}$$

Note: we see that the curl operator $\nabla \times$ behaves like a vector product. For example, it is true that $\nabla \times \mathbf{V} = -\mathbf{V} \times \nabla$. However, as we continue to emphasize, ∇ is best thought of as an operator with special properties.

A vector field that has zero curl (for example, $\nabla \times \mathbf{V} = 0$) is commonly called an *irrotational* vector field.

A.8 SUMMARY: DIV, GRAD, AND CURL

A.8.1 Integral Definitions

$$\nabla \phi = \lim_{\forall \to 0} \frac{1}{\forall} \int \mathbf{n} \phi \, dA$$

$$\nabla \cdot \mathbf{V} = \lim_{\forall \to 0} \frac{1}{\forall} \int \mathbf{n} \cdot \mathbf{V} \, dA$$

$$\nabla \times \mathbf{V} = \lim_{\forall \to 0} \frac{1}{\forall} \int \mathbf{n} \times \mathbf{V} \, dA$$

where ϕ is a scalar and $\mathbf{V}$ is a vector.

A.8.2 Differential Forms

In Cartesian coordinates,

$$\nabla \phi = \frac{\partial \phi}{\partial x} \mathbf{i} + \frac{\partial \phi}{\partial y} \mathbf{j} + \frac{\partial \phi}{\partial z} \mathbf{k}$$

$$\nabla \cdot \mathbf{V} = \frac{\partial u}{\partial x} + \frac{\partial v}{\partial y} + \frac{\partial w}{\partial z}$$

$$\nabla \times \mathbf{V} = \begin{vmatrix} \mathbf{i} & \mathbf{j} & \mathbf{k} \\ \frac{\partial}{\partial x} & \frac{\partial}{\partial y} & \frac{\partial}{\partial z} \\ u & v & w \end{vmatrix}$$

$$= \left(\frac{\partial w}{\partial y} - \frac{\partial v}{\partial z} \right) \mathbf{i} - \left(\frac{\partial w}{\partial x} - \frac{\partial u}{\partial z} \right) \mathbf{j} + \left(\frac{\partial v}{\partial x} - \frac{\partial u}{\partial y} \right) \mathbf{k}$$

In cylindrical coordinates,

$$\nabla \phi = \frac{\partial \phi}{\partial r} \mathbf{e_r} + \frac{1}{r} \frac{\partial \phi}{\partial \theta} \mathbf{e_\theta} + \frac{\partial \phi}{\partial z} \mathbf{e_z}$$

$$\nabla \cdot \mathbf{V} = \frac{1}{r} \frac{\partial r u_r}{\partial r} + \frac{1}{r} \frac{\partial u_\theta}{\partial \theta} + \frac{\partial u_z}{\partial z}$$

$$\nabla \times \mathbf{V} = \left(\frac{1}{r} \frac{\partial u_z}{\partial \theta} - \frac{\partial u_\theta}{\partial z} \right) \mathbf{e_r} + \left(\frac{\partial u_r}{\partial z} - \frac{\partial u_z}{\partial r} \right) \mathbf{e_\theta} + \left(\frac{1}{r} \frac{\partial r u_\theta}{\partial r} - \frac{1}{r} \frac{\partial u_r}{\partial \theta} \right) \mathbf{e_z}$$

A.8.3 Integral Theorems

Stokes' theorem relates the line integral of the velocity to the area integral of the vorticity. That is,

$$\int \mathbf{n} \cdot \vec{\Omega} \, dA = \int \mathbf{n} \cdot \nabla \times \mathbf{V} \, dA = \oint_C \mathbf{V} \cdot d\vec{l}$$

where the line integral is taken around the closed curve C defining the perimeter of the area A, where the direction of integration around C is positive with respect to the side of A on which the unit normals are drawn (by the right hand-screw rule).

Additional theorems relate area integrals to volume integrals. For a closed volume $\forall$ with surface area A,

$$\int \mathbf{n}\,p\,dA = \int \nabla p\,d\forall$$
$$\int \mathbf{n}\cdot\mathbf{V}\,dA = \int \nabla\cdot\mathbf{V}\,d\forall$$
$$\int \mathbf{n}\times\mathbf{V}\,dA = \int \nabla\times\mathbf{V}\,dA$$

where p and $\mathbf{V}$ are continuous and differentiable. The second of these theorems is usually called the divergence theorem.

A.8.4 Taylor-Series Expansion

The Taylor-series expansion allows us to estimate the variation of a quantity in the neighborhood of a known point in a formal and complete way. To find how the pressure p varies with z, for example, the pressure is expanded in a power series in the neighborhood of $z = \alpha$, where α is a known point (see Figure A.1). The expansion is given by

$$p(z) = p(\alpha) + (z - \alpha)\left(\frac{\partial p}{\partial z}\right)_{z=\alpha} + \frac{(z-\alpha)^2}{2!}\left(\frac{\partial^2 p}{\partial z^2}\right)_{z=\alpha} + \cdots$$

We need to assume that the pressure and its derivatives exist at $z = \alpha$ (it is obvious that if the function $p(z)$ has a discontinuity close to the point α where the derivative takes an infinite value, the Taylor-series expansion will not give a good estimate). Now if $(z - \alpha) = \delta z$, we can write this relationship as

$$p(\alpha + \delta z) = p(\alpha) + \delta z\left(\frac{\partial p}{\partial z}\right)_{\alpha} + \frac{(\delta z)^2}{2!}\left(\frac{\partial^2 p}{\partial z^2}\right)_{\alpha} + \frac{(\delta z)^3}{3!}\left(\frac{\partial^3 p}{\partial z^3}\right)_{\alpha} + \cdots$$

The dots represent higher order terms, that is, terms which contain δz in a higher order (for example, $(\delta z)^2$ is one order higher than δz and $(\delta z)^3$ is one order higher than $(\delta z)^2$). If δz is very small, $(\delta z)^2$ is much smaller still, and in the limit of δz going to very small values, higher order terms can usually be neglected. This may be illustrated by using some numerical values. For a quantity Δ that is small compared to unity, $\Delta \ll 1$, and $\Delta^2 \ll \Delta$. For example,

$$\Delta = 0.1 \qquad \Delta^2 = 0.01$$
$$\Delta = 0.01 \qquad \Delta^2 = 0.0001, \text{ etc.}$$

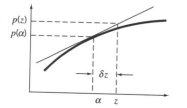

FIGURE A.1 Estimating the pressure z by using the pressure at α by applying the Taylor-series expansion.

Therefore, higher order terms rapidly become small, and in the limit of an infinitesimally small value of Δ, terms that contain Δ^2 can usually be neglected. This approach is called linearization (since quadratic terms are neglected).

A.8.5 Total Derivative

In Cartesian coordinates,

$$\frac{D\mathbf{V}}{Dt} = \left(\frac{\partial u}{\partial t} + u\frac{\partial u}{\partial x} + v\frac{\partial u}{\partial y} + w\frac{\partial u}{\partial z}\right)\mathbf{i}$$

$$+ \left(\frac{\partial v}{\partial t} + u\frac{\partial v}{\partial x} + v\frac{\partial v}{\partial y} + w\frac{\partial v}{\partial z}\right)\mathbf{j}$$

$$+ \left(\frac{\partial w}{\partial t} + u\frac{\partial w}{\partial x} + v\frac{\partial w}{\partial y} + w\frac{\partial w}{\partial z}\right)\mathbf{k}$$

In cylindrical coordinates,

$$\frac{D\mathbf{V}}{Dt} = \left(\frac{\partial u_r}{\partial t} + u_r\frac{\partial u_r}{\partial r} + \frac{u_\theta}{r}\frac{\partial u_r}{\partial \theta} + u_z\frac{\partial u_r}{\partial z} - \frac{u_\theta^2}{r}\right)\mathbf{e_r}$$

$$+ \left(\frac{\partial u_\theta}{\partial t} + u_r\frac{\partial u_\theta}{\partial r} + \frac{u_\theta}{r}\frac{\partial u_\theta}{\partial \theta} + u_z\frac{\partial u_\theta}{\partial z} + \frac{u_r u_\theta}{r}\right)\mathbf{e_\theta}$$

$$+ \left(\frac{\partial u_z}{\partial t} + u_r\frac{\partial u_z}{\partial r} + \frac{u_\theta}{r}\frac{\partial u_z}{\partial \theta} + u_z\frac{\partial u_z}{\partial z}\right)\mathbf{e_z}$$

A.9 DYADS: THE OPERATOR V·∇

In Cartesian coordinates, the total derivative of the velocity is given by

$$\frac{D\mathbf{V}}{Dt} = \frac{\partial \mathbf{V}}{\partial t} + u\frac{\partial \mathbf{V}}{\partial x} + v\frac{\partial \mathbf{V}}{\partial y} + w\frac{\partial \mathbf{V}}{\partial z}$$

By writing this as

$$\frac{D\mathbf{V}}{Dt} = \frac{\partial \mathbf{V}}{\partial t} + (\mathbf{V}\cdot\nabla)\mathbf{V}$$

we define a new operator $\mathbf{V}\cdot\nabla$. In this form, the total derivative is written in a way that is independent of the coordinate system. The expansion of $\mathbf{V}\cdot\nabla$ depends on the coordinate system. In Cartesian coordinates,

$$\mathbf{V}\cdot\nabla = u\frac{\partial}{\partial x} + v\frac{\partial}{\partial y} + w\frac{\partial}{\partial z}$$

Compare this with the divergence of $\mathbf{V}$, where

$$\nabla\cdot\mathbf{V} = \frac{\partial u}{\partial x} + \frac{\partial v}{\partial y} + \frac{\partial w}{\partial z}$$

In cylindrical coordinates,

$$\mathbf{V}\cdot\nabla = u_r\frac{\partial}{\partial r} + \frac{u_\theta}{r}\frac{\partial}{\partial \theta} + u_z\frac{\partial}{\partial z}$$

Note: the operator $\mathbf{V} \cdot \nabla$ should not be thought of as a dot product of the velocity and a "vector" ∇. As we indicated earlier, ∇ is not a true vector, although it sometimes behaves like one. It is best to think of $\mathbf{V} \cdot \nabla$ as an operator. To make this clear, it is preferable to write it in parentheses, as $(\mathbf{V} \cdot \nabla)$, although formally there is no ambiguity.

Care must be taken in expanding $\mathbf{V} \cdot \nabla$ in other than Cartesian coordinates. In fact, $\mathbf{V} \cdot \nabla$ is an example of a *dyad,* and it is more easily handled in tensor notation. It can be expressed in true vector form using the vector identity

$$(\mathbf{V} \cdot \nabla)\mathbf{V} = \nabla(\tfrac{1}{2}V^2) - \mathbf{V} \times (\nabla \times \mathbf{V}) \tag{A.3}$$

A.10 INTEGRAL AND DIFFERENTIAL FORMS

Instead of deriving the differential forms of the continuity and momentum equations using elemental control volumes as we did in Sections 6.2 and 6.3, we can obtain them directly from their integral forms.

The continuity equation for a fixed control volume (equation 5.3) states

$$\frac{\partial}{\partial t} \int \rho \, d\forall + \int \mathbf{n} \cdot \rho \mathbf{V} \, dA = 0$$

Since the control volume is fixed, the derivative can be moved inside the integral, so that

$$\int \frac{\partial \rho}{\partial t} \, d\forall + \int \mathbf{n} \cdot \rho \mathbf{V} \, dA = 0$$

The area integral can be converted to a volume integral using the divergence theorem (see Section A.8.3). Hence,

$$\int \left(\frac{\partial \rho}{\partial t} + \nabla \cdot \rho \mathbf{V} \right) d\forall = 0$$

Since the control volume is arbitrary, the integrand itself must be zero, so that

$$\frac{\partial \rho}{\partial t} + \nabla \cdot \rho \mathbf{V} = 0$$

or

$$\nabla \cdot \mathbf{V} = -\frac{1}{\rho} \frac{D\rho}{Dt}$$

as before (equations 6.7 and 6.8).

The momentum equation for a fixed control volume (equation 5.16) states

$$\mathbf{F}_{ext} + \mathbf{F}_v - \int \mathbf{n}\rho \, dA + \int \rho \mathbf{g} \, d\forall = \frac{\partial}{\partial t} \int \rho \mathbf{V} \, d\forall + \int (\mathbf{n} \cdot \rho \mathbf{V})\mathbf{V} \, dA$$

If we examine the interior of the fluid and ignore viscous effects, the forces $\mathbf{F}_{ext}$ and $\mathbf{F}_v$ can be neglected. Since the control volume is fixed, the derivative can be moved inside the integral, so that

$$-\int \mathbf{n}p \, dA + \int \rho \mathbf{g} \, d\forall = \int \frac{\partial \rho \mathbf{V}}{\partial t} \, d\forall + \int (\mathbf{n} \cdot \rho \mathbf{V})\mathbf{V} \, dA$$

Changing the area integrals to volume integrals using the divergence theorem and collecting terms,

$$-\int (\nabla p + \rho \mathbf{g})\, d\forall = \int \left(\frac{\partial \rho \mathbf{V}}{\partial t} + (\nabla \cdot \rho \mathbf{V})\mathbf{V} + (\rho \mathbf{V} \cdot \nabla)\mathbf{V} \right) d\forall$$

$$= \int \left(\rho \frac{\partial \mathbf{V}}{\partial t} + \rho(\mathbf{V} \cdot \nabla)\mathbf{V} + \mathbf{V}\frac{\partial \rho}{\partial t} + \mathbf{V}(\nabla \cdot \rho \mathbf{V}) \right) d\forall$$

Using the continuity equation yields

$$\int \left(\rho \frac{\partial \mathbf{V}}{\partial t} + \rho(\mathbf{V} \cdot \nabla)\mathbf{V} + \nabla p - \rho \mathbf{g} \right) d\forall = 0$$

Since the control volume is arbitrary, the integrand itself must be zero, so that

$$\rho \left(\frac{\partial \mathbf{V}}{\partial t} + (\mathbf{V} \cdot \nabla)\mathbf{V} \right) + \nabla p - \rho \mathbf{g} = 0$$

That is,

$$\rho \frac{D\mathbf{V}}{Dt} = -\nabla p + \rho \mathbf{g}$$

as before (equation 6.11).

CONVERSION FACTORS

Length	$1\ m = 1000\ mm = 39.37\ in. = 3.281\ ft$
	$1\ ft = 0.3048\ m$
	$1\ km = 0.6214\ mile$
	$1\ mile = 5280\ ft = 1609.3\ m$
Velocity	$1\ m/s = 3.281\ ft/s = 3.60\ km/hr = 2.28\ mph$
	$1\ ft/s = 0.3048\ m/s$
	$1\ mph = \frac{22}{15}\ ft/s = 0.447\ m/s$
	$1\ knot = 1.151\ mph = 5144\ m/s$
Volume	$1\ m^3 = 1000\ liters = 35.31\ ft^3$
	$1\ ft^3 = 0.02832\ m^3 = 28.32\ liters = 7.48\ gal\ (U.S.)$
	$1\ gal\ (U.S.) = 3.785\ liters$
Mass	$1\ kg = 1000\ g = 2.205\ lb_m = 0.06852\ slug$
	$1\ slug = 32.174\ lb_m = 14.59\ kg$
Density	$1\ kg/m^3 = 0.001940\ slug/ft^3 = 0.06243\ lb_m/ft^3$
	$= 36.13 \times 10^{-6}\ lb_m/in.^3$
	$1\ slug/ft^3 = 515.4\ kg/m^3$
Force	$1\ N = 0.2248\ lb_f$
	$1\ lb_f = 4.448\ N$
Pressure	$1\ pascal = 1\ N/m^2 = 10^{-5}\ bar = 10\ dyne/cm^2$
	$= 0.14504 \times 10^{-3}\ psi\quad or\quad lb_f/in.^2$
	$= 0.02088\ psf\quad or\quad lb_f/ft^2$
	$1\ psi = 6895\ Pa$
	$1\ atm = 101{,}325\ N/m^2\quad or\quad Pa$
	$= 1.01325\ bar$
	$= 760\ torr$
	$= 14.70\ psi\quad or\quad lb_f/in.^2$
	$= 2116\ lb_f/ft^2\quad or\quad psf$
	$= 29.92\ in.\ Hg = 760.0\ mm\ Hg$
	$= 10.33\ m\ H_2O = 33.90\ ft\ H_2O$
Viscosity	$1\ Pa \cdot s = 1\ N \cdot s/m^2$
	$= 10\ poise = 1000\ cp\ (centipoise)$
	$= 0.02088\ lb_f \cdot s/ft^2$
	$1\ lb_f \cdot s/ft^2 = 32.174\ lb_m/s \cdot ft^2$

Kinematic viscosity	$1\ m^2/s = 10^4\ stokes$
	$= 10.76\ ft^2/s$
	$1\ ft^2/s = 0.09290\ m^2/s$
Energy	$1\ N \cdot m = 1\ joule$
	$= 0.7375\ ft \cdot lb_f = 0.000948\ Btu$
	$= 0.2388\ cal$
	$1\ Btu = 1055\ joule = 252\ cal$
Power	$1\ W = 1\ J/s$
	$= 0.7375\ ft \cdot lb_f/s$
	$1\ kW = 1.341\ hp = 737.5\ ft \cdot lb_f/s = 3412\ Btu/hr$
	$1\ hp = 745.7\ W = 550\ ft \cdot lb_f/s = 2545\ Btu/hr$
Temperature	$1°C = 1°K = \frac{9}{5}°R$
	$0°C = 273.15°K$
	$= 32°F = 459.67°R$
	$= 0.02088\ lb_f/ft^2 \quad or \quad psf$
Gas constant R	$1\ m^2/s^2K = 1\ J/kg\ K$
	$= 5.980\ ft^2/s^2\ R$

FLUID AND FLOW PROPERTIES

TABLE C.1 Physical Properties of Air at Standard Atmospheric Pressure (SI Units)[a]

Temperature (°C)	Density, ρ (kg/m³)	Dynamic viscosity, μ (N·s/m²)	Kinematic viscosity, ν (m²/s)	Specific heat ratio, γ
−40	1.514	1.57 E − 5	1.04 E − 5	1.401
−20	1.395	1.63 E − 5	1.17 E − 5	1.401
0	1.292	1.71 E − 5	1.32 E − 5	1.401
5	1.269	1.73 E − 5	1.36 E − 5	1.401
10	1.247	1.76 E − 5	1.41 E − 5	1.401
15	1.225	1.80 E − 5	1.47 E − 5	1.401
20	1.204	1.82 E − 5	1.51 E − 5	1.401
25	1.184	1.85 E − 5	1.56 E − 5	1.401
30	1.165	1.86 E − 5	1.60 E − 5	1.400
40	1.127	1.87 E − 5	1.66 E − 5	1.400
50	1.109	1.95 E − 5	1.76 E − 5	1.400
60	1.060	1.97 E − 5	1.86 E − 5	1.399
70	1.029	2.03 E − 5	1.97 E − 5	1.399
80	0.9996	2.07 E − 5	2.07 E − 5	1.399
90	0.9721	2.14 E − 5	2.20 E − 5	1.398
100	0.9461	2.17 E − 5	2.29 E − 5	1.397
200	0.7461	2.53 E − 5	3.39 E − 5	1.390
300	0.6159	2.98 E − 5	4.84 E − 5	1.379
400	0.5243	3.32 E − 5	6.34 E − 5	1.368
500	0.4565	3.64 E − 5	7.97 E − 5	1.357
1000	0.2772	5.04 E − 5	1.82 E − 4	1.321

[a] Based on data from R. D. Blevins, *Applied Fluid Dynamics Handbook,* Van Nostrand Reinhold Co., Inc., New York, 1984.

TABLE C.2 Physical Properties of Air at Standard Atmospheric Pressure (BG Units)[a]

Temperature (°F)	Density, ρ (slug/ft³)	Dynamic viscosity, μ (lb$_f$·s/ft²)	Kinematic viscosity, ν (ft²/s)	Specific heat ratio, γ
−40	2.939 E − 3	3.29 E − 7	1.12 E − 4	1.401
−20	2.805 E − 3	3.34 E − 7	1.19 E − 4	1.401
0	2.683 E − 3	3.38 E − 7	1.26 E − 4	1.401
10	2.626 E − 3	3.44 E − 7	1.31 E − 4	1.401
20	2.571 E − 3	3.50 E − 7	1.36 E − 4	1.401
30	2.519 E − 3	3.58 E − 7	1.42 E − 4	1.401
40	2.469 E − 3	3.60 E − 7	1.46 E − 4	1.401
50	2.420 E − 3	3.68 E − 7	1.52 E − 4	1.401
60	2.373 E − 3	3.75 E − 7	1.58 E − 4	1.401
70	2.329 E − 3	3.82 E − 7	1.64 E − 4	1.401
80	2.286 E − 3	3.86 E − 7	1.69 E − 4	1.400
90	2.244 E − 3	3.90 E − 7	1.74 E − 4	1.400
100	2.204 E − 3	3.94 E − 7	1.79 E − 4	1.400
120	2.128 E − 3	4.02 E − 7	1.89 E − 4	1.400
140	2.057 E − 3	4.13 E − 7	2.01 E − 4	1.399
160	1.990 E − 3	4.22 E − 7	2.12 E − 4	1.399
180	1.928 E − 3	4.34 E − 7	2.25 E − 4	1.399
200	1.870 E − 3	4.49 E − 7	2.40 E − 4	1.398
300	1.624 E − 3	4.97 E − 7	3.06 E − 4	1.394
400	1.435 E − 3	5.24 E − 7	3.65 E − 4	1.389
500	1.285 E − 3	5.80 E − 7	4.51 E − 4	1.383
750	1.020 E − 3	6.81 E − 7	6.68 E − 4	1.367
1000	8.445 E − 4	7.85 E − 7	9.30 E − 4	1.351
1500	6.291 E − 4	9.50 E − 7	1.51 E − 3	1.329

[a] Based on data from R. D. Blevins, *Applied Fluid Dynamics Handbook*, Van Nostrand Reinhold Co., Inc., New York, 1984.

TABLE C.3 **Physical Properties of Water (SI Units)**[a]

Temperature (°C)	Density, ρ (kg/m³)	Dynamic viscosity, μ (N·s/m²)	Kinematic viscosity, ν (m²/s)	Specific tension[b], σ (N/m)	Vapor pressure, p_v [N/m²(abs)]
0	999.9	1.787 E − 3	1.787 E − 6	7.56 E − 2	6.105 E + 2
5	1000.0	1.519 E − 3	1.519 E − 6	7.49 E − 2	8.722 E + 2
10	999.7	1.307 E − 3	1.307 E − 6	7.42 E − 2	1.228 E + 3
20	998.2	1.002 E − 3	1.004 E − 6	7.28 E − 2	2.338 E + 3
30	995.7	7.975 E − 4	8.009 E − 7	7.12 E − 2	4.243 E + 3
40	992.2	6.529 E − 4	6.580 E − 7	6.96 E − 2	7.376 E + 3
50	988.1	5.468 E − 4	5.534 E − 7	6.79 E − 2	1.233 E + 4
60	983.2	4.665 E − 4	4.745 E − 7	6.62 E − 2	1.992 E + 4
70	977.8	4.042 E − 4	4.134 E − 7	6.44 E − 2	3.116 E + 4
80	971.8	3.547 E − 4	3.650 E − 7	6.26 E − 2	4.734 E + 4
90	965.3	3.147 E − 4	3.260 E − 7	6.08 E − 2	7.010 E + 4
100	958.4	2.818 E − 4	2.940 E − 7	5.89 E − 2	1.013 E + 5

[a] Based on data from *Handbook of Chemistry and Physics,* 69th Ed., CRC Press, 1988.
[b] In contact with air.

TABLE C.4 **Physical Properties of Water (BG Units)**[a]

Temperature (°F)	Density, ρ (slugs/ft³)	Dynamic viscosity, μ (lb$_f$·s/ft²)	Kinematic viscosity, ν (ft²/s)	Specific tension[b], σ (lb$_f$/ft)	Vapor pressure, p_v [lb$_f$/in.²(abs)]
32	1.940	3.732 E − 5	1.924 E − 5	5.18 E − 3	8.854 E − 2
40	1.940	3.228 E − 5	1.664 E − 5	5.13 E − 3	1.217 E − 1
50	1.940	2.730 E − 5	1.407 E − 5	5.09 E − 3	1.781 E − 1
60	1.938	2.344 E − 5	1.210 E − 5	5.03 E − 3	2.563 E − 1
70	1.936	2.037 E − 5	1.052 E − 5	4.97 E − 3	3.631 E − 1
80	1.934	1.791 E − 5	9.262 E − 6	4.91 E − 3	5.069 E − 1
90	1.931	1.500 E − 5	8.233 E − 6	4.86 E − 3	6.979 E − 1
100	1.927	1.423 E − 5	7.383 E − 6	4.79 E − 3	9.493 E − 1
120	1.918	1.164 E − 5	6.067 E − 6	4.67 E − 3	1.692 E + 0
140	1.908	9.743 E − 6	5.106 E − 6	4.53 E − 3	2.888 E + 0
160	1.896	8.315 E − 6	4.385 E − 6	4.40 E − 3	4.736 E + 0
180	1.883	7.207 E − 6	3.827 E − 6	4.26 E − 3	7.507 E + 0
200	1.869	6.342 E − 6	3.393 E − 6	4.12 E − 3	1.152 E + 1
212	1.860	5.886 E − 6	3.165 E − 6	4.04 E − 3	1.469 E + 1

[a] Based on data from *Handbook of Chemistry and Physics,* 69th Ed., CRC Press, 1988.
[b] In contact with air.

TABLE C.5 Properties of the U.S. Standard Atmosphere (SI Units)[a]

Altitude (m)	Temperature (°C)	Acceleration of gravity, g (m/s²)	Pressure, p [N/m²(abs)]	Density, ρ (kg/m³)	Dynamic viscosity, μ (N·s/m²)
−1,000	21.50	9.810	1.139 E + 5	1.347 E + 0	1.821 E − 5
0	15.00	9.807	1.103 E + 5	1.225 E + 0	1.789 E − 5
1,000	8.50	9.804	8.988 E + 4	1.112 E + 0	1.758 E − 5
2,000	2.00	9.801	7.950 E + 4	1.007 E + 0	1.726 E − 5
3,000	−4.49	9.797	7.012 E + 4	9.093 E − 1	1.694 E − 5
4,000	−10.98	9.794	6.166 E + 4	8.194 E − 1	1.661 E − 5
5,000	−17.47	9.791	5.405 E + 4	7.364 E − 1	1.628 E − 5
6,000	−23.96	9.788	4.722 E + 4	6.601 E − 1	1.595 E − 5
7,000	−30.45	9.785	4.111 E + 4	5.900 E − 1	1.561 E − 5
8,000	−36.94	9.782	3.565 E + 4	5.258 E − 1	1.527 E − 5
9,000	−43.42	9.779	3.080 E + 4	4.671 E − 1	1.493 E − 5
10,000	−49.90	9.776	2.650 E + 4	4.135 E − 1	1.458 E − 5
15,000	−56.50	9.761	1.211 E + 4	1.948 E − 1	1.422 E − 5
20,000	−56.50	9.745	5.529 E + 3	8.891 E − 2	1.422 E − 5
25,000	−51.60	9.730	2.549 E + 3	4.008 E − 2	1.448 E − 5
30,000	−46.64	9.715	1.197 E + 3	1.841 E − 2	1.475 E − 5
40,000	−22.80	9.684	2.871 E + 2	3.996 E − 3	1.601 E − 5
50,000	−2.50	9.654	7.978 E + 1	1.027 E − 3	1.704 E − 5
60,000	−26.13	9.624	2.196 E + 1	3.097 E − 4	1.584 E − 5
70,000	−53.57	9.594	5.221 E + 0	8.283 E − 5	1.438 E − 5
80,000	−74.51	9.564	1.052 E + 0	1.846 E − 5	1.321 E − 5

[a] Data from *U.S. Standard Atmosphere,* 1976, U.S. Government Printing Office, Washington, D.C.

TABLE C.6 Properties of the U.S. Standard Atmosphere (BG Units)[a]

Altitude (ft)	Temperature (°F)	Acceleration of gravity, g (ft/s²)	Pressure, p [lb_f/in.²(abs)]	Density, ρ (slugs/ft³)	Dynamic viscosity, μ (lb_f·s/ft²)
−5,000	76.84	32.189	17.554	2.745 E − 3	3.836 E − 7
0	59.00	32.174	14.696	2.377 E − 3	3.737 E − 7
5,000	41.17	32.159	12.228	2.048 E − 3	3.637 E − 7
10,000	23.36	32.143	10.108	1.756 E − 3	3.534 E − 7
15,000	5.55	32.128	8.297	1.496 E − 3	3.430 E − 7
20,000	−12.26	32.112	6.759	1.267 E − 3	3.324 E − 7
25,000	−30.05	32.097	5.461	1.066 E − 3	3.217 E − 7
30,000	−47.83	32.082	4.373	8.907 E − 4	3.107 E − 7
35,000	−65.61	32.066	3.468	7.382 E − 4	2.995 E − 7
40,000	−69.70	32.051	2.730	5.873 E − 4	2.969 E − 7
45,000	−69.70	32.036	2.149	4.623 E − 4	2.969 E − 7
50,000	−69.70	32.020	1.692	3.639 E − 4	2.969 E − 7
60,000	−69.70	31.990	1.049	2.256 E − 4	2.969 E − 7
70,000	−67.42	31.959	0.651	1.392 E − 4	2.984 E − 7
80,000	−61.98	31.929	0.406	8.571 E − 5	3.018 E − 7
90,000	−56.54	31.897	0.255	5.610 E − 5	3.052 E − 7
100,000	−51.10	31.868	0.162	3.318 E − 5	3.087 E − 7
150,000	19.40	31.717	0.020	3.658 E − 6	3.511 E − 7
200,000	−19.78	31.566	0.003	5.328 E − 7	3.279 E − 7
250,000	−88.77	31.415	0.000	6.458 E − 8	2.846 E − 7

[a] Data from *U.S. Standard Atmosphere*, 1976, U.S. Government Printing Office, Washington, D.C.

TABLE C.7 Densities of Some Common Solids and Fluids (kg/m³) at Atmospheric Pressure and 20°C

Gold	19,300	Lead	11,370
Silver	10,510	Copper	8906
Steel	7850	Aluminum	2770
Plexiglass	1180	Water (20°C)	998
Oak (red)	660	Ice	920
Pine (Eastern, white)	370	Sea water	1025
Mercury	13,550	SAE 30 oil	917
Kerosene	809	Gasoline	680
Ethyl alcohol	789	Air (sea level)	1.204
Carbon dioxide	1.85	Argon	1.679
Methane (natural gas)	0.677	Propane	1.854
Hydrogen	0.0851	Helium	0.169

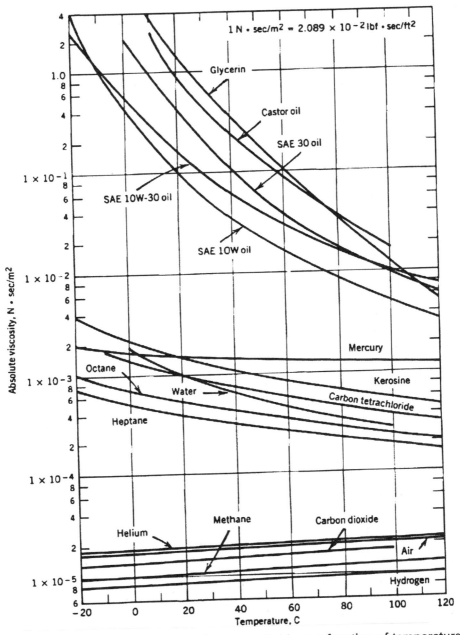

FIGURE C.1 Dynamic viscosity of common fluids as a function of temperature (at one atmosphere). Adapted from Fox and Macdonald, *Introduction to Fluid Mechanics,* published by John Wiley & Sons, 4th ed., 1992.

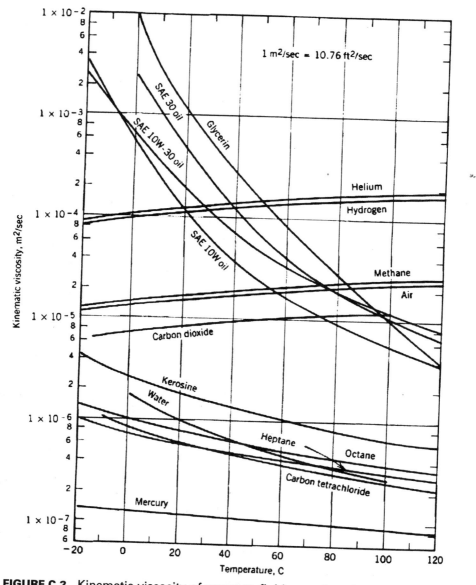

FIGURE C.2 Kinematic viscosity of common fluids as a function of temperature (at one atmosphere). Adapted from Fox and Macdonald, *Introduction to Fluid Mechanics,* published by John Wiley & Sons, 4th ed., 1992.

TABLE C.8   Properties of Common Gases at Atmospheric Pressure and 20°C

Gas	Molecular weight	R [m²/(s²·K)]	ρ (kg/m³)	μ [(N·s)/m²]	Specific heat ratio γ
H_2	2.016	4124	0.0838	9.05 E − 6	1.41
He	4.003	2077	0.166	1.97 E − 5	1.66
H_2O	18.02	461	0.749	1.02 E − 5	1.33
Ar	39.944	208	1.66	2.24 E − 5	1.67
Dry air	28.96	287	1.20	1.80 E − 5	1.40
CO_2	44.01	189	1.83	1.48 E − 5	1.30
CO	28.01	297	1.16	1.82 E − 5	1.40
N_2	28.02	297	1.16	1.76 E − 5	1.40
O_2	32.00	260	1.34	2.00 E − 5	1.40
NO	30.01	277	1.23	1.90 E − 5	1.40
N_2O	44.02	189	1.83	1.45 E − 5	1.31
Cl_2	70.91	117	2.95	1.03 E − 5	1.34
CH_4	16.04	518	0.667	1.34 E − 5	1.32

Adapted from White, *Fluid Mechanics,* McGraw-Hill, 1986.

TABLE C.9   Properties of Common Liquids at Atmospheric Pressure and 20°C

Liquid	ρ (kg/m³)	μ [(N·s)/m²]	σ (N/m)	p_v (N/m²)	Bulk modulus, (N/m²)
Ammonia	608	2.20 E − 4	2.13 E − 2	9.10 E + 5	
Benzene	881	6.51 E − 4	2.88 E − 2	1.01 E + 4	1.05 E + 9
Carbon tetrachloride	1,590	9.67 E − 4	2.70 E − 2	1.20 E + 4	9.65 E + 8
Ethanol	789	1.20 E − 3	2.28 E − 2	5.7 E + 3	8.96 E + 8
Gasoline	680	2.92 E − 4	2.16 E − 2	5.51 E + 4	9.58 E + 8
Glycerin	1,260	1.49	6.33 E − 2	1.4 E − 2	4.34 E + 9
Kerosene	804	1.92 E − 3	2.8 E − 2	3.11 E + 3	1.43 E + 9
Mercury	13,550	1.56 E − 3	4.84 E − 1	1.1 E − 3	2.55 E + 10
Methanol	791	5.98 E − 4	2.25 E − 2	1.34 E + 4	8.27 E + 8
SAE 10 oil	917	1.04 E − 1	3.6 E − 2		1.31 E + 9
SAE 30 oil	917	2.90 E − 1	3.5 E − 2		1.38 E + 9
Water	998	1.00 E − 3	7.28 E − 2	2.34 E + 3	2.19 E + 9
Seawater	1,025	1.07 E − 3	7.28 E − 2	2.34 E + 3	2.28 E + 9

Adapted from White, *Fluid Mechanics,* McGraw-Hill, 1986.

**TABLE C.10 Isentropic Flow Functions
(One-Dimensional, Ideal Gas, $\gamma = 1.4$)**

M	T/T_0	p/p_0	ρ/ρ_0	A/A^*
0.00	1.0000	1.0000	1.0000	∞
0.02	0.9999	0.9997	0.9998	28.94
0.04	0.9997	0.9989	0.9992	14.48
0.06	0.9993	0.9975	0.9982	9.666
0.08	0.9987	0.9955	0.9968	7.262
0.10	0.9980	0.9930	0.9950	5.822
0.12	0.9971	0.9900	0.9928	4.864
0.14	0.9961	0.9864	0.9903	4.182
0.16	0.9949	0.9823	0.9873	3.673
0.18	0.9936	0.9777	0.9840	3.278
0.20	0.9921	0.9725	0.9803	2.964
0.22	0.9904	0.9669	0.9762	2.708
0.24	0.9886	0.9607	0.9718	2.496
0.26	0.9867	0.9541	0.9670	2.317
0.28	0.9846	0.9470	0.9619	2.166
0.30	0.9823	0.9395	0.9564	2.035
0.32	0.9799	0.9315	0.9506	1.922
0.34	0.9774	0.9231	0.9445	1.823
0.36	0.9747	0.9143	0.9380	1.736
0.38	0.9719	0.9052	0.9313	1.659
0.40	0.9690	0.8956	0.9243	1.590
0.42	0.9659	0.8857	0.9170	1.529
0.44	0.9627	0.8755	0.9094	1.474
0.46	0.9594	0.8650	0.9016	1.425
0.48	0.9560	0.8541	0.8935	1.380
0.50	0.9524	0.8430	0.8852	1.340
0.52	0.9487	0.8317	0.8766	1.303
0.54	0.9449	0.8201	0.8679	1.270
0.56	0.9410	0.8082	0.8589	1.240
0.58	0.9370	0.7962	0.8498	1.213
0.60	0.9328	0.7840	0.8405	1.188
0.62	0.9286	0.7716	0.8310	1.166
0.64	0.9243	0.7591	0.8213	1.145
0.66	0.9199	0.7465	0.8115	1.127
0.68	0.9154	0.7338	0.8016	1.110
0.70	0.9108	0.7209	0.7916	1.094
0.72	0.9061	0.7080	0.7814	1.081
0.74	0.9013	0.6951	0.7712	1.068
0.76	0.8964	0.6821	0.7609	1.057
0.78	0.8915	0.6691	0.7505	1.047
0.80	0.8865	0.6560	0.7400	1.038
0.82	0.8815	0.6430	0.7295	1.030
0.84	0.8763	0.6300	0.7189	1.024
0.86	0.8711	0.6170	0.7083	1.018
0.88	0.8659	0.6041	0.6977	1.013
0.90	0.8606	0.5913	0.6870	1.009

TABLE C.10 Isentropic Flow Functions
(*Continued*)

M	T/T_0	p/p_0	ρ/ρ_0	A/A^*
0.92	0.8552	0.5785	0.6764	1.006
0.94	0.8498	0.5658	0.6658	1.003
0.96	0.8444	0.5532	0.6551	1.001
0.98	0.8389	0.5407	0.6445	1.000
1.00	0.8333	0.5283	0.6339	1.000
1.02	0.8278	0.5160	0.6234	1.000
1.04	0.8222	0.5039	0.6129	1.001
1.06	0.8165	0.4919	0.6024	1.003
1.08	0.8108	0.4801	0.5920	1.005
1.10	0.8052	0.4684	0.5817	1.008
1.12	0.7994	0.4568	0.5714	1.011
1.14	0.7937	0.4455	0.5612	1.015
1.16	0.7880	0.4343	0.5511	1.020
1.18	0.7822	0.4232	0.5411	1.025
1.20	0.7764	0.4124	0.5311	1.030
1.22	0.7706	0.4017	0.5213	1.037
1.24	0.7648	0.3912	0.5115	1.043
1.26	0.7590	0.3809	0.5019	1.050
1.28	0.7532	0.3708	0.4923	1.058
1.30	0.7474	0.3609	0.4829	1.066
1.32	0.7416	0.3512	0.4736	1.075
1.34	0.7358	0.3417	0.4644	1.084
1.36	0.7300	0.3323	0.4553	1.094
1.38	0.7242	0.3232	0.4463	1.104
1.40	0.7184	0.3142	0.4374	1.115
1.42	0.7126	0.3055	0.4287	1.126
1.44	0.7069	0.2969	0.4201	1.138
1.46	0.7011	0.2886	0.4116	1.150
1.48	0.6954	0.2804	0.4032	1.163
1.50	0.6897	0.2724	0.3950	1.176
1.52	0.6840	0.2646	0.3869	1.190
1.54	0.6783	0.2570	0.3789	1.204
1.56	0.6726	0.2496	0.3711	1.219
1.58	0.6670	0.2423	0.3633	1.234
1.60	0.6614	0.2353	0.3557	1.250
1.62	0.6558	0.2284	0.3483	1.267
1.64	0.6502	0.2217	0.3409	1.284
1.66	0.6447	0.2152	0.3337	1.301
1.68	0.6392	0.2088	0.3266	1.319
1.70	0.6337	0.2026	0.3197	1.338
1.72	0.6283	0.1966	0.3129	1.357
1.74	0.6229	0.1907	0.3062	1.376
1.76	0.6175	0.1850	0.2996	1.397
1.78	0.6121	0.1794	0.2931	1.418
1.80	0.6068	0.1740	0.2868	1.439

TABLE C.10 **Isentropic Flow Functions**
(*Continued*)

M	T/T_0	p/p_0	ρ/ρ_0	A/A^*
1.82	0.6015	0.1688	0.2806	1.461
1.84	0.5963	0.1637	0.2745	1.484
1.86	0.5911	0.1587	0.2686	1.507
1.88	0.5859	0.1539	0.2627	1.531
1.90	0.5807	0.1492	0.2570	1.555
1.92	0.5756	0.1447	0.2514	1.580
1.94	0.5705	0.1403	0.2459	1.606
1.96	0.5655	0.1360	0.2405	1.633
1.98	0.5605	0.1318	0.2352	1.660
2.00	0.5556	0.1278	0.2301	1.688
2.02	0.5506	0.1239	0.2250	1.716
2.04	0.5458	0.1201	0.2200	1.745
2.06	0.5409	0.1164	0.2152	1.775
2.08	0.5361	0.1128	0.2105	1.806
2.10	0.5314	0.1094	0.2058	1.837
2.12	0.5266	0.1060	0.2013	1.869
2.14	0.5219	0.1027	0.1968	1.902
2.16	0.5173	0.09956	0.1925	1.935
2.18	0.5127	0.09650	0.1882	1.970
2.20	0.5081	0.09352	0.1841	2.005
2.22	0.5036	0.09064	0.1800	2.041
2.24	0.4991	0.08784	0.1760	2.078
2.26	0.4947	0.08514	0.1721	2.115
2.28	0.4903	0.08252	0.1683	2.154
2.30	0.4859	0.07997	0.1646	2.193
2.32	0.4816	0.07751	0.1610	2.233
2.34	0.4773	0.07513	0.1574	2.274
2.36	0.4731	0.07281	0.1539	2.316
2.38	0.4689	0.07057	0.1505	2.359
2.40	0.4647	0.06840	0.1472	2.403
2.42	0.4606	0.06630	0.1440	2.448
2.44	0.4565	0.06426	0.1408	2.494
2.46	0.4524	0.06229	0.1377	2.540
2.48	0.4484	0.06038	0.1347	2.588
2.50	0.4444	0.05853	0.1317	2.637
2.52	0.4405	0.05674	0.1288	2.687
2.54	0.4366	0.05500	0.1260	2.737
2.56	0.4328	0.05332	0.1232	2.789
2.58	0.4289	0.05169	0.1205	2.842
2.60	0.4252	0.05012	0.1179	2.896
2.62	0.4214	0.04859	0.1153	2.951
2.64	0.4177	0.04711	0.1128	3.007
2.66	0.4141	0.04568	0.1103	3.065
2.68	0.4104	0.04429	0.1079	3.123
2.70	0.4068	0.04295	0.1056	3.183

TABLE C.10 Isentropic Flow Functions
(*Continued*)

M	T/T_0	p/p_0	ρ/ρ_0	A/A^*
2.72	0.4033	0.04166	0.1033	3.244
2.74	0.3998	0.04039	0.1010	3.306
2.76	0.3963	0.03917	0.09885	3.370
2.78	0.3928	0.03800	0.09671	3.434
2.80	0.3894	0.03685	0.09462	3.500
2.82	0.3860	0.03574	0.09259	3.567
2.84	0.3827	0.03467	0.09059	3.636
2.86	0.3794	0.03363	0.08865	3.706
2.88	0.3761	0.03262	0.08674	3.777
2.90	0.3729	0.03165	0.08489	3.850
2.92	0.3697	0.03071	0.08308	3.924
2.94	0.3665	0.02980	0.08130	3.999
2.96	0.3633	0.02891	0.07957	4.076
2.98	0.3602	0.02805	0.07788	4.155
3.00	0.3571	0.02722	0.07623	4.235
3.10	0.3422	0.02345	0.06852	4.657
3.20	0.3281	0.02023	0.06165	5.121
3.30	0.3147	0.01748	0.05554	5.629
3.40	0.3019	0.01512	0.05009	6.184
3.50	0.2899	0.01311	0.04523	6.790
3.60	0.2784	0.01138	0.04089	7.450
3.70	0.2675	0.009903	0.03702	8.169
3.80	0.2572	0.008629	0.03355	8.951
3.90	0.2474	0.007532	0.03044	9.799
4.00	0.2381	0.006586	0.02766	10.72
4.10	0.2293	0.005769	0.02516	11.71
4.20	0.2208	0.005062	0.02292	12.79
4.30	0.2129	0.004449	0.02090	13.95
4.40	0.2053	0.003918	0.01909	15.21
4.50	0.1980	0.003455	0.01745	16.56
4.60	0.1911	0.003053	0.01597	18.02
4.70	0.1846	0.002701	0.01463	19.58
4.80	0.1783	0.002394	0.01343	21.26
4.90	0.1724	0.002126	0.01233	23.07
5.00	0.1667	0.001890	0.01134	25.00

Adapted from Fox and Macdonald, *Introduction to Fluid Mechanics,* published by John Wiley & Sons, 4th ed., 1992.

TABLE C.11 Normal Shock Flow Functions
(One-Dimensional, Ideal Gas, $\gamma = 1.4$)

M_1	M_2	p_{0_2}/p_{0_1}	T_2/T_1	p_2/p_1	ρ_2/ρ_1
1.00	1.000	1.000	1.000	1.000	1.000
1.02	0.9805	1.000	1.013	1.047	1.033
1.04	0.9620	0.9999	1.026	1.095	1.067
1.06	0.9444	0.9998	1.039	1.144	1.101
1.08	0.9277	0.9994	1.052	1.194	1.135
1.10	0.9118	0.9989	1.065	1.245	1.169
1.12	0.8966	0.9982	1.078	1.297	1.203
1.14	0.8820	0.9973	1.090	1.350	1.238
1.16	0.8682	0.9961	1.103	1.403	1.272
1.18	0.8549	0.9946	1.115	1.458	1.307
1.20	0.8422	0.9928	1.128	1.513	1.342
1.22	0.8300	0.9907	1.141	1.570	1.376
1.24	0.8183	0.9884	1.153	1.627	1.411
1.26	0.8071	0.9857	1.166	1.686	1.446
1.28	0.7963	0.9827	1.178	1.745	1.481
1.30	0.7860	0.9794	1.191	1.805	1.516
1.32	0.7760	0.9757	1.204	1.866	1.551
1.34	0.7664	0.9718	1.216	1.928	1.585
1.36	0.7572	0.9676	1.229	1.991	1.620
1.38	0.7483	0.9630	1.242	2.055	1.655
1.40	0.7397	0.9582	1.255	2.120	1.690
1.42	0.7314	0.9531	1.268	2.186	1.724
1.44	0.7235	0.9477	1.281	2.253	1.759
1.46	0.7157	0.9420	1.294	2.320	1.793
1.48	0.7083	0.9360	1.307	2.389	1.828
1.50	0.7011	0.9298	1.320	2.458	1.862
1.52	0.6941	0.9233	1.334	2.529	1.896
1.54	0.6874	0.9166	1.347	2.600	1.930
1.56	0.6809	0.9097	1.361	2.673	1.964
1.58	0.6746	0.9026	1.374	2.746	1.998
1.60	0.6684	0.8952	1.388	2.820	2.032
1.62	0.6625	0.8876	1.402	2.895	2.065
1.64	0.6568	0.8799	1.416	2.971	2.099
1.66	0.6512	0.8720	1.430	3.048	2.132
1.68	0.6458	0.8640	1.444	3.126	2.165
1.70	0.6406	0.8557	1.458	3.205	2.198
1.72	0.6355	0.8474	1.473	3.285	2.230
1.74	0.6305	0.8389	1.487	3.366	2.263
1.76	0.6257	0.8302	1.502	3.447	2.295
1.78	0.6210	0.8215	1.517	3.530	2.327
1.80	0.6165	0.8127	1.532	3.613	2.359
1.82	0.6121	0.8038	1.547	3.698	2.391
1.84	0.6078	0.7947	1.562	3.783	2.422
1.86	0.6036	0.7857	1.577	3.870	2.454
1.88	0.5996	0.7766	1.592	3.957	2.485
1.90	0.5956	0.7674	1.608	4.045	2.516

TABLE C.11   Normal Shock Flow Functions
(*Continued*)

M_1	M_2	p_{0_2}/p_{0_1}	T_2/T_1	p_2/p_1	ρ_2/ρ_1
1.92	0.5918	0.7581	1.624	4.134	2.546
1.94	0.5880	0.7488	1.639	4.224	2.577
1.96	0.5844	0.7395	1.655	4.315	2.607
1.98	0.5808	0.7302	1.671	4.407	2.637
2.00	0.5774	0.7209	1.687	4.500	2.667
2.02	0.5740	0.7115	1.704	4.594	2.696
2.04	0.5707	0.7022	1.720	4.689	2.725
2.06	0.5675	0.6928	1.737	4.784	2.755
2.08	0.5643	0.6835	1.754	4.881	2.783
2.10	0.5613	0.6742	1.770	4.978	2.812
2.12	0.5583	0.6649	1.787	5.077	2.840
2.14	0.5554	0.6557	1.805	5.176	2.868
2.16	0.5525	0.6464	1.822	5.277	2.896
2.18	0.5498	0.6373	1.839	5.378	2.924
2.20	0.5471	0.6281	1.857	5.480	2.951
2.22	0.5444	0.6191	1.875	5.583	2.978
2.24	0.5418	0.6100	1.892	5.687	3.005
2.26	0.5393	0.6011	1.910	5.792	3.032
2.28	0.5368	0.5921	1.929	5.898	3.058
2.30	0.5344	0.5833	1.947	6.005	3.085
2.32	0.5321	0.5745	1.965	6.113	3.110
2.34	0.5297	0.5658	1.984	6.222	3.136
2.36	0.5275	0.5572	2.002	6.331	3.162
2.38	0.5253	0.5486	2.021	6.442	3.187
2.40	0.5231	0.5402	2.040	6.553	3.212
2.42	0.5210	0.5318	2.059	6.666	3.237
2.44	0.5189	0.5234	2.079	6.779	3.261
2.46	0.5169	0.5152	2.098	6.894	3.285
2.48	0.5149	0.5071	2.118	7.009	3.310
2.50	0.5130	0.4990	2.137	7.125	3.333
2.52	0.5111	0.4910	2.157	7.242	3.357
2.54	0.5092	0.4832	2.177	7.360	3.380
2.56	0.5074	0.4754	2.198	7.479	3.403
2.58	0.5056	0.4677	2.218	7.599	3.426
2.60	0.5039	0.4601	2.238	7.720	3.449
2.62	0.5022	0.4526	2.259	7.842	3.471
2.64	0.5005	0.4452	2.280	7.965	3.494
2.66	0.4988	0.4379	2.301	8.088	3.516
2.68	0.4972	0.4307	2.322	8.213	3.537
2.70	0.4956	0.4236	2.343	8.338	3.559
2.72	0.4941	0.4166	2.364	8.465	3.580
2.74	0.4926	0.4097	2.386	8.592	3.601
2.76	0.4911	0.4028	2.407	8.721	3.622
2.78	0.4897	0.3961	2.429	8.850	3.643
2.80	0.4882	0.3895	2.451	8.980	3.664

TABLE C.11   Normal Shock Flow Functions
(*Continued*)

M_1	M_2	p_{0_2}/p_{0_1}	T_2/T_1	p_2/p_1	ρ_2/ρ_1
2.82	0.4868	0.3829	2.473	9.111	3.684
2.84	0.4854	0.3765	2.496	9.243	3.704
2.86	0.4840	0.3701	2.518	9.376	3.724
2.88	0.4827	0.3639	2.540	9.510	3.743
2.90	0.4814	0.3577	2.563	9.645	3.763
2.92	0.4801	0.3517	2.586	9.781	3.782
2.94	0.4788	0.3457	2.609	9.918	3.801
2.96	0.4776	0.3398	2.632	10.06	3.820
2.98	0.4764	0.3340	2.656	10.19	3.839
3.00	0.4752	0.3283	2.679	10.33	3.857
3.10	0.4695	0.3012	2.799	11.05	3.947
3.20	0.4644	0.2762	2.922	11.78	4.031
3.30	0.4596	0.2533	3.049	12.54	4.112
3.40	0.4552	0.2322	3.180	13.32	4.188
3.50	0.4512	0.2130	3.315	14.13	4.261
3.60	0.4474	0.1953	3.454	14.95	4.330
3.70	0.4440	0.1792	3.596	15.81	4.395
3.80	0.4407	0.1645	3.743	16.68	4.457
3.90	0.4377	0.1510	3.893	17.58	4.516
4.00	0.4350	0.1388	4.047	18.50	4.571
4.10	0.4324	0.1276	4.205	19.45	4.624
4.20	0.4299	0.1173	4.367	20.41	4.675
4.30	0.4277	0.1080	4.532	21.41	4.723
4.40	0.4255	0.09948	4.702	22.42	4.768
4.50	0.4236	0.09170	4.875	23.46	4.812
4.60	0.4217	0.08459	5.052	24.52	4.853
4.70	0.4199	0.07809	5.233	25.61	4.893
4.80	0.4183	0.07214	5.418	26.71	4.930
4.90	0.4167	0.06670	5.607	27.85	4.966
5.00	0.4152	0.06172	5.800	29.00	5.000

Adapted from Fox and Macdonald, *Introduction to Fluid Mechanics*, published by John Wiley & Sons, 4th ed., 1992.

TABLE C.12   Prandtl–Meyer Function (one-dimensional, ideal gas, $\gamma = 1.4$)

ν (degrees)	M	α_M (degrees)	ν (degrees)	M	α_M (degrees)
0.0	1.000	90.000	17.5	1.689	36.293
0.5	1.051	72.099	18.0	1.706	35.874
1.0	1.082	67.574	18.5	1.724	35.465
1.5	1.108	64.451	19.0	1.741	35.065
2.0	1.133	61.997	19.5	1.758	34.673
2.5	1.155	59.950	20.0	1.775	34.290
3.0	1.177	58.180	20.5	1.792	33.915
3.5	1.198	56.614	21.0	1.810	33.548
4.0	1.218	55.205	21.5	1.827	33.188
4.5	1.237	53.920	22.0	1.844	32.834
5.0	1.256	52.738	22.5	1.862	32.488
5.5	1.275	51.642	23.0	1.879	32.148
6.0	1.294	50.619	23.5	1.897	31.814
6.5	1.312	49.658	24.0	1.915	31.486
7.0	1.330	48.753	24.5	1.932	31.164
7.5	1.348	47.896	25.0	1.950	30.847
8.0	1.366	47.082	25.5	1.968	30.536
8.5	1.383	46.306	26.0	1.986	30.229
9.0	1.400	45.566	26.5	2.004	29.928
9.5	1.418	44.857	27.0	2.023	29.632
10.0	1.435	44.177	27.5	2.041	29.340
10.5	1.452	43.523	28.0	2.059	29.052
11.0	1.469	42.894	28.5	2.078	28.769
11.5	1.486	42.287	29.0	2.096	28.491
12.0	1.503	41.701	29.5	2.115	28.216
12.5	1.520	41.134	30.0	2.134	27.945
13.0	1.537	40.585	30.5	2.153	27.678
13.5	1.554	40.053	31.0	2.172	27.415
14.0	1.571	39.537	31.5	2.191	27.155
14.5	1.588	39.035	32.0	2.210	26.899
15.0	1.605	38.547	32.5	2.230	26.646
15.5	1.622	38.073	33.0	2.249	26.397
16.0	1.639	37.611	33.5	2.269	26.151
16.5	1.655	37.160	34.0	2.289	25.908
17.0	1.672	36.721	34.5	2.309	25.668

TABLE C.12 Prandtl–Meyer Function (*Continued*)

ν (degrees)	M	α_M (degrees)	ν (degrees)	M	α_M (degrees)
35.0	2.329	25.430	52.5	3.146	18.532
35.5	2.349	25.196	53.0	3.176	18.366
36.0	2.369	24.965	53.5	3.202	18.200
36.5	2.390	24.736	54.0	3.230	18.036
37.0	2.410	24.510	54.5	3.258	17.873
37.5	2.431	24.287	55.0	3.287	17.711
38.0	2.452	24.066	55.5	3.316	17.551
38.5	2.473	23.847	56.0	3.346	17.391
39.0	2.495	23.631	56.5	3.375	17.233
39.5	2.516	23.418	57.0	3.406	17.076
40.0	2.538	23.206	57.5	3.436	16.920
40.5	2.560	22.997	58.0	3.467	16.765
41.0	2.582	22.790	58.5	3.498	16.611
41.5	2.604	22.585	59.0	3.530	16.458
42.0	2.626	22.382	59.5	3.562	16.306
42.5	2.649	22.182	60.0	3.594	16.155
43.0	2.671	21.983	60.5	3.627	16.005
43.5	2.694	21.786	61.0	3.660	15.856
44.0	2.718	21.591	61.5	3.694	15.708
44.5	2.741	21.398	62.0	3.728	15.561
45.0	2.764	21.207	62.5	3.762	15.415
45.5	2.788	21.017	63.0	3.797	15.270
46.0	2.812	20.830	63.5	3.832	15.126
46.5	2.836	20.644	64.0	3.868	14.983
47.0	2.861	20.459	64.5	3.904	14.840
47.5	2.886	20.277	65.0	3.941	14.698
48.0	2.910	20.096	65.5	3.979	14.557
48.5	2.936	19.916	66.0	4.016	14.417
49.0	2.961	19.738	66.5	4.055	14.278
49.5	2.987	19.561	67.0	4.094	14.140
50.0	3.013	19.386	67.5	4.133	14.002
50.5	3.039	19.213	68.0	4.173	13.865
51.0	3.065	19.041	68.5	4.214	13.729
51.5	3.092	18.870	69.0	4.255	13.593
52.0	3.119	18.701	69.5	4.297	13.459

TABLE C.12   Prandtl–Meyer Function (*Continued*)

ν (degrees)	M	α_M (degrees)	ν (degrees)	M	α_M (degrees)
70.0	4.339	13.325	87.5	6.390	9.003
70.5	4.382	13.191	88.0	6.472	8.888
71.0	4.426	13.059	88.5	6.556	8.774
71.5	4.470	12.927	89.0	6.642	8.660
72.0	4.515	12.795	89.5	6.729	8.546
72.5	4.561	12.665	90.0	6.819	8.433
73.0	4.608	12.535	90.5	6.911	8.320
73.5	4.655	12.406	91.0	7.005	8.207
74.0	4.703	12.277	91.5	7.102	8.095
74.5	4.752	12.149	92.0	7.201	7.983
75.0	4.801	12.021	92.5	7.302	7.871
75.5	4.852	11.894	93.0	7.406	7.760
76.0	4.903	11.768	93.5	7.513	7.649
76.5	4.955	11.642	94.0	7.623	7.538
77.0	5.009	11.517	94.5	7.735	7.428
77.5	5.063	11.392	95.0	7.851	7.318
78.0	5.118	11.268	95.5	7.970	7.208
78.5	5.174	11.145	96.0	8.092	7.099
79.0	5.231	11.022	96.5	8.218	6.989
79.5	5.289	10.899	97.0	8.347	6.881
80.0	5.348	10.777	97.5	8.480	6.772
80.5	5.408	10.656	98.0	8.618	6.664
81.0	5.470	10.535	98.5	8.759	6.556
81.5	5.532	10.414	99.0	8.905	6.448
82.0	5.596	10.294	99.5	9.055	6.340
82.5	5.661	10.175	100.0	9.210	6.233
83.0	5.727	10.056	100.5	9.371	6.126
83.5	5.795	9.937	101.0	9.536	6.019
84.0	5.864	9.819	101.5	9.708	5.913
84.5	5.935	9.701	102.0	9.885	5.806
85.0	6.006	9.584			
85.5	6.080	9.467			
86.0	6.155	9.350			
86.5	6.232	9.234			
87.0	6.310	9.119			

Adapted from Liepmann and Roshko, *Elements of Gasdynamics,* John Wiley & Sons, 1957.

WEB RESOURCES

1. *http://www.princeton.edu/ ˜asmits/fluids_text/intro.html*
   Updates and corrections to the text, links to new sites, and Picture of the Week.

2. *http://www.princeton.edu/ ˜gasdyn/fluids.html*
   A listing of current links to sites of interest to students and researchers in fluid dynamics.

3. *http://ate.cc.vt.edu/fluids/links/edulinks.htm*
   Links to educational sites maintained by Virginia Tech.

4. *http://www.engapplets.vt.edu*
   Engineering Applets. Interactive Java programs for Fluid Dynamics and Solid Mechanics
   Ideal Flow Machine
   The Virtual Shock Tube
   The Compressible Aerodynamics Calculator
   Thermodynamics of Air
   Boundary Layer Applets + Convection
   Heat Conduction Applet
   Vortex Panel Method

5. *http://aero.stanford.edu/onlineaero/onlineaero.html*
   "Applied Aerodynamics: A Digital Textbook" by Professor Ilan Kroo of Stanford.

6. *http://www.ma.adfa.oz.au/Teaching/Subjects/VFD/VFD.html*
   Viscous Fluid Dynamics course material by Geoffrey K. Aldis at the Australian Defence Force Academy.

7. *http://raphael.mit.edu:80/Java*
   The Java Virtual Wind Tunnel. An interactive two-dimensional inviscid CFD simulation of a channel with a bump. You can experiment with different inlet conditions and change 2nd- and 4th-order damping, while you see how the solution evolves. The numerical method is a standard Jameson explicit Runge–Kutta scheme. Written by David Oh.

8. *http://ourworld.compuserve.com/homepages/anima/second.htm*
   Computer simulations for MSDOS, demonstrating the Kelvin–Helmholtz instability and the von Kármán vortex street, by David Porthouse.

9. *http://wings.ucdavis.edu*
   The K-8 Aeronautics Internet Textbook. A very attractive site that is worth a visit. Some of the material is useful even at a high-school level.

10. *http://www.mame.syr.edu/simfluid*
    SimScience Fluid Dynamics Modules. Three levels of instruction.

11. *http://quest.arc.nasa.gov/aero/index-new.html*
    Aero Design Team Online. Middle school level introduction to aeronautics.

Excellent source material, and an extensive list of internet links on similar topics.

12. *http://www.princeton.edu/~asmits/Bicycle_web/bicycle_aero.html*
Bicycle Aerodynamics.

13. *http://www.eng.vt.edu/fluids/msc/bike.htm*
Links to sites on bicycle aerodynamics.

14. *http://observe.ivv.nasa.gov/nasa/aero/exhibits/planes_0.html*
NASA Observatorium: How Planes Fly.

15. *http://www.seattleweb.com/aeroworks/macairplane/macairplane.html*
MacAirplane: Graphics-oriented, conceptual aircraft design program developed at Notre Dame and on display at the Smithsonian Air and Space Museum.

16. *http://www.ts.go.dlr.de/sm-sk_info/HGAinfo/NACA-java.html*
NACA 4-Digit and 5-Digit airfoil data provided by the German Aerospace Center

17. *http://spot.colorado.edu/~dziadeck/airship.html*
Lighter-than-Air Craft information source.

18. *http://www.pitt.edu/~maarten/work/soapflow/soapflow.html*
Soap film visualization: A How To Guide.

19. *http://www.civeng.carleton.ca:80/Exhibits/Tacoma_Narrows*
Tacoma Narrows Bridge Failure. Includes mpeg movie of the disaster.

20. *http://walrus.wr.usgs.gov/docs/tsunami/PNGhome.html*
A graphic animation of the July 17, 1998, Papua New Guinea Tsunami (produced by the U.S. Geological Survey). Note the generation of multiple waves by a single earthquake.

21. *http://www.vki.ac.be/welcome.htm*
von Kármán Institute for Fluid Dynamics (Brussels). The Institute conducts extensive educational programs.

ANSWERS TO SELECTED PROBLEMS

CHAPTER 1

1.1 400 *kg*, 27.4 *slug*, 882 *lb*$_m$

1.2 11.39 *ft*^{3}, 0.821 *ft*^{3}, 1.027 *ft*^{3}, 0.001206 *ft*^{3}, 0.000169 *ft*^{3}

1.3 1204 *lb*$_f$, 102.7 *lb*$_f$, 62.5 *lb*$_f$, 0.0764 *lb*$_f$, 0.00379 *lb*$_f$

1.4 417 *N*, 69.5 *N*

1.5 19.33, 2.775, 1.027, 0.001398, 0.001682

1.6 +2.20%, −25.6%, +5.96%, −25.4%

1.7 34.8 *min.*

1.8 18,900 *psi*

1.9 $\Delta\forall = -4.79 \times 10^{-7}$ *m*^{3}

1.10 21,886 *N*, 4920 *lb*$_f$

1.11 0.645 *psi*, 4448 *Pa*

1.12 255,340 *N*, 57,400 *lb*$_f$

1.13 1.76×10^{21} *kg*

1.14 −0.158 *psi*

1.15 −1.1%

1.16 0.733

1.17 1.343

1.18 0.798 *N*

1.19 $\mu = T\delta/(2\pi\omega R^3 H)$

1.20 0.127 *N·s/m*^{2}

1.21 3.27 *m/s*

1.22 **(a)** 10,016, **(b)** 77.5×10^{-5} *lb*$_f$/*ft*^{2}, **(c)** 0.0413 *lb*$_f$

1.23 5.9 *μm*, 106 *μm*, 78 *km*

1.24 **(a)** 222,000, **(b)** 21×10^6, **(c)** 475×10^6, **(d)** 74×10^6, **(e)** 1.9×10^6, **(f)** 667

1.25 **(a)** 240 *m/s*, **(b)** 2360 *m/s*, **(c)** 11.8 *m/s*

1.26 $Re = 16,670$ (turbulent)

1.27 9.8 *mm*

1.28 0.033 *N/m*

1.29 104,327 *Pa*

506

1.30   584 *Pa*

1.31   107 *mm*

CHAPTER 2

2.1   (a) 36.3 *psi*, (b) 62.4 *psi*, (c) 15.2 *psi*, (d) 8.67 *psi*

2.2   (a) 2.07×10^3 *Pa*, (b) 4.3×10^5 *Pa*, (c) 105×10^3 *Pa*, (d) 78.5×10^3 *Pa*

2.3   (a) -80.7×10^3 *Pa*, -11.7 *psig*, (b) 149×10^3 *Pa*, 21.6 *psig*, (c) 3.66×10^3 *Pa*, 0.53 *psig*, (d) 329×10^3 *Pa*, 47.7 *psig*, (e) -41.5×10^3 *Pa*, -6.03 *psig*

2.4   2000 *N*

2.5   1820 *N*

2.6 $\rho g(a + b)wc$

2.7   (a) 2.81 *m*, (b) 3.48 *m*, (c) 0.208 *m*

2.8   0.072 *m*

2.9   (a) 23.1×10^3 *Pa*, (b) >2.5%, (c) >0.12%

2.10 $\rho_B/\rho_A = 2.8$

2.11 69.9×10^3 *Pa*

2.12 $p_g = (\rho_o z + mz^3/3)g$

2.13 $R_1 = 3496$ *lb$_f$/ft*, $R_2 = 3700$ *lb$_f$/ft*

2.14   (a) $\rho g H^2 w/2$, (b) $2H/3$

2.15   3.12 *in.*

2.16 $h_1/h_2 = (\rho_2/\rho_1)^{1/3}$

2.17 $h_1/h_2 = \sqrt{\rho_2/\rho_1}$

2.18   (a) $\rho_1 g z$, $\rho_1 g D + \rho_2 g(z - D)$, (b) $F/(\rho_1 D g L w) = 1 + (\rho_2/2\rho_1)(L/D)$, (c) $(L/2) (1 + (2\rho_2 L)/(3\rho_1 D))/(1 + (\rho_2 L)/(2\rho_1 D))$ from top of gate

2.19 $p_w = 2M_h/(WB^2) - \rho g h + \rho g B/3$

2.20 $\rho_2/\rho_1 = (3H_1 - b)/(3H_2 - b)$

2.21   (a) $2\rho w g h^2 \sqrt{2}/9$, (b) $8\rho w g h^2/81$

2.22 $D^3 - 3HD^2 + 2H^3 = 3MH \cos \theta \sin \theta/\rho b$

2.23 $M = 2\rho w(HL/2 \cos \theta + L^2 \tan \theta/3)$

2.24 $\rho g b w(3h - b)/(6 \cos \theta)$, normal to gate

2.25 $\rho_2/\rho_1 = (H_1/H_2)^3 \cos^2\theta$

2.26 $14d/9$ from surface along wall

2.27 $F = (\rho_1 H_1^2 - \rho_2 H_2^2)wg/2$, $(\rho_1 H_1^3 - \rho_2 H_2^3)/3(\rho_1 H_1^2 - \rho_2 H_2^2)$ above apex

2.28 $\rho_2/\rho_1 = 2$

2.29   58,586 *N*

2.30 $(6M/\pi\rho)^{1/3}$

2.31   (a) 0.08, (b) sinks, (c) no

2.32 volume $= 2.24\ m^3$

2.33 $389\ N,\ 87.3\ lb_f$

2.34 2

2.35 $d = \rho_w(H - D)/\rho_o$

2.37 $2.06\rho_w gA$

2.38 $0.255D$

2.40 (a) 0.85

2.41 $2b/3$ below interface

2.42 $3\rho_w gV(1 - \rho_f/\rho_w)$

2.43 $D^2(h - D/3) - 2abL(0.86 - (h - D)/b) = 0$

2.44 $\rho H^3 - 3\rho HL^2 - 3mL^2 = 0$

2.45 $d = 2h^3/(3b^2)$

2.46 $b = 5.2H$

2.47 $B = 2h$

2.48 (a) $16,564\ lb_f$, (b) $3.93\ ft$ along the plate from its upper edge, (c) $10.6\ ft$ from surface

2.49 $2MgL\cos\theta/(wd^2) - 2\rho gd\sin\theta/3$

2.50 $M = (\rho wD/\cos\theta)(H/2 + D\sin\theta/3)$

2.51 $\rho g(D + 2L\cos\theta/3)$

2.52 $h^3 = 6ML\sin^2\theta/(\rho w)$

2.53 (a) $\frac{1}{2}\rho gwh^2\sqrt{(b/h)^2 + 81}$, (b) $b/h = 9$

2.54 $2\rho gH^2 a/3$ at $3H/4$

2.55 $\rho_2/\rho_1 = 2.5$

2.56 $\rho gab(D - 2a/3);\ D = 3ML/(\rho ba^2) + a/2$

2.57 (a) $\pi\rho ghR^2/2$, (b) $\pi\rho gR^3/8$

2.58 $80\ lb_f$

2.59 $\rho(a + g)H$

2.60 (a) $0.547\ ft$, (b) $69.1\ mph$

2.61 $0.134\ g$

2.62 $39.6\ rad/s$

CHAPTER 3

3.5 $y - 2 = \ln x$

3.6 $y = 2/x;\ 0.25$ time units

3.9 $0.382\ lb_m/s;\ 10\ ft/s$

3.11 (a) $499\ kg/s$, (b) $63.7\ m/s$, (c) $39,800\ N$

3.12 $V_2 = 1.33\ m/s,\ V_3 = 2.08\ m/s$

3.13 8.33 *ft/s;* 6.67 *ft/s*

3.14 119 *m/s;* 215 *m/s*

3.15 0.8 *m/s*

3.16 0.0736 m^3/s or 1167 *US gpm;* 12 hoses

3.17 1011 *s*

3.18 level falls at 0.0026 *m/s*

3.19 2

3.20 0.00610 m^3/s

3.21 (a) $5V_1A_1/4$; $5\rho V_1A_1/4$, (b) $\rho V_1^2A_1((1 - 25\cos\theta/32)\mathbf{i} + (1/8 - 25\sin\theta/32)\mathbf{j})$

3.22 $\pi\rho V^2D^2/2$

3.23 $U_2 = U_1 + Q/2D$; $F/(2\rho U_1^2Dw) = (p_2 - p_1)/(\tfrac{1}{2}\rho U_1^2) + (Q/DU_1)/(Q/4DU_1 + 1)$

3.24 $V_B = 0.6V_1$; $p_B - p_A = \rho V_1^2/25$

3.25 4275 lb_f

3.26 $\rho V^2\pi R^2$

3.27 $\theta = \pi/4$

3.28 $U_2 = 8U_1/5$; $F_D = \rho h^2(g\Delta - 3U_1^2/25)$

3.29 7°

CHAPTER 4

4.5 (d) $p_{2g} = -8\rho V_1^2/9$, (e) $p_{1g} = 11\rho V_1^2/18$

4.7 10.16 *psia;* 8.64 *psia*

4.8 81.2 *m/s;* $C_p = -0.833$

4.9 0.517 *psi/ft*

4.10 (a) $p_{1g} = 15\rho V_1^2/2$, (b) 25.2° from horizontal

4.11 $F_x/(\rho U_1^2A_1) = 4.5$

4.12 795 lb_f

4.13 $-\rho V^2\pi R^2((1 + \cos\theta)\mathbf{i} + \sin\theta\mathbf{j}$

4.14 $-\rho V^2\pi R^2((17/2 + 4\cos\theta)\mathbf{i} + 4\sin\theta\mathbf{j}$

4.15 (a) $\tfrac{15}{2}\rho V_1^2$, (b) $\rho A_1V_1^2 (-11.96\mathbf{i} - 2\mathbf{j})$

4.16 $-9\rho V_1^2 A_1/2$

4.18 8.58 ft^3/s

4.19 $h = 8\dot{q}^2/(g\pi^2 d^4)$

4.20 $V = \sqrt{2gL}$

4.21 8.21 *ft*

4.22 (a) $A/A_t = \sqrt{H_1/H_1 - y)}$, (b) $H_3 = H_1\sin^2\theta$, (c) vol $= 2H_1A_t\sin\theta$

4.23 (a) $V_e = 4.95$ *m/s,* (b) $V'_e = 4.89$ *m/s*

4.24 (a) 7.67 m/s, (b) 2.67 m/s, 25,600 Pa, (c) 7.75 m

4.25 $H \sin^2 \theta$; $-2\rho g H A \cos \theta$

4.26 (a) $\sqrt{2gH}$, (b) $-\rho 2gHA \cos \theta$, (c) $H \sin^2 \theta$

4.27 $H \sin^2 \theta$; $\rho g H (A_T + 2A_j \sin \theta)$

4.28 $H_3 = H_1 \sin^2 \theta$; vol $= 2H_1 A_2 \sin \theta$

4.29 (a) 255 $\rho V^2/2$, (b) 289 $\pi \rho V^2 D^2/8$ (vertical)

4.31 $F = \rho V_1^2 H_1 w/2$ (to right)

4.32 (a) $\rho_2 gD$, (b) $(A_1/A_2)^2 = 2g(\rho_2 D - \rho_1(z_2 - z_1))/(\rho_1 V_1^2) + 1$

4.33 $V^2 = (2\rho_m gh/\rho_a)/((D/d)^4 - 1)$

4.34 $409 \rho_1 U_1^2/16$

4.35 $\rho_2/\rho_1 = 16/3$; $C_p = 4$

4.36 (a) $A_3/A_2 = \sqrt{h_1/(h_1 + h_2)}$, (b) vol $= 2\sqrt{h_1} A_2(\sqrt{h_1 + h_2} - \sqrt{h_1})$

4.37 $(d^3 V_i/8D) \sqrt{(\rho \pi/2Mg)}$

4.39 $\Delta T = -15°K$

4.40 26.8 m/s, 448 m/s

4.41 325°K

4.42 (a) 5.25 × 10^5 Pa, (b) 5881 N·m

CHAPTER 5

5.3 $4\rho w V_m h/3$, $2V_m/3$

5.4 $\dot{m}_3 = 4\rho QW/9$

5.5 $U_3 = 3(U_1 + U_2)/4$

5.6 $2U_1 W(h_2 - h_1)$

5.7 $U_m/U_o = 3b/2h$

5.8 $U_0 = 2V_t$; $\rho DV_t^2/(2g)$

5.9 $3(2U_1 + U_m)/4$; $16\rho WbU_m^2/15$

5.10 $U_m = 3aU_1/2b$; $6\rho Wa^2 U_1^2 \sin \theta/5b$

5.11 $U_m = 2U_1 a/b$; $(4\rho Wa^2 U_1^2/3b)\mathbf{j}$

5.12 $\rho_2 = 3\rho_1/2$; $2\rho_2 WU_2^2(\cos \theta \mathbf{i} + \sin \theta \mathbf{j})/3$

5.13 $(\rho_1 U_1 - \rho_2 U_m/2)/2L$

5.14 (a) $U_2/U_1 = (1 - \delta/2H)^{-1}$, (b) $1 - (U_2/U_1)^2$, (c) $1 - 1/(3(1 - \delta/2H)^2)$

5.15 (a) $b = 3a/2$, (b) $2\rho WU_{av}^2 a(\mathbf{i} - 0.8\mathbf{j})$

5.16 $U_2 = 2U_1$; $F_x = -\rho U_1^2 Dw/3$

5.17 $T/(\tfrac{1}{8}\rho U_1^2 \pi D^2) = (p_2 - p_1)/(\tfrac{1}{2}\rho U_1^2) + (U_m/U_1)^2 - 2$

5.18 (a) $\delta = 0.25D_1$, (b) $\tfrac{1}{15}$

5.19 (a) $V_2 = V_1 h_1/h_2$, (b) $F = (p_1 - p_2)Wh_2 + 8\rho W(V_1^2 h_1 - V_2^2 h_2)/15$

5.20 $U_m/U_{av} = 1.5$; $C_D = C_p - 0.4$

5.21 (a) $U_2 = 3U_1$, $U_m = 9U_1/4$, (b) $p_1 - p_2 = 4\rho U_1^2$, (c) $F/(\rho U_1^2 hW) = -3p_1/(\rho U_1^2) + \frac{12}{5}$

5.22 (b) $C_p = (U_2/U_1)^2 - 1$

5.23 (a) $U_0 \delta W/3$, (b) $\rho U_0^2 \delta W/3$

5.24 (a) $U_\infty H/2L$, (b) $F = \rho U_\infty^2 HW/6$

5.25 $C_D = \frac{1}{3}$

5.26 $F_y/F_x = \sin\theta/(1 + \cos\theta)$

5.27 (a) $\rho V^2 WD((\cos\theta - 1)\mathbf{i} + \sin\theta\mathbf{j})$, (b) $\rho V^2 WD((\cos\theta - 1)\mathbf{i} + \sin\theta\mathbf{j})/4$

5.28 $\mathbf{F} = 998\mathbf{i} - 1729\mathbf{j}\ N$

5.29 $3210\ lb_f$

5.30 (a) $F = 0$, (b) $113N$, (c) $0.41m$

CHAPTER 6

6.8 $4/3$

6.9 (a) $-\rho(2yt + y^2x)$, (b) $2xy + 4xy^2t^2 + 2y^3x^2t/3$

6.10 (a) $-z + 4xy + x^4y/3 - ztx^2$

6.11 $-3/x$; $-2(6x + 3yx + 2t + yt)/xt$

6.12 no

6.13 $2at + 18xy + 16x^2$; $-18(at^2 + xy)$

6.14 yes

6.15 compressible

6.16 $\nabla \cdot \mathbf{V} = 2 + 5z^2$

6.17 (a) 2, (b) $-2t - yx$, (c) $2x(1 + 2t^2)$

6.18 $z = 3$

6.19 (b) $8(\mathbf{i} + 3\mathbf{j})$, (c) 24, (d) L^3/T

6.20 (a) $-4xy$, (b) $(8x^3y^2 + 6x^2)\mathbf{i} + 12\mathbf{k}$

6.21 (c) $D\mathbf{V}/Dt = 18x^3\mathbf{i} + (4zx + 12x^2zt + 16zx^2t^2)\mathbf{k}$, (d) $\nabla \cdot \mathbf{V} = 6x + 4xt \neq 0$

6.22 (d) $a_x = 2xy + 4xy^2t^2 + 2x^2y^3t/3$

6.23 (d) $\theta = \pi/4$, (e) $-2xy^2\mathbf{i} + 18y^3\mathbf{j}$

6.24 (d) $-z^3/3 + z^5t^2/3$

6.25 (c) no, (d) $a_x = xy^2z^2 + xzt^2 + 3xy$

6.26 (c) $2\mathbf{i} - 2\mathbf{j}$, (d) 0.5

6.27 (d) $4z + 5xy^3t$, (e) $-\rho(4z + 5xy^3t)$

6.28 (b) $a_x = x^2(x^3/3 - 2)$

6.29 $-y^3/3$

6.30 y^2

6.31 $(4/r^2 - 1) \cos \theta$

6.32 $\rho = c/x^2$

6.33 $a + e + j = 0$

6.34 $u = -2yx - 2x + f(y)$

6.35 $v = Ay/x^2$

6.36 (a) $u_r = \bar{v}r/2h$, (b) $a_r = 2.22 \times 10^5 \ m/s^2$

6.38 $-3\rho V_0^2/2$

6.39 $40 \ ft/s; \ 8 \ ft/s^2$

6.40 $5 \ m/s^2; \ 31.1 \ m/s^2$

CHAPTER 7

7.3 $\psi = -(A/k)e^{ky} \cos kx$

7.4 (a) yes, (b) irrotational, (c) $p = \rho k^2 A^2 \sin^2 kx$

7.5 (a) yes, (b) rotational, (c) $p = -2\rho k^2 A^2 \sin^2 kx$

7.6 (b) $u = V_0, v = 0$, (c) V_0

7.7 $V_0(-\sin \alpha \ x + \cos \alpha \ x)$

7.8 (b) $u = x, v = -y$, (c) 1

7.9 $\psi = r^3 \sin 3\theta; \ |\mathbf{V}| = 3r^2$

7.10 (a) V_0x, (b) $V_0(\cos \alpha \ x + \sin \alpha \ y)$, (c) $x^2 + y^2$, (d) $r^3 \cos 3\theta$

7.11 $\psi = 2xy$

7.12 (c) $\phi = c(x^2y + a^2y - y^3/3)$, (d) $\psi = c(y^2x - a^2x - x^3/3)$

7.13 $u_r = 0; \ u_\theta = 20 \ m/s$

7.14 (a) $ky^2/2$, (b) yes

7.15 $\psi = 1.04y^2 + 1.5y \ m^2/s$

7.16 $\theta - \log r$

7.17 (b) $u_r = 0, u_\theta = A/r$, (c) $-A \log r$

7.18 (b) $0, -\sqrt{2}$ and $1.225, -1.225$, (c) $(2r^{3/2} \cos 3\theta/2)/3$

7.19 (a) $\phi = V_0x + \dfrac{q}{2\pi}(\ln \sqrt{x^2 + y^2} - \ln \sqrt{(x - b)^2 + y^2})$, (b) $b/2 - \sqrt{(2qb + \pi b^2 V_0)/4\pi V_0}$

7.20 $\theta = \pi/6$

7.21 (a) $Q/2\pi U_\infty, \pi$, (b) $2 \ m; \ 2.21 \ Pa$

7.23 $q = 0.6 \ m^2/s$

7.24 $q = 3.75 \ m^2/s$

7.25 $\Gamma = -21 \ m^2/s$

7.26 $1.592 \ m/s$ for $\Gamma = -10 \ m/s, \ s = 4 \ m$

CHAPTER 8

8.7 $0.0129 \times 10^{-6} \, m^2/s$

8.8   12.5 *mm*

8.9   **(b)** $30 \times 10^8 \, lb_f/in^2$, **(c)** 0.05 *Hz*

8.10   **(b)** $\omega_m = 4\omega_p$, $T_m = T_p/2$

8.11   **(a)** 4, **(c)** $P_p = 4P_m$, $V_m = V_p$

8.12   **(b)** 1794 *rpm*, 0.776 *N/m*

8.13   same

8.14 $2gL(\rho_f a/\dot{m}_f)^2 = 1 - (\rho_f/\rho_a)(a/A)^2(\dot{m}_a/\dot{m}_f)^2$

8.15   **(a)** 50 *ft/s*, **(b)** 3.75 *rps*

8.16   **(b)** $V_m = V_p/2$, $f_m = 2f_p$

8.17   **(b)** $d_2 = d_1/\sqrt{2}$, $N_2 = N_1\sqrt{2}$, **(c)** none

8.18   **(b)** $V_m = V_p$; $F_m = F_p/100$

8.19   **(b)** $D_m = D_p/2$, $\omega_m = 4\omega_p$

8.20   **(c)** $V_{1m} = V_{1p}/2$, $\nu_{1m} = \nu_{1p}/8$

8.21   **(b)** 0.0179

8.22   **(b)** 15,600 *N*

8.24   **(c)** $V_m = 2m/s$, $\nu_m = \nu_p/125$

8.25   **(b)** drag increases by 10%

8.26   **(b)** $V_2 = 0.183V_1$, $\nu_2 = 0.0061\nu_1$

CHAPTER 9

9.5   600 (laminar)

9.6   10,000 (turbulent)

9.7 $\bar{V} < 0.065 \, ft/s$

9.8   0.0028 *ft/s*, 0.0141 *ft/s*

9.10   **(a)** 519 (laminar), **(b)** 173,000 (turbulent), **(c)** 5.19×10^6 (turbulent)

9.11   64.2; 0.0787

9.12   22.0

9.13   4.50 *psi*

9.14   87 *min*

9.15 $p_{Bg} = 77,800 \, Pa$

9.16   5.10 *m/s*

9.17 $\dot{q} = 0.022 \, m^3/s$

9.18 $p_{1g} = 24.7\rho V^2$

9.19   480 *m*

9.20 $H = (V^2/2g)(1 + fL/16D + C_{D1}/16 + 2fL/D + C_{D2})$

9.21 0.22 *m/s*, 33 × 10⁻⁶ *m³/s*

9.23 33.9 *ft³/s*, 0.043 *psig*

9.24 0.456 *ft³/s*, −6.74 *psig*

9.25 0.0748 *m³/s* (0 *yrs*); 0.0738 *m³/s* (5 *yrs*); 0.0728 *m³/s* (10 *yrs*); 0.0716 *m³/s* (20 *yrs*)

9.26 13.5°

9.27 6.5°

9.28 $h_{l_1}/h_{l_2} = 7.9$

9.29 2.5 *in.*

9.30 $V_e = \sqrt{80gd}$

9.31 4.7 *hp*

9.32 +200 *kW* (pump)

9.33 −768 *W* (turbine)

9.34 131 *kW*

9.35 **(a)** 0.0924 *Pa*, **(b)** 3.70 *Pa*, **(c)** *Re* = 865

9.36 **(a)** $\bar{V} = U_{CL}/2$, **(c)** $C_f = 16/Re$

9.37 **(b)** $-8U_0\mu/D^2$, **(c)** $-8U_0y/D^2$

9.38 **(c)** $\bar{V} = U_m/3$, $C_f = 12/Re$

9.39 **(a)** $\tau_\omega = \rho gD/4$, **(b)** $\nu = gD^2/(16U_c)$

9.40 **(a)** $u = (h^2g \sin \theta/\nu)(y/h - \frac{1}{2}(y/h)^2)$; $\dot{q}/\text{width} = gh^2 \sin \theta/(3\nu)$;
(b) $\dot{q}/\text{width} = h\sqrt{2gh \sin \theta/f}$

9.42 **(a)** $\rho g\delta \sin \theta$, $\mu V_s/\delta$; **(b)** $2 \sin \theta/F_s^2$, $2/Re_\delta$

CHAPTER 10

10.4 $Re_L = 43{,}000$ (laminar)

10.9 $\mu U_e/\delta$; $2/Re_\delta$; $\delta/2$; $\delta/6$

10.12 $\delta/10$; $9\delta/110$; $11/9$

10.13 **(a)** $Re_x = 1.28 \times 10^6$ (turbulent); **(b)** 22 *mm;* **(c)** 0.00444; **(d)** 1.05 *N*

10.14 **(a)** $U_e = 1.153$ *m/s;* **(b)** $\delta = 22$ *mm;* **(c)** 0.00444; **(d)** 2.94 *N*

10.15 **(a)** 1.32 *ft,* **(c)** 0.462 *lb_f*

10.16 1.35

10.17 **(a)** $(0.037 U_e/Re_x^{0.2})(y/\delta)^{8/7}$, **(b)** 0.0067°, 0.0268°, 0.107°

10.18 193 *hp*

10.19 **(b)** $p_1 - p_2 = \frac{1}{2}\rho U_1^2((U_2/U_1)^2 - 1$

10.20 **(a)** $\tau_w = 2\mu U_{max}/h$, **(b)** $2U_{max}/h$

10.21 $\pi\mu U_e/2\delta$; π/Re_δ

10.22   4/3

10.24   (a) $d\theta/dx = (d\delta/dx)/6$, (b) $C_f = 2\nu/U_e\delta$

10.25   (b) $\tau_w = 3\mu U_e/2\delta$

10.29   1.47 N

10.30   83 lb_f

10.31   406 lb_f

10.32   72 lb_f

10.33   (a) 1.20 m/s, (b) 35.6 N

10.34   (a) 138 lb_f, (b) 92 mph, 62 mph

10.35   31.3 m/s

10.36   6.95 m/s

10.37   0.026 m/s

10.38   28.6 m/s; 3.16 m/s

10.39   (a) 1035 Hz, 1922 Hz, (b) 690 Hz, 1281 Hz, (c) 517 Hz, 961 Hz

10.40   (a) 1.76 Hz, 25 ft, (b) 422 Hz, 2.5 in.

CHAPTER 11

11.2 $U \pm \sqrt{gy}$

11.8   4.64 m

11.10   15.6 ft^3/s

11.11   8.3 ft

11.12 $F_1 = 0.386$; $F_2 = 0.932$

11.13 $F_1 = 0.192$; $H/Y_1 = 0.519$

11.14 $F_1 = 0.316$; $F_2 = 2.53$

11.15 $F_1 = \sqrt{2h_1/Y_1}$; $Y_2 = 2(h_1 - h_2)/3$

11.16   0.990; 0.316

11.17   (a) $F_1 \approx \sqrt{2H/h_1}$, (b) $F_1 = 2.53$, (c) $b = 1.41h_1$

11.18 $h/H_1 = 1 - W_1/W_2$

11.19   (a) $Y_2 = 2Y_1/3$, (b) $F_3 = 2$, (c) $Y_4 = 0.791Y_1$, (d) $F_4 = 0.547$, (e) $F_5 = 0.30$

11.20   (a) $Y_2 = 3Y_1$, (b) $Y_3 = 7.12Y_1$, (c) $F_3 = 0.547$, (d) $F_4 = 0.872$, (e) $B_4 = 0.318B_1$

11.23 $V_1 = 5.43$ m/s, $V_2 = 2.71$ m/s

11.24   6.07 ft

11.25   (d) $H_2/H_1 = (\sqrt{1 + 8F_2^2} - 1)/4F_2^2$, or $H_1/H_2 = \frac{1}{2}(\sqrt{1 + 8F_2^2} - 1)$ (e) 2.26 m

11.26   2.67 ft

11.27 $U_2/U_b = 0.5$

11.28 **(d)** 6.43 m/s

11.29 1.813 m

11.30 2.65 m

11.31 1.86 ft; 2.23 ft/s

11.33 $F_1 = 4.0$; $Y_2 = 12.9$ mm

11.34 **(a)** $Y_2/Y_1 = 2(F_1^2/2 + 1 - H/Y_1)/3$, **(b)** $F_3 = 2\sqrt{2}\,F_1$, **(c)** $Y_4 = Y_1$

11.35 $h_3/h_2 = 1.863$

11.36 $Y_2/Y_1 = 7/12$; $Y_3/Y_1 = 0.19$

11.37 **(a)** $F_2 = 2.55$, $B_2/B_1 = 1.66$, **(b)** $Y_3/Y_1 = 1.57$

11.38 **(a)** $F_3 = 2.83$, **(b)** $B_3 = 2B_1$, **(c)** $B_4 = 0.883B_1$

11.39 **(c)** 0.5; 0.77, **(d)** 2/3

11.40 $F_1 = \sqrt{2(H - Y_1)/Y_1}$

11.41 **(a)** $F_2 = 1.51$, **(b)** $3.38Y_1$

11.42 0.39

11.45 **(a)** $F_2 = 1.98$, **(b)** $F = -0.115\rho g w Y_1^2$, **(c)** $F = 0$ (ignoring friction), **(d)** $F_3 = 0.552$

11.46 **(a)** $F_2 = 1$, **(b)** $Y_2 = 0.63Y_1$, $h = 0.18Y_1$, **(c)** $C_D = 0.069$

11.47 **(a)** 0.152, **(b)** 0.529

11.48 **(a)** $15\rho g h^2/32 - 3\rho V^2 h$, **(b)** $Y_d = (h/8)(\sqrt{1 + 512F^2} - 1$, $F_d = (h/Y_d)^{1.5}$

11.49 **(a)** $F_1 = 0.193$, **(b)** $H = 0.519Y_1$, **(c)** $0.298\rho g w Y_1^2$

CHAPTER 12

12.1 347.2 m/s, 1017 m/s

12.2 1.823, 1.856, 1.929, 2.010

12.3 subsonic

12.4 60.8°F (520.5°R), 14.78 $psia$

12.5 122 kPa, 29.9°C (303.1°K)

12.6 128 kPa, 34.9°C; 102 kPa, 21.9°C; 35.7 kPa, -30.0°C

12.7 0.437, 2.033

12.8 3444 ft/s (2348 mph)

12.9 2.13, 93 × 10⁶; 2.03, 113 × 10⁶

12.11 266 kPa, 3.74 kg/m^3, -25.6°C (247.6°K)

12.12 255 m/s, 551 kPa, 881 kPa

12.13 6100 Pa, 0.236 kg/m^3, 104 kPa, 475 m/s

12.14 247 m/s, 670°K, 320 $J/kg \cdot$°K

12.15 a factor of 6

12.16 68.6°K, 668 Pa, 664 m/s, 0.91 kg/s

12.17   3.40, 0.0247 m^3, 0.0040 m^3

12.18   **(a)** 1.10×10^6 *Pa*, **(b)** 196°*K*, **(c)** 19.6 *kg/m³*, **(d)** 30 *kg/s*

12.19   0.120 *kg/s*, 3.73 *kPa* $< p_b <$ 33.5 *kPa*

12.20   **(a)** 4.05 *kg/s*, **(b)** 0.147, 2.94, **(c)** 670 *kPa*, 20.3 kPa

12.21   14.74°

12.23   2.68, 2.24, 1.93

12.24   **(a)** 2.00, 183 *kPa*, 1.97 *kg/m³*, **(b)** 6.45 *J/kg·°K*

12.25   **(a)** 1.64, 312 *kPa*, 379°*K*, **(b)** 4.07 *J/kg·°K*, **(c)** 15.5°

12.26   1.67, 0.779 p_1

12.27   23.7°, 20.6

12.28   5.92°, 41,400 *N*

12.29 $L = 779\ N, D = 99.5\ N$

CHAPTER 13

13.1 $N'_{sd} = 53$, Francis

13.2   303 *rpm*

13.3 $\dot{q} = 50.5\ ft^3/s, r = 9.25\ ft$

13.4 $\dot{q} = 3.34\ m^3/s$, 31.8 *MW*

13.5 $2r = 2.2\ m, N'_{sd} = 2.4$

13.6 $D = 0.82\ m, P = 5.4\ MW, N = 18\ rads/s$

13.7 $N_s = 1650$

13.8 $N_s = 6552$

13.9   centrifugal, mixed (respectively)

13.10   1280 *rpm*

13.11 1.85×10^6 *gpm*

13.12   4 in series

13.13   **(b)** 1370

13.14   6.5 *ft*

13.15   347 *rpm*; 73,000 *gpm*

13.16   1.86 *ft*, 172 *hp*, 121 *ft*

13.17   .076 *hp*

13.18   92%

13.19   38,800 $ft \cdot lb_f$; 1480 *hp*

13.20   393 *rpm*

13.21   25.1 *kw*

13.22   1580 *rpm*; 30.4 *hp*

13.23 (a) 61.6°, (b) 1.05 *MW*

13.24 0.71, 39 *hp*

13.25 154 *kW*, 25 *rpm*, 5.6 *m/s*

13.26 0.74, 85%

13.27 82%, 170 *mph*, 1180 *hp*, 2609 *lb_f*, 0.36 *psi*

13.28 420 *km/hr*, 10,220 *N*, 1.02 *MW*, 851 *kW*, 83%, 4020 *Pa*

13.29 4 shafts, 244,000 *lb_f · ft*

13.30 314 *kW*, 11 *m/s*, yes

13.31 1.94 *ft³/s*; 1.40 *ft³/s*

CHAPTER 14

14.4 (a) very stable, (b) stable, (c) unstable

14.5 (a) neutral, (b) unstable, (c) stable

14.6 (a) 2.83 *m/s*, 3.33 *m/s*, (b) 5.02 *m/s*, 7.80 *m/s*

14.7 13.60 *m/s*

14.8 (a) 6952 *ft*, (b) 4861 *ft*

14.9 (a) 2176 *m*, (b) 948 *m*

14.10 378 *m*

14.11 (a) 3.0×10^{-3} *m/s*; 0.3 *m/s*, 4.3 *m/s*, (b) 6.0×10^{-3} *m/s*, 0.6 *m/s*, 6.0 *m/s*

14.12 (a) 5.4×10^{-5} m/s, 5.4×10^{-3} m/s, 0.13 *m/s*, (b) 1.1×10^{-4} *m/s*, 1.1×10^{-2} *m/s*, 0.23 *m/s*

14.13 0.37 *m/s*

14.14 77 *ft*

14.15 44 *μm*

INDEX

10.15

a) $Re_{crit} = 5 \times 10^5 = \dfrac{x_t U_e}{\nu}$

$x_t = $ distance to transition

$x_t = 5 \times 10^5 \dfrac{\nu}{U_e} = 1.32$

b) $Re_x = \dfrac{x U_e}{\nu} = \dfrac{60 x}{1.58 \times 10^{-4}}$

$(Re_x)_{max} = \dfrac{6 \times 60}{1.58 \times 10^{-4}} = 2.28 \times 10^6$

For laminar flow $(Re_x < Re_{crit})$

$C_f = \dfrac{.664}{\sqrt{Re_x}}$

For Turbulent flow $(Re_x > Re_{crit})$

$C_f = \dfrac{.0576}{Re_x^{.2}}$

$Re_x \times 10^{-6}$	0	.25	.5	1	1.5	2
laminar $C_f \times 10^3$	∞	1.33	.94			
turb $C_f \times 10^3$				4.18	3.63	3.35	3.16

c) Treat the laminar portion Separate from the turbulent portion

for $x \le 1.32$ $C_f = \dfrac{1.328}{\sqrt{Re_{crit}}} = \dfrac{1.328}{\sqrt{5 \times 10^5}} = .00188$

∴ Drag of laminar part $= 2 F_r = \rho U_e^2 x_t W \times .00188$

$= .0613$

for $x \ge 1.32$ $2 F_r = 2 \int_{1.32}^{6} \tau_w W dx$

$= \int_{1.32}^{6} \rho U_e^2 w C_f dx$

$= \rho U_e^2 w \int_{1.32}^{6} \dfrac{.0576}{Re_x^{.2}} dx = \dfrac{\rho U_e^2 w}{\left(\frac{U_e}{\nu}\right)^{.2}} \times .0576 \int_{1.32}^{6} x^{-.2} dx$

$= \dfrac{\rho U_e^2 w}{\left(\frac{U_e}{\nu}\right)^{.2}} \times \dfrac{.0576}{.8} \left[x^{.8} \right]_{1.32}^{6}$

$2 F_r = .4008$

Total drag (entire plate both sides) $= .0613 + .4008 = .462$

$$\frac{U}{Ue} = \left(\frac{y}{\delta}\right)^{\frac{1}{7}} \quad \text{and} \quad \frac{\delta}{x} = \frac{.37}{Rex^{.2}}$$

1) Two dimensional constant density Flow: $\frac{du}{dx} + \frac{dv}{dy} = 0$, $v = -\int_0^{y}\frac{\partial u}{\partial x}dy$

now $\frac{\partial u}{\partial x} = \frac{du}{dx} = -\frac{U_e y^{1/7}}{7}\delta^{-\frac{8}{7}} = -\frac{U_e}{7\delta}\left(\frac{y}{\delta}\right)^{1/7}\frac{d\delta}{dx}$

$$\frac{d\delta}{\delta x} = \frac{d}{du}\left(\frac{.37 x}{Rex^{.2}}\right) = \frac{d\left(.37 x^{.8}\right)}{du\left(\frac{Ue/\nu}{}\right)^{.2}} = \frac{.37 x^{.8}}{\left(Ue/\nu\right)^{.2}} \times \frac{1}{x^{.2}} = \frac{.37 x^{.8}}{Rex^{.2}}$$

$V = \int_0^{y}\frac{Ue}{7\delta}\times\frac{.37 x^{.8}}{Rex^{.2}}\left(\frac{y}{\delta}\right)^{1/7}dy = \frac{.037 Ue}{Rex^{.2}}\left(\frac{y}{\delta}\right)^{8/7}$

b) $\frac{V}{U} = \frac{.037 Ue}{Rex^{.2}}\left(\frac{y}{\delta}\right)^{8/7}\frac{1}{Ue}\left(\frac{y}{\delta}\right)^{-1/7} = \frac{.037}{Rex^{.2}}\left(\frac{y}{\delta}\right)$

$\therefore \alpha = Tan^{-1}\left[\frac{.037}{Rex^{.2}}\left(\frac{y}{\delta}\right)\right] \rightarrow (\alpha)_{.5} = .0067°$

$(\alpha)_{.2} = .0268°$

$(\alpha)_{.8} = .107°$

Note
Streamlines are almost perfectly parallel

Problem 10.18

$C_F = \frac{Fv}{\frac{1}{2}\rho U_e^2 A} = \frac{.074}{Re_L^{.2}}$

$\therefore Fv = \frac{.037\rho U_e^2 A}{Re_L^{.2}} = 1.037\times1.94\times25^2\times5000\times\left(\frac{200\times25}{1.2\times10^{-5}}\right)^{-.2}$

10.24) $\frac{U}{Ue} = \frac{y}{\delta}$ for $y \leq \delta$

1) $\theta = \int_0^{\infty}\frac{U}{Ue}\left(1-\frac{U}{Ue}\right)dy = \int_0^{\delta}\left(\frac{y}{\delta}\right)-\left(\frac{y}{\delta}\right)^2 dy = \left[\frac{y^2}{2\delta} - \frac{y^3}{3\delta^2}\right]_0^{\delta} = \frac{\delta}{2} - \frac{\delta}{3}$

$\therefore \theta = \frac{\delta}{6}$

$\therefore \frac{d\theta}{dx} = \frac{1}{6}\frac{d\delta}{dx}$

b) $Cf = \frac{Tw}{\frac{1}{2}\rho Ue^2} = \frac{2}{\rho Ue^2}\mu\frac{du}{dy}\Big|_{y=0} \Rightarrow \frac{2\mu}{\rho Ue^2}\frac{Ue}{\delta} = \frac{2\mu}{\rho Ue\delta}$

5) Momentum integral equation for zero pressure gradient, flate plate boundary layer

$\frac{d\theta}{dx} = \frac{Cf}{2} \Rightarrow \frac{1}{6}\frac{d\delta}{dx} = \frac{\mu}{\rho Ue\delta}$

$\int \delta \, d\delta = \int \frac{6\mu}{\rho Ue}dx$

$\frac{\delta^2}{2} = \frac{6\mu}{\rho Ue}x \Rightarrow \delta = \sqrt{\frac{12\mu x}{\rho Ue}}$

$\theta = \frac{1}{6}\delta = \sqrt{\frac{\mu x}{3\rho Ue}} \sim \sqrt{x}$